Biogeography: natural & cultural

Biogeography

natural & cultural

I G Simmons

Duxbury Press
North Scituate, Massachusetts

© I. G. Simmons 1979

First published 1979 by
Edward Arnold (Publishers) Ltd
41 Bedford Square, London WC1B 3DQ

This edition first published in the
United States of America in 1980 by
Duxbury Press, a division of Wadsworth, Inc.,
Belmont, California 94002

ISBN 0-87872-260-2

All rights reserved. No part of this publication may be
reproduced, stored in a retrieval system, or transmitted
in any form or by any means, electronic, photocopying,
recording or otherwise, without the prior permission
of Edward Arnold (Publishers) Ltd.

Library of Congress Cataloging in Publication Data

Simmons, Ian Gordon.
   Biogeography, natural and cultural.

   Bibliography: p.
   Includes index.
   1. Geographical distribution of animals and plants.
2. Man—Influence on nature. 3. Man—Influence of
environment. I. Title.
QH84.S55 1980   574.9   80-11106
ISBN 0-87872-260-2

Printed in the United States of America
1 2 3 4 5 6 7 8 9 — 84 83 82 81 80

Contents

Preface
Acknowledgements
List of Plates
Framework   1

Part I: Natural Biogeography   6

1 The organism scale   7
2 The ecosystem scale   54
3 The biome scale   90

Part II: Cultural biogeography   149

4   Pre-agricultural man's effects on plants and animals.   151
5   The domestication of plants and animals.   163
6   Man and biota in pre-industrial times.   192
7   Industrialization and biota.   204
8   The inadvertent creation of new biotic patterns.   230
9   Man's deliberate creation of new biotic patterns.   256
10 Productive resource processes underlain by biota.   273
11 Protective resource processes underlain by biota.   314
12 The ecology of man in his environment.   331

Glossary   347
Bibliography   353
Index   379

Preface

This book is intended specifically for undergraduate students of geography, and in particular for those involved with biogeography for the first time at anything other than the patchy treatment given in most schools; I realize that such people may not have taken much biology either. My aim here is to give an overview of much of the content of the field as seen from within geography and is deliberately planned as a 'once over lightly' treatment. The subject matter is therefore rather wider (it includes animals, for instance) than has been the case with most books of this kind written by geographers recently, though such was not the case with the biologist N. Polunin's *Introduction to Plant Geography* (1960). Inevitably, selection for breadth has meant sacrifices of depth of detail and explanation and some readers may therefore find the general level of treatment too superficial: they ought to go directly to more specialized material. My hope is that it will be read by the users rather than copiously noted, and that the students, aided by their teachers, will follow up some of the topics in greater depth. To that end I have buttressed the text with rather more references than is usual in books at this level, with a small selection of them at the end of each chapter. Apart from a few references to the late Mesolithic period in upland Britain, all the material has been culled from other people's work and I have doubtless misunderstood some of it, so that I shall be glad to have the errors pointed out; some Latin binominals may have been changed since the source in which I read about them was printed and again I should be pleased to be corrected by any systematist who happens to come across the book.

The background to the present work can be seen as a set of stages. I was fortunate to be taught as an undergraduate by a number of outstanding lecturers in the Plant Geography course of the University of London, and they strengthened an interest in the natural world which stemmed from parts of my childhood. So Palmer Newbould, David Harris, Francis Rose and Stanley Woodell introduced me to some of the material and attitudes of ecological and biogeographical thinking in Britain during the post-war years, a tradition into which my Ph.D. thesis on the ecological history of Dartmoor fitted quite well. A year at the University of California's Berkeley campus showed me another type of biogeography, which I found strange in its emphasis on human culture in an almost ethnographic sense as it related to particular groups of plants and animals. Its value, though, was made convincing to me in the course of conversations and field trips with men like Jim Parsons and the late Carl Sauer. So this book is an outcome of having been in contact with two traditions although it is not to be considered as an attempt at a synthesis with methodological intent. In more recent years, concern over environmental issues and the future of threatened species has become a concern of many students with biological interests and this view too will be apparent in this book. Lastly, I agree with David Watts that a humanistically-oriented biogeography might well be in the vanguard of a resurgence of the man-environment relations tradition in geography and it seems to me that this view, bringing together both the natural and the cultural systems of the world, might now be something which our discipline can offer mankind in its search for a *modus vivendi* with nature. The way I have tried to weave all these strands into one book is explained in the section entitled 'Framework' beginning on page one.

I have had a great deal of help in the preparation of this work. Michael Tooley took part in many discussions about format and content when we were both colleagues together at Durham. My wife Carol prepared much of the material for Part I and I have benefited greatly from her experience as a biology teacher. My thanks are due to the unknown Dr X who commented on a draft of Part II, and to Sarah Ayling and Annie Berry who helped with the rather boring tasks of checking and compilation. Part II especially was improved during a spell in the Berkeley library in 1976, and I am grateful to the Department of Geography there for lending me a desk, and to Lois and Dan Luten for their always generous hospitality; Dan's personal library yielded a fair amount of material, too, including his unpublished diagram which appears here as Fig 11.2. Early drafts were typed by the secretarial staff at Durham and the final version here in Bristol by Mrs D. Macey. My secretary, Mrs Mary Southcott, helped in very many ways and especially by re-typing tables, captions and other bits at short notice. Much of Part II was written while I was at Durham University and the time in which to do it was made possible by the generosity and flexibility of my colleagues there (and especially the Head of Department, Professor W. B. Fisher) in not insisting that I always gave particular courses; Part I was written after moving to Bristol University early in 1978 and here I am grateful to Professor Peter Haggett and his colleagues for not loading me too heavily with new courses and administrative duties.

Catherine and David are the dedicatees of this book: in small recompense for all the times the study door has been firmly shut.

I. G. SIMMONS
Bristol, June 1979

Acknowledgements

The author and publishers gratefully acknowledge permission granted by the following to reprint or to modify copyright material:

AAG Bijdragen for Fig. 6.1;*Academic Press Inc. for Fig. 8.6; the Agricultural Society of Newcastle upon Tyne for Figs. 10.2–3; Aldine Press Ltd for Fig. 4.4; Allen & Unwin Ltd for Fig. 1.2; the American Association for the Advancement of Science for Figs. 4.3 and 10.11; Anchor Books Inc. for Figs. 7.5; Blackie & Sons Ltd for Figs. 10.6–7; Blackwell Ltd. for Figs. 2.12 and 3.13; Blackwell Scientific Publications for Figs 1.1, 1.3, 7.7 and 8.1; Butterworth Ltd for Fig. 3.12; Cambridge University Press for Figs. 4.2 and 5.9; the Clarendon Press for Fig. 10.8; Collier-Macmillan Inc. for Fig. 2.7; Croom-Helm Ltd for Fig. 3.6; Elsevier Publishing Co. for Fig. 10.1; Freeman & Co. Inc. for Figs. 2.6 and 2.8; Harper & Row Inc. for Figs. 1.17–20 and 1.22–3; Heinemann Ltd for Fig. 11.1; Her Majesty's Stationery Office for Fig. 7.1; Hutchinson & Co. Ltd for Figs 7.3 and 7.4; IPC Publications for Figs. 9.1 and 10.4; the Institute of Biology for Fig. 7.6; Junk Publishing Co. for Fig. 3.1; the Longmans Group Ltd for Figs. 1.12–15 and 5.4–8; Macmillan Publishing Co. Inc. for Fig. 1.11; Mills & Boon Ltd for Figs. 1.8–10; Methuen & Co. Ltd for Fig. 7.2; Natural History Press for Fig. 8.5; National Academy of Sciences for Fig. 2.15; Nelson & Sons Ltd for Figs. 1.4, 1.7 and 8.7; Oliver & Boyd Ltd for Fig. 2.11; the Open University Press for Figs. 2.2, 3.7, 3.14, 3.15 and 10.10; Prentice-Hall Inc. for Figs. 1.5, 1.6, 1.21, 2.5 and 5.1; Reidel Publishing Co. for Figs. 8.2, 12.2 and 12.3; W. B. Saunders & Co. for Figs. 2.3, 2.13, 3.3, 3.4, 3.8, 3.11 and 8.8; *Scientific American* for Figs. 2.6, 2.8, 5.2 and 5.3; Springer-Verlag for Figs. 2.9, 2.10, 3.2 and 3.5; the Swedish National Research Council for Fig. 3.9; the United Nations Organization for Fig. 12.1; the University of Toronto Press for Fig. 12.4; the University of Wisconsin Press for Figs. 5.10 and 5.11 and John Wiley & Sons Inc. for Figs. 1.16 and 8.4.

*Figure numbers refer to this book. Detailed citations may be found by consulting captions and bibliography.

List of Plates

		page
1	Three species of the genus *Quercus*	9
	(a) *Q. gambelli* (Courtesy of Ardea, London, photo Ake Lindau)	
	(b) *Q. Virginiana* (Courtesy of Grant Heilman)	
	(c) *Q. Petraea* (Courtesy of Grant Heilman)	
2	Three genera of the *Felidae*	10
	(a) the lion, *Panthera leo* (Courtesy of L. Hugh Newman, NHPA, photo Andrew M. Anderson)	
	(b) the lynx, *Lynx lynx* (Courtesy of Ardea, London, photo Ake Lindau)	
	(c) the ocelot, *Felis pardalis* (Courtesy of Ardea, London, photo Adrian Warren)	
3	Phytoplankton (Courtesy of D.T. Boalch, Marine Biological Association)	14
4	The peppered moth, *Biston betularia* (Courtesy of L. Hugh Newman, NHPA, photo M. F. W. Tweedie)	26
5	The restored skull of *Australopithecus* (Courtesy of the British Museum, Natural History Collection)	33
6	*Tyrannosaurus rex* (Courtesy of the American Museum of Natural History)	34
7	A fossil Lycopod (Courtesy of Ardea, London, photo P. J. Green)	35
8	Members of the order Marsupiala	38–9
	(a) the red kangaroo, *Macropus rufus* (Courtesy of Ardea, London, photo F. Collet)	
	(b) the greater glider possum, *Schoinobates volans* (Courtesy of the Australian Information Service, photo Harry Fracua)	
	(c) *Sminthopsis* sp., a nocturnal desert mouse (Courtesy of the Australian Information Service)	
	(d) the numbat, *Myrmecobius fasciatus*, an anteater (Courtesy of the Australian Information Service, photo Mike Brown)	
9	Different types of ecosystems according to their energy characteristics	64–5
	(a) tropical forest in Guyana (Courtesy of Ardea, London, photo Adrian Warren)	
	(b) a river estuary in west Wales (Courtesy of Aerofilms)	
	(c) modernized crop agriculture in Rhodesia (Courtesy of Ardea, London, photo Alan Wearing)	
	(d) Pittsburgh by night (Courtesy of Grant Heilman)	
10	A pond undergoing succession (Courtesy of J. K. St Joseph)	67

xii List of Plates

11 Warthogs, *Phacochoerus aethiopicus* (*Courtesy of L. Hugh Newman, NHPA, photo Andrew M. Anderson*)   83
12 A characteristic rain-forest tree with buttress roots and lianas (*Courtesy of Ardea, London, photo Adrian Warren*)   94
13 East African savanna (*Courtesy of Ardea, London, photo John Wightman*)   101
14 Redwood chaparral, Big Sur (*photo I. G. Simmons*)   103
15 Short-grass prairie in the Sand Hills of Nebraska (*Courtesy of Grant Heilman*)   105
16 Desert plants in the southwest USA (*Courtesy of Grant Heilman*)   109
17 Deciduous forest in England (*Courtesy of the Forestry Commission*)   113
18 Coniferous forest in North America (*Courtesy of the USDA*)   118
19 A tundra scene in Greenland (*Courtesy of the Danish Tourist Board*)   122
20 The Trans-Alaska pipeline raised off the ground so as not to melt the permafrost (*Courtesy of BP*)   126
21 The unusual biota of the Galapagos (*Courtesy of Sally Anne Thompson Animal Photography Ltd*)   129
22 A fringing reef of coral and algae (*Courtesy of John Small*)   139
23 Agricultural scenes from the *Book of the Dead* (*Courtesy of the British Museum*)   165
24 A picture of the last living specimen of the aurochs, *Bos primigenius* (*Courtesy of Hutchinson & Co. Ltd, photo A. C. Cooper*)   180
25 A yak bull in Nepal, *Bos grunniens* (*Courtesy of FAO, photo W. Schulthess*)   187
26 A simple farming system in Ethiopia (*Courtesy of Barnaby's Picture Library, photo Peter Larsen*)   201
27 Severe defoliation of native tree species in New Zealand by the opossum (*Courtesy of the New Zealand Forest Service, photo J.H.G. Johns*)   212
28 A herd of reindeer in the Cairngorm mountains of Scotland (*Courtesy of Aviemore Photographic, photo David Gowans*)   218
29 The roof garden of the Kaiser Building in Oakland, California (*Courtesy of Kaiser Graphic Arts, photo Bill Wasson*)   225
30 The red kite (*Courtesy of the RSPB*)   230
31 Dead fish killed by pollution in an Ontario lake (*Courtesy of Keystone Press Agency Ltd, photo Peter L. Gould*)   240
32 Contemporary ecosystem types   252–3
   (a) an Australian rain forest (*Courtesy of L. Hugh Newman, NHPA, photo M. Morcombe*)
   (b) agriculture in Thailand (*Courtesy of Aerofilms*)
   (c) Woodland and grassland in the English Lake District (*Courtesy of Aerofilms*)
   (d) a biologically inert system: near Chicago, Illinois (*Courtesy of Grant Heilman*)
33 An agricultural experiment station (*Courtesy of Rothamsted agricultural experiment station*)   258
34 An eland undergoing domestication (*Courtesy of the South African Embassy*)   263

List of Plates   xiii

35	An irrigation canal in Florida clogged with water hyacinth (*Courtesy of Grant Heilman*)	268
36	A beef feed-lot operation in the USA (*Courtesy of Grant Heilman*)	280
37	Forest fires	292
	(a) a crown fire in Australia (*Courtesy of Associated Press*)	
	(b) a forest fire in southern France (*Courtesy of Keystone Press Agency*)	
38	An experiment in fish-and-duck culture (*Courtesy of FAO, photo Peyton Johnson*)	305
39	A goat browsing on thorn trees (*Courtesy of FAO, photo by F. Botts*)	308
40	A pair of California Condor (*Courtesy of Frank Lane, photo William L. Finley*)	315
41	A water-hole at Mkuse in Zululand (*Courtesy of the South African Tourist Corporation*)	316
42	Erosion from recreational use in the Chilterns, England (*Courtesy of Ardea, London, photo A. C. Cooper*)	318
43	A piece of primary woodland in England (*Courtesy of the Nature Conservancy Council*)	321
44	California Coast Redwoods (*Courtesy of Grant Heilman*)	323

Names of organizations abbreviated in the text

AAAS	American Association for the Advancement of Science (Washington, DC)
ACS	American Chemical Society
CAST	Council for Agricultural Science and Technology
CIA	Central Intelligence Agency (of the USA)
FAO	The Food and Agriculture Organization of the United Nations (Rome)
GMAG	Genetic Manipulation Advisory Group
IBP	International Biological Programme (London)
IBPGR	International Board for Plant Genetic Resources
IRRI	International Rice Research Institute
IWC	International Whaling Commission
IUCN	International Union for the Conservation of Nature and Natural Resources (Morges, Switzerland)
NAS	National Academy of Sciences (Washington, DC)
UCFTF	University of California Food Task Force
UNESCO	United Nations Educational, Social and Cultural Organization (Paris)

Framework

Students of geography are usually introduced to the study of the plant life (and, less often, the animals) of the world and the environmental factors which control their distribution; frequently their work includes a consideration of how human activities have affected the distribution and growth of the organisms. This sub-field of the subject is usually called biogeography: this is a term which is not wholly satisfactory since the word is already in use by biologists to whom (either as 'biogeography' or its subdivisions 'phytogeography' and 'zoogeography', dealing with plants and animals respectively) it has a rather specialized meaning. In biology it is used for the study of the distributions of plants and animals as a key to their long-term evolution, and for looking at the effects of large-scale environmental factors such as continental land-form, continental-scale climate and world soil groups. Much of the material normally used by geographers would be considered as 'ecology' rather than 'biogeography' although the boundary is not rigid and biologists as a group are not particularly demarcation-minded. But terms like 'ecological geography', 'geographical ecology' and 'geoecology' have not found common acceptance and so, for all its problems, within geography the term 'biogeography' has stayed firmly put to describe our study of the biosphere and of man's effects on its plants and animals and the ecological systems of which they are a part.

But there is not much doubt that within geography, the biogeography teaching has lacked any crystallizing focus. Instead there has been something of a supermarket approach: according to the interests of the teacher a number of packages have been brought off the shelves and put in the course trolley. The major world soil-climate vegetation units; soils; the 'natural' vegetation units of a region or country; some vegetation history; and the consideration of energy and chemical element flows through ecological systems, have been put together and presented as 'biogeography' at the checkout point of the examination paper. This writer is not about to suggest that such an approach is 'wrong', but for those teachers and students who would prefer to have a stronger conceptual framework for their learning, then it is suggested that one can be found. The source, unaccountably neglected by many recent authors in this field, is the North American geographer of distinction, Pierre Dansereau, in his book *Biogeography* (1957). His concepts are, simply, that man creates new **genotypes** and man creates new **ecosystems**. To expand slightly, man changes the genetic makeup of plants and animals so that they pass on the alterations to future generations. Some of this transformation is done deliberately by processes of selection in which human needs and perceptions influence the desired characteristics of the organisms, along with or instead of the 'natural' environment. Other modifications may happen by accident, as a by-product of some other change wrought by man. As examples, consider the domestication of plants and animals by selecting for breeding those individuals which showed desired characteristics like easy harvestability and heavy yield in plants, or docility and tender meat in animals (p. 182), as a deliberate change. Inadvertently, industrial smoke selectively enhanced the survival of a black form of a normally light-coloured moth found in temperate regions, so that this darker form progressively became commoner than its lighter relatives (p. 25). Similar changes have happened to the interactive combinations of the non-living environment with plants and animals which are called ecosystems. Deliberate changes include, for example, the replacement of forests by croplands, where the complex natural ecosystems of the woodland is replaced by the relatively simple field of a single crop organism and its associated plants and animals. Inadvertent alteration might be typified by the gradual changes in the flora of a grassland during decades of grazing by domesticated beasts where the system changes gradually but not necessarily the genetic structure of the individual organisms. Another example is the effect of agricultural and industrial wastes in poisoning plankton, fish and birds.

However, in order to study the effects of the human species, it is necessary to consider nature without man, as a datum-line. Part I of this book is largely designed to do this and so it presents a distillation of a great deal of biological science. First, the individual organism is considered particularly in relation to environmental factors in its distribution at various scales (e.g. global, continental, local); this is more or less the biogeography *sensu stricto* referred to above. Initially, though, there is some basic material on classification since geography students may not have studied much biology, some schools having conceived the curious idea that a number of their pupils might be allowed to leave without a working knowledge of the biological and physical systems of the planet which keep them alive. We then move to the ecosystem idea and consider the functional relations of plants, animals and their non-living environments, hanging the study on such pegs as energy flow and the cycling of chemical elements, and ending with a brief look at how energy and matter come together as the population of a species. Lastly in Part I, the major world natural associations of climate, plants, animals and soils (usually called biomes) are discussed individually. Like all distillations, the end-product can be a bit rough, especially when new, and so this attempt to compress so much into so small a space has a lot of unsatisfactory qualities especially in the form of generalizations which are prone to large numbers of exceptions. The suggestions for Further Reading will, it is hoped, help clear up some of the inevitable questions raised by the more alert readers. In one respect, this section is particularly unsatisfactory, *viz.* in trying to separate the natural from the man-made it imposes a difficult-to-find boundary: it is not known, for example, to what extent the remaining 'primeval' forests and grasslands of the world have been affected by pre-agricultural peoples. Some investigators think that man, fire, grasslands and animal communities have all co-evolved in Africa at least since the Early Pleistocene, so that the vegetation and fauna there have not been 'natural' at any time when scientific study could have been carried out on them.

In Part II, the precepts of P. Dansereau are used as a guide, though not followed in precisely the form in which he stated them. Instead, a historical approach is followed in Chapters 4–7 in which the effects of man at various technological levels on genotypes and ecosystems is considered; we turn then to the particular consequences of modern industrialization in the creation of new patterns, both by accident (Ch. 8) and design (Ch. 9). Chapters 10 and 11 are given to the thesis that the Earth's biota are man's most important resource and so discuss some of the resource systems which are fundamentally biological in nature. Only contemporary biota are considered, so that fossil fuels *per se* are excluded, although the effects of their use are central to the themes of Chapters 7–10. The final chapter comes as a future-oriented extension of Chapters 10 and 11 and considers the possible outcome of man–biota relationships if the present trends are extrapolated, and also whether there are any alternative types of relationship which might benefit both our own species and all the rest with whom we live on this planet. A narrower focus is adopted here than would be customary in books on 'The Environment' or on population–resource–environment interactions (anybody wanting a broader treatment is quite welcome to buy a copy of the present author's (1974) book which covers that field) but the discussion is deliberately widened out beyond the scientific and biological at this point. Whatever the virtues of scientific 'objectivity', it is unreal to dismiss such phenomena as human values where the future of the world's plants and animals is concerned; consider only the differential reaction of most people to pandas and hairy spiders. Although no firm answer is attempted here, no intelligent student (or citizen for that matter) ought to avoid the question, 'how much more creation of new genotypes and new ecosystems ought we to have?'

Bibliographic note; Units; Names

At the end of each chapter, 'Further reading' supplies a short list of summary sources or major works on relevant topics. These bibliographies, together with the text references, should provide a beginning for readers wanting to follow up a topic in greater depth. The items in the 'Further reading' also appear in the Bibliography, along with the bibliographic citations of the text references. A glossary of some of the scientific terms is provided at the back of the book; words defined in it are printed in **bold** when discussed in the text.

In general, the units used by an original author have been used here, with a conversion to the metric system where Imperial units were employed. Not all the units are SI, however, since quite a lot of ecological energetics is expressed in calories rather than joules. These two units are both used, therefore, rather than converting all to one system or the other. The conversion factors:

$$1 \text{ calorie} = 4.184 \text{ joules}$$
$$1 \text{ joule} = 0.239 \text{ calories}$$

may be useful. Sometimes productivity statistics are given in terms of grammes of carbon. To convert to the more commonly used dry-weight measurement, multiply $\times 2.2$. This is an approximation but is widely used. Following the usage of many IBP studies on production ecology, the conventions g/m^2/yr or kg/ha/yr have been used instead of the more strictly correct gm^{-2} yr^{-1} or kg ha^{-1}yr^{-1}.

Latin binominals of most organisms referred to are given, except where the common name is considered an adequate identification or where groups of animals (e.g. 'hawks') or plants (e.g. 'mosses') are referred to.

Further Reading

Dansereau, P. 1957: *Biogeography. An ecological perspective.* New York: Ronald Press.

Part I
Natural Biogeography

This part of the book seeks to establish a set of datum lines about how the world of living things would be distributed and would function if man were not present. Most of our knowledge has been gained since the nineteenth century and unless there is clear evidence to the contrary we have to make the assumption that the characteristics being studied, whether for instance of physiology or ecology, have not been affected by human activity. In order to simplify the very complex set of interactions between organisms and their environment which result in the distributions and dynamics of organisms, local ecological systems, and the major world biomes, we shall study them in succession, noting that it is only in the laboratory that an organism can be seen in isolation from its non-living or abiotic environment and from other plants and animals.

1
The organism scale

All living things have in common certain characteristics. They are made up of a complex material called protoplasm which is usually divided up into units called cells. At the simplest level of organization, plants and animals consist of a single cell; at the most complex, several million, of which many will have a specialized function such as nervous tissue, hormone secretion, or hair. The functioning of both plants and animals also exhibits common characteristics: they feed, grow, reproduce, move, respire, are sensitive to their surroundings, and excrete waste substances. Differences obviously exist between these two types of life: plants make their own food from inorganic substances and solar energy and hence are called **autotrophic** ('self-feeding') whereas nearly all animals rely on food already in plant or animal form and so are called **heterotrophic** ('outside feeding'). The key difference lies in the ability of the plant to synthesize organic material from inorganic molecules, incorporating the energy from the sun which is the basis for all life. This process is called **photosynthesis** and at the simplest level of explanation involves the use of solar radiation to bind together carbon dioxide and water to make sugars and starches, with the giving off of oxygen as a concomitant. The addition of chemicals taken up by the plant from its environment makes possible the synthesis of proteins, which are the basis of the complex molecules at the heart of living matter.

Some groups of living things are not clearly either animals or plants, e.g. bacteria and fungi, and viruses show by turns that they appear to be on the border of living and non-living things. The intricacies of their classification need not be elaborated here but we need to note that the simplified systems as presented in this book are not universally agreed by all biologists.

Classification of organisms

The 'natural' system of classification, which implies some evolutionary relationship between groups, depends on arranging organisms in a series from the more simple to the more complex forms, with the latter often showing special adaptations to their mode of life, as with various types of parasite adapted to life inside another animal, or the giraffe adapted to feed off the top of savanna trees. Both plants and animals are classified according to their natural characteristics (usually morphological although biochemical are often now used), and the basic unit is the species, of which about two million are currently recognized. A definition of this classificatory or **taxonomic** unit is not easy but can perhaps be summarized as consisting of those individuals which are able to reproduce among themselves but unable to breed at all freely with members of other

8   The organism scale

groups. (The individuals of a species may well not be identical but form a variable population in which some subgroups are recognizable as subspecies, written spp, or even more subordinate units called varieties, written var.) Species which are closely related are collected into the next highest unit, the genus (plural genera) and most plants and animals are known by their generic and specific names. This is the binominal system invented by the Swedish biologist Karl von Linne (C. Linnaeus) in the eighteenth century who, following the custom of the time, decreed Latin to be the proper language for such naming. Thus the oaks of the temperate zone are collected into the genus *Quercus*, with many species, e.g. *Quercus robur*, the deciduous English oak; *Q. ilex*, the evergreen live-oak of southern Europe; *Q. dumosa* an evergreen oak from California, and many more (Plate 1). Several species of cats form the genus *Panthera*, including *Panthera leo*, the lion; *P. pardus*, the leopard and *P. tigris*, the tiger (Plate 2). Genera are collected into families (e.g. *Panthera* with genera like *Felis* and *Acinonyx* is collected into the Felidae, the cats) then to orders (Carnivora), classes (Mammalia) and phyla, singular phylum (Chordata). All members of a phylum have the same basic plan in their structure whereas the lower taxonomic groups show variations on the basic plan and adaptations to different modes of life. Table 1.1 shows a simple classification of living things with some notes on the major groups.

There are also 'artificial' classifications based on some aspect of the functioning of the organism but implying no evolutionary relationship. An example is the division into autotrophic and heterotrophic organisms, with the further subdivision of heterotrophs into **herbivores, carnivores, diversivores** (or **omnivores**), **saprovores** and **parasites.** Plants can also be classified according to their water relations (**xerophytes, hydrophytes, mesophytes**), or their life-form, based on the location and nature of the perennation mechanism (e.g. buds on aerial shoots, underground bulbs, seeds of annual plants). But none of these, which have their uses when describing a population or an ecosystem, seem to take away the need for an internationally recognized system of putting a name to an individual plant or animal.

One point must be stressed here. Especially in geographical works it is common to lay emphasis on the study of the more complex plants and animals like the Angiosperms and Mammals, largely because they are highly visible and often of economic or aesthetic importance. But the role of two groups of smaller organisms should not be overlooked. The first of them are the micro-organisms such as bacteria and unicellular algae and fungi. They are very important in the soil, especially in the circulation of nutrients where, for example, they bring about the decay of dead plants and animals. In this way, they break down complex organic molecules into simpler inorganic compounds available once again for uptake by plants. It is micro-organisms, too, which disintegrate many man-induced contaminants, such as sewage in fresh water and crude oil in the sea. Also, we might note that it is algae, bacteria and fungi which constitute the only organisms in the most extreme conditions where life is found: in the windswept dry valleys of Antarctica, which are both extremely cold and extremely dry, and where they live just below the surface layers of the rocks. The second group are the microscopic plants and animals of the sea, referred to as the plankton. The plant members of this group (**phytoplankton,** mostly algae) are the basis of life in the sea (Plate 3) since they form the

Plate 1 Three species of the genus *Quercus*, the oaks: (a) *Q. gambelli* (Western USA); (b) *Q. virginiana* (Southeast USA); *Q. petraea* (West and Central Europe).

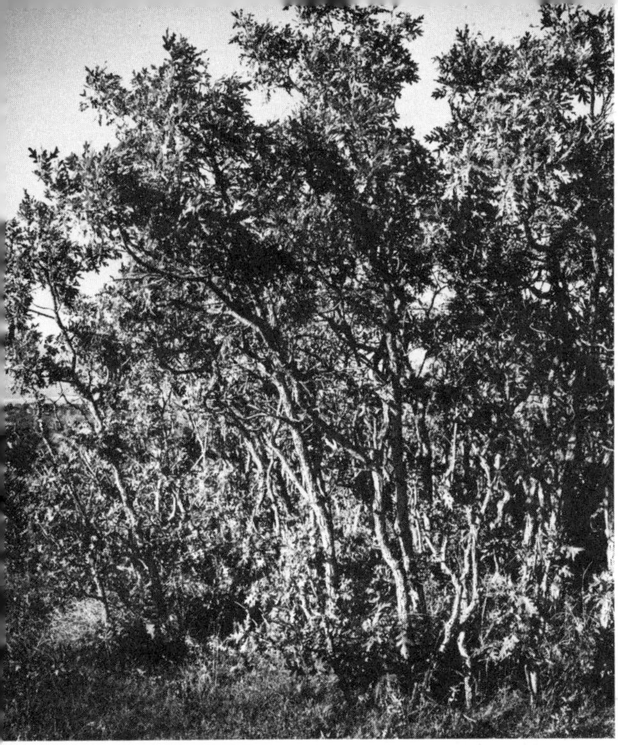
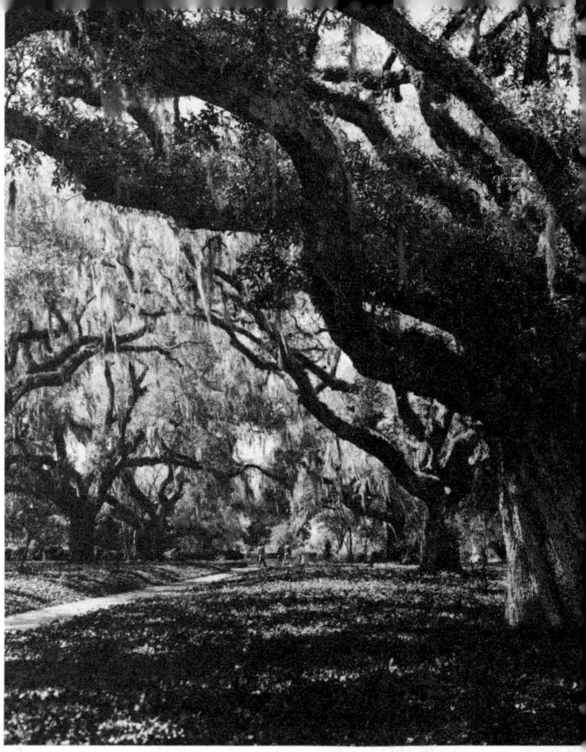
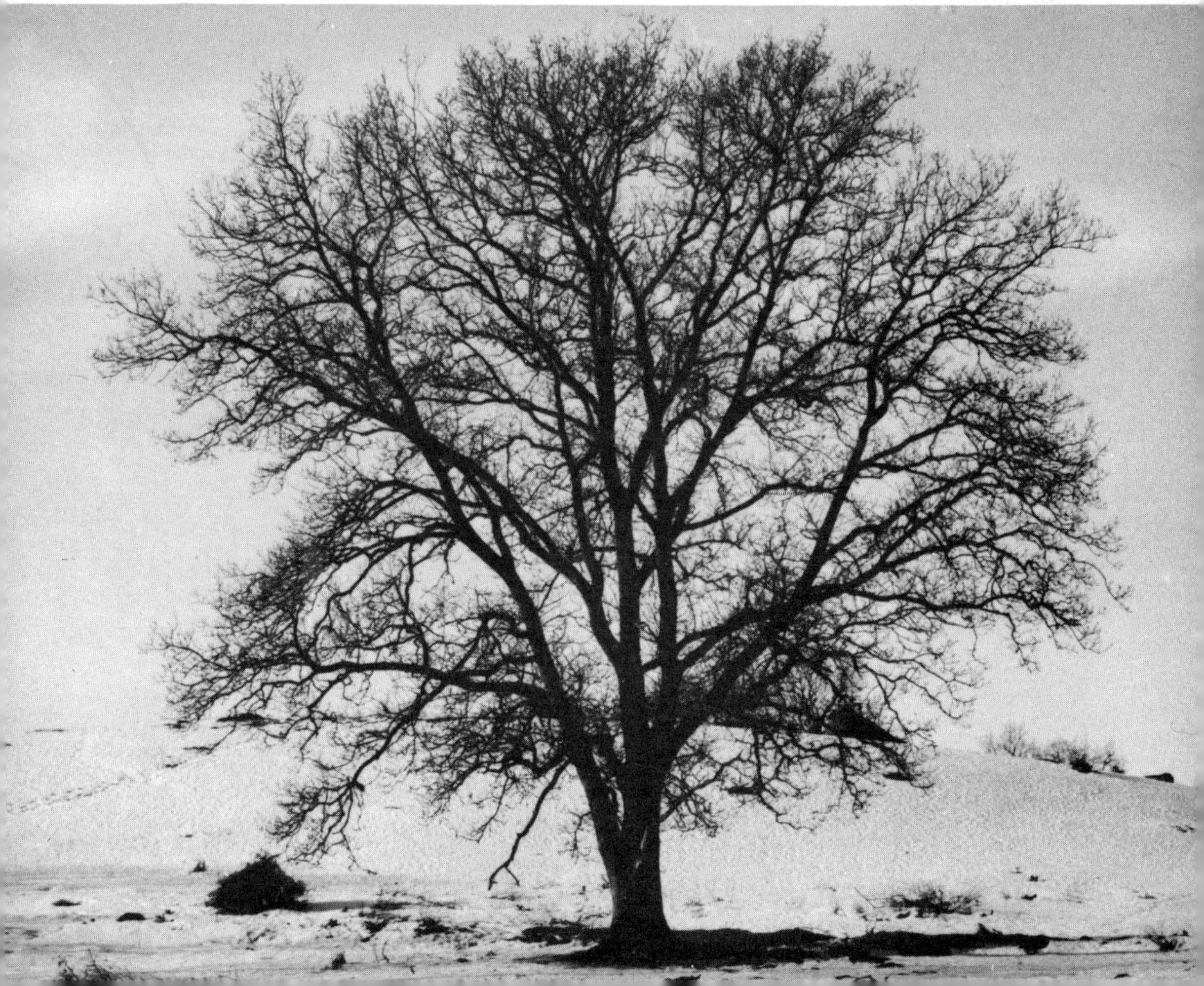

TABLE 1.1   Classification of living things

1. Protista
   Unicellular animals and plants (micro-organisms) classified by complexity.

1a. Bacteria	Microscopic, unicellular organisms. Saprophytic, or parasitic (rarely photosynthetic), e.g. *Staphylococcus aureus*.	
1b. Blue-green Algae	Unicellular or filamentous. Simplest plants which contain chlorophyll, e.g. *Arabaena*, *Nostoc*.	
1c. Unicellular green Algae	Microscopic algae containing chlorophyll, e.g. diatoms, *Pleurococcus*.	
1d. Protozoa	Unicellular organisms living in water, moist soil or parasitically, e.g. *Amoeba*, malarial parasite.	
1e. Unicellular fungi	Microscopic fungi without chlorophyll. Saprophytic or parasitic, e.g. yeasts.	
1f. Viruses	Organisms smaller than bacteria, which can only reproduce inside living cells. On the boundary between the living and non-living world. Cause many plant and animal diseases, e.g. common cold, measles, smallpox.	

2. Thallophyta
   Plants without stem, roots or leaves. Reproduce by spores.

2a. Algae	Contain chlorophyll and sometimes other pigments. Unicellular members are often classified now as Protista (see above). Other Algae have more complex tissues, e.g. brown seaweeds such as *Fucus vesiculosus* (bladderwrack). Are also filamentous type, e.g. *Spirogyra*.	1 900 spp
2b. Fungi	Without chlorophyll; saprophytic or parasitic. Body usually composed of branching threads, e.g. pin mould, potato blight, *Penicillium*, mushrooms and toadstools. Sometimes Fungi are classified separately from both plants and animals.	42 000 spp
2c. Lichens	Organisms where an algae and a fungus live together symbiotically. Structure varies from an encrusting growth on bare rock to a dense, branching structure, e.g. Iceland moss, *Rhizocarpon*.	

3. Bryophyta
   Small green plants without true roots, but usually with stem and leaves. Reproduce by spores.

3a. Liverworts	Body either a simple, flat thallus which may be branched, or leafy, with leaves arranged in 2 rows, e.g. *Pellia*.	9 000 spp
3b. Mosses	All have leaves which have mid-ribs and are arranged spirally on stem. Often grow in compact cushions or carpets. e.g. *Sphagnum* (bog moss) *Polytrichum* (hair moss)	14 000 spp

4. Pteridophyta
   Green plants with true roots, stems and leaves. Reproduce by spores.

4a. Clubmosses and Horsetails.	Present day species are small and not very numerous, but in the Carboniferous this group produced large trees which are fossilized in coal measures. e.g. *Lycopodium* (Clubmoss) *Equisetum* (horsetail)	1 025 spp
4b. Ferns	Spore-producing plants of moderate size, which bear spores on certain areas of lower leaf surface, e.g. bracken (*Pteridium*).	9 500 spp

5. Spermatophyta

5a. Gymnosperms (conifers)	Ovules and seeds are naked, no carpels. Seeds usually produced in cones. e.g. Pine, yew, larch, spruce.	720 spp

Plate 2 Three genera of the *Felidae*, the cats: (a) the lion, *Panthera leo*; (b) the lynx, *Lynx lynx*; (c) the ocelot, *Felis pardalis*.

12 The organism scale

TABLE 1.1 contd.   Classification of living things

5b. Angiosperms (flowering plants)	Ovules enclosed in an ovary of one or more carpels. Seeds enclosed in a fruit. Group may be further subdivided into:
	i) Monocotyledenous
	   e.g. grass family, lily family
	        (Gramineae) (Liliaceae)
	ii) Dicotyledenous
	   e.g. Rose family, Cabbage family,
	        (Roseaceae) (Cruciferae)
	        Daisy family
	        (Compositae)
		250 000 spp

Classification of animals.

1. Invertebrates — animals without a backbone or internal skeleton. May have many pairs of limbs or none. About 1 million species.
   N.B. The term, 'invertebrates' is not used in scientific classification, but is in common usage.

Phylum	Characteristics
2. Coelenterates	Body of 2 layers of cells round a central cavity. Only opening is the mouth. e.g. *Hydra*, sea anemones, jellyfish, corals.
3. Platyhelminthes	Small, flattened worms, no segments. Parasitic or free-living. e.g. *Planaria*, tapeworms, liverflukes.
4. Nematoda (threadworms)	Small, cylindrical worms, no segments. Mostly parasites. e.g. hookworms in man, eelworms in potato.
5. Annelida (segmented worms)	Body divided into segments, usually with bristles on each segment. e.g. earthworms, leeches.
6. Arthropoda	Animals with jointed limbs and a hard external skeleton.
6a. Crustacea	Animals who are mostly aquatic and have 2 pairs of antennae. Have a number of paired limbs used for a variety of purposes, e.g. feeding, walking, swimming. e.g. Crabs, lobsters, waterfleas, woodlice, barnacles.
6b. Myriapoda	Terrestrial animals with many segments each bearing similar limbs. e.g. centipedes, millipedes.
6c. Insecta	Segmented body divided into 3 parts: head, thorax, abdomen. 3 pairs of limbs on thorax. Often have 2 pairs of wings. e.g. locusts, houseflies, wasps, mosquitoes, termites, green-fly, fleas, ants.
6d. Arachnida	Body divided into 2 parts. 4 pairs of limbs attached to anterior part of body. e.g. spiders, ticks, scorpions.
7. Mollusca	Soft-bodied animals without segments. Usually have calcareous shells. e.g. snails, limpets, octopus, squids, oysters, slugs.
8. Echinodermata (spiny-skinned animals)	Unsegmented marine animals. Parts of body arranged symmetrically, usually on 5 radii. e.g. starfish, sea urchins, sea cucumbers.
9. Vertebrates (scientifically, phylum is Chordata)	Term not used in scientific classification, but commonly used of animals which have a backbone and internal skeleton of bone or cartilage. 2 pairs of limbs or fins. Well-developed head with brain protected by skull. Usually possess a tail.
9a. Pisces (fish)	Aquatic animals. 20 000 spp. Movement by muscular tail and paired and median fins. Respiration by gills. e.g. skate, shark, eel, cod, herring.
9b. Amphibia	Spend life on land and in water. Eggs must be laid in water and larvae (tadpoles) are aquatic. Moist, naked skin. e.g. frog, newt, salamander. Amphibia+reptiles=*ca* 6 000 spp.

Classification of animals.

Phylum	Characteristics
9c. Reptilia	Mostly terrestrial. Eggs are protected by a shell and are laid on land. Breathe through lungs. Dry, scaly skin. e.g. lizards, crocodiles, snakes.
9d. Aves (birds)	Terrestrial: lay eggs on land even if water-oriented rest of life. Skin covered with feathers and front limbs adapted for flight, forming wings (some vestigial in flightless birds). 9 000 spp. e.g. ducks, owls, finches (penguins).
9e. Mammalia	Mostly terrestrial. Possess hair/fur. Young are born alive and fed on milk produced by mammary glands. e.g. whales, man, apes, deer, bats. 4 000 spp.

N.B. 9d and 9e are the only groups of animals which are 'warm blooded' (**homoiothermic**) i.e. body temperature is kept more or less constant. All other animals are 'cold blooded' (**poikilothermic**) i.e. body temperature fluctuates with temperature of environment.

foundation of the food for all the higher forms of life in the oceans, including fish and whales.

The uneven distribution of individual plants and animals

It is a simple fact that even in a world without man's influence, plant and animal species would be distributed unevenly. Some of the reasons are obvious and easily recognized intuitively: a plant adapted to live under water is hardly likely to survive if one of its seeds is excreted by a migratory duck somewhere on a sand dune. Similarly, a small rodent adapted to survive in the intense heat and dry atmosphere of a desert might stand little chance if transported to the conditions of a tundra winter. Factors governing the distributions of species fall into two broad groups, the **abiotic** or environmental factors such as temperature, moisture, geological history, soil type and nutrient availability, and **biotic** factors such as food, cover, relationships with other members of the same species and relationships with other species. These groups of factors are not mutually exclusive and any factor may act either directly or indirectly upon a particular individual. For example, the drainage of a marsh by a farmer might not affect his chickens directly, but could lead to increased killing of them by foxes which had hitherto eaten large quantities of frogs from the marsh. Again, in a time of climatic change, species A might grow perfectly well in the new climate but if its seedlings where shaded out by those of species B which grew faster in the altered conditions, then A could lose its place in the plant community.

A species will exhibit different scales of difference of distribution (see p. 35), e.g. a plant species may grow throughout the north temperate zone of the earth but only in acid bogs; an animal subspecies may be found only on one isolated island. In every case, the species' distribution is bounded by barriers where it cannot grow. The nature of these barriers, as suggested above, may be either environmental or biotic. The environmental conditions usually work through a set of **limiting factors** which tend to

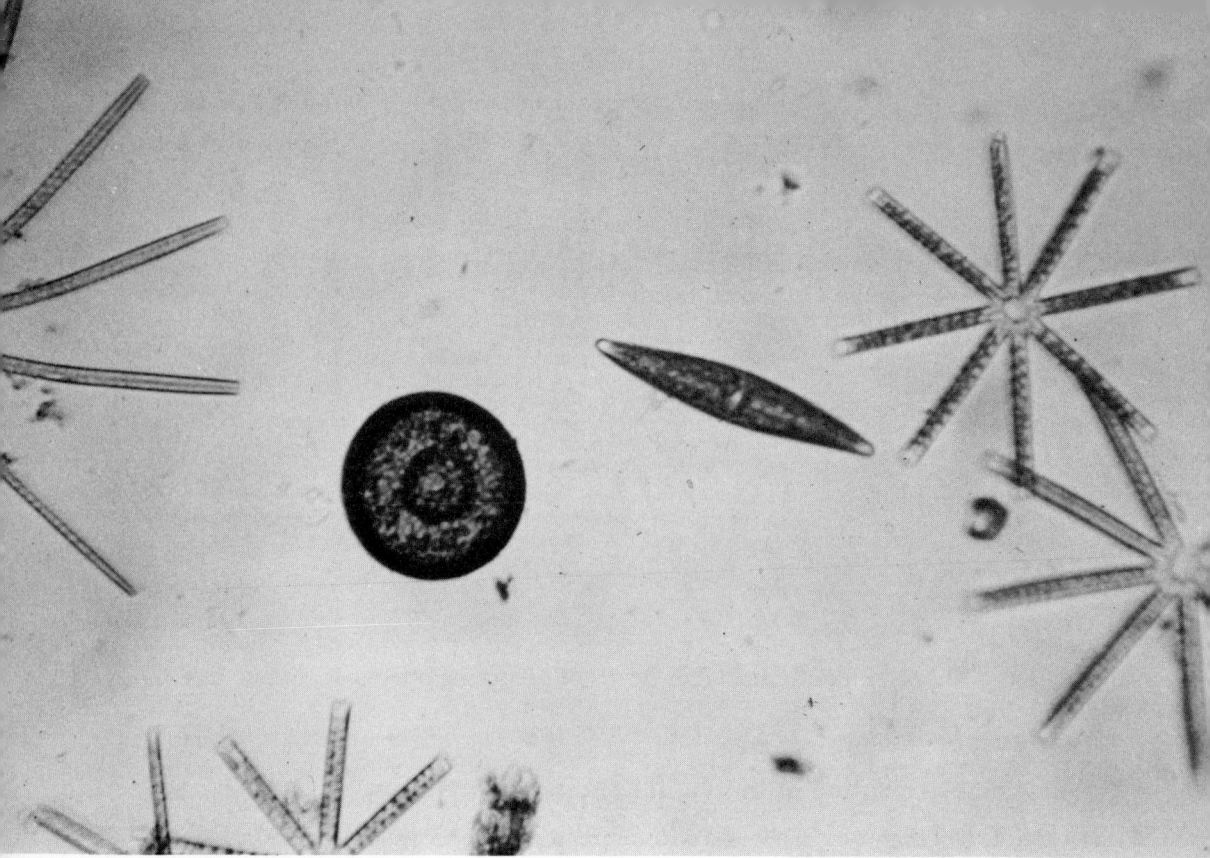

Plate 3 Diatom components of a phytoplankton. The star-shaped organisms are about 50 μ across.

make it more difficult for a species to live, grow or reproduce in a particular place. They need not necessarily be lethal but make its physiology or behavioural patterns less 'efficient' so that the organism is less able to reproduce or to compete with other species for food or living space. Often there is not just one such limiting factor but a complex of them, all interacting. Most environmental factors which affect an organism do so along a gradient, and the tolerance of the gradient varies slightly from individual to individual and more widely from species to species: some species are tolerant of a wide range of environmental conditions (**eurytopic**), others only a narrow range (**stenotopic**) but each species can function most efficiently over only a limited part of each environmental gradient, its **optimal range** (Fig. 1.1). But a species may not attain its full potential range in terms of abiotic factors because of competition with other organisms (Fig. 1.2).

One example of an environmental gradient is temperature, where there is a global transition from poles to Equator which is intersected by altitudinal changes and complicated by the occasional occurrence of relative extremes in any one place (Table 1.2). Animals and plants adapted to live in, for example, cool temperate conditions find that northwards temperatures are too cold and that southwards they are too hot, although there may be small populations of cool temperate species at the southern fringe of the cold end of the gradient in particularly favourable years and likewise for the warmer

The uneven distribution of individual plants and animals   15

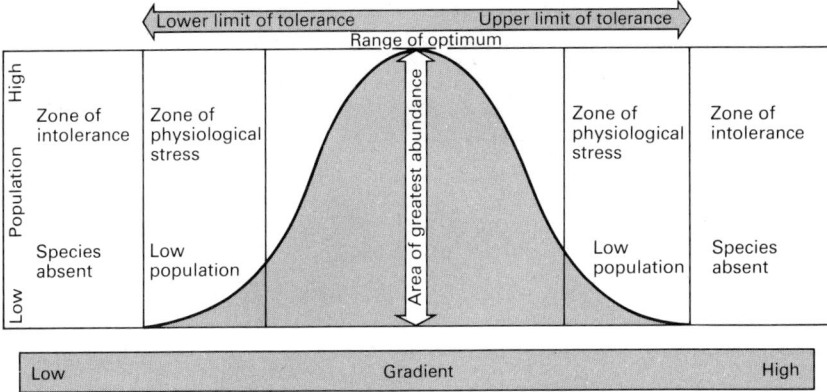

Fig. 1.1 A graphical representation of the abundance of an organism along the gradient of a physical factor in its environment.
Source: Cox et al., 1976.

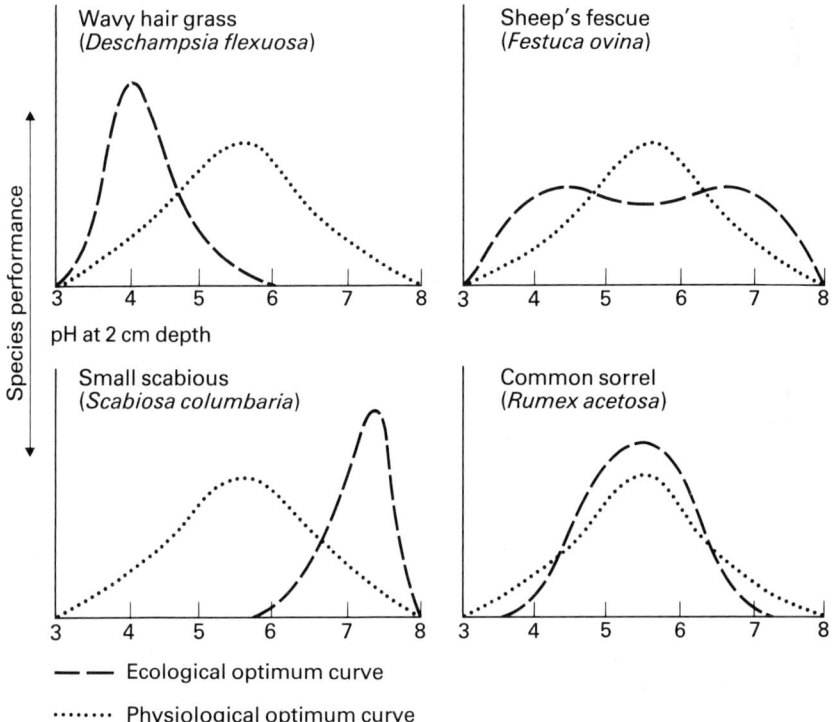

Fig. 1.2 The difference between the behaviour of four plant species growing in the field ('ecological optimum' curve) and under non-competitive conditions in controlled laboratory plots ('physiological optimum' curve).
Source: Collinson, 1978.

TABLE 1.2   Species richness along a latitudinal gradient

Taxon	Approximate number of species			
Approximate latitude	Florida (27°N)	Massachusetts (42°N)	Labrador (54°N)	Baffin Island (70°N)
Beetles	4000	2000	169	90
Land snails	250	100	25	0
Intertidal molluscs	425	175	60	*
Reptiles	107	21	5	0
Amphibia	50	21	17	0
Freshwater fishes	*	75	20	1
Coastal marine fishes	650	225	75	*
Flowering plants	2500	1650	390	218
Ferns and club mosses	*	70	31	11

* Data lacking.
Source: Clapham 1973.

TABLE 1.3   Number of species of mayflies along a stream

Collecting station	Water temperature (°C)	Number of species
1	9·0	7
2	16·3	15
3	19·5	16
4	21·5	22
5	20·5	21
6	24·0	29

Source: Ricklefs 1973, amended.

edge. The English daisy (*Bellis perennis*) will grow in high daytime temperatures but will not survive unless the night temperature drops below 10°C. Such a plant cannot therefore persist in constantly warm climates. Several instances have been shown where the distribution of a plant coincides with a particular environmental value, although such correlations are often hypotheses which lack experimental proof. The northern boundary of the wild madder *Rubia peregrina* in Europe coincides with the January 4·5°C isotherm (Fig. 1.3) and so it was suggested that this temperature is critical for the early phases of the growth of new shoots. At a more restricted spatial scale, Table 1.3 gives a simple example of how the number of mayfly species varied along a single stream according to the temperature. Another way in which temperature can be critical in distribution at a small scale is shown by the behaviour of an Australian species of locust *Chortoicetes terminifera* whose nymphal stage sought out microclimatic situations which were at a preferred temperature. This caused them to become crowded and hence apparently precipitated the migratory or swarming phase of their life-cycle (Collier *et al.* 1973).

Another critical abiotic factor in determining distributions is water. This is essential for all living organisms and so its virtual absence or its superabundance would clearly preclude the presence of all species not adapted to those particular conditions. The amount of water available to land plants from a given amount of precipitation is a function of such factors such as the ability of the soil to hold water, the ambient temperature, the wind velocity and features like rainfall interception (and subsequent evaporation)

Fig. 1.3 The distribution of wild madder (*Rubia peregrina*) and the location of the January isotherm for 4·5°C.
Source: Cox *et al.*, 1976.

by other plants. Where the effective moisture is decreasing then a plant adapted to dry **xeric** conditions may extend its range, as when overgrazing of semi-arid ranges by domesticated animals reduces the mulch cover of the soil and produces a drier microclimate. Plants growing in habitats subject to stress from aridity often show adaptations which enable them to store water or reduce the rate of its loss from leaves.

Light is of fundamental importance for the photosynthesis in the green plants but the actual periodicity of light incidence (daylength) can affect the ability of many plants to flower successfully; often a daylength must be within particular limits for flowering to occur in a species. In latitudes where daylength is outside these limits during the flowering period then the plant cannot reproduce sexually. Thus many southern species fail to flower in the north and vice-versa. Daily and seasonal fluctuations also regulate the lives of many animals, and especially major cycles such as migration and reproduction.

The concentration of essential gases such as CO_2 and O_2 is clearly important as well. The mixing of CO_2 into the surface layers of the ocean is one of the limiting factors in the photosynthetic rate of phytoplankton and is a consideration in the possible growth of algae in tanks where the soup of algae would have to be stirred or have CO_2 bubbled through it if it were not to be quickly self-limiting due to the lack of CO_2 penetration through the layers of cells at the water surface. Aquatic animals like fish are often very sensitive to O_2 concentrations in the water, and so many species are not found in water where O_2 levels are depleted because of pollution (Fig. 1.4).

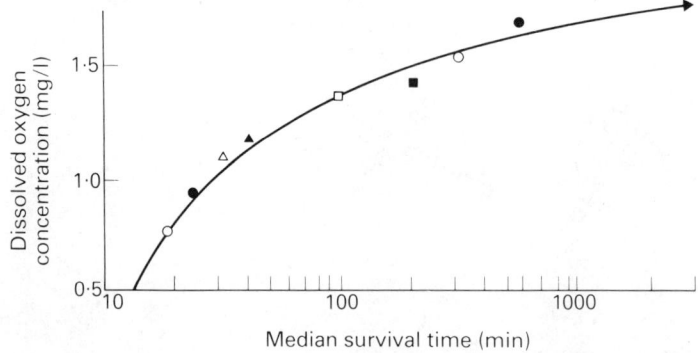

Fig. 1.4 The median survival time of young brook trout (*Salvelinus tontinalis*) in waters with different concentrations of dissolved oxygen. The trout had previously been raised in water at 10·5 mg/l. Source: Ricklefs, 1973.

In marine environments, environmental factors such as salinity are also powerful influences, sometimes providing barriers as where the very high salinities of the Red Sea have prevented the migration of Indian Ocean species into the Mediterranean. Conversely the low salinity of the Baltic allows freshwater species of plants and animals to appear off the islands of the Stockholm archipelago where there are fringes of common reed and freshwater fish such as pike and bream are found. The amount of desiccation of which marine plants and animals are tolerant may be a primary determinant of their location within an inter-tidal zone, other factors being equal. So a rocky shore may exhibit a zonation of brown algae according to the number of hours out of water each species can withstand: *Fucus spiralis* is much more tolerant of desiccation than *F. serratus* and *Laminaria* spp.

It is worth re-emphasizing that many of these environmental factors do not act in isolation to provide limiting factors in the distribution of a species. They interact among themselves and, in addition, the biotic relationships of organisms are clearly important in their distribution, a complexity which means that no one factor is likely to change without affecting the others (Billings 1974, Mooney 1974). Two things follow: it is not surprising that discontinuity of presence is more common than continuity, given the diversity of the habitats present on this planet; and the complexity of interactions is such that man-directed management aimed at modifying a limiting factor sometimes fails because the complexities of its interaction with other environmental influences have not been understood.

Relationships between individuals within species

There is often competition among individuals of the same species for an environmental resource in limited supply; this is known as intraspecific competition. Competition for light, nutrients and water is very often found in plants, especially among large numbers of young individuals. Work by Harper (1961) on seedlings of *Trifolium repens* can be cited: they were grown at different densities and one set of replicates was watered freely while the other had water withheld after the first 18 days. After seven weeks, the survival of seedlings mirrored their initial density: where regular watering had taken place, mortality was not dependent on density, but in the simulated drought conditions, mortality at a seedling density of 60 000 individuals per square metre was three times that at a density of 18 000/m² (Fig. 1.5). Competition at root level for a limited water supply

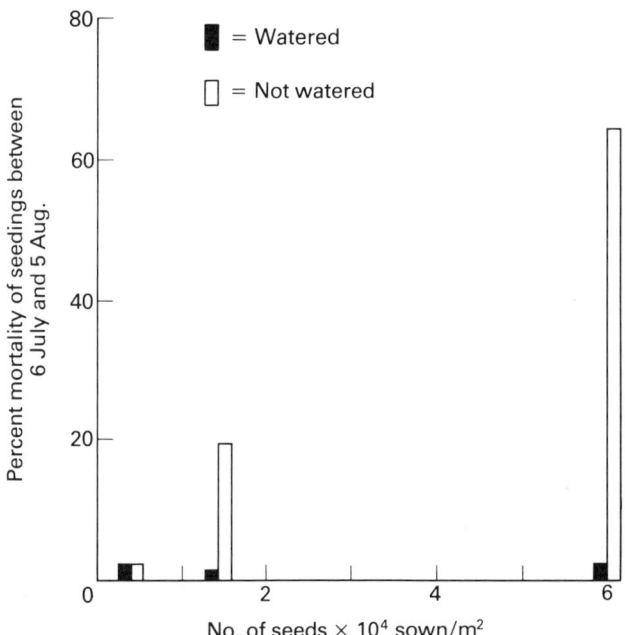

Fig. 1.5 Seed density and mortality in *Trifolium repens* after seven weeks. The unshaded bars represent the mortality of seedlings not watered after day 18.
Source: Collier *et al.* 1973.

had therefore resulted in the elimination of individual plants. A similar phenomenon occurs in desert plants of the saltbush type where the plants appear to be widely spaced but whose root systems in fact fill the underground space and are in contact laterally. The spacing of such plants may in fact reflect the rainfall over the lifetime of the individual, and desert areas with slightly higher precipitation totals have denser spacing of a particular species.

In animals, intraspecific competition may take an analogous form to that described above: a mass of maggots may feed off a piece of meat which is insufficient for them all. Each individual will try to get the resources it needs without regard for others. Many vertebrates and some invertebrates engage in contests in which an individual or group of individuals defend a territory against intrusion by other members of the same species: in birds, for example, a male will 'stake out' a volume of space for their activities which will include nesting and feeding. The number of territories in a given area is related to the total amount of, e.g., energy and nutrients available to the species and may vary in time. The size of a territory is a function of the total number of animals that can be maintained in the region and of the relative ability of individuals to defend them: a stronger individual (usually a male) can maintain a larger territory than a weak one. Individuals which cannot keep others out will be excluded from the breeding population so that territoriality seems to ensure that the population size does not exceed the resources available and also that the strongest individuals reproduce. The phenomenon has been detected most strongly in birds, ungulate mammals and wild primates and so much thought has been given in recent years by ethologists as to whether these kinds of considerations affect human behaviour or whether these 'animal' characteristics have been entirely superseded by 'cultural' behaviour; the issue is clearly not yet decided.

Relationships between individuals of different species

A number of different relationships exist here along a spectrum which ranges from outright and exclusive *competition* for space and resources, to cooperation for mutual benefit. At one end there is interspecific competition. At the early stage of establishment of a beech wood, for example, the tree seedlings coexist with other tree and shrub seedlings and a diverse ground flora. But once mature beech trees are established, their leaves cast such a dense shade that only a few species (e.g. the moss *Leucobryum glaucum*) can exist below a complete canopy. Some plants may exude substances from their roots which inhibit the growth of potential competitors; this is one case of a more complex phenomenon called allelopathy, which may involve micro-organisms as well as chemical agents. A common plant which exhibits the phenomenon is bracken fern (*Pteridium aquilinum*) which often inhibits the growth of tree seedlings by this means.

Another relationship is a differential coexistence where two species exist side by side in space but utilize different resources. Two chaparall shrubs, chamise (*Adenostoma fasciculatum*) and red shank (*A. sparsifolium*) grow together but the first has an extensive tap-root system which penetrates deeply into fractured parent material below the surface soil. The other has no tap root and indeed all its roots are concentrated within the top layer of the soil; a series of thick buttress growths help to support the shoot. Temporally, chamise flowers and sets seed earlier in the year than red shank (Fig. 1.6). These adaptations of morphology and physiology allow two closely related species to coexist: competition during their evolution seems to have produced a variety of species to take up most of the available resources.

A well recognized relationship is that of **predation** which applies to both plant–animal food chains and those involving only animal predation. The requirement for stability in this relationship is clearly the actions of the predator in controlling its numbers so as not to over-use the prey to the point where it fails to yield a regular food supply.

Relationships between individuals of different species 21

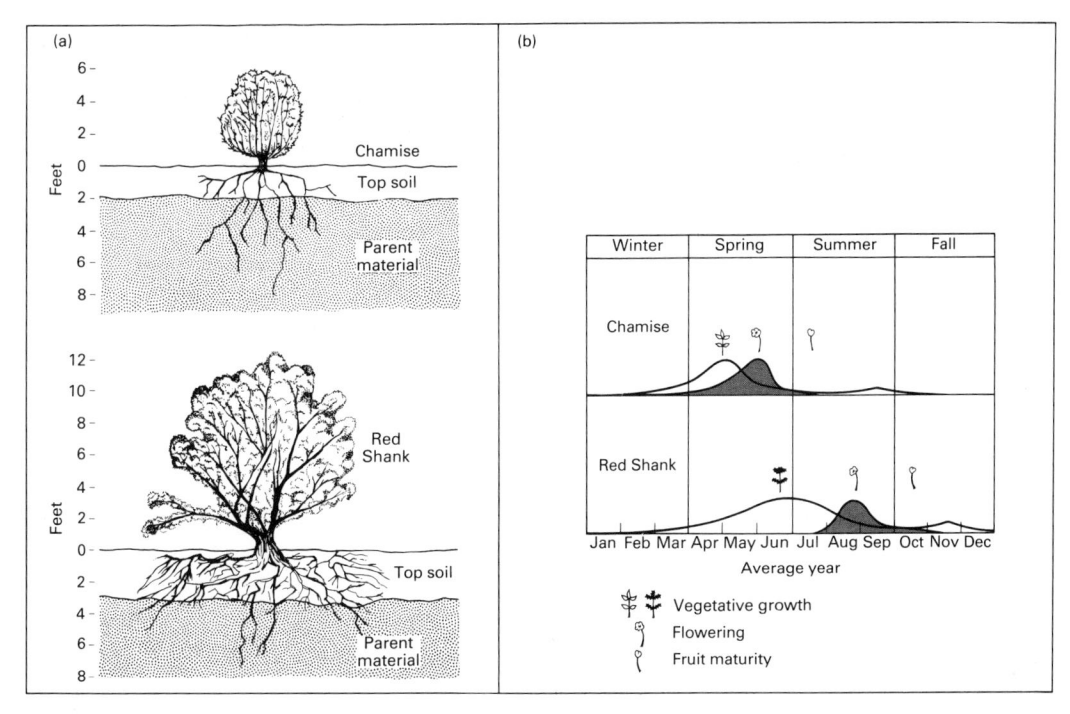

Fig. 1.6 The separation of two closely growing species, chamise and red shank. **A** shows how the root systems are different, with that of chamise penetrating the parent material and of red shank being largely of lateral roots in the topsoil; **B** shows the metachronous nature of their yearly cycle. Thus the two species minimize competition between each other.
Source: Collier et al., 1973.

The effect of predation can be beneficial to the population of prey also, as is shown by the history of Isle Royale in Lake Michigan where a moose (*Alces alces*) population became established in 1908 and by 1930 had expanded so fast that the vegetation of the forest and the small lakes of the island was being damaged beyond recovery by regeneration. Periodic die-offs of large numbers of moose occurred. About 1948, a wolf population was established. Since moose formed about 75 per cent of the wolves' diet, their population was substantially reduced to a stable level where regrowth of vegetation could occur and the die-offs have ceased. In this case, the relationship clearly stabilizes the whole ecological system.

Other relationships can include **commensalism** where one partner benefits and the other neither benefits nor suffers. A bird species may need the tree fork for its nest but no benefit accrues to the tree; epiphytes such as Spanish moss (*Tillandsia usneoides*), a bromeliad, benefit from the light and moisture conditions around the tree on which they grow, but the tree, so far as is known, gains nothing. Neither does the shark which transports the remora fish *Echeneis naucrates* around. The fish hangs on by a sucker, drops off to clean up scraps of food left by the shark and then clings on again to ride to the next meal. In the case of **parasitism** one party benefits strongly from an interaction but the other suffers; e.g. an intestinal tapeworm needs the gut in which to live but the host gains nothing and may well be debilitated by its presence. The successful parasite (in natural as in social systems) does not kill hits host, however. In a more balanced fashion, **mutualistic** interactions are found which are beneficial to both species. One of the outstanding examples is the pollination of flowering plants by insects, where an insect relies on one species of plant for its existence and the plant is pollinated by no other species of insect. The tropical orchid *Stanhopea grandiflora* and the bee *Eulaema meriana* have developed a very close relationship in which *Stanhopea* provides the male bees which visit it not with nectar but with a fragrance which it appears enables them to attract mates. The pollen-bearing and receiving parts of the flower are exactly the right size to brush against the pollen carrying parts of the males of this species of bee. In general, though, plants have evolved showy flowers to attract pollinators, nectar (a sugar solution) provides a food incentive for insects and groups like humming birds; any surplus pollen may be eaten for its high protein content, and the structure of the flowers is such that the pollinators will transfer pollen efficiently from one flower to another.

In overall perspective, it seems as if the effect of evolution has been to reduce the pressures of competition and to produce a division of food and other resources between a variety of species: competition of an intraspecific kind may result in the evolution of two different species with different tolerances; competition of an interspecific nature may result in the partitioning of the resources of the habitat often by means of bodily or behavioural adaptations. Such trends are mutually advantageous since the risk of extinction is reduced but beyond that consideration, they result in a community of many species, each with its own differing food sources and specialized adaptations to the point where no single species is so dominant (our own excepted) that it can displace all or most of the others. This makes for stability in natural ecological systems at their mature, self-maintaining stage, since the system has few resources for invaders unless they are possessed of a strong competitive ability. However, ecological systems with relatively few species, for instance at an early stage of their development or on oceanic islands removed from the mainstreams of dispersal, are particularly vulnerable to rapid change from invading organisms, as from the rats, goats and pigs which went ashore from the ships of explorers and other seamen onto islands of the Caribbean and Pacific in the seventeenth and eighteenth centuries.

The inheritance of biological characteristics

The offspring of living organisms resemble their parents in most ways, yet may not be exactly the same, so that there is some variation even within a species. The inheritance of major characteristics and the onset of variation is the field of **genetics**. The basic units of inheritance are the **genes**, which are situated on **chromosomes**, themselves thread-like structures within the nucleus of each plant or animal cell. Each species has a fixed number of pairs of chromosomes, e.g., 3 in the mosquito, 7 in the pea and 23 in man. In sexual reproduction, the male and female each contribute half the number of chromosomes so that one of each pair comes from each parent; thus an organism inherits characteristics from both parents. Not all genetic traits have equal strength, however, and some may be **dominant** over others which exist in the genetic material but are not exhibited in the measurable characteristics of the organism (i.e. are **recessive**). The genetic make-up is called the **genotype** and its outward expression the **phenotype**. For example, in Mendel's experiments with garden peas, tallness was dominant over shortness, said to be **recessive**. So if the genetic factors for tall and short were present in one individual, that plant would be tall since tallness had masked shortness. But if such **hybrid** tall plants were crossed with each other, then the offspring would be both tall and short in a 3:1 ratio.

	Phenotype	*Genotype* (T=tall; t=short)
Parent generation	Tall × short plant, both pure breeding	TT × tt
F$_1$ hybrid	Tall plants	Tt × Tt
F$_2$ offspring of hybrid cross	(× tall plants) 3 tall 1 short	TT Tt Tt tt tall short

Thus the genes for each trait were unaltered during reproduction, but there was a recombination of them. Modern techniques of artificially implanting genes onto chromosomes open up a new world of **genetic engineering** which is discussed on p. 256.

In some organisms, groups of genes are inherited together and do not assert themselves independently in the offspring: this is termed linkage. Further, many continuously varying phenotypic traits, e.g. size and colour, are determined by the cumulative contribution of many genes, a condition called multiple gene inheritance. This is one of the most important of all gene interactions in the environmental relationships of plants and animals: it is the basis for nearly all the adjustments in size of organisms and rates of physiological processes, as well as behavioural traits, by which plants and animals can respond to changes in their environments.

In a natural population there is always a degree of genetic variation and this has long been viewed by biologists as beneficial to the population although some individuals will necessarily be among the variants less fitted to survive or reproduce. The primary cause of variations in populations is **mutation**. The replication of genetic material during cell division is not always perfect and occasionally copies are produced which are not faithful to the original: the structure of the genetic material itself (dioxyribonucleic acid,

TABLE 1.4   Comparison of mutation rates, expressed per generation and per unit time

Species	Range of mutation rates per generation	Midpoint	Generation time (days)	Modal mutation rates per day
Bacteria	10^{-10} to 10^{-6}	10^{-8}	1 hour ($10^{-0.4}$)	$10^{-7.6}$
Fly	10^{-7} to 10^{-5}	10^{-6}	3 weeks ($10^{1.3}$)	$10^{-7.3}$
Mouse	10^{-6} to 10^{-5}	$10^{-5.5}$	2 months ($10^{1.8}$)	$10^{-7.3}$
Corn	10^{-6} to 10^{-4}	10^{-5}	1 year ($10^{2.6}$)	$10^{-7.6}$
Man	10^{-6} to 10^{-3}	$10^{-4.5}$	20 years ($10^{3.9}$)	$10^{-8.4}$

Source: Ricklefs 1973.

DNA) may change, or a whole chromosome may undergo gross changes. These alterations appear to be random in occurrence but under 'normal' conditions, the mutation rates for a variety of different organisms seem very similar (Table 1.4); the reasons for this are unknown.

The mutation therefore becomes established in the population's genotypic structure though since mutations are not necessarily dominant characteristics, they may not find expression in the phenotype until several generations later. This phenotypic expression usually reduces the fitness of an individual to survive and reproduce, especially if the population is already well adjusted to the conditions of its environment. Gross effects in the phenotype particularly reduce fitness, whereas smaller effects are more likely to produce a few individuals with a selective advantage. Most mutations are not likely to persist but a very few may be the means to continued existence of the species.

Genetic variation can also occur during cell division by the process of **recombination**. During the cell division in each individual's reproductive organs that precedes sexual reproduction, the number of chromosomes is halved in each daughter-cell: only one chromosome of each pair goes to each half (**gamete**) of the eventual whole formed by the union of male and female genes (the **zygote**). During this process the chromosomes cross over and genetic recombination can take place at the separation of the chromosomes; thus, the genotype of the gametes can differ from each of their parents and so the genetic structure of the zygote will be different (Fig. 1.7). As with mutation, it provides a source of variation, in terms of the number of gene combinations in a population, which enables a population to adapt to environmental change.

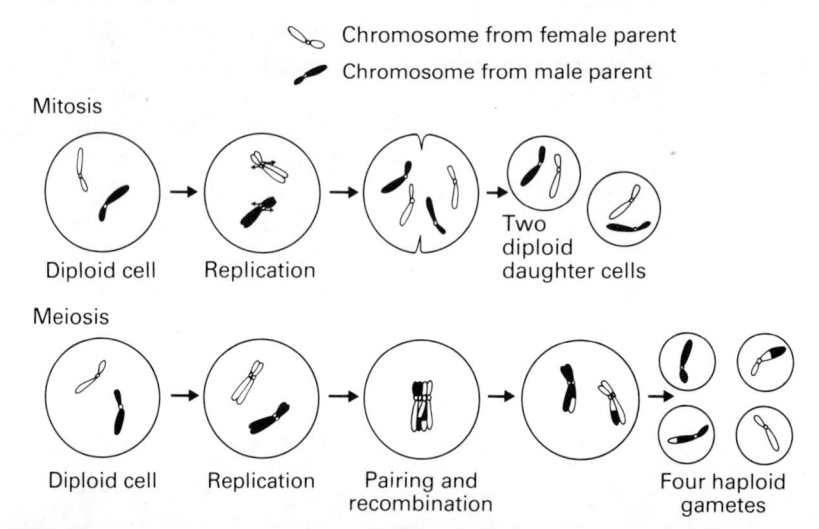

Fig. 1.7 A simplified representation of the behaviour of a pair of chromosomes during mitosis—a simple division of the cell resulting in the formation of identical daughter cells; and in meiosis—the formation of cells which will form part of a new gamete with different combinations of characteristics thus leading to variation within the species.
Source: Ricklefs, 1973.

Natural selection

This process, usually considered very important in the process of **evolution**, is defined as the differential survival and reproduction of individuals carrying varying genetic make-up. Genetically distinct individuals tend to contribute differentially to the number of progeny in future generations, thus effecting a gradual drift in the genetic composition of the population. An example given by Ricklefs (1973) is shown below:

In general
a) The reproductive potential of populations is great but
b) populations tend to remain constant in size, because
c) populations suffer a high mortality.
d) Populations exhibit variation which leads to
e) differential survival of individuals.
f) Individual traits are inherited by offspring.
g) The composition of the population changes by the selective elimination of unfit individuals.

An example
a) Rabbits should cover the earth, but
b) they don't because
c) many are caught by predators.
d) Some rabbits run faster than others,
e) and escape from predators.
f) So do their young.
g) Populations of rabbits, as a whole, tend to run faster than their predecessors; so do those of their predators!

A well known example of selection at work is the development of dark pigmentation (melanism) in British populations of the peppered moth (*Biston betularia*). Early in the nineteenth century, occasional melanistic specimens of this moth (the form called *B. b. carbonaria*) were collected. Over the next 100 years, the *carbonaria* form of this moth became increasingly common in industrial areas until it formed nearly 100 per cent of those populations. Where there was relatively little industrialization, the light form of the moth was still the most frequent. Release and recapture experiments show that in polluted areas the dark form survived better than the light, and *vice-versa* in unpolluted areas (Plate 4). Thus the moth population seems to have undergone an evolutionary adaptation to survival in polluted areas where presumably the dark form resting on grimy surfaces is less prone to predation by birds than the light form and hence contributes more of its genes (including those for melanism) to the future generations. The implementation of the Clean Air Act in the 1950s, which has reduced smoke levels in British cities, should in theory reduce the representation of the *carbonaria* form in the urban and urban-fringe populations, and recent work (Askew *et al.* 1971) has shown that this is in fact happening. These observations show that **natural selection**, operating over a period of decades in this case, resulted in genetic changes in the population of the moth in contaminated areas. The actual agents of selection are insectivorous birds, with the colour of the background against which the moth spends the day as a critical aspect of the environment.

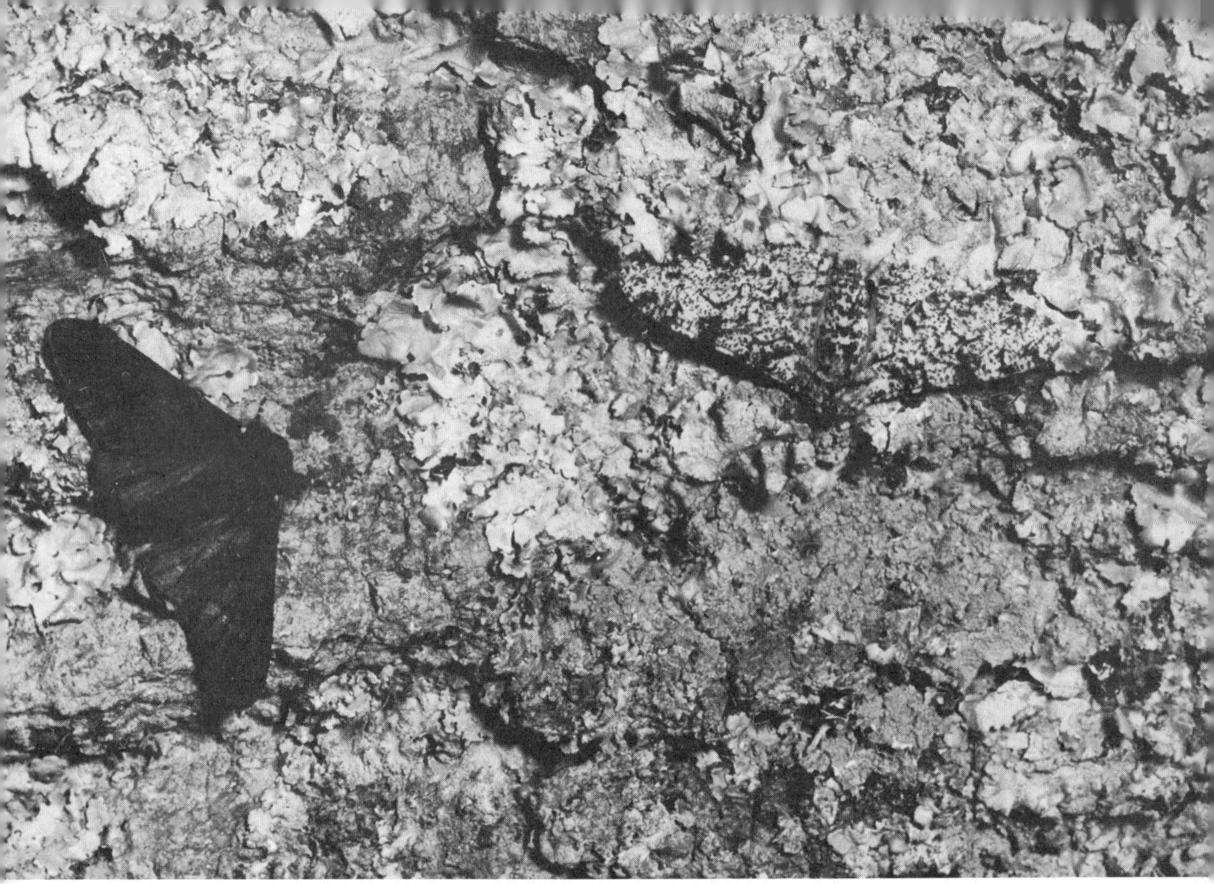

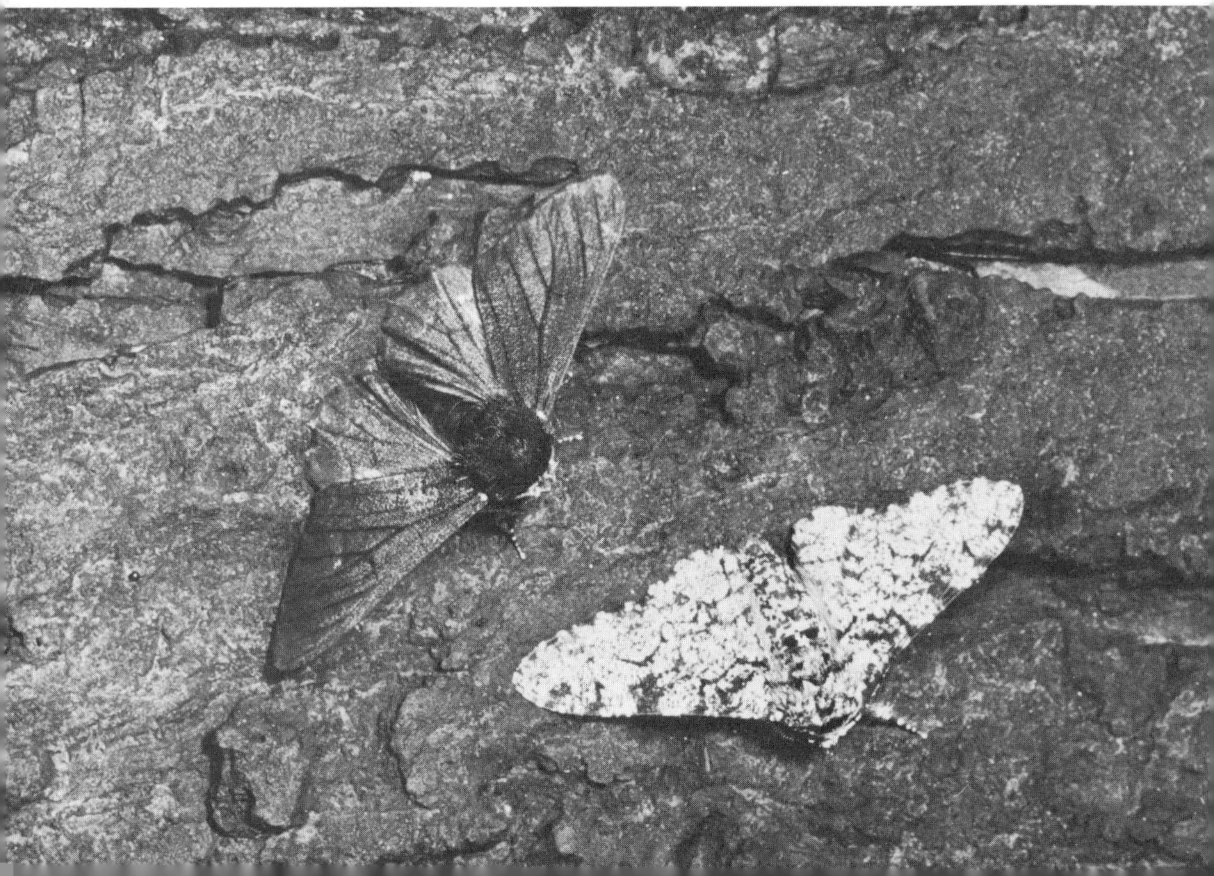

Speciation and evolution

Speciation is usually defined as the irrevocable separation of populations into independent evolutionary units. The concept of a reproductively isolated evolutionary unit though easy to grasp intuitively, is difficult to segregate in nature; however, the result of evolutionary change is normally the accumulation of genetic differences between two populations to the point where they cannot interbreed. Speciation therefore requires more than just evolutionary change within a population: it requires divergent change in two or more populations to the point of reproductive isolation. Notwithstanding, those processes may be bypassed where major chromosome mutations caused by environmental factors (e.g. cosmic-ray radiation) bring about 'instant' speciation.

Because evolutionary change takes the form of the replacement of individuals in a population, generation by generation, the evolution of, for example, a morphological, physiological or behavioural adaptation may require a long time. Admittedly, a sudden change in environment may immediately and completely alter the relative fitness of different genetic traits and change the course of evolution, but a population will continue to reflect earlier conditions (unless it is made extinct by an inability to tolerate the new conditions or is competitively displaced) until a complete evolutionary adjustment to the new environment has occurred. Thus the degree to which a population is well adapted depends partly on the rate of evolutionary change in a population and partly on the rate of environmental change. (It is interesting to consider the evolution of our own behaviour in this context: are we, in the West, now behaviourally adapted to living in an urban-industrial environment or do we still have traits derived from the 90 + per cent of our evolutionary time as hunter-gatherer?) The time taken by a population to adjust to new conditions varies greatly, but a key factor is the length of generations: a bacterial population may adapt to the addition of an antibiotic to a culture in a matter of hours or days; a long-lived and slowly reproducing organism such as man may require many millennia to make an evolutionary transition of any magnitude, if we exclude behaviour. In between these extremes, the theoretically calculated time required for the *carbonaria* form of the peppered moth to increase from a frequency of 0·05 to 0·95 in a polluted area is about 47 generations, and from a frequency of 0·01 to 0·99, about 204 generations; these estimates accord well with the observed times. Another example of temporal change comes from the trapping returns of foxes at a trading post in Labrador during the years 1834–1933 which show that the percentage of silver foxes in the catch declined from about 0·15 to 0·05 during the course of a century: clearly trapping was selecting against the survival of that variant in the population.

Long-term evolution

The complex hypotheses about the nature of evolutionary change in no way detract from the value of the record of the progress of change in plant and animal form to be found in the fossil record. During the evolution of life on this planet genetic groups have gone through periods of rapid evolution and diversification followed by quiescent periods of little alteration, or even eras of major extinctions. Thus our present flora and fauna are the outcome of perhaps 3000–3500 million years of evolutionary change

Plate 4 The peppered moth *Biston betularia*: (a) the normal and melanic (*carbonaria*) forms on a lichen-clad oak trunk; (b) both forms on a soot-blackened tree trunk.

PLANTS		Age at base in million years	ANIMALS
Holocene — Dominance of herbs			Dominance of man, modern species of mammals
Pleistocene — Increase of herbs / Decrease of trees		2	Extinction of large mammals e.g. mammoth, Early Man
⌐ AGE OF FLOWERING PLANTS			⌐ AGE of MAMMALS
Pliocene / **Miocene** / **Oligocene** / **Eocene** — Herbaceous plants especially grasslands / Tropical forests / Modernisation of flowering plants.		65	Abundant species of mammals / Modern mammals appear / Evolutionary explosion of mammals
Cretaceous — Flowering plants dominant evolving parallel with insects / Conifers decreasing		136	Extinction of large reptiles and ammonities. Bony fish abundant / Pollinating insects evolve
Jurassic — First flowering plants / Cycads dominant		190	⌐ AGE OF REPTILES / dominant on land, sea and in the air / First birds and mammals
⌐ AGE of CONIFERS			
Triassic — Conifers dominant / First cycads		225	First dinosaurs, ichthyosaurs, plesiosaurs and turtles
Permian — Decrease of ancient plants / Seed ferns extinct		280	Radiation of reptiles replacing amphibia
			⌐ AGE OF AMPHIBIA
Carboniferous — First mosses and conifers / Great tropical coal forests		345	Radiation of amphibia. First reptiles / Rise of insects / Sharks and sea lilies abundant
⌐ AGE of SPORE — BEARING FERNS			
Devonian — First liverworts and clubmosses, horsetails, ferns and seed ferns, many the size of trees		395	First amphibia / Early radiation of fishes, mostly in fresh water
			⌐ AGE of FISHES
Silurian — Plants invade land		440	Arthropods invade land. Wide expansion of invertebrates
Ordovician — Marine algae dominant		500	Appearance of jawless armoured vertebrates. Brachiopods and cephalopods dominant
⌐ AGE of ALGAE			
Cambrian — Diversity of algae / First spores		570	Appearance of all invertebrate phyla / Trilobites dominant
Precambrian — Traces of bacteria, fungi, algae back to 3000 / Blue greens 3500			Worm tracks and burrows, jellyfish / First autotrophs (no free oxygen)

Fig. 1.8 An outline chronology of major stages in the evolution of plants and animals.
Source: Archer et al., 1976.

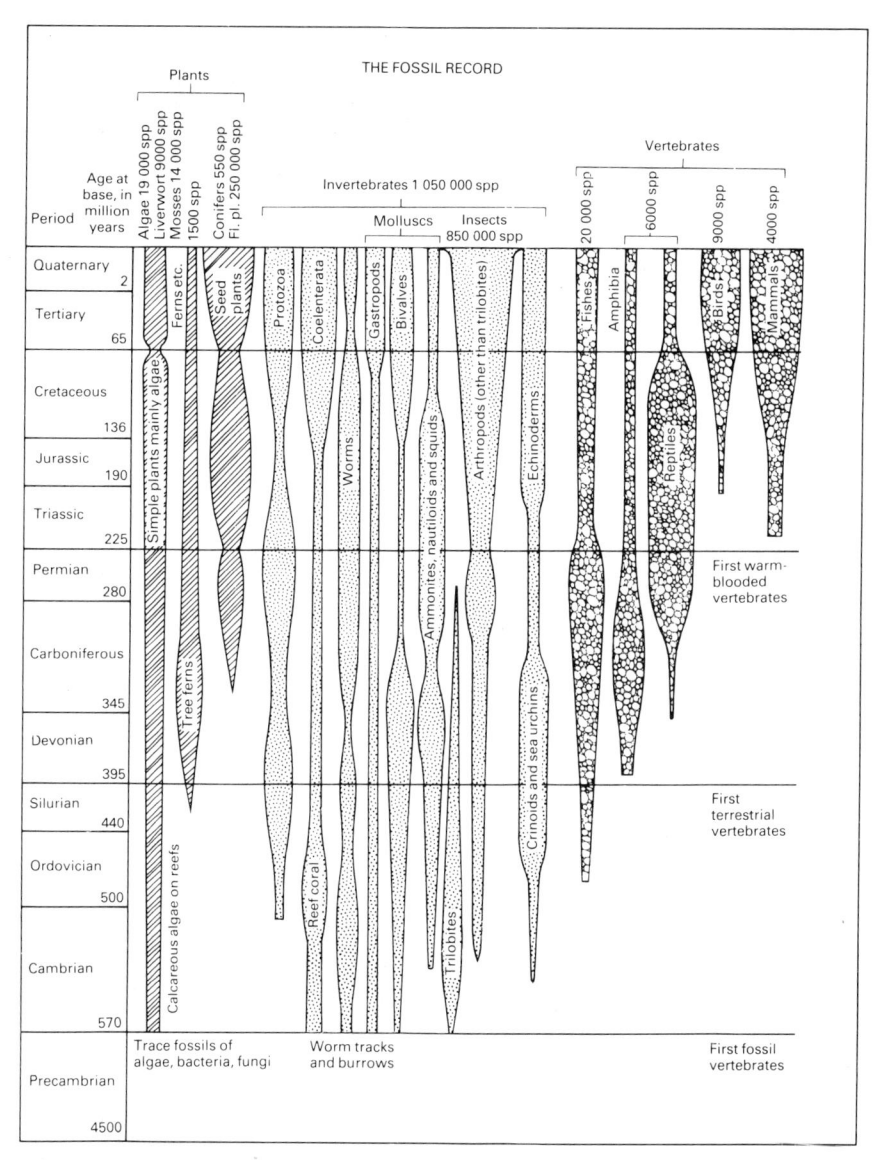

Fig. 1.9 An outline chronology of plant and animal evolution showing the relative importance of different groups at different stages and the approximate number of species now living.
Source: Archer et al., 1976.

with perhaps 1 or 2 million years of the selective effects of man superimposed upon the last years of that change (Figs 1.8 and 1.9).

In the total span of life on Earth plants have covered the land surface in some abundance for only about 400 million years. Their relatively late evolution indicates the difficulties presented by the terrestrial environment in terms of feeding, support and reproduction, although clearly other planets have presented apparently insuperable problems.

The first photosynthetic cells on earth were probably blue-green algae: they are known as successful colonizers of recently exposed volcanic terrain, for example, and their decay would add the first humus to the rock particles produced by erosion. They can also fix atmospheric nitrogen, thus adding nitrates to the soil for other plants to use later. The earliest land plants were leafless but soon leaves evolved, with the first ferns having very large fronds, and the **clubmosses** reaching the size of trees. The first **seed-ferns** appeared at the end of the Devonian (ca 345×10^6 yr ago) and by the end of the Carboniferous (ca 280×10^6 yr ago), primitive conifers began to evolve among the spore-bearing flora of the 'coal' forests. Conifers bear seeds rather than spores, but carry them exposed on the surface of the scales of the cones, not enclosed in **carpels** as in the Angiosperms. (Refer back to Table 1.1 for details of these taxonomic groups.)

The Angiosperms seem to have originated in the late Jurassic or early Cretaceous in a primitive continent which eventually became South America and Africa. Although relatively insignificant in the early Cretaceous, they had become dominant by the middle Cretaceous, when some Angiosperms characteristic of present-day southern temperate regions, e.g. the Proteaceae and *Nothofagus* (the southern Beech) had already appeared. By the end of the Cretaceous, the Angiosperms had differentiated into a vast number of forms which dominated the vegetation of most of the Earth's terrestrial habitats and much of the aquatic plant life as well. It is also interesting that a considerable proportion of the genera, and even species, which are now familiar as common trees and shrubs, seem to have persisted throughout the Tertiary and Quaternary with little or no apparent change. That is not to say that their distributions have remained constant: in early Tertiary (Eocene) times, for example, some individual species of plants tended to be much more widely distributed than today (e.g. palms were common in England and Canada) and forests much more widespread than grasslands and other open vegetation. By the last period of the Tertiary (Pliocene) there seems to have been much more relative cooling in the northern hemisphere (perhaps less so in the southern hemisphere) with the result that the vegetation became much more like that of the post-Pleistocene period, with some of the forests being replaced by natural grasslands. Nevertheless, the same major plant groups seem to have persisted through the Pleistocene, coming back after displacement by Pleistocene glaciations with only minor shifts in the species composition, a striking decline in the number of coniferous species being perhaps the most notable other characteristic.

In the case of animals, many soft-bodied invertebrates must have evolved before the land was invaded but have left few fossil traces. The conquest of the land from shallow water would have been relatively easy: earthworms for example are not very different from their aquatic relatives and can only exist under damp conditions. Successful adaptation to dry land required a number of adaptations other than the water-conserving skin or shell, and the reptiles were the first group to possess them, including a skeleton for muscle attachment, a shelled egg containing food for the developing embryo, internal fertilization, strong legs for lifting the body clear of the ground to facilitate movement, and the processing of nitrogenous waste to a less toxic form than the ammonia excreted by aquatic animals, i.e. to urea or uric acid. The reptiles evolved into a great number of species in the Mesozoic ($225-66 \times 10^6$ yr ago) and dominated the Earth for 150 million years, adapting to every possible habitat, including the air. At the end of the Cretaceous most of them became extinct and now there are only turtles, crocodiles, snakes and lizards.

The fossil reptile *Archaeopteryx* had feathers on primitive wings and could only glide but from its stock the birds may have evolved, undergoing **adaptive radiation** during the Eocene. All birds are like feathered reptiles but have hollow bones, no teeth, and are 'warm-blooded'; clearly the ability to fly enabled them to colonize a whole range of habitats hitherto unoccupied, especially at a time of forest dominance.

The end-point of animal evolution from our present stance in time is the mammal. The first true mammals appeared in the Triassic and were tiny shrew-like animals with sharp teeth, probably feeding on worms and insects and having proportionately larger brains than reptiles. Before the end of the Cretaceous, both marsupial and placental mammals had appeared: marsupial eggs develop inside the female but there is no **placenta** to supply continuous nourishment to the embryo. Young marsupials are therefore born in a very immature state and finish their development in a pouch, feeding on milk from mammary glands. Early marsupials were highly successful and world-wide in their distribution until the placental mammals underwent a period of rapid evolution during the early Tertiary which spread them over most of the world; eventually they replaced the marsupials except in Australasia (Plate 8). Many mammalian features are improvements on reptilian adaptations: they can move faster because the limbs bend underneath the body and not sideways as in reptiles; they have three kinds of teeth, for biting, tearing and chewing; being warm-blooded they can be active most of the time regardless of the external temperature; hair, subcutaneous fat and sweat glands regulate the temperature. The possession of a placenta allows the embryo to be nourished inside a womb and thus the offspring are born in a well-developed state; post-natal feeding is ensured by the production of milk from mammary glands and parental care, including the teaching of behavioural patterns, is normal; and there is an enormous increase in the size of the cerebral hemispheres so that the development of higher brain centres in a cerebral context can occur, a trend associated with an enhanced ability to learn and with 'intelligent' behaviour in the class of mammals including our own, the Primates.

Evolution in the Primates and man

The Primates are a group of mammals which include the monkeys and apes as well as our own species. In general they seem to have undergone much of their evolutionary history in tropical forests where the sedimentary environments that form fossils are scarce and so remains are infrequent. The original stock seems to have been a small insectivore of the late Cretaceous similar to the tree shrew of today but with a larger brain. These organisms could move rapidly through the trees and possessed the advantages of binocular vision, a relatively large cerebral cortex and flat nails rather than claws giving an improved grip to the finger pads and enabling opposed thumbs and big toes to grasp small objects. From these beginnings man's ancestors evolved, although the fossil record is full of gaps (especially between 25 million and 4 million years ago) so that the story is by no means complete (Fig. 1.10).

32 *The organism scale*

Approx. times of Ice Ages	Number of years ago	Species	Brain size in cm³	Skull type
Ice Age	30 000 / 40 000	Homo sapiens sapiens (Cro Magnon man)	1200/1500	Cro Magnon man
Ice Age	100 000 / 250 000	Homo sapiens neanderthalensis (Neanderthal man)	1300/1425	
	400 000	Homo erectus pekinensis (Peking man)		Peking man
	600 000	Homo erectus (Java man)	900/1200	
Ice Age				Java man
Ice Age	1 000 000		770/1000	
	1 800 000	Australopithecus		
	3 000 000		450/550	Australopithecus

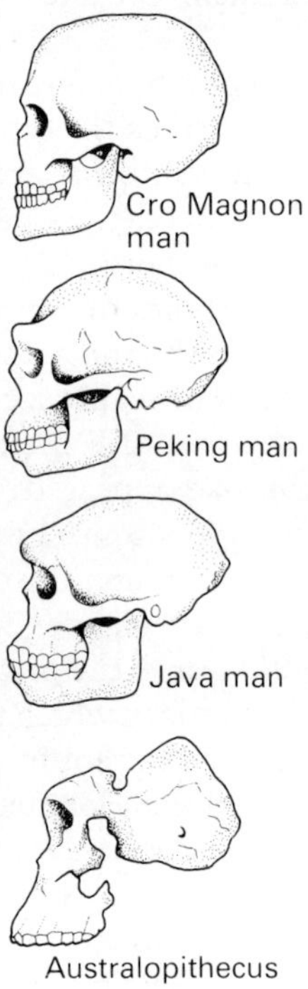

(Pleistocene / Pliocene)

Fig. 1.10 Stages in the evolution of man, with illustrations of the skull shape of each type. Others have now been found which are not portrayed in this sequence.
Source: Archer *et al.*, 1976.

Abundant evidence for the ancestry of man starts about 4 million years before the present with the fossil beds of East Africa, such as Olduvai in Tanzania. Discoveries are still frequently made from this area, and so any picture must be taken as highly provisional. Between 4 million and 1 million years BP several species of hominids existed with brain capacities of 450–550 cm³ (the chimpanzee averages 500 cm³), known usually as *Australopithecus* (Plate 5). These animals were probably nomadic hunters and food-gatherers but lacked the large canine teeth of the apes. Their hands resembled those of modern man and large pebbles were probably used as tools or in hunting. It is not clear which of these species was the direct ancestor of modern man (if indeed it was the only one, or one of these at all) but during the Pleistocene a later species evolved in the tropics of Africa and Asia which had a developed stone tool culture involving some specialized implements, was both erect and taller than earlier groups and had a striding walk. It is usually called *Homo erectus*. Its skull was ape-like with very strong jaws but had a large brain capacity (770–1200 cm³) and it is thought likely that the males formed cooperative hunting bands and could probably communicate by speech; per-

Plate 5 A restored skull of an early hominid from Olduvai Gorge, Tanzania: *Australopithecus* (*Zinjanthropus*) *boisei*.

haps they could even count, if notched sticks found with their remains may be thus interpreted. Towards the end of the Pleistocene, *ca* 250,000 yr BP, human remains are found in extra-tropical regions such as Europe and the name *H. sapiens* is usually applied to this group, with the subspecific name *neanderthalensis*. Although the skull of Neanderthal man was low-browed and apelike (a sort of archetypal punk rocker), the brain capacity at 1300–1425 cm³ was almost that of modern man. By 40 000 yr BP the first 'modern' man in biological terms was present: Cro-Magnon man, designated as *Homo sapiens sapiens*, about 1·75 m tall and with a brain capacity of 1500 cm³, a small jaw and dentition like our own, and a vertical chin and forehead. No human species have appeared since then, if we except bionic persons. The economy of these groups was hunting and gathering and until the waning of the ice from the temperate zone they

34   *The organism scale*

are, together with earlier groups, culturally classed as Palaeolithic; after the Palaeolithic, evolution in man is cultural rather than biological.

Natural extinction

Throughout the evolution of living organisms, taxonomic units have come and gone. The extinction of a species or other group has come about because it could not adapt well to change in that environment or because other species evolved which were better adapted to life in a particular environment (Plates 6 and 7). Several major extinctions happened during the pre-Pleistocene periods of geological history, for example the Pteridosperms accounted for a large part of the forest-like vegetation in the Carboniferous–mid-Jurassic periods but became extinct by the beginning of the Cretaceous; a large

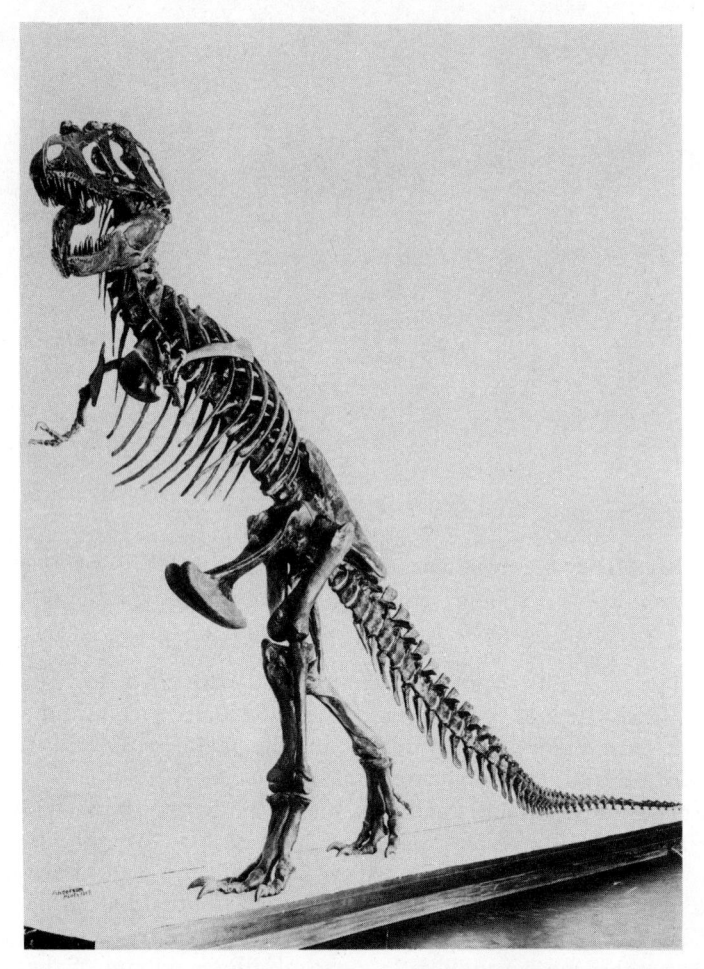

Plate 6   An extinct animal: the large reptile *Tyrannosaurus rex* from an age when reptiles dominated animal life.

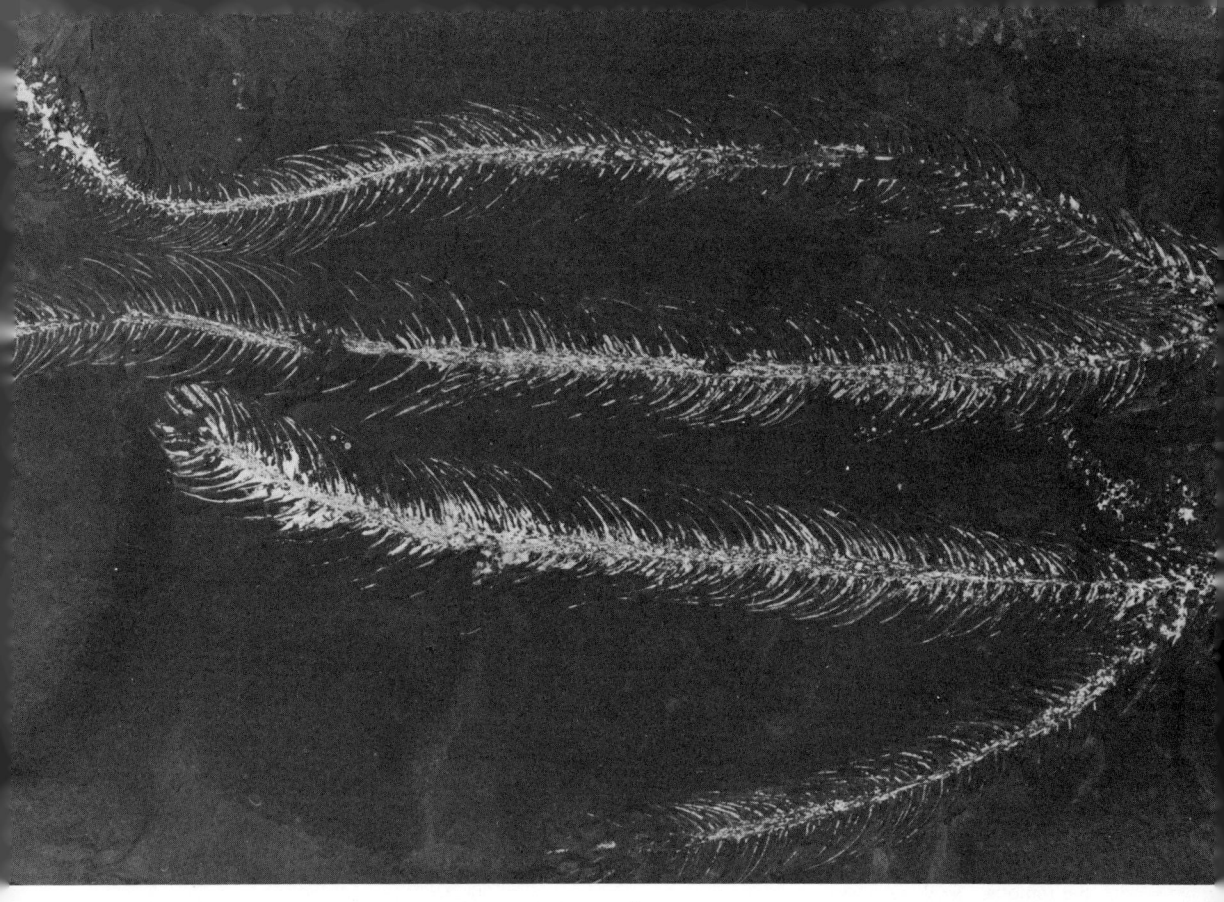

Plate 7 The leafy shoots of a now-extinct fossil Lycopod.

number of Coniferae became extinct during the Cretaceous, apparently displaced by Angiosperms. In the animal kingdom, the Permian was a major period of animal extinction, especially of marine animals such as Trilobites. Some estimates suggest that nearly half the animal families previously in existence became extinct during the 50×10^6 yr of the Permian. The dinosaurs seem to have become extinct relatively rapidly during the Cretaceous.

The conclusion to be drawn is that extinction is part of the normal process of organic evolution. Upon this process, the human species has imposed its own rate of extermination of species, e.g. in the terminal Pleistocene (p. 158) and in the modern period (pp. 218–19). The rate at which *Homo sapiens* causes extinctions is much faster than the natural rate (Myers 1976) and although scientific knowledge can produce new variants from existing material, it cannot replace what is lost entirely.

Distribution patterns of individual taxonomic units

In this section we are concerned with the patterns of distribution of classificatory groups like species, genera and families. Such distributions can be looked at on a variety of scales, from the whole globe down to an individual habitat, and different factors operate at different intensities in determining these distributions.

The world scale

The best known of the classifications applicable at this scale is the set of **zoogeographical realms** which each encompass a large terrestrial area (Fig. 1.11) and contain a fauna

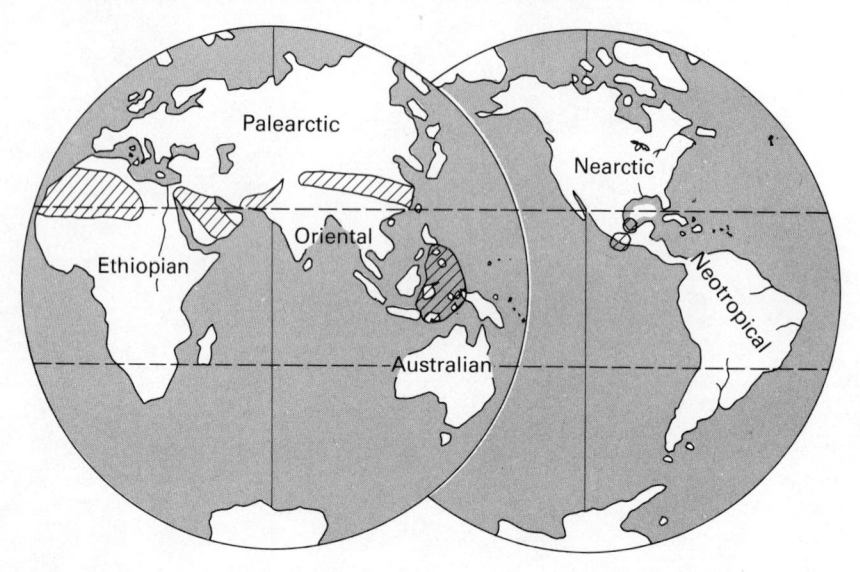

Fig. 1.11 The six major zoogeographical realms. The diagonal shading represents transitional areas between the regions.
Source: Knight, 1965.

particular to that region, although some elements may be confined to a small part of it. The basis of the realms is partly climatic and partly based on the existence of barriers which impede the movements of organisms. These realms were first proposed in the nineteenth century and have undergone relatively little modification since: basically the Palaearctic and Nearctic realms are separated from the tropical areas by climatic differences; the Neotropical and the Australasian are segregated by oceans. The Ethiopian and Oriental realms are not completely separate from their neighbours.

Considered individually, each realm shows particular characteristics. For example, the tropical part of the Ethiopian realm has a particularly diverse (i.e. many different species) fauna which diminishes to both north and south. The region is noted for its suite of game animals but some others are confined to this realm such as various species of shrews, the aardvark, hippos and some species of monkey. The Oriental realm has a fauna rather like the Ethiopian (both have a species of elephant, for example), but with differences such as a larger number of rodents. Both realms have about 66–67 families of birds, but only one of these is confined to the Oriental realm: the Irenidae, a group of leafbirds and fairy bluebirds. Among the fish, a great diversity of cypriniforms (carp-like) is characteristic.

The Palearctic realm is one of the most faunally diverse, containing deciduous and coniferous forests, deserts and tundra, and extending from Iceland to Japan. However, the variety of fauna is much lower than in the Ethiopian and Oriental realms. Reptiles, for example, are relatively few in number and there are no monkeys. Similarly, the Nearctic realm is poorer in species than its tropical neighbours. It exhibits considerable taxonomic affinity with the Palearctic realm, probably because it was at one time subject to colonization by Palearctic species across a Bering land bridge formed at periods of low sea-level during the Pleistocene. There appears to be some east–west differentiation in the Nearctic realm: more species of freshwater fish, turtles, salamanders and snakes are found east of the Rocky Mountains, whereas the western USA is the exclusive home of many lizards and rodents such as ground squirrels, prairie dogs and kangaroo rats. In the Neotropical realm, covering Central and South America, the richest faunal region is that of the tropical forests of northern South America. This is the headquarters of a very rich and distinct bird fauna, of which 30-odd families are confined to the New World.

The Australasian realm contains one of the most distinct faunas. It has relicts of ancient groups now confined entirely to this realm as well as some with affinities in the adjoining realms. The mammal fauna contains a few monotremes (primitive mammals which lay eggs), such as *Echidna* (ant-eaters) and the duckbilled platypus (*Ornithorhynchus anatinus*), many marsupials (Plate 8) and a particular richness of insectivorous and herbivorous bats. Reptiles are relatively diverse, including a turtle (*Carettochelys*) found in no other realm, and of the 58 families of birds, 12 are exclusive to this realm such as the birds of paradise, cassowaries and emus, lyre and bowerbirds, and the magpie larks. By contrast, amphibians are confined to four families of frogs.

For plants, an analogous series of distribution types has been elaborated characterizing the generally discontinuous ranges of most plant groups. These classificatory units differ from faunal realms since they enclose areas of distribution of a particular set of taxa rather than large continuous terrestrial areas containing assemblages. For example, Arctic–Alpine plants are found in the Arctic and then in the great mountain systems of the Alps, Rockies and Himalayas. (Fig. 1.12). Species common to the Arctic and the Rockies are fairly numerous, but common to the Arctic and the Himalayas much less so, partly because of present-day isolation but more so because of greater isolation during the Pleistocene. Each mountain massif has, of course, its own purely Alpine flora. Other plant groups include, for example, the North Atlantic group, distributed in Europe and eastern N. America: an example is the hooded ladies'-tresses *Spiranthes romanzoffiana*; a North Pacific group links North America and East Asia, for example with the skunk-cabbage *Symplocarpus foetidus* (Fig. 1.13). A North–South American group is distributed in both North and South America but lacks continuity between, as exemplified by some members of the Sarraceniaceae, the pitcher-plant family (Fig. 1.14). There are several other such groupings, including the Antarctic groups whose best-known member is the southern beech, *Nothofagus*, found in southern South America and in Australasia (Fig. 1.15).

The reasons behind these distributions of plants and animals at a world scale are complex and mostly historical, although some can clearly be explained by contemporary adaptations of a particular group to a particular environment, for example, those plants confined to Arctic-Alpine conditions, or birds whose sustenance is gained in the course

Plate 8 Members of the order Marsupiala: (a) the red kangaroo, *Macropus rufus*, a grassland herbivore; (b) the greater glider possum, *Schoinobates volans*, a tree-dweller; (c) *Sminthopsis* sp., a nocturnal desert mouse; (d) the numbat (*Myrmecobius fasciatus*), an ant-eater.

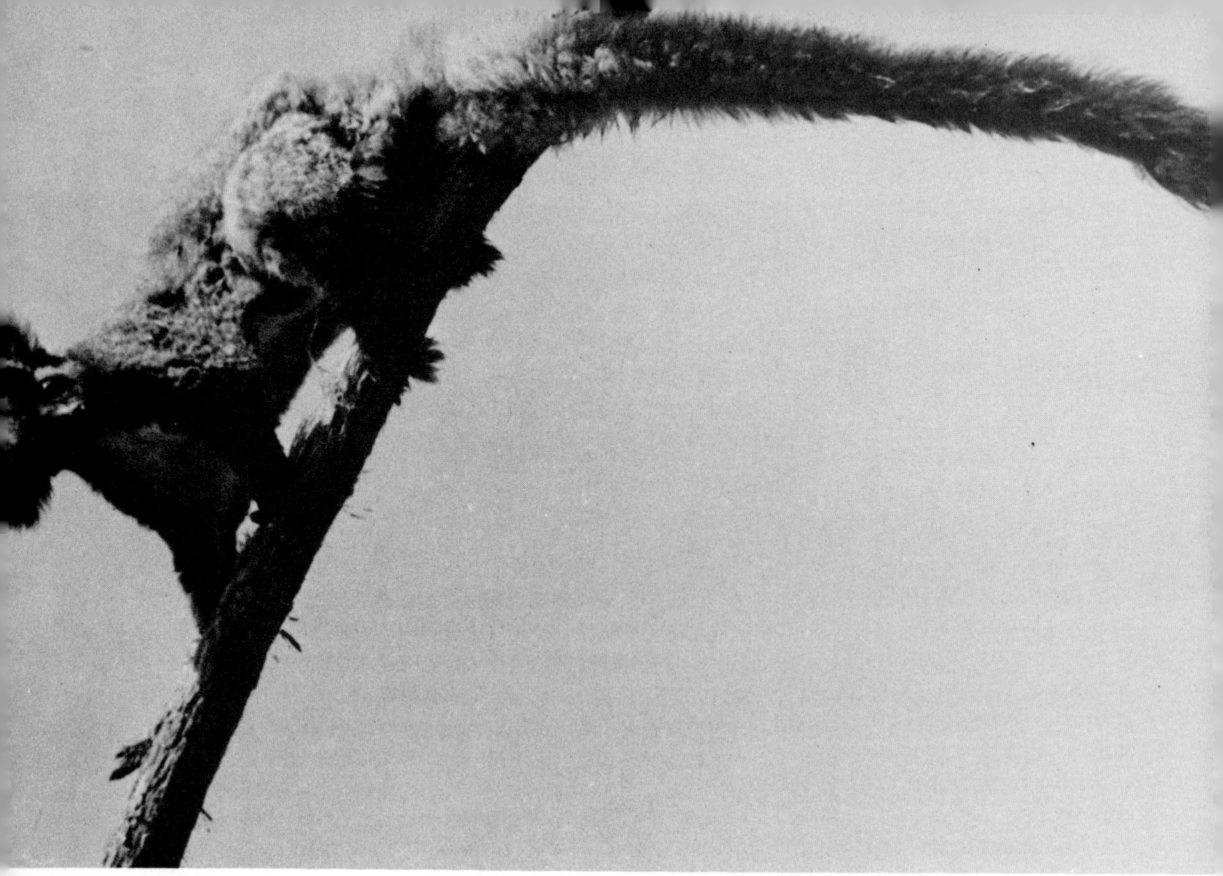

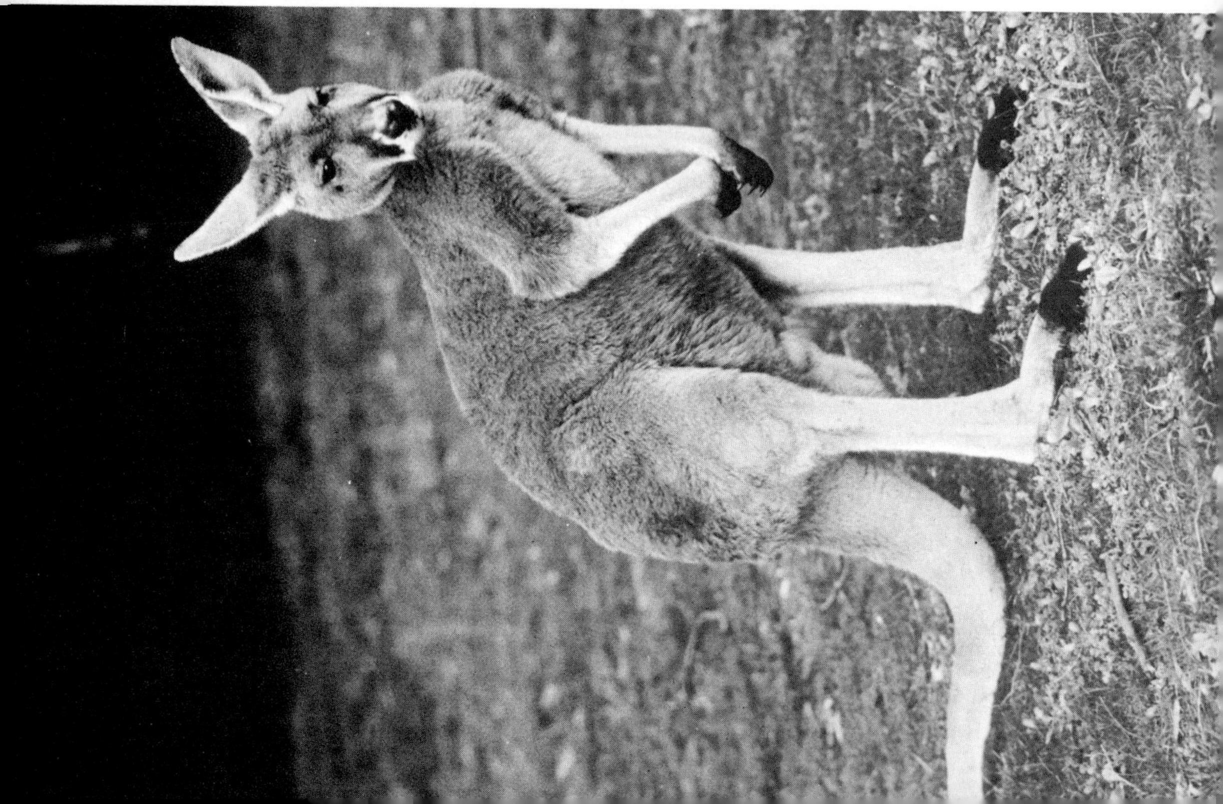

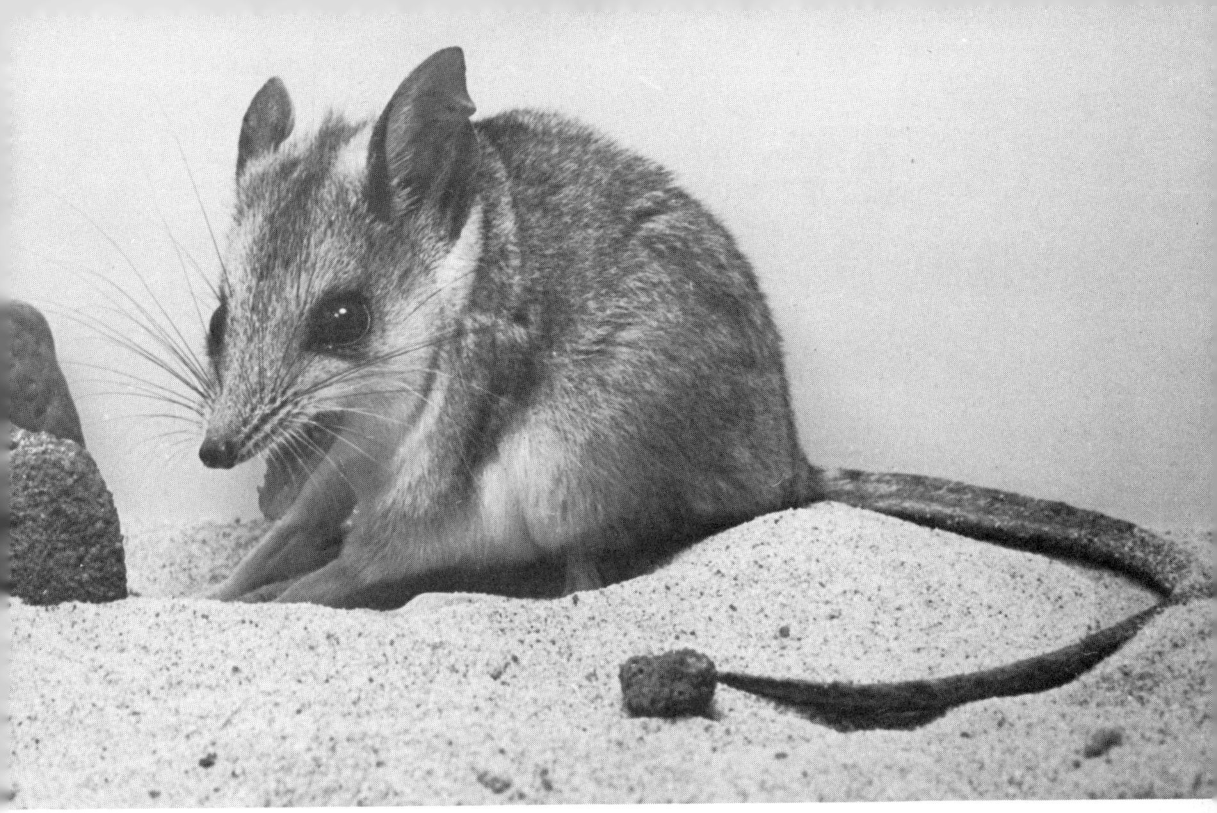

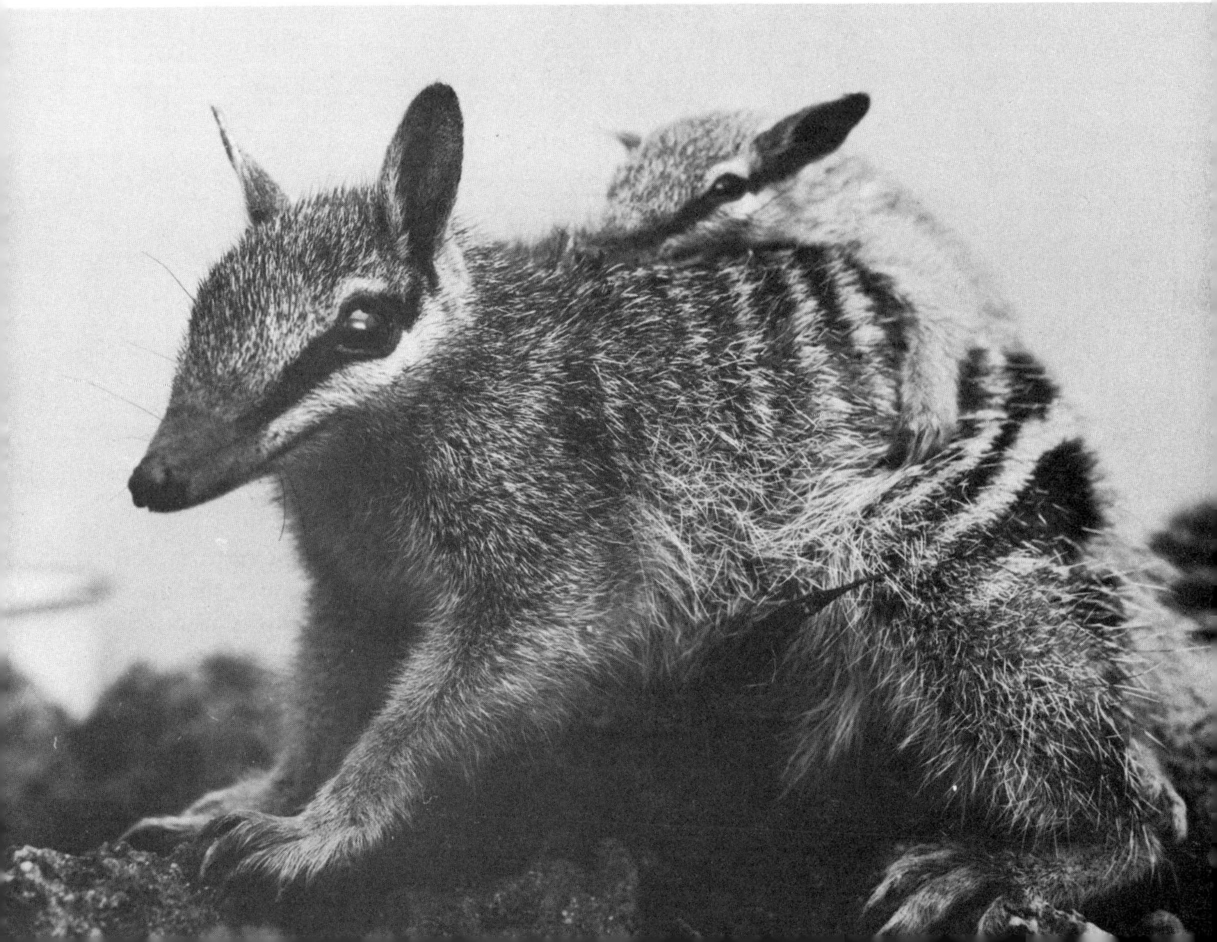

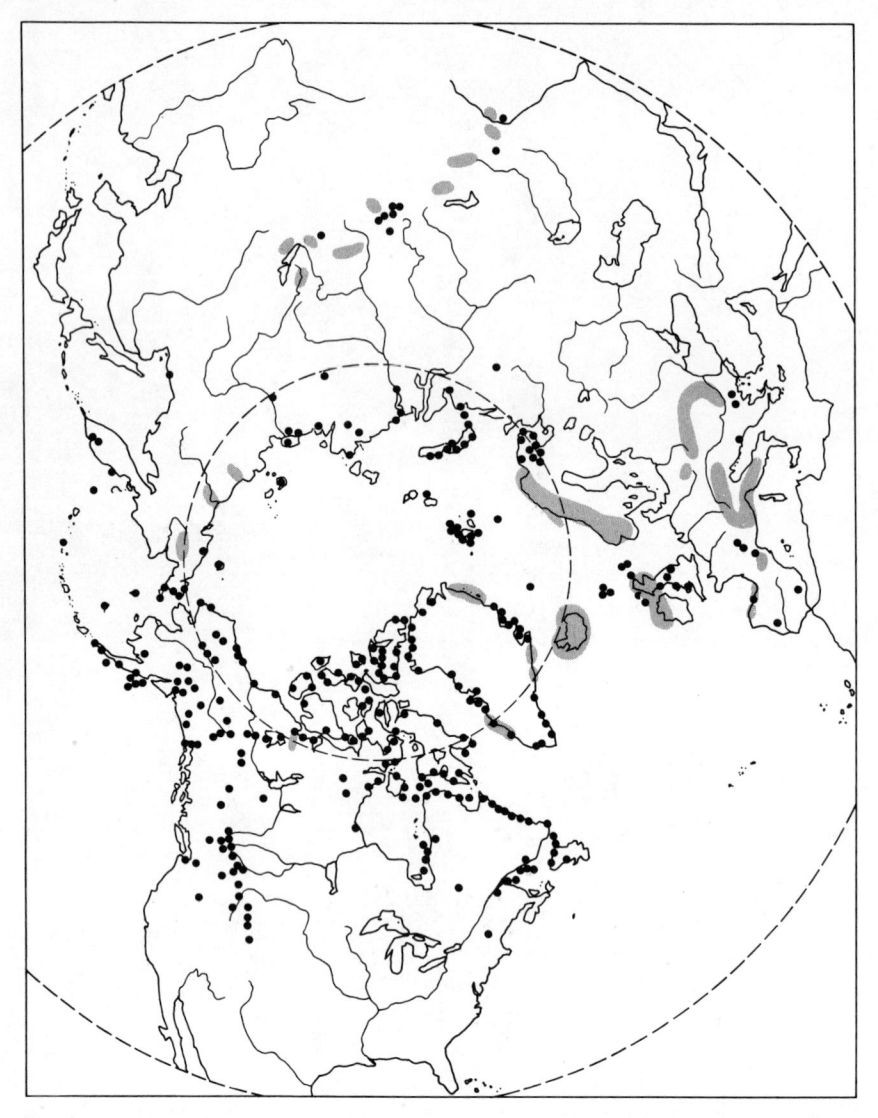

Fig. 1.12 A plant of arctic–alpine distribution (*Saxifraga oppositifolia*) in the northern hemisphere. Source: Polunin, 1960.

of pollinating tropical trees. However, the geological history of the earth is widely thought to have played a role in some of the distributions, particularly through the mechanism of continental drift. Although Wegener's original hypothesis of a single continent far to the south of the present land-masses has been much modified, the contemporary findings of the study of plate tectonics, allied to those of palaeontology, suggest that a drifting of land masses took place even though uneven in time, direction and speed. Thus in some parts of the earth there are now floras of essentially similar composi-

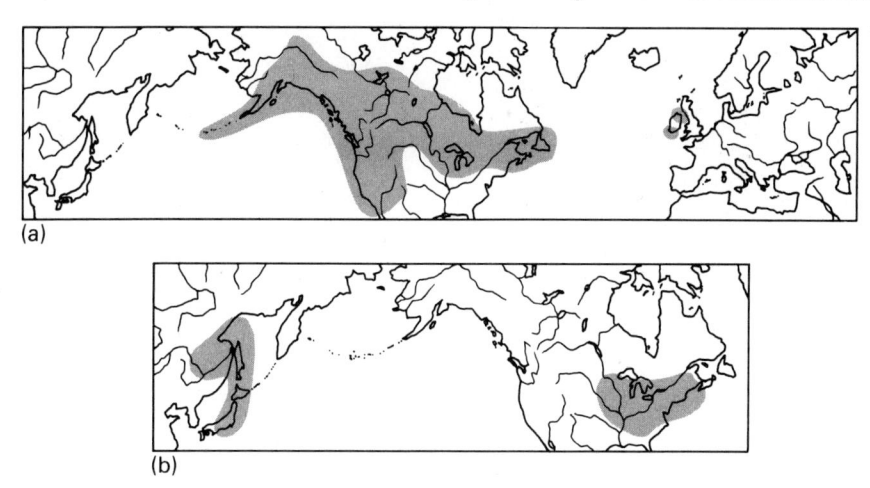

Distribution patterns of individual taxonomic units 41

Fig. 1.13 Two different types of plant distribution: **A** *Spiranthes romanzoffiana*, mostly in N. America but with a small representation in the western British Isles: **B** *Symplocarpus foetidus*, in eastern North America and eastern Asia.
Source: Polunin, 1960.

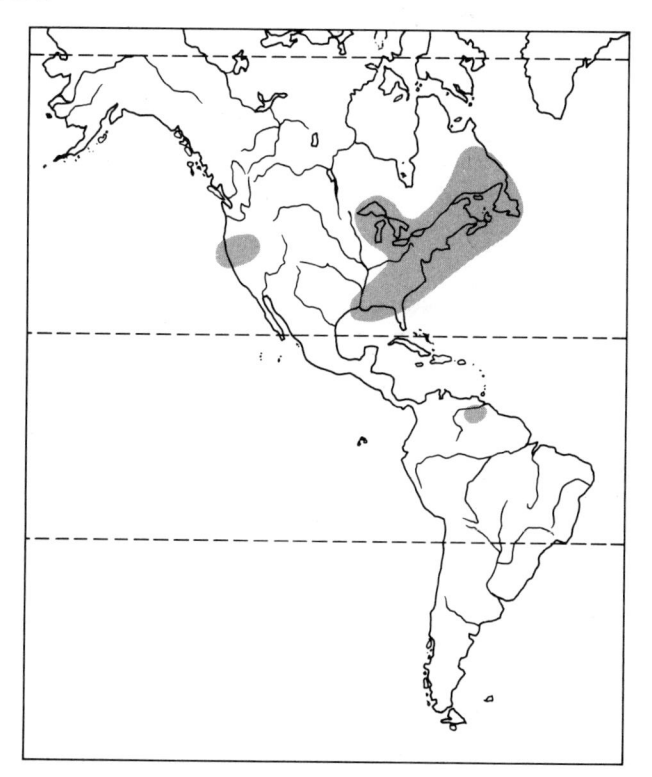

Fig. 1.14 The distribution of the pitcher-plant family (Sarraceniaceae) mostly, but not entirely, found in eastern North America.
Source: Polunin, 1960.

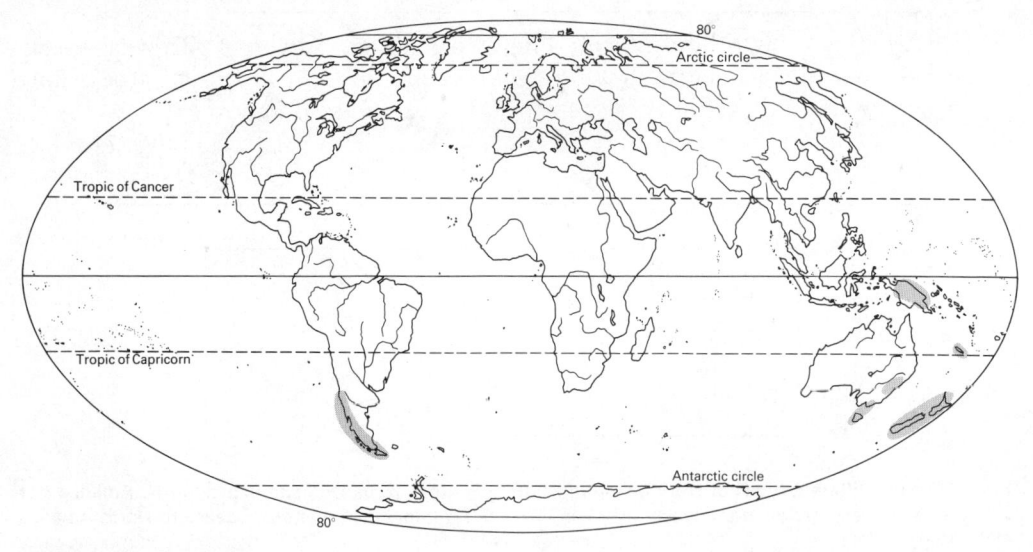

Fig. 1.15 The range of the genus *Nothofagus*, found only in widely separated continents at southerly latitudes of a temperate climate.
Source: Polunin, 1960.

tion with great spaces in between, whereas in other places essentially different floras are found in close proximity. For example, the floras of Australia, southern South America and southern Africa show considerable similarity despite separation by wide expanses of ocean. Approximately 700 species of Angiosperms are more or less entirely restricted to two or three of these regions, and six families are found in all three. The genus *Nothofagus* (although not found in South Africa) exemplifies this type of distribution (Fig. 1.15). The 'Antarctic' flora of the cold temperate regions south of 45°S is composed of groups of Angiosperms which are scarcely represented further north. Taken together, a considerable period of evolution on a single continent before its break-up is suggested.

When the factors which control the distribution and abundance of plants and animals at a world scale are considered, there seems little doubt that although the factors are complex and interactive, the most important of them is the global climate, and intuitively we recognize major world regions with a climatic basis (usually called **biomes,** see p. 90) identified by the relationship of climate, soils, vegetation and fauna. As would be expected, however, the relations between, for example, flora and climate are not always simple. It is often necessary to devise compound indices of climate in order to find a 'fit' between a plant species and a particular climatic element or elements. Some are relatively simple (e.g. the example given on p. 16), but others probably reflect a complex interaction of temperature, precipitation, competition and habitat availability. The northern tree limit in the Arctic coincides largely with the mean July isotherm of 10°C and this appears to apply to most Arctic timberlines. This may well be because the growth processes of the trees so far north are concentrated into July. The limited number of species which form the Arctic timberline also enhances the likelihood of a good fit.

Distribution patterns of individual taxonomic units 43

In the case of animals, it is difficult to find correlations between a particular species and a single climatic parameter. The northern limit of some animals appears to be defined by the direct effect of various mean temperatures, whereas others are affected by the influence of climate upon reproductive success, the survival of young individuals, food supply or the prevalence of competitors. Some groups seem to show intolerance of particular climatic regimes: reptiles, for example, are not found in the Arctic and are relatively few in number in the cool temperate zones; however, the northernmost genera of lizards and snakes (*Lacerta* and *Vipera* respectively) are also widely distributed in the Old World tropics, so no specially adapted genera seem to have arisen. The northernmost amphibian is a frog of the genus *Rana* which is not morphologically very different from its tropical relatives in the same genus (Fig. 1.16).

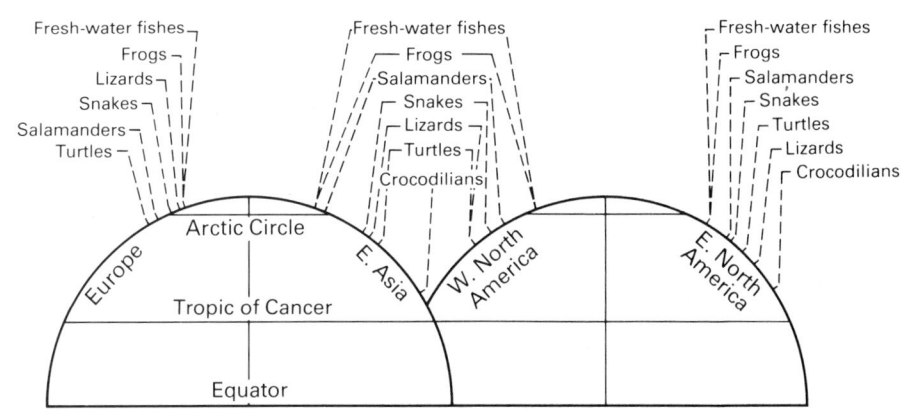

Fig. 1.16 The northern limits of various groups of cold-blooded vertebrates.
Source: Darlington, 1957.

Global distribution patterns are difficult to summarize: apart from the major patterns such as the faunal realms, a great number of types of distribution are found, with a corresponding variety of causal interconnections. As will become apparent from Part II, many global distribution patterns are altered by human activity. The dispersal of domesticates is one example, that of pathogens of domesticates and of ourselves another. Many of the patterns and regionalizations described are therefore becoming of scholarly interest only: for practical purposes, the work of man is producing whole new sets of global regions, in which one general tendency is perhaps that of a greater cosmopolitanism than in those evolved by nature.

The continental scale

Within the area of a major continent, considerable environmental contrasts will occur, leading to differential distribution patterns of biota: contrasts between mountain, floodplain and arid zone are obvious and well documented. Where there is rapid environmental change within a short distance, it will exert an effect on flora and fauna. For example, on mountains in Britain, the rush *Juncus squarrosus* is found up to about 300 ft (915 m).

44 The organism scale

Above *ca* 2200 ft (670 m) few of its flowers mature into seed capsules because the warmth needed does not normally occur since flowering is not complete until the end of August. But in warm summers, seeds will be produced at the higher altitudes and since an individual plant may be long-lived, this ensures its survival in the higher places (Fig. 1.17).

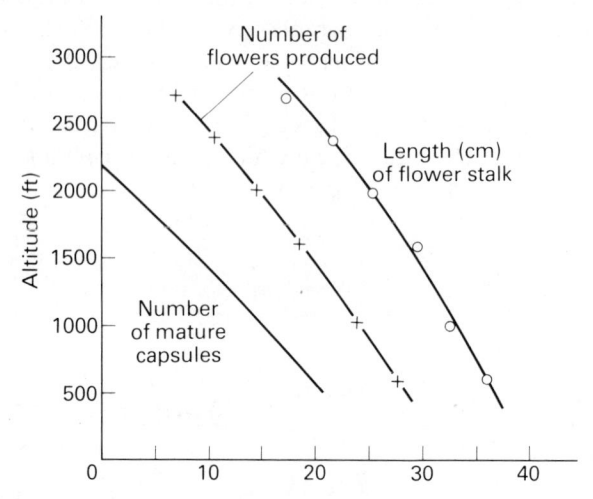

Fig. 1.17 The effect of altitude upon the growth and reproduction of the moor rush (*Juncus squarrosus*) in the British Isles.
Source: MacArthur, 1972.

Where the differentiating effect of a mountain is sharper, then species substitution may take place. In the higher mountains of New England, four species of one genus of thrushes are arranged in a more or less horizontal zonation, varying with the tree species present: in the deciduous forest up to 1700 ft (518 m). *Catharus fuscescens* is present but is replaced in the 1000–3000 ft (305–915 m) zone of pines by *C. guttata* and from 1800 ft (549 m) to the timberline in wet spruce and hardwoods, *C. ustulata* is found. In the stunted softwoods at the timberline, *C. minimus* is present. A comparable example is given in Fig. 1.18.

Even within a single species, different environments may produce variform races with inheritable characteristics, usually called **ecotypes**. In North America, the alpine sorrel (*Oxyria digyna*), an Arctic-Alpine plant, is made up of races which respond to distinctions in day length (Billings 1972). All *Oxyria* plants remain dormant at a daylength (**photoperiod**) of 12 hours; those from the alpine areas of California, Colorado and Wyoming will break dormancy at a photoperiod of 15 hours, but those from north of the Arctic circle but at lower elevations will grow and flower only at a photoperiod of 20+ hours. Arctic *Oxyria* plants have more chlorophyll per unit of biomass, higher respiration rates at the same temperatures than the Alpine race, and usually reach peak photosynthetic rates at lower temperatures. Arctic individuals are not as efficient in using CO_2 as their alpine relatives which must survive in the thin atmosphere of the high mountains. In this way, although the tolerances of an individual plant may not be very

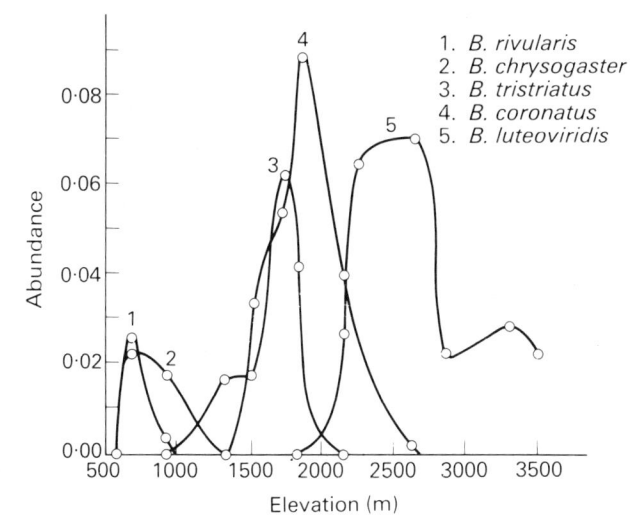

Fig. 1.18 Population-density curves for species in the genus *Basileuterus* (warblers) in Peru, showing the way in which they replace each other along an altitudinal gradient. Species 1 and 2 in fact overlap but occupy different habitats in the more diverse vegetation of lower levels.
Source: MacArthur, 1972.

wide, the species may be very widespread over a considerable range of environmental conditions: many weedy species probably share these characteristics.

Given the effects of mountains, it is not surprising that in historical terms they have been regarded as barriers to dispersal. The lighter fruits, seeds and spores can be blown over them, but conditions of soil and climate may well be different on the other side and so prevent the establishment of a plant species. By contrast, even high mountain ranges seem less of a barrier to the dispersal of animals, especially as compared with salt water. The difference between the Oriental and Palearctic faunas, for example, seems to be due to climatic zonation rather than a mountainous barrier, but the distribution of animals around and across the Himalayas now and in the past shows that in the long term they are perhaps not such important barriers as they might seem.

Within a continent, differences in the diversity of flora and fauna can be traced. Often, these are a function of latitude with few species in Arctic environments grading through to a great diversity of species in sub-tropical and tropical forests, although this is by no means universal: it appears not to apply to Australian birds and lizards, for instance (Schell and Pianka 1978). Animals exhibit this tendency quite markedly: Greenland has 56 species of breeding birds, New York State 105, Guatemala 469, and Colombia 1395. Figure 1.19 shows the poleward decrease of the number of bird species within North America. At any particular latitude the diversity in a particular biome varies from habitat to habitat: the flora of forests with 'normal' moisture conditions is generally wider than that of very wet or very dry forests and usually richer than in grasslands. The richness of the tropics is quite outstanding for in a small area of forest there may be over 100 tree species, with many of them occurring only once in 100 hectares. The number of insect species is not known but the diversity is high: one investigator collected 545

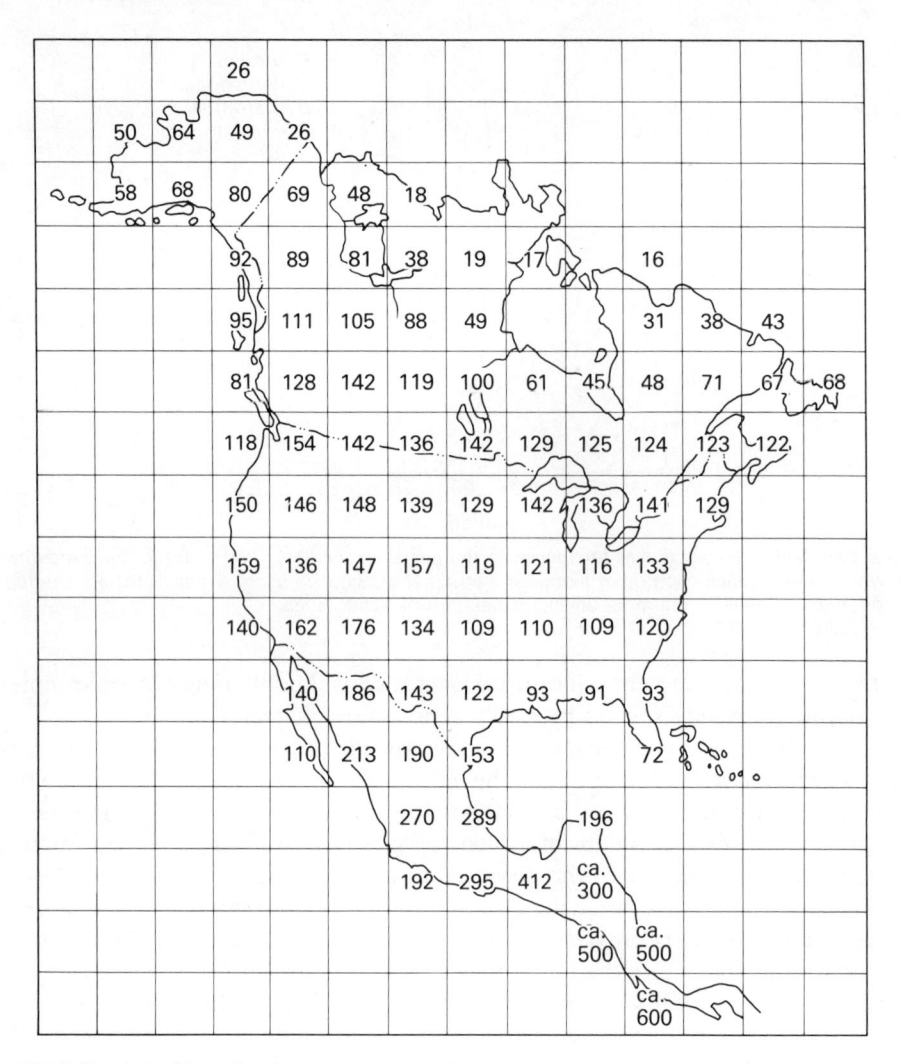

Fig. 1.19 Number of breeding land-bird species in North America.
Source: MacArthur, 1972.

species with 2000 sweeps of a net in a lowland tropical forest. The reasons for the high diversity of biota in the tropics are not universally agreed upon but Table 1.5 provides a summary of contemporary ideas.

Continents such as North America which contain a variety of topography and climate may exhibit diversity gradients other than that corresponding to latitude. A longitudinal division of the continent along the Rocky Mountain front shows distinct differences in flora and fauna between the two, controlled largely by climate and topography. The eastern half is low in genera confined solely to that half, having not more than 100 such genera of plants (e.g. *Hudsonia, Planera, Ceratiola*). By contrast, the western por-

TABLE 1.5   An outline of the basic hypotheses concerning species diversity, particularly the increased species diversity in the tropics compared to temperate and arctic regions

Nonequilibrium hypothesis
  Time — the tropics are older and more stable, hence tropical communities have had more time to develop.
Equilibrium hypotheses
 I Speciation rates are higher in the tropics.
    A. Tropical populations are more sedentary, facilitating geographical isolation.
    B Evolution proceeds faster due to
       1 a larger number of generations per year;
       2 greater productivity, leading to greater turnover of populations, hence increased selection;
       3 greater importance of biological factors in the tropics, thereby enhancing selection.
 II Extinction rates are lower in the tropics.
    A Competition is less stringent in the tropics due to
       1 presence of more resources;
       2 increased spatial heterogeneity;
       3 increased control over competing populations exercised by predators.
    B The tropics provide more stable environments, allowing smaller populations to persist, because
       1 the physical environment is more constant;
       2 biological communities are more completely integrated, thereby enhancing the stability of the ecosystem.
Source: Ricklefs 1973.

tion of the continent has at least 300 endemic genera, with the richest diversity in California: the California poppy *Eschscholzia* is an example of one such endemic. In terms of animals, the lower topographic diversity of the east is often held to be a major contributory factor to lower numbers of species: in the eastern part of North America there are only one or two species of chipmunk, whereas the innumerable basins and mountain ranges of the West seem to have produced speciation to the level of 14 species of chipmunk, mostly found in habitats distinct from each other (Fig. 1.20).

(The confinement of a species or other taxonomic unit of animal or plant to a particular region is called **endemism** and the plant or animal termed an **endemic**. The term is conventionally restricted to units having a comparatively or abnormally restricted range—that is, whose ranges are less than average for their kind, as for example with families found only on one continent. It can be a useful concept in delimiting regions where an area has a high proportion of endemics contrasted with its neighbours; for example the Cape Region can thus be delimited floristically from the rest of South Africa. Theoretically an endemic species is thought to be either near the beginning or the end of its period of existence, with a non-endemic distribution phase intervening, but unless there is abundant fossil evidence, this is impossible to verify.)

The local scale

Distribution at this scale is dominated by the concept of the **habitat**: that is, the small-scale concentration of local and microclimates, moisture availability, soil conditions, slope and similar factors which determine the exact locality in which an individual plant or animal is found and is, of course, the basic unit upon which all the other distributions are built: if the organism does not occur at this scale then it occurs not at all. So to that extent it is misleading to discuss it last, although wider (i.e. continental or global) distribution patterns may point to other interesting information about Earth history or the history of the taxon.

48 The organism scale

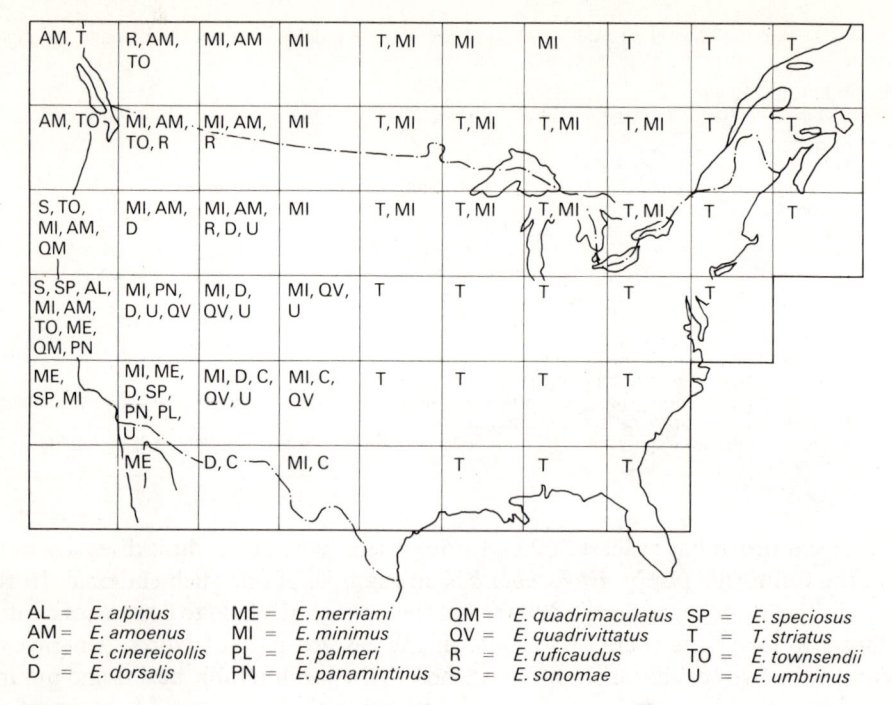

AM, T	R, AM, TO	MI, AM	MI	T, MI	MI	MI	T	T	T
AM, TO	MI, AM, TO, R	MI, AM, R	MI	T, MI	T, MI	T, MI	T, MI	T	T
S, TO, MI, AM, QM	MI, AM, D	MI, AM, R, D, U	MI	T, MI	T, MI	T, MI	T, MI	T	T
S, SP, AL, MI, AM, TO, ME, QM, PN	MI, PN, D, U, QV	MI, D, QV, U	MI, QV, U	T	T	T	T	T	
ME, SP, MI	MI, ME, D, SP, PN, PL, U	MI, D, C, QV, U	MI, C, QV	T	T	T	T		
	ME	D, C	MI, C	T	T	T			

```
AL = E. alpinus        ME = E. merriami       QM = E. quadrimaculatus   SP = E. speciosus
AM = E. amoenus        MI = E. minimus        QV = E. quadrivittatus    T  = T. striatus
C  = E. cinereicollis  PL = E. palmeri        R  = E. ruficaudus        TO = E. townsendii
D  = E. dorsalis       PN = E. panamintinus   S  = E. sonomae           U  = E. umbrinus
```

Fig. 1.20 Chipmunk species in the continental USA, showing the greater diversity in the western states. Source: MacArthur, 1972.

Most of the factors affecting distribution patterns at larger scales are present here as well, though their operation will often produce a mosaic with a much smaller tessellation than at continental or global scale. Competition and predation, for example, are probably most evident at this scale: a yellow warbler (*Dendroica petechia*) is confined to mangroves over most of the tropics, but on some islands where there are fewer competitors then the species occupies all the dry-land forests as well. The restriction to mangroves is therefore probably due to competition. Clearly all kinds of climatic factors operate as well in determining where particular plants should be located, and they in turn influence the local and microclimates around them. In the Milo River valley of Guinea, south-facing slopes exposed to the monsoon are covered with dense forest but north-facing slopes exposed to the dessicating Harmattan wind are savanna-covered. The interaction of wind and moisture can be seen in northern California, where the coast redwood *Sequoia sempervirens* is confined to areas where there are frequent wind-blown fogs. Within a formation the structure determined by the dominants may exert a vital influence in the presence or absence of a species of plant or animal. A very simple example is the distribution of bracken fern (*Pteridium aquilinum*) in woodlands: it is intolerant of dense shade and so in woods with a closed canopy, it occurs only where there are openings in the forest canopy. The layering of woodland vegetation provides a link to another aspect of local variation: structural heterogeneity within a small area (e.g. where the physiography allows different types of plant substrate which in turn provide several sources of food and cover for animals) increases the species diversity.

Where plants are concerned, one of the most common differentiating media at local level is the nature of the soil, in which the degree of acidity (pH) is often the most important facet. Where, for example, a limestone grassland adjoins an area of similar vegetation type over silicious rocks, then certain species may range over both, but some may be confined to the alkaline soils (**calcicoles**) and others to the acidic soils (**calcifuges** or **acidophiles**). The common heather (*Calluna vulgaris*) of western Europe is one example: it is unable to tolerate alkaline soils, and its seeds germinate best at pH 4·0–5·0 (Gimingham, 1975). The mechanism appears to act through the effect of pH on the availability of mineral nutrients. On acid soils, the calcicoles suffer from aluminium toxicity, whereas calcifuges are resistant; on calcareous soils the calcifuges are unable to obtain enough iron, whereas the calcicoles are adapted for its uptake. On the other hand, limestone areas often contain plants found nowhere else, for example in the Craven Pennines of northern England, the montane saxifrages are found only on thin bands of limestone which interrupt the grits and sandstones. In upland Britain, soil characteristics, the type of vegetation cover and the conditions of temperature and humidity help to produce three distinct types of grasslands which also have distinct faunal characteristics. Where base-rich rocks (e.g. limestone, basalt, dolomite) outcrop, there are grasslands of sheep's fescue (*Festuca ovina*) intermingled with many other species of grasses and herbs. The fauna contains ants, snails, earthworms and a wide variety of insect larvae. Moles and rabbits may also be present. The second type has base-deficient soils and so snails are replaced by slugs since there is too little calcium carbonate for proper shell formation. Larvae such as leather-jackets are present and also the larvae of small moths. A third type is the damp and peaty soils vegetated by grasses such as mat-grass (*Nardus stricta*) and blue moorgrass (*Molinia caerulea*). Here a visible soil fauna is virtually absent, except for some slugs and moths. Beneath the surface, the deficiency of oxygen in the peat restricts the fauna greatly, mostly to mites (Acarina) and springtails (Collembola). Under these sets of conditions, the type of soil parent material appears to be the chief factor in local faunal distributions (Pearsall 1968).

At an even more compressed spatial scale, a single plant (especially a large one like a tree) may provide different zones for other biota. A North American spruce can accommodate five species of warblers (*Dendroica* spp.) which do not compete for food since each species hunts its insect prey in different parts of the tree (Fig. 1.21). Not only does each species have its own zone but an individual hunting behaviour so that even where zones overlap, the different behavioural patterns help to minimize competition (Hanson 1964).

The actual occurrence of a plant or animal at a point in space and time is an integration of all the environmental factors discussed so far, and of its history both as an individual and as a species. What is clear is that none of these factors acts in isolation so that the investigation of the reasons for the occurrence of biota in terms of why here and not there is a complex study in which apparently simple explanations when further pursued (the example of pH discussed above is apposite) turn out to be much more intricate.

50   *The organism scale*

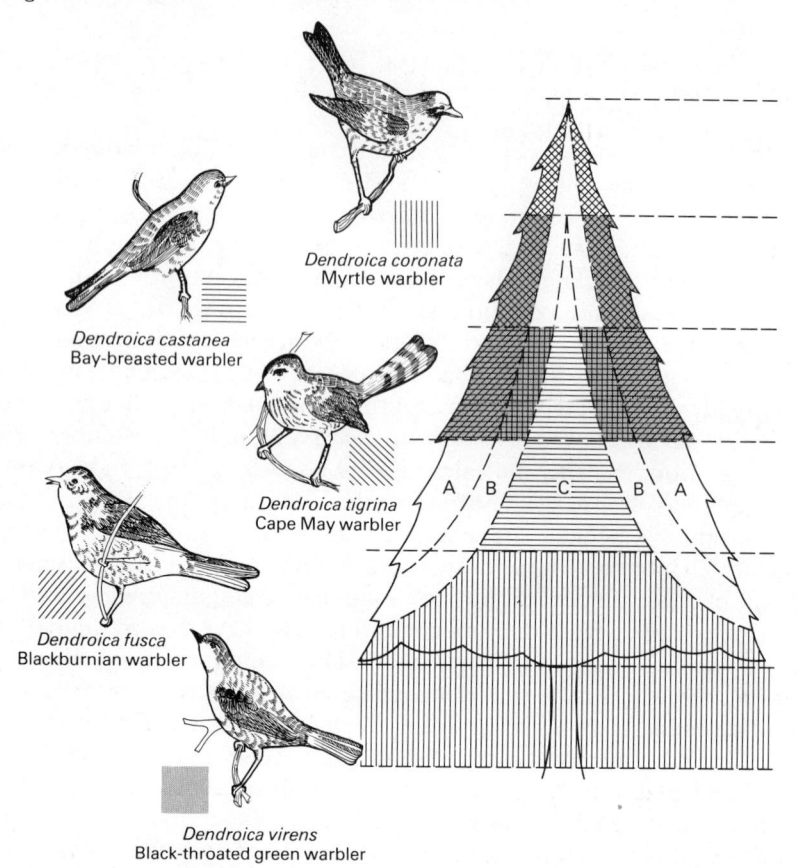

Fig. 1.21 A spruce tree occupied by five species of North American warblers, showing the feeding zones of each species. Not only are there vertical zones but concentric zones as well: **A** an outer zone of new needles and buds; **B** a middle zone of old needles; and **C** an inner zone of bare or lichen-covered branches. The shading indicates where each species spends at least half its time.
Source: Hanson, 1964.

Flora and fauna of islands

Islands present a special case in terms of their plants and animals. Their isolation makes access difficult in proportion to their distance from another land mass, and if small they may present only a limited range of habitats for incoming species, but on the other hand competition may be limited. Several biotic generalizations seem to emerge, especially from the work of MacArthur (1972):

1. No island has nearly the number of species it would have if it were part of the mainland: the salt-water barrier acts as a filter for dispersal of all types of organisms. The commonest bird of the California chaparral, the wrentit (*Chamaea fasciata*), is absent from the Channel Islands, some 20–50 km off the coast of Southern California, and common mainland birds like the brown towhee (*Pipilo fuscus*) and California thrasher (*Toxostoma redivivum*) are similarly absent. In addition, an island chain may well show

a diminution in diversity down the archipelago away from the continental mass. Where islands in a chain have become created at different times, then the diversity of their fauna seems to correlate well with their age. Some of them seem also to become 'supersaturated' with animals, some of whom then become extinct, producing an equilibrium level (Wilcox 1978).

2. A large island is likely to have a greater variety of habitats than a small one and so contain a greater diversity of species. Studies in the West Indies show that each tenfold increase in the size of island brings with it a doubling of the numbers of species of amphibians and reptiles (Fig. 1.22).

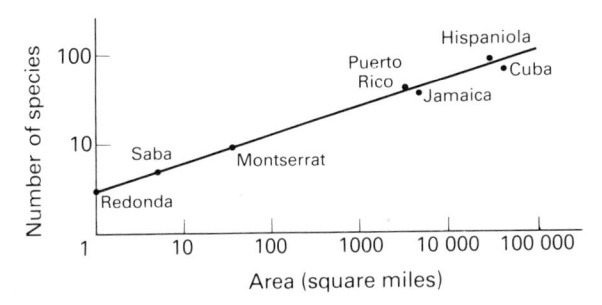

Fig. 1.22 The number of species of amphibians and reptiles plotted against the size of island in the West Indies.
Source: MacArthur, 1972.

3. Adaptation to the new environment of the island may be difficult for an immigrant. It is basically adapted to conditions on the mainland and since only a few individuals are likely to form the immigrant population, the scope for the genetic variability which might enable the species to establish itself successfully is limited. But this establishment may then bring competition for a species already present which may then decline. High rates of immigration and high rates of extinction are therefore characteristic of island biota (Fig. 1.23).

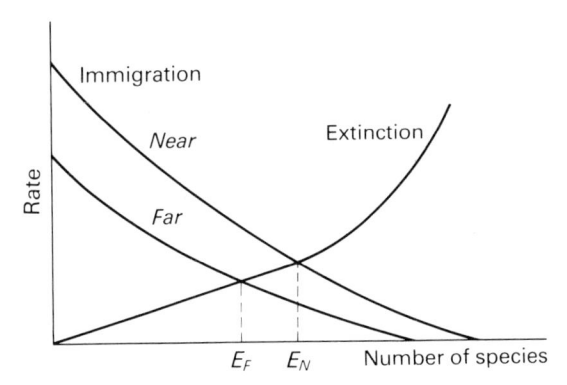

Fig. 1.23 The hypothesis that an island far from a continental mass (E_F) will equilibrate its rates of immigration and extinction with fewer species than an island (E_N) nearer the continent.
Source: MacArthur, 1972.

4. The precariousness of some of the ecosystems based on low diversities of flora and fauna makes them susceptible to rapid and sometimes catastrophic change, especially when human influences are involved as in the import of domesticated animals, weedy species, and pests such as rats. Some of the consequences are discussed on p. 130.

5. Adaptations to islands take a number of well-known forms. Firstly, the isolation produces speciation and hence islands are rich in endemic species. In terms only of birds, for instance, the Hawaiian islands are thought to have one endemic family, five endemic genera, three endemic species and four endemic subspecies (Table 1.6, Darlington 1957). Secondly, the lack of predation and competition permits the existence of groups like giant tortoises (as on Aldabra and the Galapagos Islands) and flightless birds, characteristic of, for example, New Zealand, Hawaii, Mauritius (including the much-lamented dodo), Aldabra, the Galapagos and Madagascar. It seems as if long periods of isolation have produced large flightless birds like the now extinct moa of New Zealand whereas shorter periods of time have only generated rather small flightless avians like the rails of several island groups. Lastly, the low numbers of species on an island has allowed the development of adaptive radation. The classic example are the finches of the Galapagos islands where four genera with twelve species are considered to have evolved from a single ancestral species and in so doing have filled the roles usually played by other groups: their beaks have become modified to play the parts of ground-feeders, plant-eaters, insect-eaters, cactus-feeders, warblers and woodpeckers. Another example is that of the Hawaiian honeycreepers which comprise nine endemic genera and exhibit not only the presumably original nectar-eating habit, but species which have become seed- and fruit-eaters. A roughly parallel instance in plants is shown by immigrant sunflowers on St Helena, which have given rise to five endemic species of tree of 5–7 m in height.

TABLE 1.6   Land and fresh-water birds of the Hawaiian islands (excepting Laysan)

Endemic Family
Drepaniidae (probably from America)

Endemic Genera
A rail (derivation unknown)
Two genera of honey-eaters (from Australasian ancestor)
An Old World flycatcher (from Polynesian ancestor)
A goose from American ancestor)

Endemic Species
A crow (probably from North America)
A hawk (from America)
A duck (probably North American)

Endemic Subspecies
The short-eared owl (probably from North America)
The stilt (from America)
A gallinule (from America)
The coot (from America)

Not Endemic
The black-crowned night heron (from America)

Source: Darlington 1957.

The biogeography of islands has two further and related points of interest. One is that with the restricted number of species involved it is relatively easy to quantify the biotic diversity and to make mathematical models of it. This helps with protection, for example, in showing the minimal areas necessary for successful perpetuation of a rare species and relates to the second point which is that even on continents many terrestrial habitats are *de facto* islands (e.g. a woodland surrounded by agricultural land; a nature reserve surrounded by intensively used land) and so the general models derived from islands can be applied to conservation problems. A little more discussion of this will be found on p. 317.

Conclusions

The presence of a plant or an animal in a particular locality is the outcome of a multiplicity of factors: at the local scale, where its habitat is concerned, all the tolerances of the individual organism must be satisfied by these local conditions, otherwise it will not be there at all. This scale is also the one at which most man-made changes to flora and fauna are observed. At the continental scale, an organism may reflect in its distribution major features of the Earth's surface such as broad climatic divisions or the presence of high mountain ranges. At the world scale, the presence or absence of a taxon may show features of both its evolution and of the continental masses themselves, reflecting for example the influence of continental drift upon the formation of species or the influence of the Pleistocene glaciations.

Since, however, living organisms do not occur in isolation from their environment and usually have interactive linkages with other organisms, it is now common to study them in this functional context, usually at the local scale or at an agglomeration of such small-scale units. This type of study is at the core of the next scale unit considered here, the ecosystem.

Further Reading

ARCHER, B., GRENVILLE, H. W., JAGO, M., JOHNSON, C. B. 1976: *Understanding biology.* London: Mills and Boon.
BILLINGS, W. D. 1972: *Plants, man and the ecosystem*, 2nd edn. Belmont, California: Wadsworth Publishing Company Inc.
COX, C. B., HEALEY, I. N., and MOORE, P. D. 1976: *Biogeography. An ecological and evolutionary approach*, 2nd edn. Oxford: Blackwell Scientific Publications.
DARLINGTON, P. J. 1957: *Zoogeography: the geographical distribution of animals.* New York: Wiley.
GOOD, R. 1974: *The geography of the flowering plants*, 4th edn. London: Longmans.
KREBS, C. J. 1972: *Ecology.* New York: Harper and Row. ch. 2–8.
MACARTHUR, R. H. 1972: *Geographical ecology.* New York: Harper and Row.
POLUNIN, N. 1960: *Introduction to plant geography.* London: Longmans.

2
The ecosystem scale

The individual organism represents one level of complexity of organization; the next level is the organism together with its living and non-living environment. The study of this level is called **ecology**, a term coined in the nineteenth century from the Greek *oikos*, a household. The basic concept can be applied at almost any scale: we can refer to the ecology of a drop of water with micro-organisms in it, or to the ecology of the whole planet. Different types of approach can also be adopted, of which one of the most common has been the 'inventory' where the dynamics, frequency and distribution of an organism were studied against the background of all its environmental factors, including other plants or animals, soil, and local climate. This isolation of one species as a focus for study is called **autecology**. More complicated but often nearer reality is **synecology** which tries to study all the components of a living community either in a 'static' fashion by inventory of all its components, or in a 'dynamic' way by considering the functional linkages between the various parts. This latter concept of living things interacting among themselves and with their non-living environment is analogous to that of a system and so the term **ecosystem** was coined in 1935 for this model of biota and their environment.

Modern attempts to provide a conceptual basis for the study of the functional linkages between the components of ecosystems has led to definitions like that of E. P. Odum (1971):

> Any unit that includes all of the organisms in a given area interacting with the physical environment so that a flow of energy leads to ... exchange of materials between living and non-living parts within the system ... is an ecosystem.

In this context, which emphasizes the linkages between the various components, at least two approaches are often made. The first is a synoptic approach which starts with an intuitive perception of an ecosystem, which is usually identified by a physiognomic feature, e.g. a pond, a wood, or the tundra. Studies of its ecological cohesion are then made—e.g. the relationships in a forest–deer–wolf system in which the wolves prevent the deer numbers rising to the point where they browse young trees so heavily that the regeneration of the woodland is hindered. Studies such as these may well have implications for management, and the ecosystem concept can easily accommodate man as one of its components. Another approach is more analytical: measurements are made of the flow and partitioning of energy through the ecosystem and of the cycles of mineral nutrients within the system. These are combined in studies of the rate of production of organic matter in an ecosystem (**production ecology**) and in studies of the changes in population numbers of the species of the ecosystem (population dynamics). These, too, can accommodate human actions, e.g. in altering nutrient pathways by taking a crop.

The usefulness of this dynamic approach is readily apparent: it deals with change, and such ideas as **feedback loops** and **homeostasis** derived from cybernetics fit readily into the models proposed. Likewise, many constructs are capable of mathematical modelling and of being simulated on a computer. Philosophically, the ecosystem can be viewed as just one of a class of systems and so the whole gamut of General Systems Theory can be used in ecosystem study; further, the theory of integrative levels (the atom–molecule–cell–organ–organism–ecosystem–global ecosystem–solar system progression, part of which introduced this section) puts it in a neat hierarchy of relationships with the rest of the planet and the cosmos. At a more practical level, the ecosystem concept is applicable at a variety of scales, so much so that qualifying adjectives like 'local' and 'regional' often have to be applied, though no agreed definition exists; some continental ecologists have built up elaborate spatial hierarchies with terms like 'biotope' and 'biochore' but no generally acceptable system seems to have emerged.

There are difficulties. The concept is a construct of the human mind and therefore may not always be right. The problems of scale and hierarchy are difficult to contend with, especially when a mosaic of natural or semi-natural and man-made ecosystems is being considered, and because hierarchical, nested, classifications are more suited to static phenomena than those undergoing change. Thus boundary problems exert difficulties: a pond may seem to have a clear-cut interface at the water/soil junction but consider (*a*) that due to gradual encroachment of marginal vegetation, the boundary is constantly changing its spatial location, and (*b*) that a duck spending its winter on the pond may summer in the Arctic: to what ecosystem does it belong? (Musical readers may care to ponder the varying ecosystemic relations of the unfortunate duck in Prokofiev's *Peter and the Wolf*.) Lastly, even the simplest ecosystem is in reality immensely complex and when obviously complex systems are considered, it is possible that the mind of *Homo sapiens* is not equipped to comprehend the volume of simultaneous, mutually modifying and interactive transactions that characterize most ecosystems. This will not prevent attempts to do so but at present requires the use of pegs upon which to hang the study, and in this context, parameters such as energy flow, nutrient cycling and population dynamics become especially useful. These studies have the additional advantage that they can be of practical value in environmental management and in studies such as the International Biological Programme's efforts to measure the **biological productivity** (see p. 58) of the Earth.

Energy and ecology

The topic to start this discussion is green plants which contain the pigment **chlorophyll**: these are called autotrophs because they can fix energy from inorganic sources into organic molecules (i.e. molecules containing carbon). This process is called **photosynthesis** and is the source of all the energy in ecosystems and thus in living systems entirely, with the exception of a few lower organisms which can fix energy in the absence

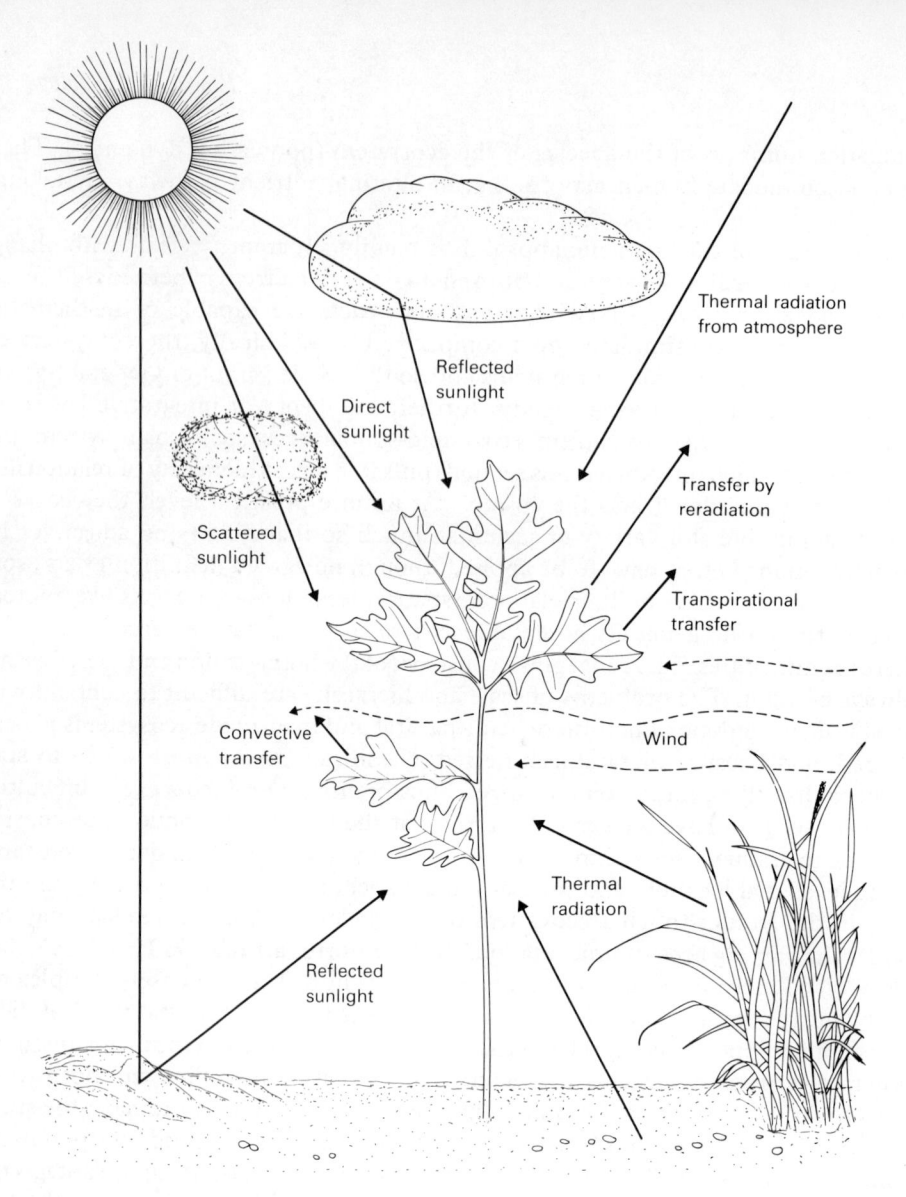

Fig. 2.1 The energy environment of a plant showing the various physical factors which influence the amount of energy it receives and the temperature which it attains.

of light. The importance of photosynthesis for life (and hence for man's economies) can scarcely be underestimated.

The process has an involved biochemistry but, simplified, occurs in two steps: the first uses light energy absorbed by chlorophyll to split a water molecule, releasing oxygen; the second, which does not require light, uses the energy in several steps to reduce carbon dioxide to carbohydrates. This can be summarized in the equation:

$$6CO_2 + 12H_2O \xrightarrow[\text{(chlorophyll)}]{2816\,\text{kJ}} C_6H_{12}O_6 + 6O_2 + 6H_2O$$

Various factors can modify the course of the reaction, e.g. the changes in the amount of energy entering the system, the concentration of any of the inputs, the efficiency of the **enzyme** systems which carry out the biochemical steps in the process, and external environmental factors such as temperature, humidity and light intensity (Fig. 2.1).

Obviously, the availability of energy as light is fundamental to photosynthesis, and light intensity varies according to many factors such as time, season, weather, topography, orientation of the photosynthetic surface (usually a leaf in higher plants), shading by other biotic or abiotic structures and even the depth of water where aquatic plants are concerned. In general the rate of photosynthesis increases linearly in relation to the intensity of light but after a certain threshold value, photosynthesis uses less and less of the light (i.e. becomes less efficient) until above an intensity of 10 000 lux, an increase in light intensity produces no increase in photosynthetic rate. This plateau is called light saturation and if the rate of photosynthesis is to be further increased then other factors have to be adjusted. In fact, except for plants growing in dense shade (e.g. below a closed forest canopy) there is normally enough light from the sun to saturate the photosynthetic capacity of most plants.

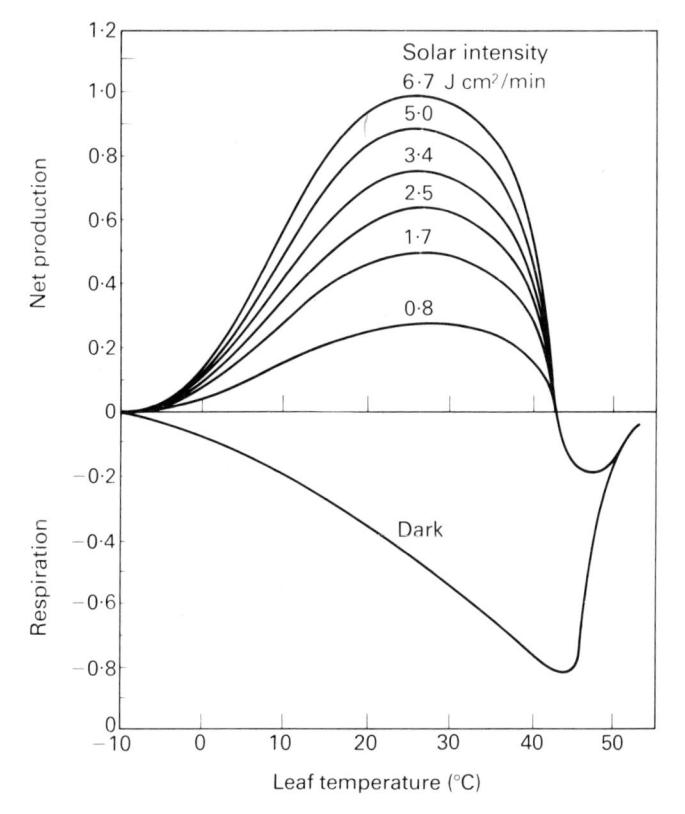

Fig. 2.2 The rates of net production at various light intensities, and respiration during darkness, for a temperate-climate plant's leaf at different temperatures.
Source: Varley, 1974.

58 *The ecosystem scale*

Through its effect on plant physiology, temperature may exert an effect on photosynthetic rates: at 25°C the rate of photosynthesis is much higher than at 15°C. For most plants, the optimum temperature is around 25°C and most function well between 10–35°C. (Fig. 2.2.) At low light intensities, the concentration of CO_2 (normally 0·03 per cent of the air) does not affect the rate of photosynthesis, but at middle and high intensities the rate is appreciably enhanced by increasing the CO_2 concentration: in some experiments with crop plants a concentration of 0·5 per cent of CO_2 was particularly effective for short periods, although prolonged exposure to this concentration injured the leaves. It seems as if an optimal concentration of CO_2 would be about 0·1 per cent and so in general plants do not have enough CO_2 to make the maximum use of the sunlight falling on them.

The end result of photosynthesis is a carbohydrate $C_6H_{12}O_6$ (see the equation on p. 56) which can be converted by the plant into starch and stored; it can be combined with other sugar molecules to make cellulose which is a basic structural material in plants; or it can be combined with elements such as nitrogen, phosphorous and sulphur to produce proteins, nucleic acids and all the other constituents of living cells. Some of the sugar produced by photosynthesis is used as an energy source by the plants themselves for growth, the maintenance of tissues, and biochemical processes. This process is called **respiration** and can be summarized as:

$$C_6H_{12}O_6 + 6O_2 \longrightarrow 6CO_2 + 6H_2O + 2830\,kJ$$

It is worth noting that this energy is converted to heat in the course of its use by the plant and so is never available for use again within the ecosystem since it is a dispersed low-grade energy rather than the concentrated high-grade chemical energy which is incorporated into plant tissues. The amount of energy available for the heterotrophs in the ecosystem (i.e. the animals and the decomposer organisms) is therefore dependent upon the balance between the rates of photosynthesis and respiration (Fig. 2.3). Put simply and symbolically,

$$N = P - R$$

where P = gross production in plants, or the total energy assimilated by the organism in a given time;

R = respiration, that part of the assimilated energy converted by the plant to heat or mechanical energy or used in life processes;

N = net production, or the increase in organic matter or total energy content in a given time. It appears as an increase in the 'living weight' of plants (and by extension it can be used of the whole ecosystem), i.e. an increase in **biomass**. The rate of accumulation of biomass (i.e. weight of living matter/unit area/unit time is called **Net Primary Productivity** (NPP).

Looked at in terms of a world scale, about 520×10^{22} joules (J) of energy strike the Earth's atmosphere every year, of which about 100×10^{22} J reaches the Earth's surface and is of a suitable wavelength for photosynthesis. Of this, some 40 per cent is reflected back into space by deserts and oceans, leaving 60×10^{22} J as the 'pool' for photosynthesis. Of this, a large quantity is respired and the remainder appears as biomass, i.e. as organic

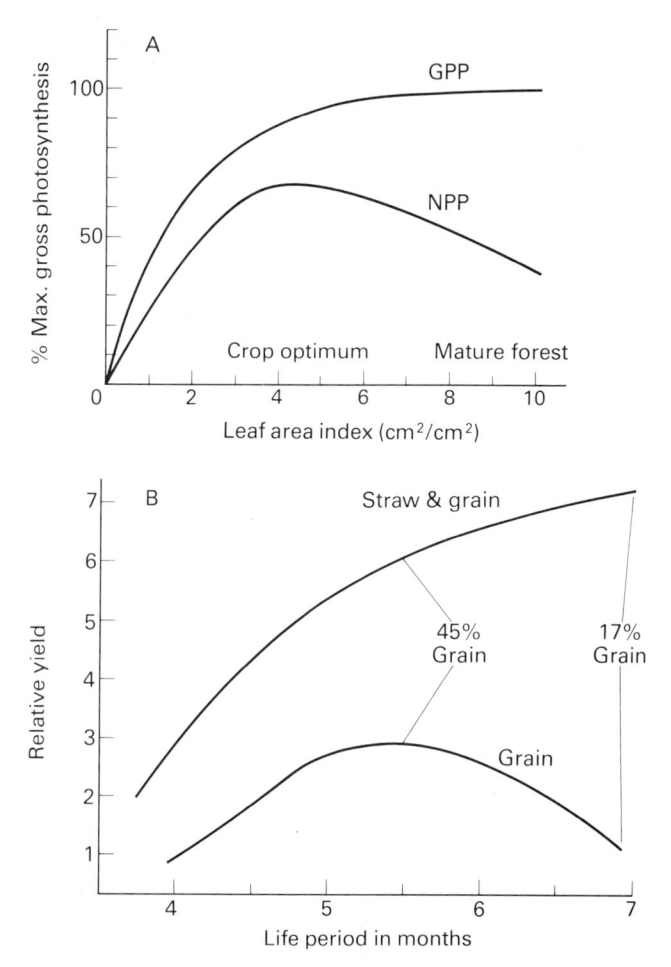

Fig. 2.3 Productivity rates for crops. **A** The effect of increasing the leaf area index (i.e. cm² of leaf surface exposed to light per cm² of ground surface) upon the relation of gross primary productivity and net primary productivity. For crops there should be as small a difference as possible. **B** The effect of the duration of the growing season on the yield of grain and straw in rice.
Source: Odum, 1971.

matter. Table 2.1 gives some data for the P/R relationship for different plants. The total amount of biomass produced annually by autotrophic plants is estimated to contain about 170×10^{19} J (much of it from phytoplankton) and so the average coefficient of utilization of the incident photosynthetically active radiation by the flora of the globe is *ca* 0·2 per cent. We may interpose the information that less than 1 per cent of that 0·2 per cent is consumed by man as food to give us a context in which to ponder one of our uses of the world's biotic resources.

TABLE 2.1 Annual primary production and respiration in 10^4 J m²/yr

	Alfalfa* fields (USA)	Maize (per 100 days)** (USA)	Young pine forest (England)	Medium aged oak-pine forest (New York)	Mature tropical rainforest (Puerto Rico)	Average for the Earth
Gross primary production (P)	10 210	3 400 (12 420)**	5 100	4 970	23 140	1 200
Respiration (R)	3 850	790 (2 900)**	1 960	2 730	17 820	600
Net primary production (NPP)	6 360	2 610 (9 520)**	3 140	2 240	5 320	600

* Alfalfa is a member of the legume family, which includes peas and beans.
† Maize production is measured during the growing season of 100 days. The figures in brackets are derived from extrapolation from 100 days to a whole year; if maize could be grown continually throughout the year (as alfalfa can) these values would be obtained.
NB The low average values for the whole Earth are due to the very low levels of primary production in the open oceans which occupy two-thirds of the Earth's surface.
Source: Varley 1974, amended.

Ecological energetics

The fate of energy which appears as net primary production gives us a way of constructing a model of an ecosystem, usually called a Lindeman model after its first (1942) articulator. In its simplest (Fig. 2.4) form, it consists of a series of boxes which represent stored energy—i.e. biomass—together with lines which indicate flows between the boxes. The area of the boxes can be made proportional to the quantity of biomass and so a quantitative model of a simple energy transfer system or **food chain** is constructed. Normally, some of the plant material continues as storage in perennial plants, some is eaten by herbivorous animals, and some dies to form a litter on the soil surface. Herbivorous animals are eaten by carnivores or else join the litter. Carnivores have only the litter layer to look forward to unless they in turn are predated upon by another or 'top' carnivore. Organisms which live in the litter layers immediately start to break down the litter, which is then chemically decomposed by the soil flora and fauna or **decomposer** organisms. The type of litter and its specialist organisms play a key role in the recycling of mineral nutrients in the system (p. 69). Each stage in the chain is called a **trophic level**: the plant producers (autotrophs) being the first trophic level, the herbivores (heterotrophs) being the second trophic level and so on. In most terrestrial ecosystems, three or at most four trophic levels are found. A few ecosystems are dominated by their **detritus** layers: a natural example would be a cave used as a roosting place by bats, where the energy source would be the bat-dung; a man-made example might be a river heavily contaminated with chemicals toxic to all but some of the detritus-feeding organisms of the river bed. In reality, there are usually many different plants and animals (as taxonomically classified rather than by trophic status) at each trophic level and so the idea of a **food web** rather than a food chain is introduced. Figure 2.5 shows an example: it will be seen at once that a multiplicity of species at each level

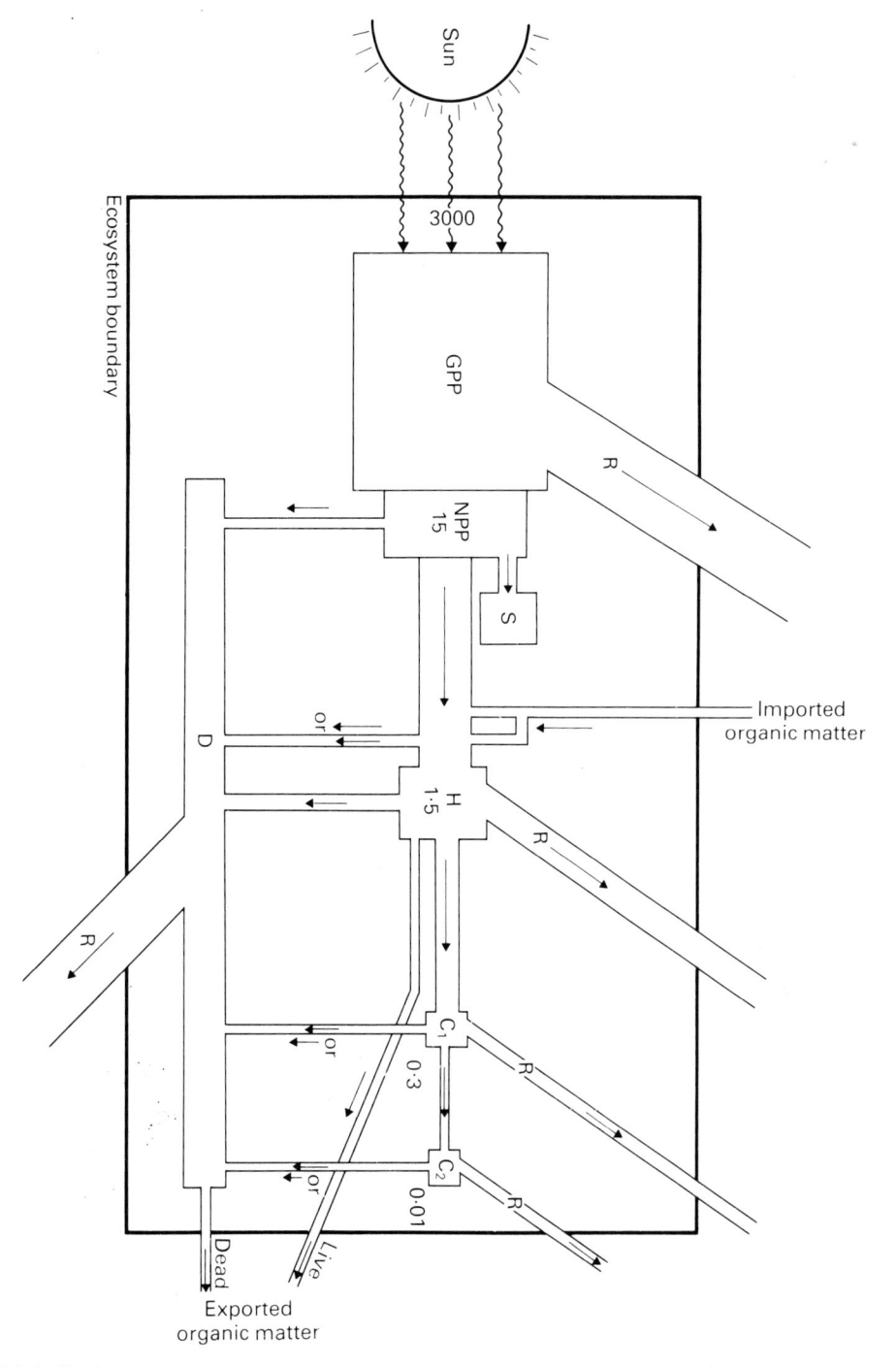

Fig. 2.4 A simple model of energy transfer in an ecosystem with imports and exports of organic matter. GPP=Gross Primary Productivity; NPP=Net Primary Productivity (i.e. GPP−Respiration, R); S=Storage; H=Herbivore animals; C₁ and C₂=Carnivores; D=Decomposers. Values are in Kcal/m²/day.

62 *The ecosystem scale*

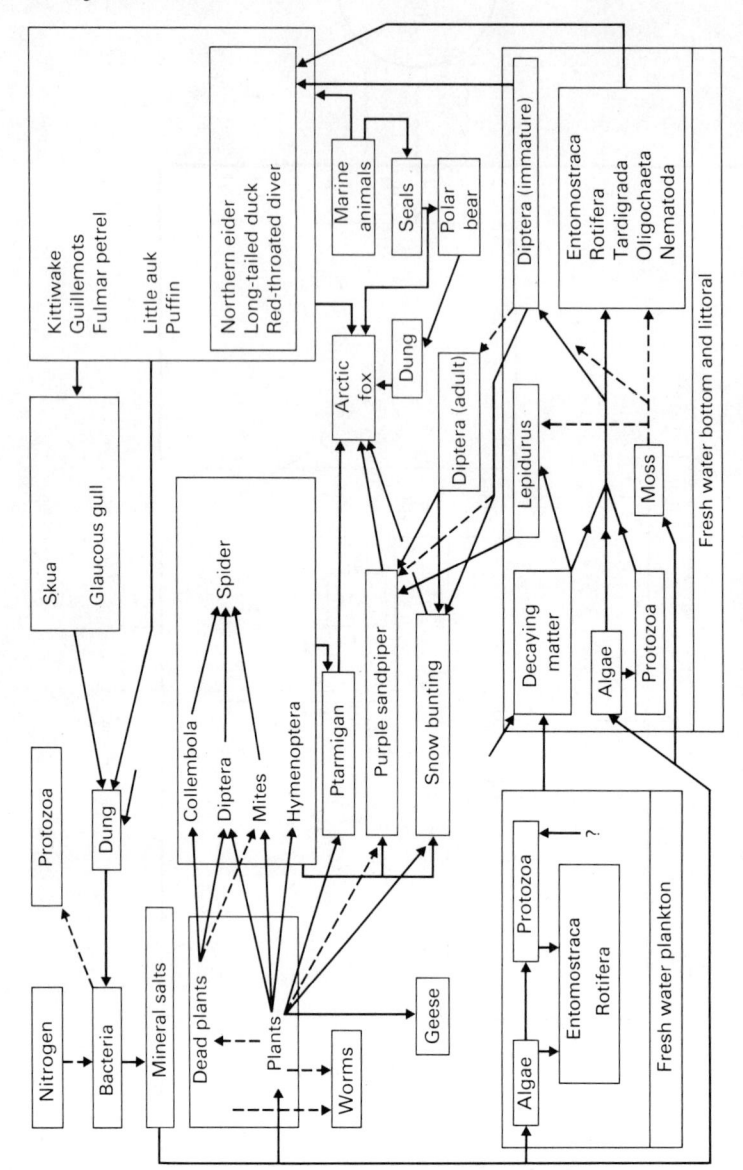

Fig. 2.5 A simplified food web for Bear Island, Spitsbergen. Originally published by Summerhayes and Elton in 1923, this is a classic early example of the concept of the web. Note the important role of the sea.
Source: Collier *et al.*, 1973.

is likely to make for greater stability in the system should one be removed. The idea of trophic levels is, however, complicated by the fact that in nature not all species feed constantly at one trophic level: some may move between levels seasonally, or at different stages of their life history, or opportunistically as food becomes available.

In their energy relations, all such chains and webs are subject to one of the basic laws of physics governing energy flow, namely the Second Law of Thermodynamics. Put simply, the law requires that at every transfer step in the ecosystem some energy will be degraded from a highly concentrated chemical form to a highly dispersed form as heat which cannot be recycled into chemical energy but must be radiated out to space and lost to the ecosystem. So at each trophic level a conversion to heat takes place which means that less energy becomes biomass at the succeeding trophic level, especially at the herbivore and carnivore levels. Consider as a simple example the foraging of a herbivore which means that a portion of its energy intake is expended in finding more energy, a condition which is exacerbated if the herbivore is warm-blooded and has to expend energy maintaining a constant body temperature. A carnivore may have to expend a great deal of energy catching its prey and many are thus adapted to eating at relatively infrequent intervals, e.g. lions and snakes. One result of this law is that in a given ecosystem, the net availability of energy gets less at levels successively away from the plant producers: if the energy content (usually given as the **calorific value**) of each trophic level is plotted then a pyramidal shape is obtained; very often the numbers of organisms also follow the same trends, although if the producers are very large (e.g. oak trees) then they may well be exceeded in number by herbivores such as defoliating caterpillars. However, a simple grass field—mice and voles, hawks and owls—ecosystem clearly follows the normal trends in this respect: the absolute number of predators is quite small compared with the number of grass plants. Contrariwise, the size of individual animals may increase up the chain since it may well be an advantage for a predator to be larger than its prey. Exceptions are not hard to find to this generalization, however. Some ecosystems may appear to have more energy at their consumer stages than in their plant production: this can be so because organic matter is imported across the ecosystem boundaries, e.g. carried by running water into a pond, by the tides into an estuary or by lunch-time office workers into a duck-inhabited park lake. These facts have obvious relevance for the human use of biota. If an ecosystem (whether natural or man-directed) is being cropped for food, then there will be much more energy available per unit area if man eats as a herbivore than as a carnivore, assuming that the plants are as edible as the animals.

The manipulating role of man may alter the energy relations of ecosystems in many ways: by simplifying food webs, for example, so as to eliminate competitors for his chosen crop (these unwanted organisms are then labelled 'weeds' and 'pests') or by clearing away a complex web like a forest and replacing it with a corn monoculture. But none of these is more important than his net additions of energy to ecosystems via the use of fossil fuels such as oil, coal and natural gas, and increasingly also of nuclear power. Fossil fuels are in effect stored photosynthesis and can be added to the ecosystem in a variety of ways but most commonly through machinery and chemicals. The bulldozer is perhaps the most crude and obvious; the appearance of a synthetic organic chemical like DDT in the body-fat of Antarctic penguins rather more subtle. There can come a point, in cities for instance, where the energy flows are dominated by those of the fossil fuels, and so the city is powered by the sun of millions of years ago, not that of today. But the idea of fossil fuel 'subsidies' to the flows of energy from contemporary solar power means that we have a way of classifying both natural and man-made ecosystems using the mix of the two sources as a diagnostic measure. Table 2.2 shows one such classification derived by E. P. Odum (1975), of which some examples are given in Plate 9. This linkage of natural and human systems carries with it the potential for

Plate 9 Different types of ecosystems according to their energy characteristics (see Table 2.2): (a) unsubsidized solar-powered system: tropical forest in Guyana; (b) naturally subsidized solar-powered system: a river estuary in west Wales; (c) a fossil-fuel subsidized solar-powered system: modernized crop agriculture in Rhodesia; (d) a non-solar fossil and nuclear-powered system: Pittsburgh by night.

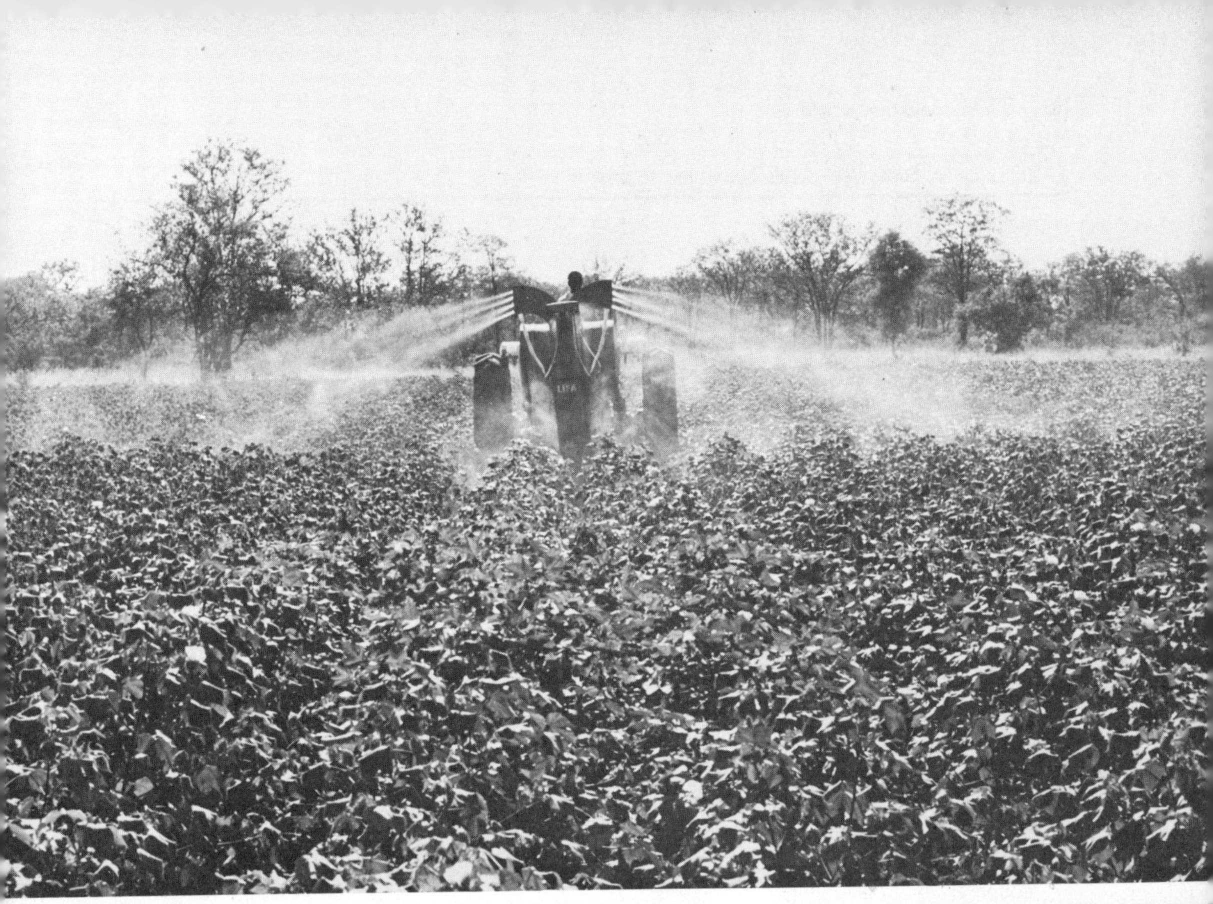

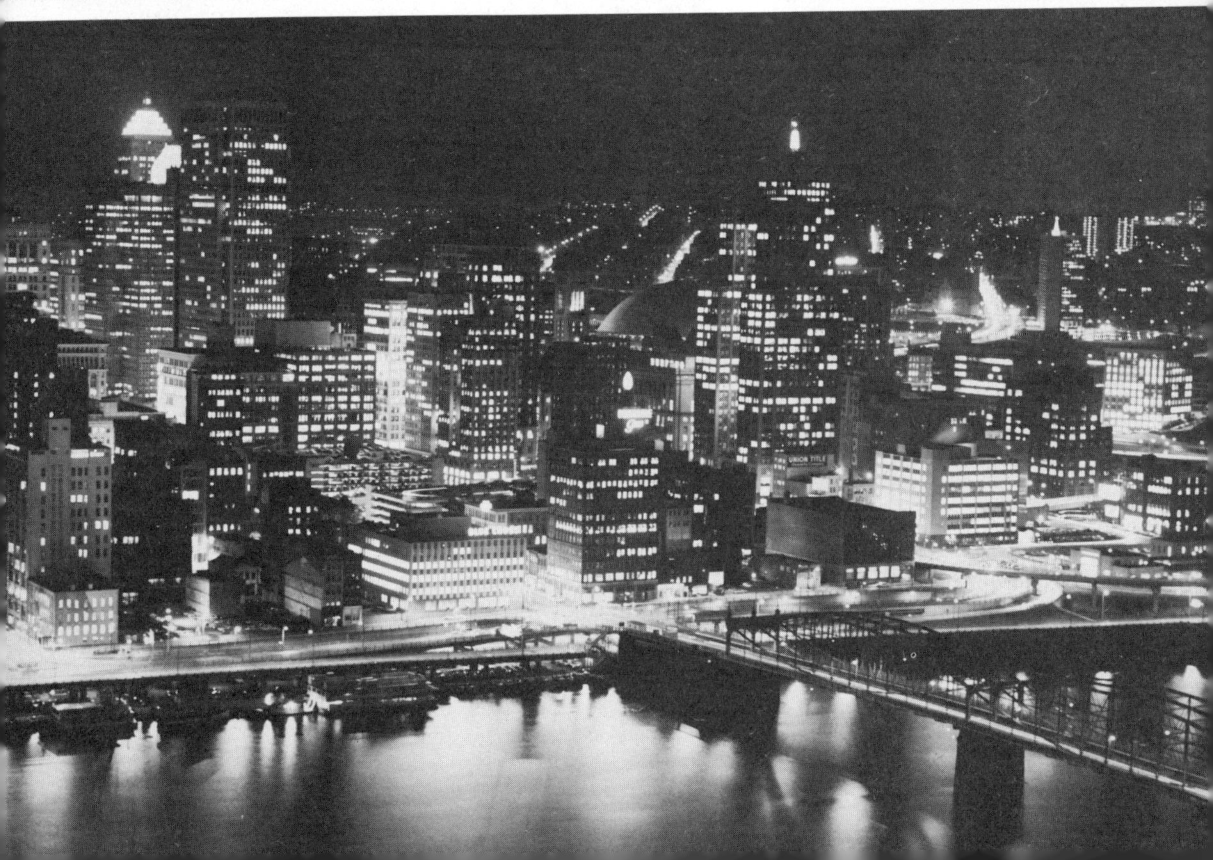
UNION TITLE

The ecosystem scale

TABLE 2.2 Ecosystem types according to energy level

Ecosystem type	Annual energy flow (kcal/m²/yr) Range	Average (estimated)
1. Unsubsidized natural solar-powered ecosystems: e.g. open oceans, upland forests. Man's role: hunter-gatherer, shifting cultivation	1 000–10 000	2 000
2. Naturally subsidized solar-powered ecosystems: e.g. tidal estuary, lowland forests, coral reef. Natural processes aid solar energy input: e.g. tides, waves bring in organic matter or do recycling of nutrients so most energy from sun goes into production of organic matter. These are the most productive natural ecosystems on the earth. Man's role: fisherman, hunter-gatherer.	10 000–50 000	20 000
3. Man-subsidized solar-powered ecosystems: food and fibre producing ecosystems subsidized by human energy as in simple farming systems or by fossil fuel energy as in advanced mechanized farming systems. E.g. Green Revolution crops are bred to use not only solar energy but fossil energy as fertilizers, pesticides and often pumped water. Applies to some forms of aquaculture also.	10 000–50 000	20 000
4. Fuel-powered urban-industrial systems: fuel has replaced the sun as the most important source of immediate energy. These are the wealth-generating systems of the economy and also the generators of environmental contamination: in cities, suburbs and industrial areas. They are parasitic upon types 1–3 for life support (e.g. oxygen supply) and for food: possibly fuel also although this more likely comes from under the ground except in LDCs where wood is still an important domestic fuel.	100 000–3 000 000	2 000 000

The most productive natural ecosystems and the most productive agriculture seem to have upper limits of ca 50 000 kcal/m²/yr.

Source: E. P. Odum 1975.

valuing natural systems in human terms, e.g. the waste-processing capabilities of a river or wetland using the current price of fossil fuel energy as a means of valuing the natural system's 'free' contribution. Much more work remains to be done but preliminary results suggest that many wild places are much undervalued in our present combination of intuition, emotion and ignorance.

Ecosystems in time

So far we have treated ecosystems as if they were static entities. This is not so for obviously they change during time: they may start from zero after an event like the eruption of a volcanic cone from the sea; and they change on smaller scales as when a senescent tree in a forest is windthrown and replaced by a series of herbs, shrubs and trees until a mature forest is present once again. The process from the establishment of life at a particular place up to an ecosystem which is stable and self-renewing is called **succession** and the mature stage is often referred to as the **climax** (Golley 1977). Some apparently mature ecosystems are so old that we do not know their original stages of

Plate 10 A pond undergoing succession: a succession of vegetation types can be seen from open water to scrub, and these zones slowly move inwards until the lake fills up entirely.

succession: the tropical rainforests for instance may have been in much the same state for 100 million years. Other ecosystems are much newer: the forests of temperate lands started as bare land after glaciation and proceeded through successional stages of lichen and moss tundra, scrubby birch and aspen, through pine forest to a stable mature forest of oaks, limes, elms and beeches. Another succession occurs when a volcanic lava flow cools and is colonized by a series of plants and animals until forest may cover it; a pond left after the decay of ice gradually fills in as the succession of plants moves inwards, filling the pond with organic debris (Plate 10). Forest may then colonize or in some places acid bog populated by plants nourished only from rainwater may become established. At each stage there may be a particular plant species or group of taxa which exert a controlling effect over all the other biota (e.g. trees in a forest, phytoplankton in the sea). These are called the **dominants**.

The self-regulating process of succession towards a stable state is called **autogenic** succession; it contrasts with **allogenic** succession where an outside influence imposes change, usually by putting back the succession to an early stage. A climatic deterioration, an unusually severe flood, or the effects of human activity are examples of allogenic processes. Overall, however, there has been sufficient time since the Pleistocene for most parts of the world to receive mature systems which are in equilibrium with climate and topography. These systems are the major world biomes which are described on p. 90. They are in many cases present only in relict form because of the transformations wrought by man.

The characteristics of ecosystems change during succession. In many cases this is obvious: mosses are replaced by trees so that the biomass per unit area increases, for example, and the number of species increases so that food webs become more complex. A model of the changes is set out in Table 2.3. (Some of the characteristics refer to population dynamics and nutrient cycles discussed later in this chapter.) We notice that

TABLE 2.3 The chief characteristics of ecosystem development

Ecosystem attribute	Ecosystem state Developmental (*successional*)	Mature (*climax*)
ENERGY FLOW		
1. Food chains	short (linear, mainly autotrophic grazing)	long (web-like mainly heterotrophic detritus feeding)
2. Community respiration: gross primary production (R/P)	1	approaches 1
3. Production per unit biomass (NPP/biomass)	high	low
4. Net community production (NPP)	high	low
STRUCTURE AND DIVERSITY		
5. Biomass	low	high
6. Species diversity (variety and equitability components)	low	high
7. Stratification	poor and simple	well-developed and complex
INDIVIDUALS AND POPULATIONS		
8. Fluctuations in numbers	marked	weak
9. Population control	external (climatic)	internal (biological)
10. Size of organisms	small	large
11. Life cycles	short (simple)	long (complex)
12. Feeding specialization	broad	narrow
NUTRIENT CYCLES		
13. Inputs from outside system.	large (open cycles)	small (closed cycles)
14. Nutrient exchange rate between organisms and environment (immobilization in biomass)	rapid	slow
15. Role of detritus in nutrient regeneration	unimportant	important
16. Nutrient conservation within system	poor	good
STABILITY		
17. Interdependence of organisms	weak	strong
18. Resistance to external disturbance	poor	good

Source: Amended from Varley 1974; originally from Odum 1969.

net production is high in early stages, as the volume of organic matter per unit area increases but low at mature stages when the increase of organic matter is virtually nil in the stable state. Biomass, however, has increased and the physiognomic structure is likely to have become more complex, for example, a single layer of lichen is replaced by a forest with tree, shrub, herb and moss layers. The question of trends in species diversity is somewhat controversial: in some ecosystems there is not a simple addition of species as succession proceeds, but a high peak in mid-succession, after which the mature state possesses fewer species. Indeed, it must be said that the whole concept of succession encompasses, as Horn (1976) puts it, 'a bewildering variety of patterns'. Succession may rapidly produce a stationary vegetational composition; it may lead to alternative stationary states dependent upon the initial composition; the vegetation may continually change in response to more or less random changes in the environment or change cyclically following an environmental factor. Small wonder also that the concept of climax is not simple, either, and that many ecologists have declined to recognize it except in terms of a description of a specific ecosystem at a specific point in time.

Looked at synoptically, the world consists of a mosaic of ecosystems at different stages. In a mostly natural world (say perhaps at *ca* 1000 AD although even by then man's impact had been strong in some places), there are large areas of mature ecosystems, smaller areas of successional systems (some filling gaps caused by catastrophies in the mature systems, some colonizing newly available areas like recently deglaciated terrain, weathered lava flows or eustatically emergent coasts), some mixed mature and successional areas at zones of contact where a small climatic change is critical for the ecosystem type, and some biologically inert areas like volcanoes, ice-caps and shifting sand. The present man-altered world shows an analogous set of systems: mature ecosystems are still present though at any rate on land they are much reduced in area; early successional ecosystems are widespread because this is the type of ecosystem which represents agriculture: a short food chain with low species diversity. Mixed systems abound where for example forestry and agriculture are intermingled, and inert systems are much increased in area where urban and industrial land use has greatly diminished if not entirely extirpated biological activity. In summary, human activity has greatly increased the inert and early successional phases of the mosaic at the expense of the mature. The question to ask, although there is not yet any clear answer, is whether there is a necessary ecological balance between the phases and whether the human species, if it continues its conversionary activity will somehow set off ecological fluctuations of an unforeseeable and possibly uncontrollable kind.

Mineral nutrients and their pathways

Apart from energy, life is sustained by a number of chemical elements which enter ecosystems via the plants. These elements come from the materials of the crust and atmosphere and if involved in the growth and maintenance of organisms they may conveniently be termed **mineral nutrients**. The elements needed for life are dominated by four: carbon, hydrogen, oxygen and nitrogen, and indeed these elements combined in various ways represent all but a tiny portion of the Earth's terrestrial vegetation and hence most of the living matter of the planet. The other elements which have been discovered as essential to life are 13 in number, comprising K, Ca, Mg, P, N, S, Fe, Cu, Mn, Zn, Mo, B and Cl. These are usually derived from the bedrock whereas the other four come ultimately from the atmosphere; thus neither soil nor atmosphere alone can support life: interaction between them nearly always occurs. Only oxygen and hydrogen are freely available in large quantities and so a characteristic feature of the pathways of the other elements is that they circulate between living organisms and non-living or abiotic pools of various scales. These pathways are generally cyclic: they involve the use and re-use of the atoms of the nutrients and since they also include an abiotic phase they are usually called **biogeochemical cycles**. Study of these cycles usually reveals two scales of rotation. The first has a slow turnover and involves a very large pool of the element stored in the atmosphere or the Earth's crust (or both in the case of carbon), and is usually non-living in character. Such an accumulation is usually called a reservoir pool. Contrasted to this is a small pool which undergoes rapid exchange between an organism and its immediate environment, both living and non-living. This lesser store is called the cycling pool. As a simple example, we may think of a molecule of calcium which weathers out from a rock (a reservoir pool) and into a soil from which it is taken up by the roots of a tree. At leaf-fall, the molecule falls to the ground, and through decay processes finds its way into the soil from which it is taken up again into the tree. The soil and the tree are both cycling pools in this case. This molecule may one year be lost from the soil pool into the runoff and end up in a marine sediment, thus being returned to the reservoir pool where it will very likely stay for billions of years until uplifted into a terrestrial context once again. In this scale of cycle, the litter layer is a very important stage in the nutrient circulation in terrestrial ecosystems.

Although much attention has focused in recent work on the dynamics of mineral nutrients in ecosystems at the local scale, it is still of interest to consider the global circulation of some of them, especially any which may be limiting factors upon the growth and dispersion of organisms or in which man may significantly interfere. One of the cycles most intensively studied has been nitrogen, because it has often been a limiting factor in agricultural productivity, and the discovery of the value of nitrogen-containing chemical fertilizers must rank as one of the major advances in the development of food resources for man. The world nitrogen cycle is summarized in Fig. 2.6. One of the paradoxes about nitrogen is that although the atmosphere is 79 per cent N_2, plant growth (especially crops) is limited more by the availability of fixed nitrogen than by any other element. This is mainly because the atmospheric nitrogen is an inert gas and it has to be fixed in a combined form before uptake by plants can occur; the largest single natural sources of such compounds are firstly terrestrial micro-organisms, and then close associations between such micro-organisms and green plants. In the latter case the Leguminosae family, which house microbes in root nodules, are probably the single greatest natural source of fixed nitrogen. As Fig. 2.6 shows, a smaller amount of nitrogen is fixed in the atmosphere by ionizing processes thought to involve the energy represented by lightning, and some 'juvenile' N_2 compounds result from volcanic action. To offset these, there are denitrification processes which return inert nitrogen to the

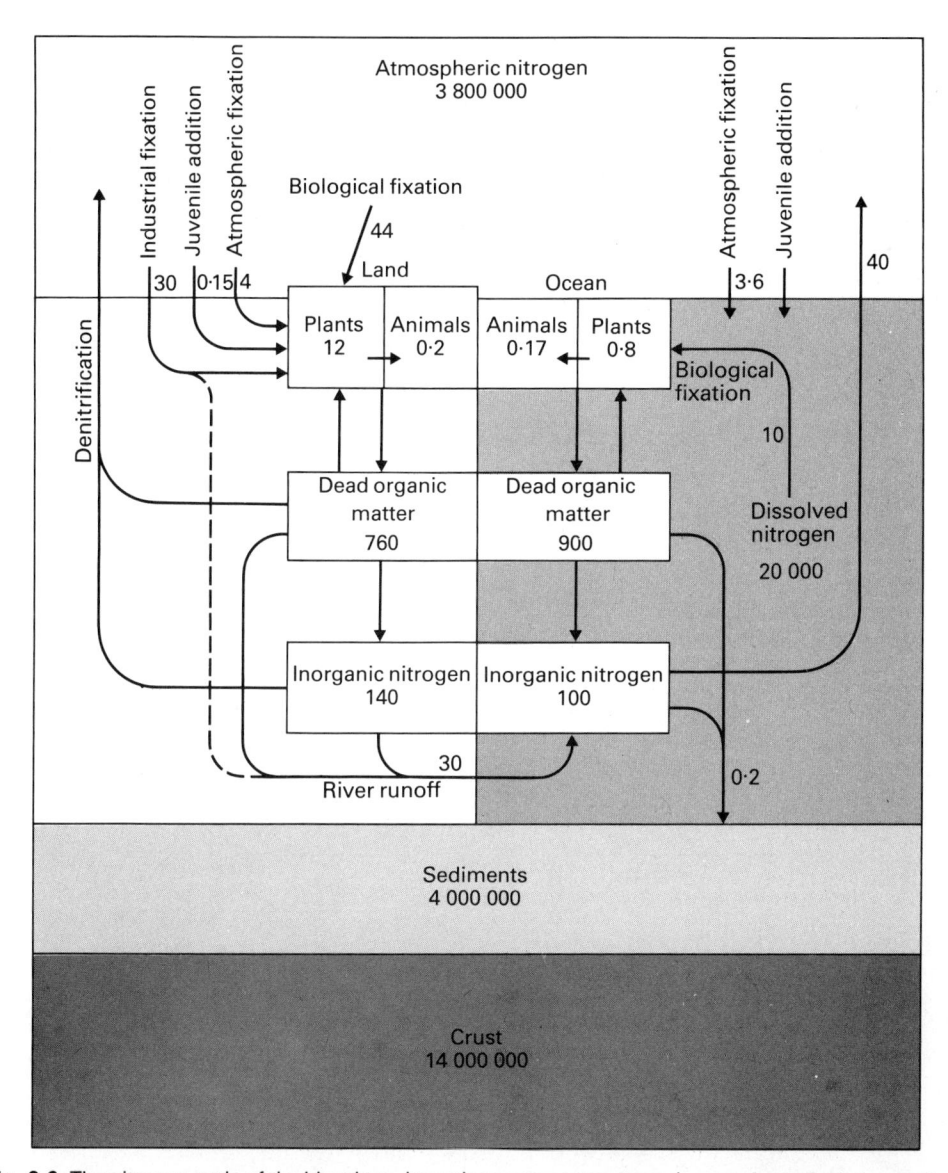

Fig. 2.6 The nitrogen cycle of the biosphere, its various compartments and rates of transfer between them. The figures are tonnes×10^9 for the compartments and tonnes×10^6 for the transfers. The assumption is made that there was an overall balance before industrial fixation became such a significant flow. Source: *Scientific American*, 1970.

atmospheric reservoir and also, although in much smaller amounts, to sedimentary rocks. In a 'natural' world, fixation and denitrification balance. Added to these natural processes is the effect of human activities; since the discovery of the value of nitrogenous fertilizers, the provision of these by industrial means has been growing to the point where it is estimated that industrially fixed nitrogen now annually amounts to 26 per cent of the present biological terrestrial fixation, including agricultural areas. At the present growth rate of the fertilizer industry the industrial fixation will equal the biological fixation by 1989 (Soderlund and Svensson 1976). The results of this are uncertain but at small scales, the addition of fixed nitrogen to ecosystems usually results in the biological shifts called eutrophication (see p. 239) resulting from fertilizer runoff and the production of organic wastes, including human sewage. The management of the cycling pools of nitrogen thus become a factor in the biogeochemical cycle of this element.

An element with a more restricted distribution than nitrogen is phosphorus, which does not appear to have an atmospheric pool: if it is present in the atmosphere (as in sea-spray) then it is only as a suspension which falls out quite rapidly. For the biosphere, the ultimate source of phosphorus is crystalline rock which makes available its nutrient content as it weathers. The phosphorus passes along the food chain and eventually comes into the detritus chain either as faeces or dead organisms. Here, it is liberated in inorganic form to be recycled to autotrophs or to enter a sedimentary phase in the soil or in the deposits of aquatic systems, including the sea floor (Fig. 2.7). The amount of phosphorus available for life depends on the rate at which it moves through the various phases of the cycle, and its movement in sedimentary cycles such as soils and lake deposits is apt to be very slow: once there it is not quickly released. An increased demand for phos-

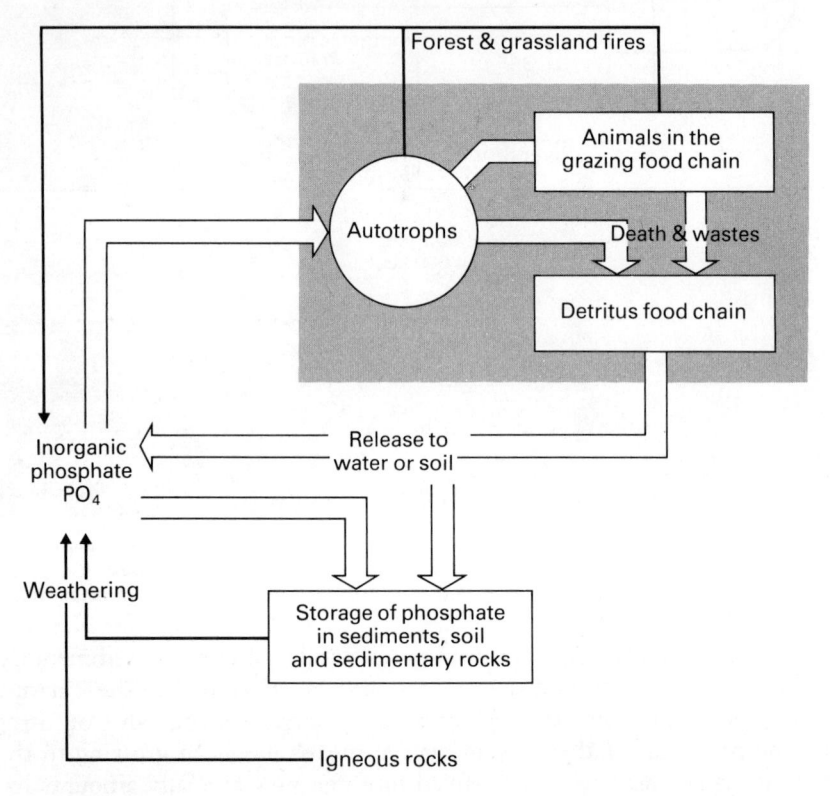

Fig. 2.7 A non-quantitative representation of the global phosphorus cycle. The organic portion of the cycle is shaded.
Source: Clapham, 1973.

phorus in an ecosystem (e.g. as succession proceeds and larger organisms colonize) is met by an increased turnover rate within the organic components rather than by faster release from sediments. Hence, phosphorus is often in short supply in an ecosystem, i.e. it is a limiting factor. Deevey (1970) says of it:

> ... if short supply is chronic, the output of the entire system could be expected to be adjusted to the rate of exploitation of one critical element, much as the performance of a bureaucracy is closely geared to the supply of paper-clips.

(We might think now that photocopying is perhaps a better analogy.) Again, human activity can become significant in this biogeochemical cycle. To increase crop yields, phosphate rock is mined and turned into chemical fertilizer. The runoff accelerates the cycle in its aquatic phase and releases the growth of organisms for which it is a limiting factor. Nitrogen may then become limiting and so blue-green algae bloom because they can fix their own nitrogen from the atmosphere. Once again, eutrophication has occurred; more seriously, increasing amounts of scarce phosphate is returned to the sea floor so that an element often limiting in nature is being pushed through the system instead of being held within it by conservation practices. In economic terms, phosphate rock is at present plentiful but its recovery depends largely upon price, and in particular on energy availability.

A last example is an element with both atmospheric, oceanic and crustal reservoirs. Carbon is the cornerstone of organic materials, yet only a few tenths of 1 per cent are present in the cycling pool: most of it is locked in the Earth's crust. The main cycle (Fig. 2.8) is from carbon dioxide to living matter and then back to carbon dioxide, and the terrestrial biota, according to Woodwell et al. (1978), are a source of CO_2 rather than a sink. Some of the carbon is bled off to the sediments of the crust where it forms the huge reservoir pool. The linkage therefore between the components of the cycling pool (oceanic and atmospheric) is the transfer of CO_2: the atmosphere and ocean exchange CO_2 at their interface, probably at the rate of 100×10^9 t/yr (Bolin 1970). The oceans appear to act as the main regulator in this process: the amount of CO_2 in the atmosphere is essentially determined by the partial pressure of CO_2 dissolved in the sea. If the CO_2 content of the oceans is stable, then so is that of the atmosphere. This balance has of course been subject to change from human activity in the form of burning fossil fuels and thus accelerating the release of CO_2 into the atmosphere from the reservoir pool of the Earth's crust. At present, the yearly rate of combustion adds some 2 ppm to the existing 320 ppm of CO_2. The observed increase is only 0·7 ppm so that much of the excess is being absorbed by the biosphere. This extra amount is nevertheless cumulative and concern has been expressed about the effect of the increased CO_2 in trapping incoming solar radiation. This is the so-called 'greenhouse effect' and is discussed further on p. 336. On the other hand, there might be a very small stimulation of photosynthesis due to the extra carbon dioxide.

The general lesson from these cycles is that although the quantities involved are large and the cycles lengthy, it is not impossible for man to intervene substantially, especially to accelerate the speed of movement of an element through a reservoir, or to concentrate elements far beyond their natural levels. Biological shifts to accommodate the new

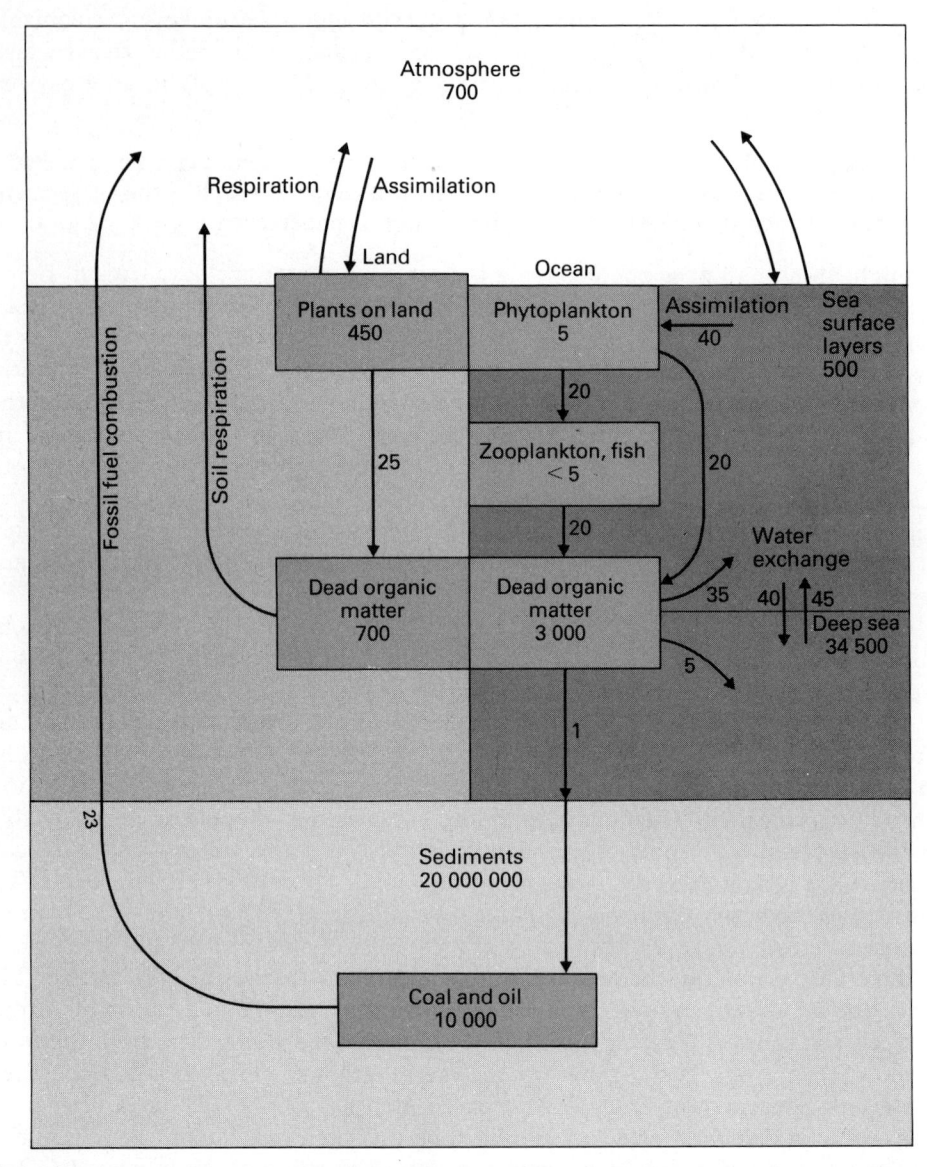

Fig. 2.8 The carbon cycle of the biosphere. It consists of two rather separate cycles, one on land and one in the sea, which are connected at the interface between the ocean and the atmosphere. All values are in tonnes×10^9. Man's intervention in the cycle is seen especially in the combustion of coal and oil: two-thirds of the extra CO_2 in the atmosphere is quickly removed either by the oceans or the terrestrial plants.
Source: *Scientific American*, 1970.

nutrient availabilities are usually found and these are only infrequently deemed useful by the societies whose activities created them.

Recent research has also greatly extended our knowledge of the cycling pool of mineral nutrients at the scale of the local ecosystem. The principal result of ecosystem development seems to be the creation of 'tight' or 'closed' nutrient pathways which recycle essential nutrients within the system. Any losses are balanced by inputs from the abiotic parts of the cycling pool. The two major pathways for recycling nutrients are firstly a return by way of animal excretion and secondly by way of microbial decomposition of plant and animal detritus. Both function in most ecosystems, but the first dominates in systems where most energy transfer follows a grazing chain, as with plankton; the second is dominant in, for example, temperate forests and grasslands where more energy is transferred via the litter or detritus chain. A third cycle is possible where there is direct transfer from plant to plant via symbiotic micro-organisms; such a method is thought to be particularly important in tropical forests. All these processes need energy for their metabolism and this is one way in which the energy content of an ecosystem is 'used'. An ecosystem which is to persist (i.e. be relatively stable through time) must therefore have adequate storage of both energy and nutrients within its cycling pool and a regulated mechanism for mobilization of stored nutrients and energy: with mineral elements the activities of decomposers are especially important here. The residence time of elements in their level of the ecosystem is, however, usually short compared with the biomass, so that the loss by runoff in terrestrial systems is minimized. For example the turnover time in one temperate deciduous forest of nitrogen in the biomass was 88 yr, whereas in the litter layer it was 5 yr (Reichle *et al.* 1975).

One of the most thorough investigations ever of the biogeochemistry of an ecosystem has been carried out on a forested area in New Hampshire in the USA (Likens and Bormann 1975; Likens *et al.* 1977). Not only were the pathways of chemical elements studied under natural conditions, but experimental disforestation was carried out to investigate what happened to the biogeochemical cycles when the ecosystem was disrupted and then allowed to recover. The detail of the study defies easy summary, but some idea of one part of it may be gained from Fig. 2.9 which shows the biogeochemistry of calcium within an experimental watershed on the Hubbard Brook under undisturbed conditions. In this case we note that the cycling pool is mostly in the soil and forest floor and that much less is actually in the vegetation; but the major feature of interest is that the magnitude of the cycle within the forest-soil system is much greater than the inputs from precipitation and outputs to the runoff. The system has a clearly developed mechanism for retaining calcium within the plant-soil components of the ecosystem: the 'intra-system cycle' is said to be 'tight'. The effects of deforestation and subsequent recovery are shown in Fig. 2.10. After the trees were felled and regrowth suppressed for three growing seasons, considerable differences in the biogeochemistry were found. The virtual elimination of transpiration caused the volume of liquid water in the runoff to increase, and the concentrations of dissolved substances were much higher because of *inter alia* the acceleration of the decomposition of organic matter on the forest floor and the absence of the uptake of nutrients by vegetation. Regrowth of vegetation shows a rapid recovery of the forest, with the regulatory mechanisms for

76 *The ecosystem scale*

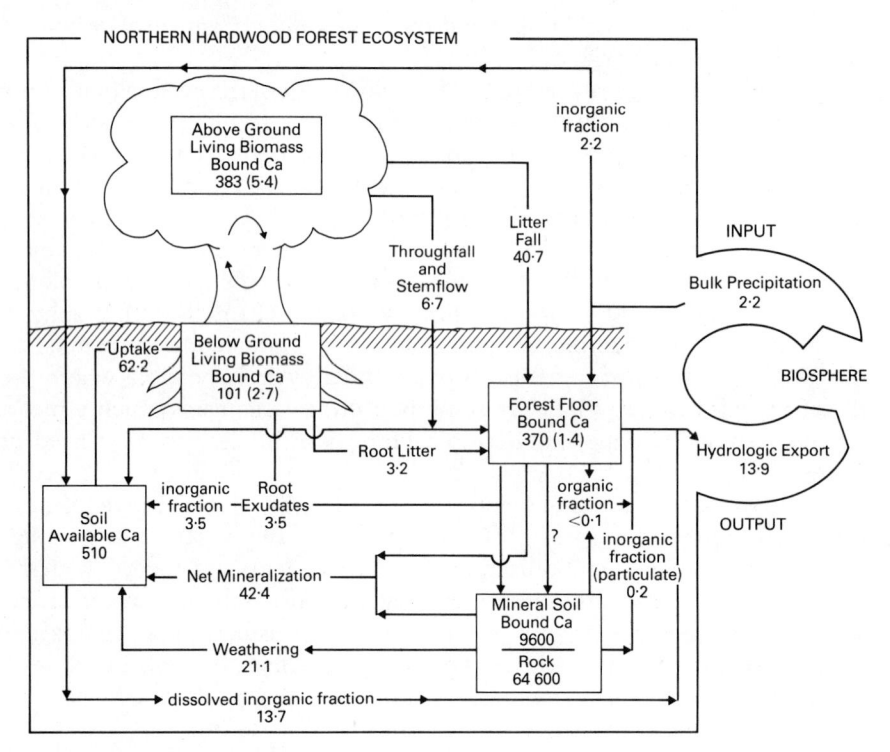

Fig. 2.9 The annual calcium budget for a deciduous forest at Hubbard Brook, New Hampshire. The compartments represent the content of the biomass in kg/ha and the transfer lines are in kg/ha/yr. Because the system is still accumulating biomass (i.e. is not yet ecologically mature), each box has a figure (in brackets) for the annual accretion.
Source: Likens *et al.*, 1977.

the retention of minerals and particulate matter being provided by the vegetation of the pioneer phase of recolonization, especially shrub species such as the pin cherry, *Prunus pensylvanica*. By 1976, however, not all the measured items had returned to normal and calculations of the losses during periods of deforestation suggest that the replacement of all the nutrients lost and the build-up to the original biomass may take as long as 60–80 yr (Likens *et al.* 1978).

One of the results of the Hubbard Brook study has been to point out that different species of plant accumulate minerals selectively, and so the type of box model which lumps together all the vegetation of an ecosystem may misrepresent the biogeochemical pattern to some extent. In the temperate zone, for example, the flowering dogwood *Cornus florida* accumulates and recycles calcium; a New Zealand endemic *Pimelea suteri* has unusual accumulations of chromium and nickel. These concentrations may sometimes merely reflect high levels in the soils (and can be a clue to mineral deposits of economic significance) or may in fact represent the accumulation by a plant of all the available supply of a mineral even of a low level of concentration in the soil. The role of the vegetation in holding minerals in intrasystem cycles is especially marked in the

Mineral nutrients and their pathways 77

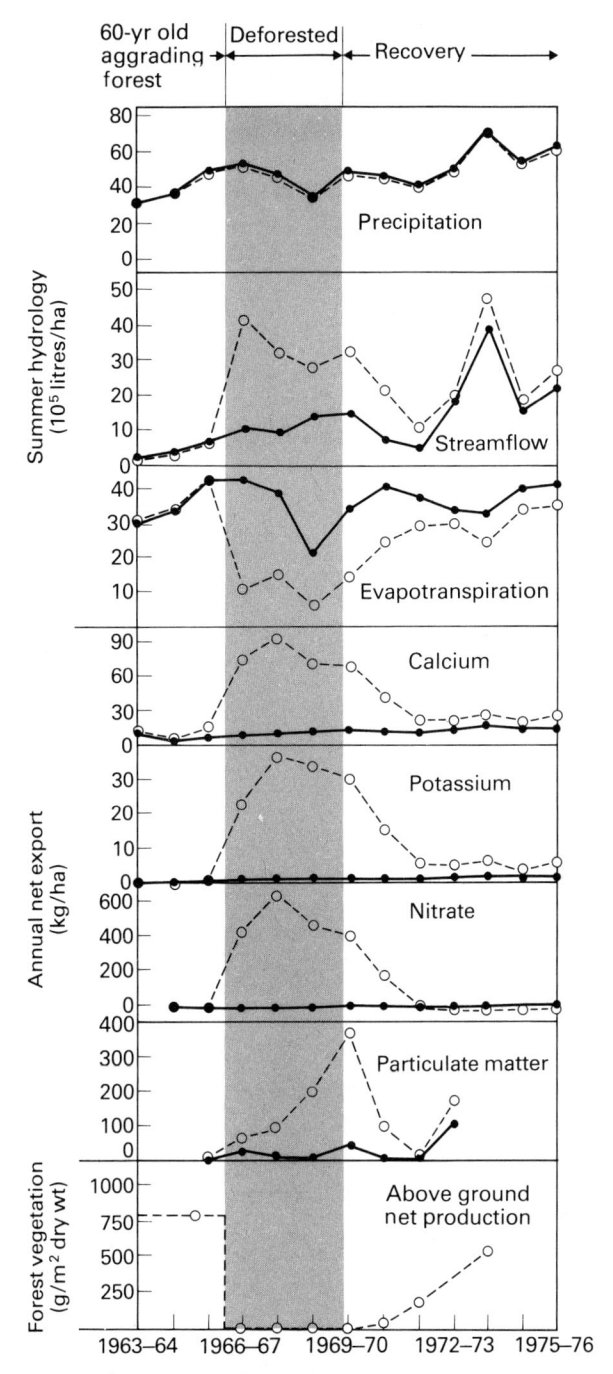

Fig. 2.10 Effects of deforestation on various parameters of the Hubbard Book forest. The open circles represent the experimentally deforested watershed, the black circles a control area. Hydrological effects were minor during the winter period because of continuous snow cover.
Source: Likens *et al.*, 1978.

78 *The ecosystem scale*

tropics where abundant rainfall would quickly leach minerals from soils and weathered rock. One of the mechanisms hypothesized is the transfer of minerals from a short residence time in quickly decomposing litter directly to the uptaking roots by means of mycorrhiza which selectively take up the minerals required by the tree (Fig. 2.11). At

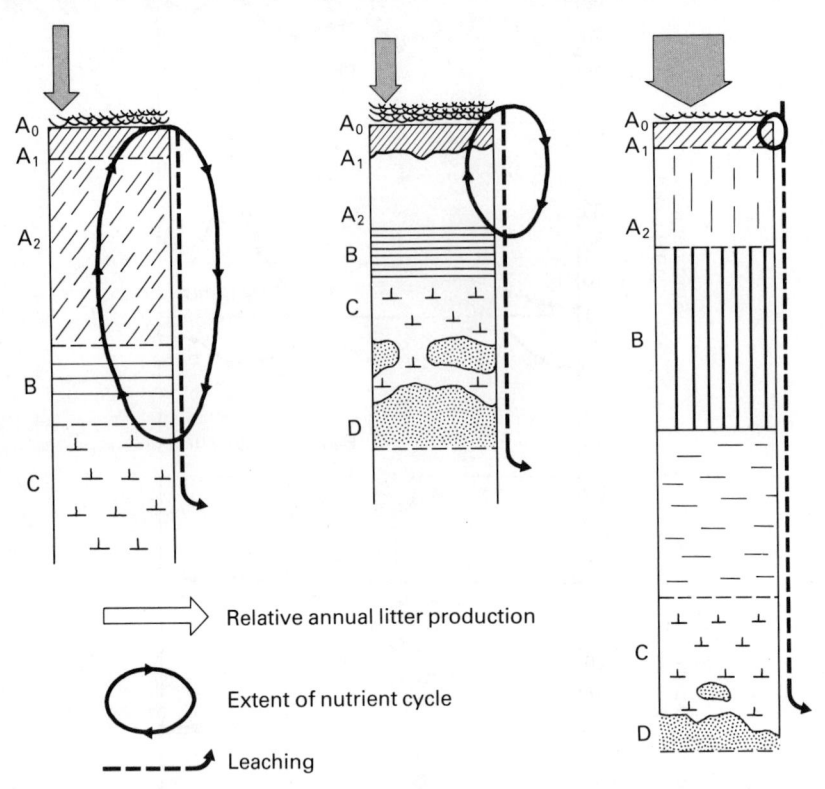

Fig. 2.11 Three types of forest soils showing the distribution of organic matter and the depth of nutrient circulation. From left to right, there are a brown earth soil as found under deciduous forest, a podzol as found under coniferous forests, and a latosol typical of tropical lowland forests.
Source: Tivy, 1971.

any one time, most of the nutrients are in the biomass and the mineral soil has a very low nutrient content compared with, for example, the Hubbard Brook calcium distribution.

Another feature of tropical forest vegetation is that the cycling time of mineral nutrients is short; the turnover time for minerals in the intrasystem cycles seems to rise with latitude which presumably exerts its efforts through the length of the growing season and the activity season for the soil flora and fauna. A world pattern of cycling times, driven by the sun's energy differentially according to solar input, seems to emerge (Jordan and Kleine 1972).

The biogeochemical cycle concept may also be used in habitats disturbed by man. The uptake of nutrients such as nitrogen and phosphorus by trees has led to suggestions that forests might be the most effective sewage treatment works yet discovered, since they might deal with the sludge as if it were a litter layer, and absorb the potential eutrophicating elements as well, which even three-stage treatment does not at present effect. Another example is the selection of management practices for colliery spoil heaps which encourage the vegetation so as to lower the rate of particular matter runoff and acid water drainage. 'Seeding' with chemical nutrients raises the level of biomass and organic matter turnover and helps reduce the erosion of and acid runoff from the tip (Chadwick 1975).

As with energy flow, the models based on biogeochemical cycles can tell us a great deal about the functioning of an ecosystem (Trudgill 1977). We do need to remember that they are not separate from energy flow but that an ecosystem has to use part of its energy in maintaining them, as distinct from adding to biomass, for example. The living tissues made possible by mineral nutrients are also the chemical repositories of the energy in the concentrated form in which it can be utilized by other organisms, so that the study of energy flow and mineral cycling must of necessity proceed together.

Population dynamics

The flows of energy and mineral nutrients through an ecosystem manifest themselves as actual animals and plants of a particular species. Groups of the same taxonomic unit which exist within definable limits of space and time are called populations and the characteristics of primary interest are the absolute size of the population (which can be very small: the Javan rhino may have a population of less than 50 individuals, or can be very large: we can only guess at the number of, for example, *Salmonella* bacteria in the world), and its density, i.e. its relation to the volume of space in its habitat.

The main biological criteria that affect population size and density can be summed up in graphical form as:

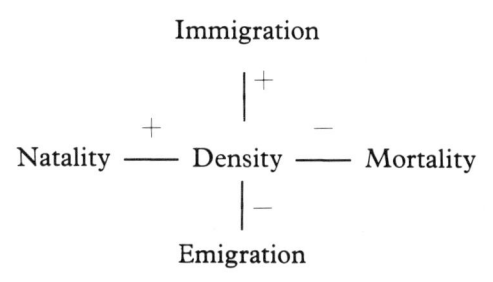

$$
\begin{array}{c}
\text{Immigration} \\
\big| + \\
\text{Natality} \xrightarrow{+} \text{Density} \xrightarrow{-} \text{Mortality} \\
\big| - \\
\text{Emigration}
\end{array}
$$

And this scheme can be applied at any scale and to any taxonomic group including our own. The concepts of immigration and emigration need little explanation except perhaps to stress the difficulty in measuring them that exists in real ecosystems; and that as with the ecosystem concept itself the placing of boundaries across which an individual becomes part of one population rather than another, can be problematical. The demographic measures of **natality** and **mortality** are more complex: in natality, two aspects of reproduction must be distinguished: **fertility**, which is the actual level of reproductive performance, which is a proportion of the **fecundity**, the potential level of

80 *The ecosystem scale*

reproduction in the population. The natality rate, therefore, may be expressed as the numbers of new individual organisms produced per unit time, and is highly variable. Some species breed once per year, others continuously; some produce prolific quantities of seed or eggs, others a very few. Similarly, mortality comprises two concepts: physiological mortality, which is the average longevity of individuals in a population living under optimum conditions and which therefore die of senescence; and ecological mortality which measures the average life-time of individuals under 'normal' conditions in which few die of senescence but fall to predators, disease and accident before they reach old age. Generation time is another variable: in the animal kingdom this ranges from a matter of minutes in the case of bacteria, to about 80 years in redwood trees (Fig.

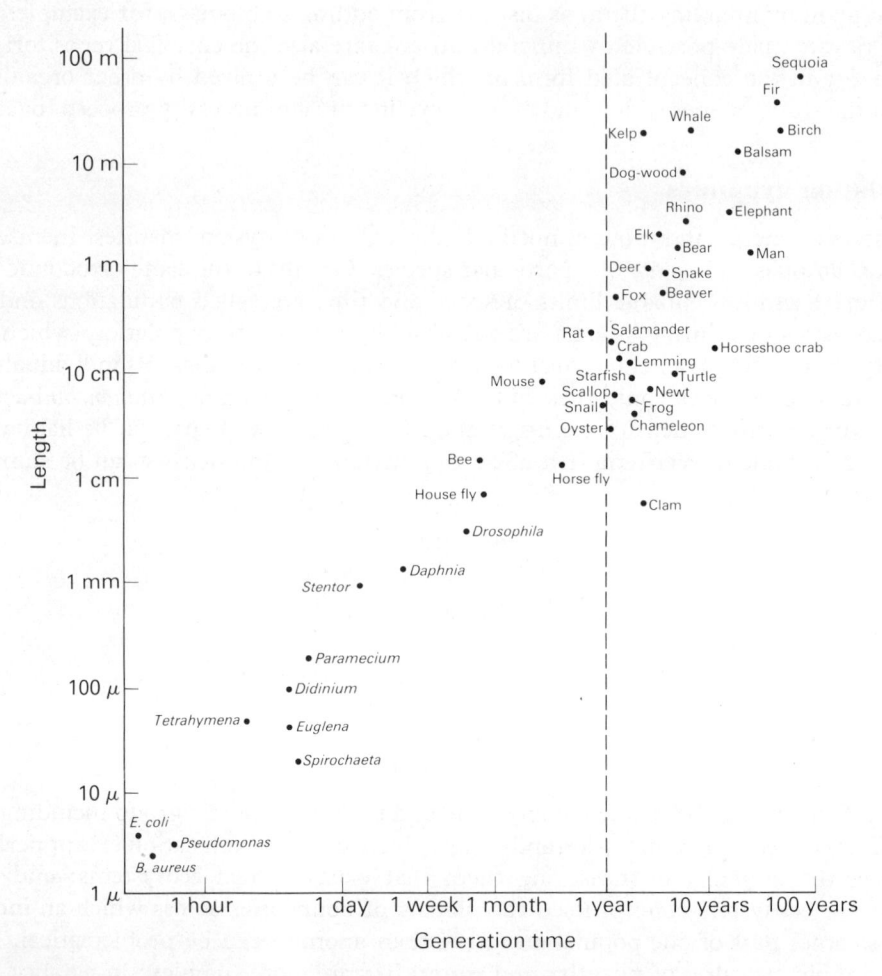

Fig. 2.12 Relationships in organisms between size (as measured by length) and generation time, on a log-log scale.
Source: Southwood, 1976.

2.12). The potential expansion of a population into a habitat must obviously vary considerably with the value of this parameter.

The factors which actually determine population size are very complex and are often inseparable from those which govern the life and growth of the individual itself, as discussed on p. 15. In plants, competition between individuals of the same species is often a regulator of density; indeed some plants grow better in a mixture than in single-species stands. Predation by herbivores may also exert a strong effect upon density: St John's Wort (*Hypericum perforatum*) is a weed of grazing land in many parts of the world, displacing forage palatable to stock and even poisoning domesticated animals if eaten in quantity. In California, its density has been reduced by the introduction of leaf-beetles of the genus *Chrysolina* to about one-hundredth (i.e. 1 per cent) of its former level, an example of biological control (see p. 264). That larger herbivores may also have an effect can often be seen by the erection of exclosures which keep them out of various types of vegetation, or when a large herbivore itself undergoes a population decline. In Europe, for example, the decline of rabbit populations following the spread of myxomatosis allowed many grasses and woody species to flourish, and species unpalatable to the rabbit such as *Senecio jacobea* (ragwort) diminished in abundance in the face of the new competition. With the return of high densities of rabbits, it became an effective competitor once more.

The regulation of animal populations is similarly complex, and short-cycle fluctuations of considerable amplitude are quite common in short-lived and rapidly breeding groups. Nevertheless, over a period, and given a stable environment, most taxa preserve characteristic levels of abundance so that we can think of a species as 'rare' or 'common'.

Fig. 2.13 Changes in the abundance of the heron (*Ardea cinerea*) in Great Britain 1933–63, implying a connection between cold winters and a decline in abundance.
Source: Odum, 1971.

82 The ecosystem scale

There may be fluctuations in abundance but if the habitat remains stable, then the populations will regain their characteristic level. As abundance increases, then the population is generally affected by various influences which bring about the stabilization of its numbers and density. The simplest form of this effect is increased mortality due to starvation since there is insufficient food (Fig. 2.13), although V. C. Wynne-Edwards (1962) has argued that this is relatively uncommon in animal populations. Predation may increase: as lemming numbers rise in the tundra, so their predators are said to immigrate from adjacent areas. Alternatively, at higher densities the proportion of individuals affected, e.g. by a parasite, may be increased because of easier transmission of the parasite from host to host. Solomon (1969) quotes the example of the parasitism of a moth by a fly: when the moth's caterpillars were at a density of 50/unit space, then 5 individuals were parasitized; at 150/unit space, some 28–29 were parasitized: the proportion affected had risen markedly. All such effects, where the regulatory factor operates more strongly at high densities than low, are called **density-dependent** factors. Some of these are very closely interrelated, as with the oscillations of predator and prey numbers: the predator regulates the numbers of its prey and in turn is itself regulated by the prey numbers. The relation can be quite complex and varies with the organisms involved: in general, however, the predators will only take higher numbers of prey from an increasing density up to a certain level, an analogue perhaps of satiation (Fig. 2.14).

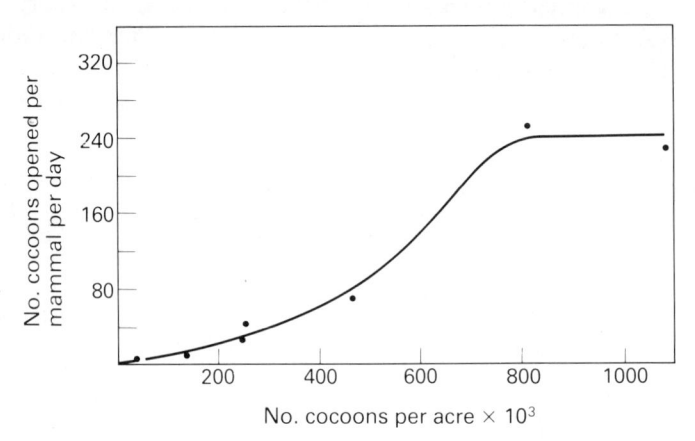

Fig. 2.14 Numbers of pine sawfly cocoons destroyed per day per individual predator in relation to the abundance of cocoons.
Source: Solomon, 1969.

In the course of coping with increasing abundance, certain 'functional responses' will become apparent in predator behaviour as adaptations to the smaller amount of time it has to spend in capturing its prey. If an organism is a rapid breeder, it may be able to increase its own population size in tune with that of the prey, but with slow breeders, a lag effect may involve the risk of a high predator density at a time when the prey numbers have fallen away.

At the other end of the density spectrum, we should mention that if some organisms become particularly rare, then their density may be too low for effective continuance of the population. It is feared that some species of whales, for example, have been thinned to the point where it becomes very difficult for males and females to locate each other for mating; as a rare plant gets rarer, the chances of long-distance pollination must inevitably fall. It is sometimes said, as well, that a minimum number of individual animals must be present to provide the correct behavioural 'environment' for reproduction and that is why some species will not breed in zoos: if a male does not have to fight off other males in order to serve the female, then apparently, she isn't worth bothering to impregnate.

Following his conviction that mortality from starvation seems rare in wild animals, and that other density-dependent mechanisms 'would in most cases be incapable of serving to impose the ceilings found in nature', Wynne-Edwards (1962) has hypothesized that many species develop social mechanisms which prevent over-breeding beyond the resources available for a particular population. The establishment of territories defended by males, for example, ensures not only that there will be enough food for ensuing offspring, but the occupation of all the space by territories will keep down the number of breeding pairs as well as confining breeding to the dominant animals: a form of selection for the best genes (Plate 11). Certain animals develop displays which affect reproduction: flock birds like the starling are said to display by wheeling about and thus the existing population density is revealed and a system of social dominance set up, which are the main prerequisites for controlling the breeding rate. Similarly, the acorn-storage

Plate 11 Fighting within species is usually a method of ensuring that the 'best' genes — i.e. those most fitted for survival — are passed on to the next generation. These are warthogs (*Phacochoerus aethiopicus*), an African herbivore with a well-defined family territory.

habits of the California woodpecker (*Melanerpes formicivorus*), where acorns are ceremonially placed in a ceremonial tree to an accompaniment of self-display and social competition, but not usually afterwards eaten, is probably a manifestation of a social mechanism which brings about population homeostasis.

The concept which emerges from all these considerations is that of **carrying capacity.** This is the total number of individuals of a species that will live in an ecosystem under certain conditions, and although it clearly applies to plants, is generally a term used of animal populations. The carrying capacity is achieved when the regulatory processes acting on the population density bring it into line with the available resources for the species after a period of rapid growth. These resources are inevitably finite and though they exert their effects in different ways, one or more of them will prove limiting for a species. The limiting resources may be invariant in time and space (analogous to the non-renewable resources of human economics), such as space itself. If this only is limiting, then the carrying capacity is the number of organisms that can be packed into the space: for example, barnacles on a rock or the number of nesting sites for seabirds on a very precipitous cliff. More likely, the limiting resources may be variable over space and time (analogous to renewable resources) and an equilibrium has to be reached that can accommodate the variations in its supply. This means that a population may be able to 'over-exploit' a resource at a time of its particular abundance but that it will fall back when the supply diminishes. Once a limiting factor has been identified in an ecosystem, it may be possible for a human manager to manipulate the system to alleviate it. In the California chaparral, for example, fire can be used to keep down the proportion of woody vegetation and increase the leafy vegetation within the reach of deer. The carrying capacity of burnt chaparral for deer becomes much higher than unburnt areas, not only because there is more browse, but because the leafy material is higher in protein as well. A similar process is thought to have happened in Britain in late Mesolithic times (p. 160).

A population dynamic of considerable interest at present is that of the human species. After a long period of very slow growth and with definite checks and recessions, the human population entered on a rapid phase of growth in about the seventeenth century AD, with a growth rate in recent years of around 2 per cent p.a. which means that the absolute size of the population doubles at 35-year intervals. Given our degree of consciousness, this rapid rate of growth has evoked concern over the future, especially as by no means all the existing members of the species have adequate food, clean water, shelter and clothing. We would expect that an equilibrium level will be achieved at the carrying capacity of the planet for our species but two highly contentious questions emerge: how is that carrying capacity to be defined, and how is the equilibrium to be achieved? The former is not merely biological and nutritional, but subject to cultural and aesthetic preferences as well; the latter could be simple biological (war, famine and starvation, disease, spontaneous abortion from stress, low-level violence short of war) but can be cultural also through the means of population-control programmes. There is some feeling that the latter are bringing down the rate of increase a little, but that they will only have a major effect among the poorest peoples of the world (whose populations are mostly the fastest growing) when the carrying capacity has been increased through 'development' of various kinds. Yet development could mean manipulation of ecosystems to the point of instability (p. 339) and thus be self-defeating and so ways of increasing resource availability without long-term ecological destabilization must be found. As we would expect, energy availability is often crucial in matters of development for it is access to stored energy as fossil fuels that sustains the dense populations of the industrial nations. Pessimists would say that *Homo sapiens* has already exceeded its long-term carrying capacity by using the fossil fuels and we must lower our population to come into equilibrium with the life-support made available by renewable sources of energy like the sun, wind, tides and green plants. Technologically based optimists, by contrast, envisage replacing and indeed exceeding the fossil fuel energy with nuclear power and alleviating shortages of all kinds using such energy to lift limiting factors in, for example, food supply by pumping water to arid areas, retaining the remaining oil as a feed-stock for single-cell protein growing and other technological approaches to human problems. More discussions of these matters will be found near the end of the book (p. 331).

Biological productivity

Another way in which the energy and matter present in living organisms is made manifest is in the rate of production of organic matter, of both autotrophs and heterotrophs. As explained earlier, the fundamental process is the production of organic tissues incorporating solar energy by green plants, and its most important practical expression is **net primary productivity** (NPP) which is the material actually available for harvest by animals and for decomposition by the soil fauna and flora or their aquatic equivalents. NPP therefore integrates abiotic phenomena such as the non-uniform distribution of incident radiant energy and the different conditions of moisture supply, with biotic features such as the genetic properties of the plants which are the primary producers. It is usually measured as dry organic matter synthesized per unit area per unit time and expressed either as $g/m^2/yr$ or $kg/ha/yr$. It can also be expressed as the calorific value of the dry matter, in kilocalories or joules (1 cal = 4·2 J); or in grams of carbon in the dry matter: this figure can reasonably be multiplied by 2·2 to obtain a dry-matter equivalent. At any one time, the standing crop of living organisms present per unit area is the biomass.

A first consideration with NPP is its efficiency. Photosynthesis may reach an efficiency of conversion of 20–30 per cent of the light absorbed, i.e. of photosynthetically active radiation (PAR), yet NPP rarely exceeds 1–2 per cent of PAR over an entirely growing season because of respiration losses. Exceptionally, higher values occur over short periods of rapid growth, as with the 7–9 per cent efficiency of sugar beet in the middle of the season and 8 per cent for mass algal cultures with an excess supply of CO_2 (Wassink 1975).

The recent work of the IBP has resulted in more accurate estimates of NPP on a world scale, for both the continents and the oceans. (Lieth 1975; Rodin *et al.* 1975). A set of these data is given as Table 2.4 and a distribution in Fig. 2.15. The ranking of the average NPP of the various biome types (whose ecology is further treated on

TABLE 2.4   NPP for the world, ca 1950

Ecosystem type	Area (10^6 km²)	Mean NPP (g/m²/yr)	Total production (10^9 t/yr)
Tropical rainforest	17·0	2200	37·4
Tropical seasonal forest	7·5	1600	12·0
Temperate forest: evergreen	5·0	1300	6·5
Temperate forest: deciduous	7·0	1200	8·4
Boreal forest	12·0	800	9·6
Woodland and shrubland	8·5	700	6·0
Savanna	15·0	900	13·5
Temperate grassland	9·0	600	5·4
Tundra and alpine	8·0	140	1·1
Desert and semidesert scrub	18·0	90	1·6
Extreme desert: rock, sand, ice	24·0	3	0·07
Cultivated land	14·0	650	9·1
Swamp and marsh	2·0	3000	6·0
Lake and stream	2·0	400	0·8
TOTAL CONTINENTAL	149	782	117·5
Open ocean	332·0	125	41·5
Upwelling zones	0·4	500	0·2
Continental shelf	26·6	360	9·6
Algae beds and reefs	0·6	2500	1·6
Estuaries (excl. marsh)	1·4	1500	2·1
TOTAL MARINE	361	155	55·0
WORLD TOTAL	510	336	172·5

Source: Whittaker and Likens 1975.

p. 90) yields few surprises. The list is headed by the tropical rainforests which have a year-round growing season and a high biomass, so that they would perhaps be expected to produce the most organic matter in the course of a year. Not generally appreciated perhaps is that limited parts of the oceans such as estuaries and coral reefs reach the NPP of tropical forests. In these data they are outweighed in absolute terms by the immense areas of open ocean whose NPP is more like that of the tundra. Tropical grasslands (including grass-dominated savanna) overlap with some of the woodlands of less favourable climates, and there is a big gap between these ecosystems and those of tundra, desert scrub and desert. The position of agricultural land is of some interest. At a world average of 650 g/m²/yr, it exceeds the average figure for grasslands but falls well below most of the forests. In one sense this is a misleading figure because in Western style agriculture it is brought about by large inputs of energy derived from fossil fuels as well as solar energy. The column for total production emphasizes the role of the forests in providing the bulk of the NPP of the world (62·8 per cent of the continental area, 42·8 per cent of the total), whereas cultivated lands produce only about 7·7 per cent of the terrestrial and freshwater NPP.

Within some ecosystems, various groups of plants stand out in terms of their NPP characteristics: within deserts for example the lichen *Ramalina maciformis* attains a NPP of ca 6 g/m²/yr, twice the average. Its water content is so low that even the smallest amount of humidity allows metabolic activity, but also enables it to return easily to a virtually non-living state without a great loss of substance. Some large freshwater plants

Biological productivity 87

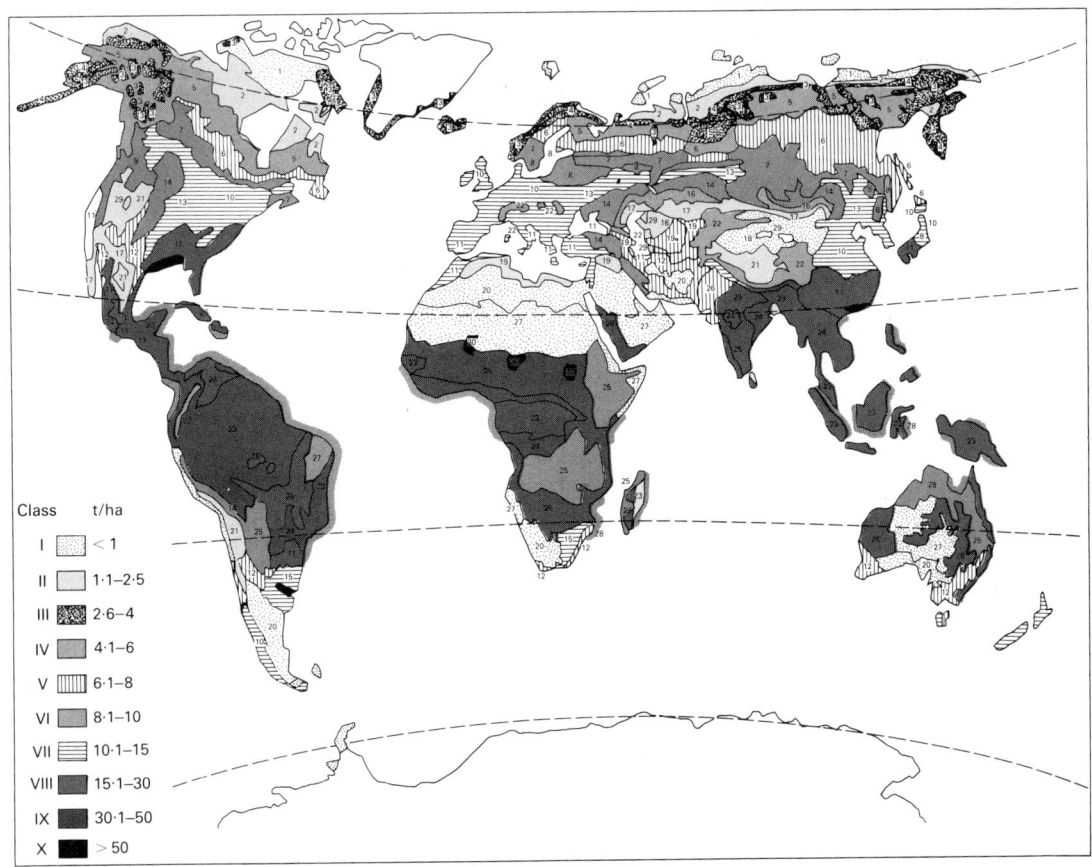

Fig. 2.15 Distribution of NPP on the continents in t/ha of dry matter. This map comes from work in the USSR which arrives at slightly different values from the work of Lieth as used, e.g., in Table 2.4. But the relative values and their spatial distribution are not very different.
(For comparison with Table 2.4, etc., g/m² = ×0·01 t/ha; t/ha = ×100 g/m²).
Source: Rodin et al., 1975.

have a very high NPP: *Cyperus papyrus* grows up to 6 m high and can produce 20–40 g/m²/day; in Malaysia some similar plants produce 60 t/ha/yr of dry matter. Similarly marine plants, especially large brown algae of the genera *Laminaria* and *Macrocystis* have very high rates of NPP. On the whole, the primary production of the oceans is less than the land because of the low availability of nutrients and CO_2 and the limited penetration of PAR into the water mass. Phytoplankton production over shelf areas is 3·0–7·5 t/ha/yr but in the open ocean no more than 2·5 t/ha/yr. This limitation has considerable consequences for the management of the ocean's living resources.

Primary producers appear to dominate ecosystems in a very direct fashion. Heterotrophs exert influences upon the primary producers (e.g. by grazing or tree defoliation) but their NPP usually appears very small compared with the autotrophs, partly because the efficiencies of energy transfer from plant to herbivore are often low. Thus in the tundra, the biomass of plants exceeds that of animals by $\times 15$, in deciduous forests $\times 300$, and in coniferous forests $\times 1200$. These data are to be expected given the metabolic needs of animals with regard to temperature and the operation of the thermodynamic laws within ecosystems (p. 55). More to be emphasized at this point is that secondary production has two branches: (a) the herbivore-carnivore-top carnivore chain, where secondary production represents successive reorganization of the same molecules, and (b) the decomposer chain which begins with dead organic matter, where the sum of new organisms in the decomposer chain (e.g. fungi-bacteria-protozoa) is the secondary production. A measure often used to describe the efficiency of heterotrophic production is Production/Assimilation (P/A). For invertebrates a representative value is 0·40; for warmblooded vertebrates 0·02 and for coldblooded vertebrates 0·10 (Heal and Maclean 1974). The low value for warmblooded animals reflects the high energy cost of maintaining a constant temperature, even where, for example, ruminants used their multiple stomachs to extract a greater fraction of nutrients and energy from their forage than their single-stomached companions. Even though their P/A ratios are low, the efficiency of ingestion of food may be quite high: vertebrate herbivores may consume up to 25 per cent of available NPP of grasslands, and 5 per cent in forests. Invertebrate herbivores usually consume less than 5 per cent of above-ground primary production although during population 'explosions' of, for example, locusts or caterpillars all the annual leaf production may be consumed: this still represents less than 50 per cent of NPP in grasslands and 20 per cent in forests although the overall system effect may be somewhat catastrophic. Predation by vertebrates upon invertebrates usually accounts for 26 per cent of prey mortality; the invertebrate prey of titmice and shrews in a wood suffered only 5 per cent of its mortality from these its predators. By contrast nearly all the mortality of wildebeest in East Africa resulted from predation by the spotted hyaena: 100 per cent of the herbivore production was in this case harvested by predators; only about 60 per cent of the hyaenas themselves fell to predation.

Although the large secondary producers of the herbivore-carnivore chain are the most obvious and the most discussed, the decomposer (or **saprovore**) system must be given its due importance. Heal and Maclean (1974) quote a typical grassland when 86·5 per cent of the NPP energy passed to the saprovore system whereas only 13·5 per cent went to the herbivores. No wonder therefore that the productivity of the micro-organisms exceeds that of the animals: in many ecosystems there is more life below the surface than above it.

If it is now estimated with reasonable accuracy that total world NPP is about 170×10^9 t/yr, of which 50–60 t/yr is from the oceans, does this have a significance for our use of biota as a resource, bearing in mind for example the low proportion of world NPP contributed by agriculture (7·7 per cent of terrestrial NPP), and the low productivity of much of the oceans? First of all, it puts in perspective the activities of man. The energetic magnitude of world primary production is estimated at $6·9 \times 10^{17}$ kcal/yr, whereas man's use of fossil fuels and other industrial energy in 1970 was $4·7 \times 10^{16}$ kcal/yr, *ca* 7 per cent of NPP. The concern is that the latter figure has been doubling every 10 years whereas the former may diminish from human impacts, and that the impact of the industrial energy is not uniform: we need to know which high natural productivities are being affected by it: estuaries are one example. Man's harvest of food also looks small compared with biosphere production: about 0·72 per cent of the energy of global NPP. From this figure to a world of milk and honey or beer and steak is a simple step: in the mind. In reality, the limitations are quite severe. Apart from economics, politics and other cultural manifestations, the great abundance of wood, grass tissue and phytoplankton cannot be used without incurring heavy energy expenditure in the form of industrial fuels or wasteful animals, although radical experimenters are working on this problem (p. 267). Much of any extra food will probably have to come from more or less conventional agriculture.

What is disturbing is the effect of human societies in reducing all forms of productivity, by replacing highly productive systems like rainforests with grassland and scrub and by toxification of ecosystems with agricultural, industrial and urban wastes. Together these reduce the production of organic matter which is man's most important resource. Indeed, since it is infinitely renewable, the NPP of an area might be said to be, in S. R. Eyre's (1978) phrase, 'the real wealth of nations'.

Further Reading

BOUGHEY, A. S. 1973: *Ecology of populations*, 2nd edn. New York: Maxmillan.
DOBBEN, W. H. VAN, and LOWE-McCONNELL, R. H. (eds.) 1975: *Unifying concepts in ecology*. Hague: Junk.
EYRE, S. R. 1978: *The real wealth of nations*. London: Edward Arnold.
GOLLEY, F. B. (ed) 1977: *Ecological succession*: Benchmark Papers in Ecology 5. London: Wiley.
HALL, D. O. and RAO, K. K. 1974: *Photosynthesis*. Studies in Biology 37. London: Edward Arnold.
JORDAN, C. F. and KLEIN, J. R. 1972: Mineral cycling: some basic concepts and their application in a tropical rain forest. *Ann. Rev. Ecol. System.* **3,** 33–50.
KRUMBEIN, W. E. (ed) 1978: *Environmental biochemistry and geomicrobiology*. London and New York: Wiley. 3 vols.
LIETH, H. and WHITTAKER, R. H. (eds.) 1975: *Primary productivity of the biosphere*. Ecological Studies 14. Berlin, Heidelberg and New York: Springer-Verlag.
LIKENS, G. E., BORMANN, F. H., PIERCE, R. S., EATON, J. S. and JOHNSON, N. M. 1977: *Biogeochemistry of a forested ecosystem*. New York, Berlin and Heidelberg: Springer-Verlag.
ODUM, E. P. 1969: The strategy of ecosystem development. *Science* **164,** 262–70.
ODUM, E. P. 1971: *Fundamentals of ecology*, 3rd edn. Philadelphia, London and Toronto: W. B. Saunders Company.
ODUM, E. P. 1975: *Ecology*, 2nd edn. New York and London: Holt, Rinehart and Winston.
PHILLIPSON, J. 1966: *Ecological energetics*. Studies in Biology 1. London: Edward Arnold.
Scientific American 1970: *The biosphere*. San Francisco: Freeman.
SOLOMON, M. E. 1969: *Population dynamics*. Studies in Biology 18. London: Edward Arnold.

3
The biome scale

Following E. P. Odum (1971), we may designate the biome as the largest land community or ecosystem unit which it is convenient to designate. This recognition depends largely on the life-form of the mature vegetation stands, so that although the species which dominate deciduous forests vary on different continents, the visual dominance of deciduous tree life-form enables the recognition of a deciduous forest biome. This life-form exercises control over the rest of the ecosystem, both in terms of the structure of the habitat for non-dominant plants and for animals, and in terms of the energy supply for heterotrophs (Fig. 3.1). Together, the vegetation and the climate are major influences

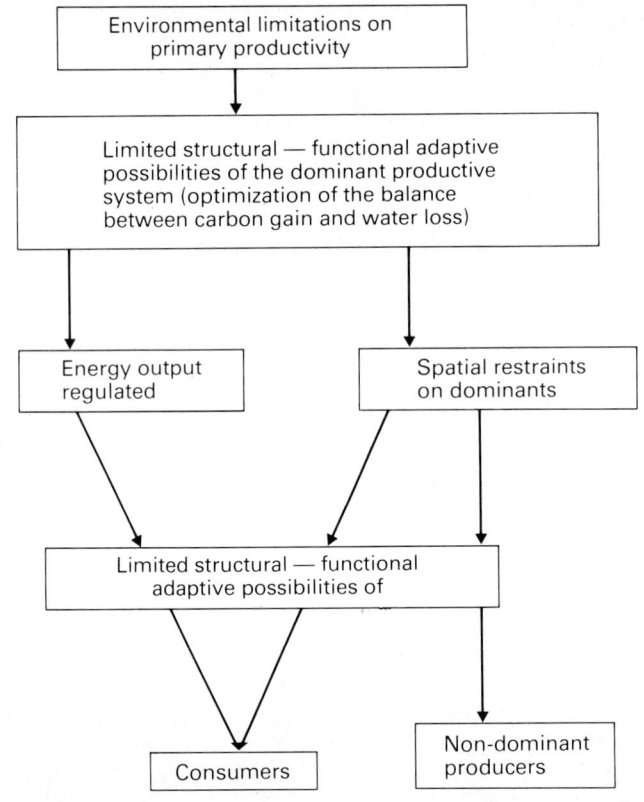

Fig. 3.1 The influence of the physical or abiotic environment upon the structure of an ecosystem. Source: Mooney, 1974.

The biome scale 91

in creating soils, and so the large-scale soil regions of the world correspond to a great extent with the vegetation units. The biomes of the world (and we can add the oceans as another distinctive type) are thus major world regions which integrate a number of factors into a recognizable but complex whole.

World climate is nonetheless the overriding control in the distribution of biomes, and Fig. 3.2 shows the relation of some major biome types to the distribution of radiation

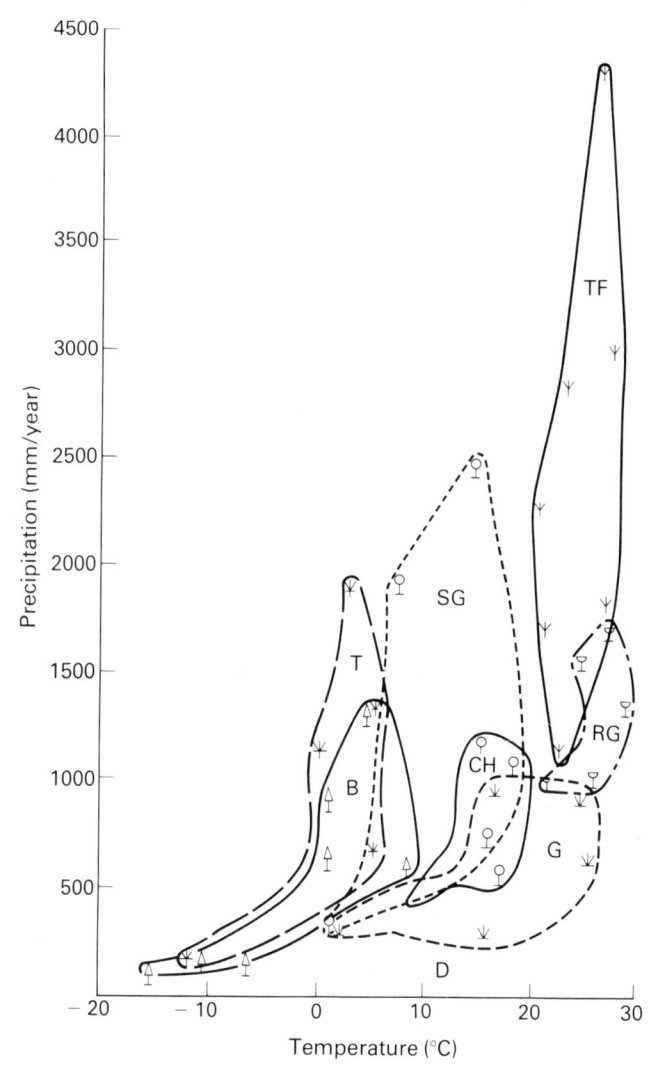

Fig. 3.2 Relationships of temperature, precipitation and biome type. Continuous lines enclose evergreen types: B=Boreal forest; CH=chaparral; TF=tropical rainforest. Pecked lines enclose seasonally green vegetation types: T=tundra; SG=deciduous forest; RG=temperate rainforests; G=grassland; D=desert and semi-desert.
Source: Lieth, 1975.

as measured by mean annual temperature, and of precipitation. Some characteristics may be noted: that the tropical forests cover a wide range of precipitation but are stenothermic, whereas the grassland type can stretch over a wide variety of temperature conditions but is delimited by moisture. In general, the shape of the distributions suggests that temperature, reflecting solar energy input, becomes a limiting factor far more often than precipitation. The distribution of the biomes is shown on Fig. 3.3 which is basically a vegetation map and does not indicate that there are soils and animals which are characteristic of each biome type as well. No extensive comment is offered on this map except that comparison with maps of climatic factors such as temperature and rainfall, and great soil groups will show very similar patterns. The recent work on primary productivity is also producing similar patterns: compare the terrestrial portions of Figs. 2.15 and 3.3. Such maps conceal certain complexities: in this case, variations of the biome characteristics brought about by, for example, differences in topography within it or by the presence of different stages in the natural succession. Another complicating factor is that the boundaries are often gradual; there is often a gradient of change rather than an abrupt break and so rather arbitary lines of definition have to be chosen when maps are made.

It remains to be said that the map of biomes is partly conjectural. It is largely a map of what the world would be like if man's activities were suddenly removed and all the ensuing successions were telescoped in time; alternatively it might be a map of some time in the past before human manipulation had become significant but after the major post-Pleistocene climatic changes and subsequent vegetational adjustments had largely taken place: perhaps about 0 BC. But even then, large areas of southwest Asia must have been altered by agriculture and pastoralism and large tracts of southeast Asia converted to cereal-growing. It is not a map of contemporary reality. Most of the tundra, for example, is still there and little altered; the deciduous forests of Eurasia have largely gone; the lowland tropical forests of the Congo and Amazon basins are shrinking fast. This leads to one more *caveat* about the reality of the biomes; in the case of those which have been altered by agriculture and pastoralism and large tracts of southeast Asia conbeen done on relict areas either accidentally or deliberately preserved. We do not know exactly to what extent their ecology differs from that of the biome in its pristine state.

A set of accounts of the ecology of the major biomes is now given, for which the main general sources are Odum 1971; Clapham 1973; Müller 1974; Collinson 1977; Eyre 1968, 1971. More detailed material is given its own attribution.

Tropical rainforests

It looks as if tropical lowlands with year-round precipitation are conducive to the development of high biomes and high rates of NPP and high rates of litter breakdown. Also, these forests suffered little disturbance during Pleistocene times and so have been *in situ* at least since the Tertiary and so have had a long time to evolve efficient mechanisms for the partition of energy and for the cycling of nutrients.

The tropical rainforests are found in the lowland areas of the Congo and Amazon basins and in archipelagic and peninsular Southeast Asia with extensions into lowland

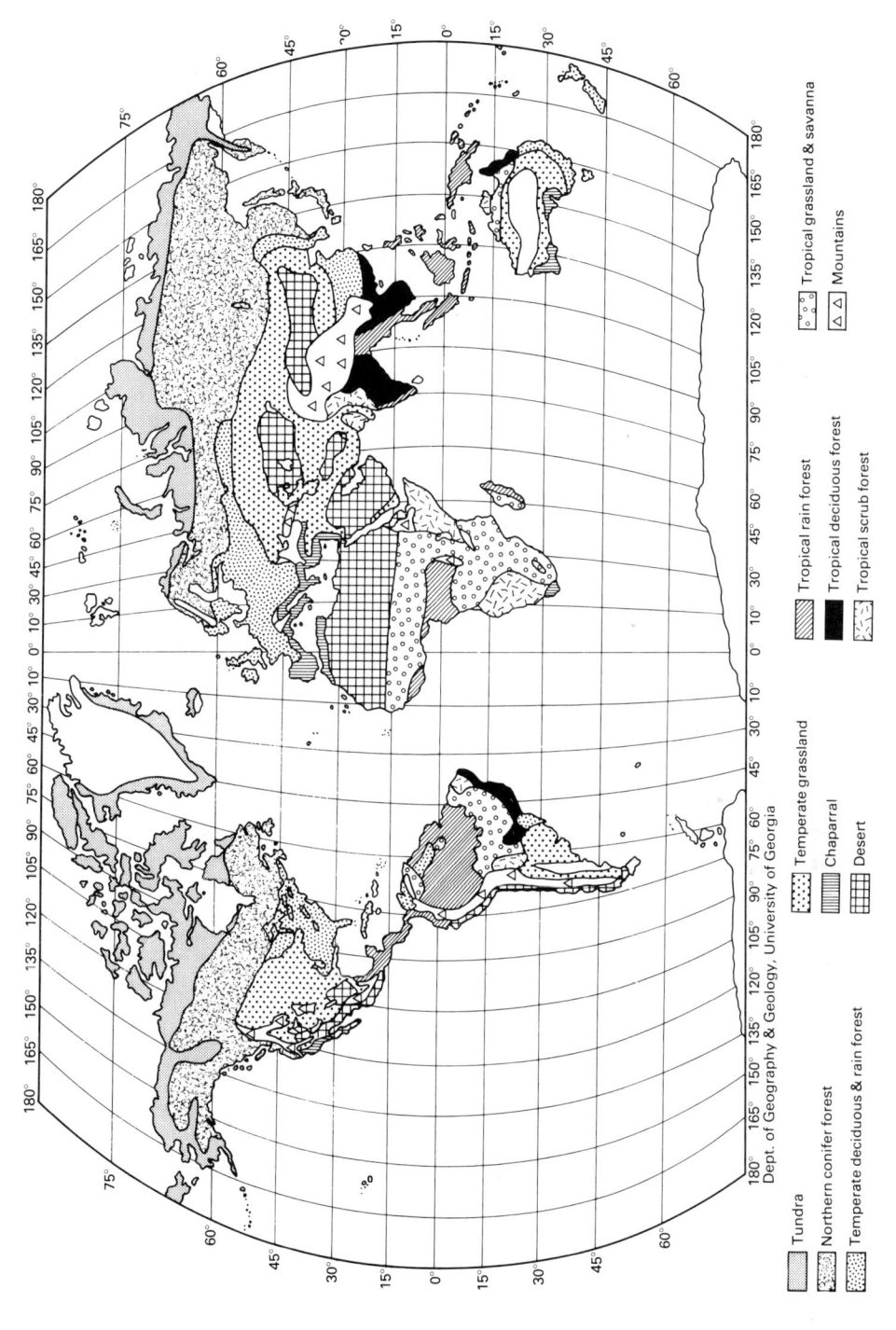

Fig. 3.3 Major biomes of the world as if unaffected by human activity.
Source: Odum, 1971.

94 *The biome scale*

Meso-America and Madagascar (Fig. 3.3). Climatically, they are distinguished by both high and constant temperature and humidity, with precipitation of over 2000 mm/yr and at least 120 mm in the driest months. The vegetation is dominated by trees of a great variety of species: in some parts of Brazil there are 300 different species of tree in 2 km^2. The trees are typically tall and are structured into three layers. The highest or emergent layer consists of the tallest trees (45–50 m high) which are scattered but project through the lower canopy layer (25–35 m high) which forms an almost continuous cover, and which absorbs some 70-80 per cent of the incident light. When there are gaps in this lower layer, then a normally sparse understorey layer may become dense. The effect of all these strata of trees is to absorb all but a few per cent of the incoming light and so shrubs and ground vegetation are not normally found except in gaps in the forest and at its edges, as near rivers. The higher energy level of the upper layers is reflected

Plate 12 A characteristic rainforest tree with buttress roots and lianas.

in the epiphytes and lianas which are densest there. The trees themselves typically have buttressed bases for mechanical support, are evergreen with leathery dense green leaves, and are seasonal in that they both flower and lose leaves all year round (Plate 12). They are rarely wind-pollinated because of the species diversity but rely instead on a variety of insects, bats, birds and even mammals: in Madagascar, lemurs appear to be important in pollinating various plants, including trees, and the same seems to be true of the bush babies of continental Africa (Sussman and Raven 1978).

Examples of the tree species in this biome are the rubber tree (*Hevea brasiliensis*), wild banana (*Musa* spp) and cocoa (*Theobroma cacao*). Dominance may be difficult to detect since family dominance may replace species dominance in areas of high species diversity: in the lowlands of Malaya the Dipterocarpaceae provide perhaps 25 per cent of the mature individuals, and in Borneo the large Dipterocarps may constitute 90 per cent of the emergent layer. But single-species dominance is found as well: in South America genera like *Mora*, *Eperua* and *Dimorphandra* form such stands, and their equivalents are found in southeast Asia and Africa. This type of dominance (common in temperate zones) was thought to be a feature of poor soils, but more recent work (summarized by Connell 1978) has put forward the idea that very high tree diversity in rain forests is a disequilibrium state (i.e. successional) due to past disturbances and that in an equilibrium or mature state the tree diversity of the forest is lower.

Animal life, though very low in productivity compared with the trees, exhibits the greatest taxonomic variety of any biome. The richness of food resources available and the relative constancy of environmental conditions seem conducive to such a state. Like the trees, the animal communities are stratified: the emergent layer is inhabited mostly by birds and insects which live their whole lives in this arboreal habitat. Below them, the canopy layer houses the highest variety of animals in the forms of tree-dwelling monkeys, sloths, anteaters and small carnivores. They rarely descend to the ground, but in the understorey layer the animals may range down from the trees to the forest floor. Ground-dwellers are less diverse than arboreal types but include deer, rodents, peccaries and wild pigs. The pigs are known to be scavengers on the carcasses of other animals (e.g. apes and monkeys) that fall from the trees when dead. The proportions of different adaptations are shown by studies in Guyana in which 31 out of 50 species of mammals were arboreal, 5 amphibious and the rest mainly ground-dwellers. The species diversity is perhaps most obvious among birds, where the abundance of fruits, seeds, buds, nectar and insects allow dense populations of groups such as hornbills and toucans and the presence of birds-of-paradise.

The soils of tropical rain forests are typically latosols or podsols. The latosols are thick, red in colour with a high content of iron and aluminium sesquioxides, whereas the podsols have lost their iron oxides and are mostly developed over siliceous parent materials. Both have in common a low availability of mineral nutrients, to the point where in one of these forests some 58 per cent of the nitrogen in the plant-soil system was in the plants (44 per cent above ground), compared with 6 per cent in a temperate pine forest (Fig. 3.4). The productivity of the rain forests seems to depend upon mechanisms which keep the nutrients in the organic components of the cycle so that they are not leached out of the inorganic phase by the abundant rainfall. One such strategy is

96 The biome scale

Distribution of organic carbon
(Ca 250 tons/ha)

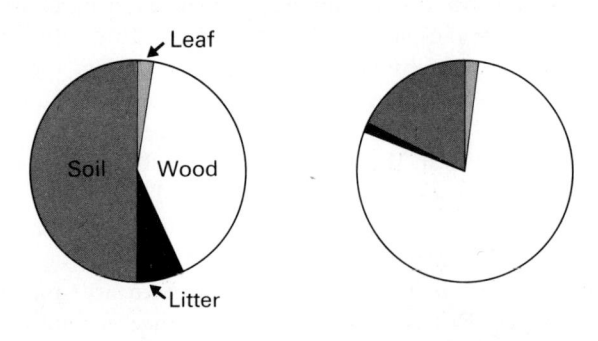

Leaf

Soil Wood

Litter

Northern coniferous forest   Tropical rain forest

Fig. 3.4 Comparison of a northern coniferous (Boreal) forest and a tropical rainforest in terms of the distribution of organic carbon in the different compartments of the ecosystems. The dominance of the biomass in the tropical system is emphasized.
Source: Odum, 1971.

the extensive development of fungal mycorrhizae along the shallow, spreading, root systems of the trees (Went and Stark 1968). The rate of litter fall from the forest canopy is high (typically 11 t/ha/yr in the Amazon basin) but there is a humus turnover of 1 per cent per day so that litter does not accumulate: if mean temperatures are above 30°C litter is broken down faster than it is supplied, at 25–30°C supply and breakdown are about equal. The main agents of litter breakdown seem to be fungi in mycorrhizal associations, so that mineral nutrients are passed directly from the decaying litter to the roots of the trees for uptake. Further, the roots of tropical trees may selectively grow towards a nutrient source (H. T. Odum 1970). Thus loss of minerals to the runoff is minimized, even to the point where soil animals are forced to feed on the fungi rather than the litter. Earthworms, for example, do little mixing of the soil and so the organic upper horizon is sharply marked off from the mineral soil beneath. However the energy-rich litter does sustain a dense population of bacteria and blue-green algae which are nitrogen-fixers and these are essential maintainers of a nitrogen cycle that plays a key role in sustaining the high biomass, though the circulation of large quantities of silicon, calcium and potassium are important as well. Input from precipitation is by no means negligible for nutrients either: near Manaus, Brazil, the annual average input from rainfall was 0·3 kg/ha of phosphorus, 2·0 kg/ha of iron, 10 kg/ha of nitrogen and 3·6 kg/ha of calcium (Richards 1973).

The nitrogen budget is also increased by fixation by lichens containing blue-green algae in the crowns of the trees (Forman 1975), and the tight circulation of nutrients may also be aided by larger animals. On Barro Colorado Island (Panama), a few trees lost 7–20 per cent of their annual leaf production to cropping by sloths (*Choloepus hoffmanni* and *Bradypus infuscatus*) which made those individual trees the focal points of their home ranges. But the sloths climb down those particular trees and defecate selectively near the base of the tree off which they have fed. About half the material eaten by the sloth is returned to the tree as a slowly decomposing (>6 months) source of nutrients which contrast with the rapid return from decomposing leaves (*ca* 1·5 months). This is called a long-term stable source by Montgomery and Sunquist (1975).

The outcome of high solar input, abundant rainfall and rapid nutrient cycling is a very high NPP, with the mean for rain forests estimated at 2200 g/m^2/yr of dry matter; multiplied by the area of the biome, we get $37\cdot4 \times 10^9$ t/yr which is far higher than for any other terrestrial biome. The average subsumes the fact that there are large areas of rain forest on podsol soils which fall at the lower end of the productivity range (1000 g/m^2/yr) for this forest type.

Nutrient-cycling studies show that the clearance of forest for agriculture robs the system of its store of nutrients and hence its fertility in agricultural terms. Once the trees are gone, nutrients are lost in runoff after being rapidly mobilized in the higher temperatures of the now unshaded soil surface. It is notable that traditional shifting cultivation adapted to these conditions by planting a miniature forest of crops which covered the clearing surface: tidy row crops were avoided. The plot was abandoned after a few years to enable the forest to recolonize and rebuild the nutrient cycles (Geertz 1964). Such a strategy is not, however, the aim of large-scale clearance schemes which involve keeping the forest down, once felled, by grazing domesticated stock. Therefore the tropical rain forests may be ecologically unsuitable for cropping methods based originally on the characteristics of European ecosystems, a fact which does not seem to prevent various governments and lumber companies from removing them at a current rate of 11 million ha/yr so that of an original mid post-glacial area of 16 million km^2 of rain forests only two-fifths are now left: at this rate of destruction, another thirty years will see the demise of the biome (Huxley 1978). Apart from its scientific interest (for example there are estimated to be over 25 000 species of flowering plants in the rain forests of Southeast Asia), such a biotic profusion must inevitably be a reservoir of great economic potential. (For detailed ecological studies see Richards 1952; H. T. Odum 1970; Whitmore 1975; accounts of ecosystem structure and function can be found in Fittkau and Klinge 1973; Lugo 1974, Klinge *et al.* 1975; man–forest interactions in Reynolds and Wood 1977.)

Tropical seasonal forests

Where the total annual rainfall is high but segregated into pronounced wet and dry periods, for example where the driest month has less than 50 mm of precipitation, even though the annual total in monsoon areas may be higher than in the equatorial lowlands, then seasonally-adapted forests are found. Typical locations are India, the Southeast Asian mainland, Central and South America, North Australia, western Africa and the tropical islands of the Pacific. The dominant trees, like the equatorial rain forests, may reach 40 m but are more commonly 20–30 m high and the stratification is relatively simple: only one single understorey tree layer is usually found beneath the main canopy. The adaptation to the moisture regime is found particularly in the deciduous habit although the form this takes may exhibit variation. For example, the canopy may be deciduous but the understorey may remain evergreen; or there may be a mosaic of evergreen/deciduous species according to the water-holding properties of the soils beneath. At all events, these forests are likely to be more variable in their physiognomy than the rain forests, as in the presence of virtually single-species stands, a condition less common in the rain forests. One such tree is teak (*Tectonia grandis*) in India and Southeast Asia, though doubtless its selection has been increased by its value as a commercial wood. Other species of these forests include *Shorea robusta*, *Terminalia* spp and the leguminous pyinkado tree *Xylia xylocarpa*.

Beneath these forests, the soils also exhibit differences from evergreen trees. The latosols rarely form where annual precipitation is below 1500–2000 mm and the clay-humus content appears to be more stable, holding greater amounts of available mineral nutrients. Iron is often mobilized, however, and can form concretions but the typical laterite seems not to form. Such studies as have been carried out suggest that mineral cycling has more in common with temperate forests than with the tropical rain forest.

In spite of the somewhat more restrictive environmental conditions, the biomass of these forests ranks high on the world scale, at an average of about 205 t/ha. This average, however, conceals a set of wide deviations between the wetter and drier ends of the climatic spectrum within which these forests develop; and the wet–dry variation also affects other aspects of the forests' functions. For example, the breakdown of the large quantities of litter which accummulate at the beginning of the dry season seems to be dependent upon that season's length—if short, then breakdown soon resumes with the onset of wet conditions; if long, then litter will accummulate on the forest floor. If the material is resistant to attack then the quantities can be considerable: in one investigation in Thailand as much as 19–20 t/ha of dead bamboo was recorded. In the montane rain forests of New Guinea, the decomposition rate of litter has been shown to be about half that of lowland rain forests (Edwards 1977).

An accummulation of dry matter in a dry season provides a considerable potential for fire: another contrast with the lowland equatorial forests where fire is not reckoned as an ecological factor. But in the seasonal forests an occasional fire (not even an annual burning appears necessary) may exclude some species of trees, especially those with a thin bark and unprotected buds. Experimental exclosures in West Africa and India have shown that some species of evergreen rain forest will colonize the seasonal forests in the absence of fire. This suggests that fire may have played a role in differentiating the two types of forest at their **ecotone** or tension zone; even where lightning-set fires are a possibility, the hand of man must always be suspected in such processes. Even more than in the lowland tropics, this tool has been one of man's chief aids in the clearance of forest and its conversion to grassland and cropland (Bennett 1968; J. J. Parsons 1976).

Tropical savannas

This biome type is in many ways intermediate between a forest and a grassland and indeed the term has been applied to a variety of vegetation formations from a nearly closed canopy woodland to a grassland with thinly scattered bushes. Common to them all is a continuous ground layer dominated by grasses. The biome is formed (Fig. 3.3) in a wide belt on either side of the equator in areas with a tropical temperature regime. Total rainfall will vary from 250 mm/yr on the desert fringes of the savanna, to 1300 mm where it abuts true tropical forests, but characteristically there is at least one dry season; it used to be thought that a long dry season was the cause of the physiognomy of the vegetation but this feature is now recognized as the outcome of a wider complex of environmental factors. However the dry season exerts direct effects on, for example, large herbivorous mammals, causing them to migrate in search of forage and free water. Found mainly in Africa, Australia and South America, the savanna biome covers about 20 per cent of the world's land surface.

The characteristic soil is a latosol with more humus than in the tropical forests but one which nevertheless may become solid if cleared. Variations in the soil, such as the proportions of gravel, clay or sand, may apparently influence the mix of grass and trees in some places. Many savannas are formed on land surfaces of great antiquity and so the soils are likely to be depleted of many mineral nutrients except on steep slopes where downslope flushing may occur and denser woodland is then found, especially if such slopes are less often burned. The obvious characteristics of typical savanna vegetation are trees and grasses. The former exhibit a great taxonomic variety and are usually 6–12 m in height, strongly rooted and with flattened crowns. They exhibit drought-resisting features, including partial or total deciduousness, water-storage modifications, and reduced leaves, and in addition are usually fire-resistant (pyrophytic) in having a thick bark and thick bud-scales. In Africa, savanna trees include species of *Isoberlinia*, the Baobab (*Adansonia digitata*), and the dom palm (*Hyphaene thebaica*); in Australia, species of *Eucalyptus* such as *E. marginata* and *E. calophylla* form the tree layer; in South America a great variety of species is involved, and in Honduras a savanna vegetation dominated by pine trees has been described by Johannasson (1963). Even where the savanna is cultivated, the savanna trees may be retained because of their usefulness to the farmers, e.g. for shade, food, materials and firewood (Pullan 1974). The grasses are often long, reaching up to 3·5 m in height and thus providing ample fuel for dry-season fires. In Africa the elephant grass *Hyparrhenia* grows to 5 m in height. Other typical savanna grass genera are *Panicum*, *Pennisetum*, *Andropogon*, and the species *Imperata cylindrica* now widespread throughout the tropics.

The biological productivity of the vegetation is seasonal, with a flush following the rains. The range of NPP can be from 150 t/ha/yr in closed savanna down to 2 t/ha/yr where the savanna type is more akin to a desert scrub. In grassy areas the productivity appears to be closely related to annual rainfall but this relationship is less close in wooded regions. In an Ivory Coast savanna, the productivity of the palms was 23 t/ha/yr, other trees 14 t/ha/yr and the herbaceous layer 23 t/ha/yr, but, as Fig. 3.5 shows, the turnover time of each component was radically different (Lamotte 1975). Investigations of nutrient cycling suggest that the nutrient cycled in greatest quantity is silica but that nitrogen is also returned to the soil in high amounts.

Where man has not over-hunted them, and where fire maintains a variety of habitats, the savanna can support a very diverse fauna. The savannas of East Africa, for example, support the greatest diversity of grazing vertebrate life in the world, with over 40 species of large herbivorous mammals and up to 16 species grazing together in apparently the

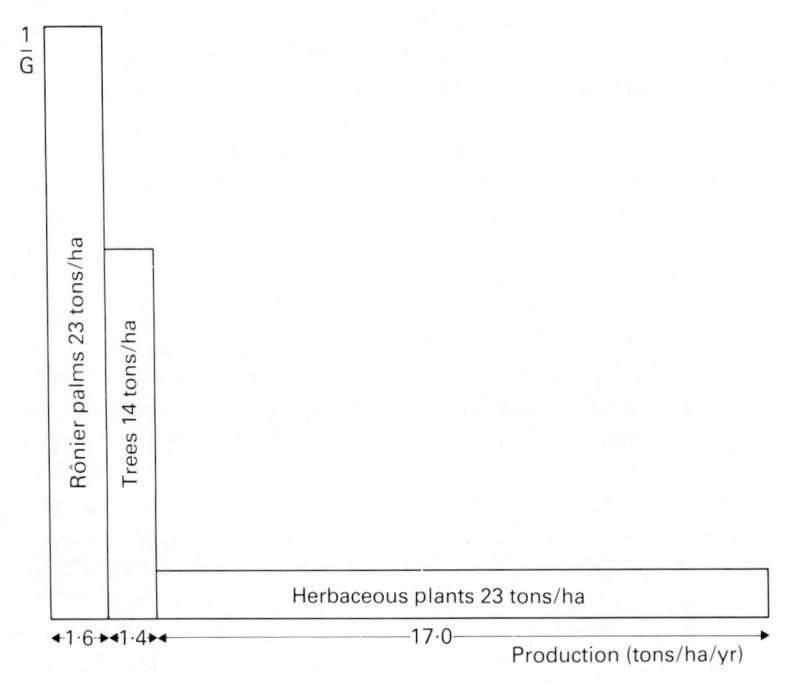

Fig. 3.5 NPP structure of a West African savanna. The surface of the rectangles represent biomass, the width production, and the height is the inverse (1/G) of the turnover rate.
Source: Lamotte, 1975.

same habitat. A wide variety of scavengers and predators is supported by this fauna (Plate 13). Other savannas are less rich: in many parts of Africa and South America, hunting has depleted the fauna although the known occurrence of 50 species of kangaroo (*Macropus*) in Australia suggests an evolution towards a similar faunal richness. Where a rich fauna still exists as in East and Central Africa, it may achieve a yearlong vertebrate biomass of 100–350 kg/ha and the animals avoid overgrazing by various behavioural adaptations: they may be separated by small variations in the habitat, for example, as where one species of small antelope lives in permanent swamps, another in seasonal swamps. The same plant may provide food for different species during its yearly growth cycle; and in parallel with this, some animals may be very selective in their foraging. Different beasts will use different levels of vegetation: the giraffe has sole rights to the tops of, for example, *Acacia* trees, but lower down the two species of African rhinoceros are separated by food habits since one is a browser off twigs and foliage, and the other a grazer from the ground layer. Over-use of the vegetation is also avoided by seasonal migrations of the commonest herbivores, and the seasonality of climate is also seen in the breeding cycles of the vertebrates, which usually give birth when food is abundant for their young. (It is said that lions eat men more often in the rainy season when other animals are more dispersed). The interaction of the herbivores and their forage is illustrated by careful measurements such as those from grasslands in Uganda which show

Plate 13 East African savanna showing trees set in a grassy plain. The diversity of large mammals in this biome is represented by giraffe, zebra, impala and wildebeest.

that grazing mammals remove half of the above-ground NPP (Strugnell and Pigott 1978).

Other animals are also plentiful in savannas: birds can make good use of a variety of food species and can move quickly to evade unfavourable conditions. Year-round ground-living species and scavengers are joined by many migrant species avoiding colder winters elsewhere on the globe who often fly considerable distances. The insects are dominated by the Orthoptera (locusts and grasshoppers), Hymenoptera (ants) and Isoptera (termites); the savannas are the inception zones of the outbreaks of migratory locusts which are so devastating to agriculture. Locally more important are the termites which are a very important component of the decomposer chain and include (a) foragers which gather in dead and some living herbage; (b) humivorous termites which decompose cellulose underground; and (c) fungus-growing termites which feed a fungus with dead wood and dead grass and then themselves eat the fungus. The importance of termites in recycling nutrients can be gauged from estimates in the Ivory Coast (Lamotte 1975) which suggest that the biomass of foraging termites is 6 kg/ha live weight, and of the humivorous termites 12 kg/ha and that these latter consume 30 kg/ha/yr of cellulose and so yearly rearrange several dozen tonnes of surface soil, improving the structure and aeration for plant growth. The mounds built by some species of savanna termite are important landscape features, with up to 600 hills/ha, and their frequency seems to be positively correlated with the rate of litter decomposition (Pomeroy 1978). The mounds may also become the foci of trees and shrubs not palatable to the termites and such thickets are often pockets of vegetation which is not burnt, thus adding to the diversity of habitats. Some termites, however, move underground during the fire season.

The occurrence of fire in this biome has been mentioned several times. It is probably responsible for the maintenance of savannas as grassy communities, for when absent the tree cover increases and there are fewer ruminants. It is clearly of evolutionary significance for the ecosystems, for in Africa at least, evidence of fire goes back to 60 000 yr BP. In general regular fire favours perennial grasses with underground stems which can regenerate after the fire has passed. A fire later in the dry season will consume about 84 per cent of the herbaceous layer, which is some 75 per cent of the annual production of above-ground shoots of the vegetation; burning of a plot for 5 consecutive seasons reduced the tree population by 32 per cent (Hopkins 1965). The burning also actuates growth by removing dead material and mineralizing litter, and the stimulation of perennial grasses in middle or late dry-season burns is particularly significant for savanna vegetation, for if the burn is early in the dry season then the growth may be initiated but there may not be sufficient water available to last the plant through to the rainy season (San Jose and Medina 1975). Where large mammal herbivores are absent, then fire consumes what they would otherwise have eaten. Thus fire appears to be a normal part of the savanna biome and one of the major factors in its nature, as it is in many tropical forests and grasslands other than the lowland rain forests (Batchelder and Hirst 1966).

Whether fire is the major determinant of the savanna biome's ecological characteristics is more controversial. There are areas of South America where savanna-type vegetation is found without fire, and where pollen analysis indicates spells of open grassland before the coming of man. The arguments for and against fire and geomorphological history as major determinants of the nature of the savanna ecosystems have been summarized by Eyre (1968) and by Sarmiento and Monasterio (1975) and need not be repeated here. In Africa, however, opinion seems to favour the idea of the savanna as a delicate balance of the outcome of climate, soils, vegetation, animals and fire, with fire as the key agent whereby men have created the biome; as it now stands this biome in Africa cannot be regarded as a climatic climax but as a product of human activity (Hills 1976). It is less certain that this diagnosis applies elsewhere and the savanna may be an example of a similar ecosystem being the end-product of different ecological histories.

Temperate sclerophyll woodland and scrub

This biome includes the vegetation type characteristic of the lands having a Mediterranean-type climate with warm and rainless summers and the rainfall of 500–1000 mm/yr concentrated into a cool but often frost-free winter. The woodland facies of the biome is best represented by the *Eucalyptus* forests of the southeast and southwest corners of Australia, and it is not surprising that this tree has been so successfully introduced into the Mediterranean basin and into California. In other areas of the biome, the adaptation to the summer drought has been via the coniferous evergreen habit and much of the land surrounding the Mediterranean was once covered with pine and cedar forest. In California, too, several species of pine and cypress, for example, are found within the biome, though not usually as forest stands.

Common to the Mediterranean, Australia, South Africa, Chile and the Pacific coast of North America are various forms of evergreen scrub with a similar vegetation structure (Parsons and Moldenke 1975). Known in California as chaparral, in Europe names such as maquis, garrigue and encinar are found; in Australia, mallee scrub; and in Chile, matorral (Plate 14). The dominants are trees and shrubs either with needle-like leaves or with leathery evergreen leaves of sclerophyll character (i.e. having a thick epidermis with a waxy cuticle). The dominants are usually 3–4 m high and form a close-set and sometimes impenetrable scrub. In the Mediterranean, the wild olive (*Olea europea*), carob (*Ceratonia siliqua*), evergreen oaks (e.g. *Quercus ilex*), pines such as *P. pinea* and *P. pinaster*, arbutus or strawberry tree (*Arbutus unedo*), heathy shrubs of the genera *Erica*, *Ulex* and *Genista*, and herbs of the Labiatae family and the genus *Thymus*, are the commonest. In California, many species of bush-like oak are found, along with species of *Ceanothus*, the chamiso bush (*Adenostoma*), manzanita (*Arctostaphylos* spp) and several pines. In Australia the equivalent mallee consists of *Eucalyptus* scrub from 2–3 m high. In an average area of the biome, a biomass of *ca* 6 kg/m^2 results from a

Plate 14 Redwood Chaparral, Big Sur

NPP of *ca* 700 g/m²/yr: one sample site had an above-ground biomass of 315 t/ha under *Quercus ilex* and 518 t/ha organic matter in the soil, connected by 3·8 t/ha/yr of leaf fall. (Lossaint 1973.) It appears that many of the plants grow quite fast on soils which are quite low in phosphorus and so conserving mechanisms exist: a fine mat of roots penetrates the litter and phospholate-storing enzymes within the roots accumulate the phosphorus until it is needed for rainy-season growth (Specht 1973). The litter does not penetrate far into the soil by way of humus and so the unobscured iron gives the name of terra rossa to the characteristic soils in the Mediterranean.

The abundant food and good cover of these scrub lands allow a plentiful animal life. At the San Dimas Experimental Forest in Southern California, 201 species of vertebrates were inventoried, 75 per cent of which were birds. Mammals tend now to be dominated by ground squirrels, the wood rate and the mule deer (*Odocoileus hemionus*) although before the heavy impact of man, predator species such as the wolf and mountain lion, and diversivores like the grizzly bear (*Ursus arctos horribilis*) were more common. The role of small mammals in the ecosystem is hinted at by the fact that the wood rat (*Neotoma fuscipes*) has been known to consume the entire acorn crop of one species of oak in a particular year.

In such climates, it is not surprising that fire is a normal occurrence in the biome. In the San Dimas forest, lightning set 8 fires in 75 years and burning by the Indians is well documented. Most of the species are adapted to fire (e.g. Eucalypts stump-sprout after it) and no doubt have been selected by many thousands of years of its occurrence. The fires seem to stimulate the germination of some seeds, to reduce to ashes much vegetation and litter and hence speed up the process of mineralization of organic matter; and also to destroy phytotoxic compounds secreted by plant roots which interfere with litter decomposition and the processes of nitrogen fixation in the soil. An Australian study showed that an unburned *Eucalyptus* forest was experiencing no regeneration of the Eucalypts but that they were being replaced by *Casuarina*, *Banksia* and *Acacia* species (Withers and Ashton 1977). On the other hand, fire increases erosion rates: in the San Gabriel mountains the pre-burn silt loss was 7980 kg/ha; in the first year after the burn it was 230 000 kg/ha, the second 52 700 kg/ha and the third 26 600 kg/ha (Mooney and Parsons 1973).

Chaparral and similar forms of vegetation can clearly be regarded as biomes in which natural fire has played a key role, even though its frequency has been increased by human activity. Fire, though, is not the only form of human impact: centuries of grazing of goats and sheep in the Mediterranean, terracing for cultivation, management for high deer populations, conversion to grassland using biocides, agricultural use including irrigation, coppicing, and urbanization have all hit the sclerophyll scrubs at various times, some of them quite recent (e.g. home-building in the Santa Maria mountains near Los Angeles), others quite ancient (e.g. pastoralism in the Levant since time out of mind). In Chile, the vegetation has a greater diversity of species and growth forms than in California, and D. J. Parsons (1976) has suggested that this is due to a more intensive land-use history as well as areas of naturally more open vegetation. But it is no surprise that some of the scrublands of the Mediterranean can be regarded as degraded communities in need of careful management (Naveh and Dan 1973) and that everywhere else these ecosystems are thought to have a potential for rapid change often hazardous to human life and property (Aschmann 1973).

Temperate grasslands

Although large areas of tropical forests and wooded savannas have been converted to grassland, there exists outside the tropics large areas of grassland (called 'steppe' by some authors) which have been thought to be largely natural. Usually the rainfall and snowfall (250–750 mm/yr) are too low to yield enough water to support forest but are above the level of desert, and so the grasslands are seen as an intermediate life-form between the forest and the desert. The prairies, Great Plains and arid grasslands of North America, the Eurasian steppe, the veldt of Africa and the South American pampas are the main areas of this type of grassland. The idea of their origin as climatically determined is not always acceptable, however. Some grasslands occur in forest climates where perhaps a high water table favours grasses in competition with trees, or where fire prevents the regeneration of trees and produces a grass-dominated plant community. Although lightning-set fires may play a part in producing such mosaics, it is more often the case that

Plate 15 Short-grass prairie in the Sand Hills of Nebraska. Note how broken and more steeply sloping areas have trees; this is taken by some authors to be indicative of the role of fire in the ecology of the grasslands.

the fire results from human activity: early European explorers in the northeast woodlands of North America noticed large grassy openings in the forest due to fires which the Indians had set, largely to aid hunting. Such instances were extended, particularly in the case of the North American grassland, to develop the idea that all the grassland was in fact a fire climax and that in post-Pleistocene times the true natural vegetation had been forest. As with the origin of the savanna, there has been some controversy, but there is now perhaps an acceptance of the concept of a multiple origin in which some parts of the North American grassland are derived from forest, especially in the wetter eastern zone, and others are products of low precipitation (Dix 1964). In both cases, a history of grazing by domesticated animals and by populations of wild animals subject to human manipulation had altered the vegetation from a 'natural' state.

The grassland vegetation can be classified on the basis of the height of the dominant species although the grasslands are not composed entirely of the family Gramineae but a number of other herbaceous groups (notably the Leguminosae and Compositae) are present (Plate 15). The moister parts of the biomes have grass species which grow more than 1 m high (in some places as high as 2·5 m), whereas the drier parts only achieve much shorter grasses. The North American grassland has been further classified into tall grass, mixed grass and short grassbunch grass prairies, following an east–west trend of declining average precipitation and forming a gradient of declining NPP. The tall grasses (1·5–2·5 m) include the big bluestem, *Andropogon gerardi* and the switchgrass *Panicum virgatum*; the mixed grasses (0·6–1·2 m) include the little bluestem *Andropogon scoparius*, the needlegrass *Stipa spartea*, and the June grass *Koeleria cristata*; and the short and bunch grasses (0·15–0·45 m) include the buffalo grass *Buchloe dactyloides* and the blue grama, *Bouteloua gracilis*. The average biomass of these grasslands, taken together, is 1·6 kg/m² and the average NPP is 600 g/m²/yr. But the shoots are only the obvious part of the vegetation, for the roots of most species will penetrate deeply (up to 2 m) into the soil and so the majority of the plant biomass is in fact below ground: 2 kg/m² in some places. Some of the species have underground rhizomes which help to form a very tightly knit sod which was for long resistant to ploughing. This sod also comprises the 50–55 per cent of the total plant biomass which annually forms part of the litter fall: there is usually more organic matter in the litter and upper soil horizons than in the green parts of the vegetation. The growth habits of the grasses are tuned to achieve a stable NPP: some genera like *Andropogon*, *Stipa* and *Poa* renew their growth early in spring and reach their maximum development in late spring or early summer, when they set their seed. In the hot weather of high summer they become semi-dormant but resume growth in the autumn and remain green despite frosts. Other genera, such as *Buchloe* and *Bouteloua* renew their growth late in spring but grow continuously during the summer, reaching their maximum standing crop in late summer or autumn, with no further growth thereafter. The management of the grasslands to produce a less diverse stand of grasses is therefore very likely to bring about oscillations in productivity.

The animal ecology of the grassland has some distinctive characteristics. Large mammals are by no means as diverse taxonomically as in savannas: a few species tend to be dominant, as with the buffalo and pronghorn antelope in North America, the wild horse and saiga antelope in Eurasia, antelopes in Southern Africa and guanaco in South

America. These large herbivores tend to be herd animals, which affords them some protection from predators (e.g. wolf and coyote) in the open terrain. They are usually migratory in order to avoid over-use of their forage: buffalo, for instance, appear to be rather unselective grazers (compared with cattle or sheep) although able to digest very poor quality forage (Peden *et al.* 1974). The other mammals are small, often burrowing and quite frequently nocturnal: gophers, prairie dogs, jackrabbits and small birds such as meadowlarks and prairie chickens are predated upon by weasels, foxes, badgers, owls and rattlesnakes.

All these components interact to produce the grassland ecosystem: removal of predators, for example, allows burrowing animals to create very large areas of open ground where forage for large animals is absent; the selective grazing of large mammals largely controls the composition of the plant community, making it more likely that an individual plant will survive if it grows steadily rather than rapidly to the point where it is an obvious target for a grazer. The grassland sod holds the high quantities of nutrients and the organic matter which helps to retain moisture during long periods of drought and prevents the erosion which occurs when the turf mat is broken. The nature of these and other interactions led Sims and his co-workers (1978) to suggest that grasslands and herbivores have co-evolved and that it is 'unnatural' to study an ungrazed grassland. The effect of man has been, in general, to break open these tight linkages either by pastoralism to the point of grazing out the palatable species and then increasing the representation of drought-tolerant or **xeric** species (see p. 108), or by ploughing and exploiting the rich stores of mineral nutrients and humus, sometimes with irrigation as an aid. Stable agriculture can be achieved although both the Great Plains and Khazakstan steppe, for example, have seen crop failure and soil loss due to unsuitable agricultural methods. So virgin grasslands are rare since most of them have been altered by pastoralism of domesticated animals, replaced by agricultural ecosystems, or converted to a different species composition through the use of biocides or mechanical processes like brush removal, seeding with leguminous species, or simply through the invasion of new (including **exotic**) species following utilization by man, as described by Harris (1966) for the arid and semi-arid lands of the southwest USA. The management of grazing resources in various biomes is further discussed on p. 307.

Deserts

This biome is found in areas of the world where the average precipitation is below 250 mm/yr and is often irregular. The biotic response is firstly in scantiness of vegetation so that there is usually more bare ground than plant cover, and secondly in marked adaptations of both plants and animals to enable them to survive long periods of drought or the lack of access to free water. The aridity is usually compounded by very high daytime temperatures so that evaporation greatly exceeds precipitation in the soils and water-loss potential from individual organisms is very high. For example, the ground temperature may reach 60°C during the early afternoon, when the air temperature is around 40°C. The relative humidity is at its lowest, perhaps lower than 20 per cent, at the same time, having fallen from an 0600 h peak of 60 per cent.

Deserts are found in a number of topographic-climatic situations. Sub-tropical deserts are the results of semi-permanent belts of high pressure in the tropics from which air is warmed by compression on descent and becomes very hot and dry; the Sahara is an example of this type. Cold currents offshore may produce cool coastal deserts such as the Namib, Atacama and the coastal desert of Baja California. The other type of cool desert occurs in the interior of great continental masses, such as the Gobi Desert north of the Himalayan massif which help to keep precipitation away just as in the rain-shadow deserts such as those of the Great Basin and Mojave (southwestern USA) and Chihuahua (northern Mexico), which are in the lee of mountain ranges like the Sierra Nevada, the San Gabriel Mountains and the Sierra Madre Occidental.

In the hot deserts, the combination of aridity and high temperatures is probably the most adverse for life except regions of permanent snow and ice and their margins. Precipitation is erratic and may consist of occasional torrential rain breaking long (sometimes over a year) periods of drought. Dew-fall during the comparatively cold nights is the only other source of water.

Since vegetation is sparse, the soils reflect climate, parent material and relief to a much greater extent than where biomass is less sparse and so they are formed almost entirely by the mechanical and chemical weathering of rocks. Processes such as abrasion by wind-blown sand may produce scree-like formations of rocky waste, especially if finer material is also removed by the rare rainstorms. At the other extreme, the fine sand may form the entire substrate in the form of dune systems. Depressions receive much of the washed-out material and this accumulates in the form of flat plains of sediment with a high content of the sulphates and carbonates of sodium and potassium, so that the depression glistens with the white salts which may form a crust at the surface or just beneath it.

As said above, plants exhibit special adaptive mechanisms for survival in these conditions (Plate 16). The amount of water which the cells can lose before irreversible dehydration sets in is very variable, but for example the leaves of the creosote bush (*Larrea divaricata*) may lose a large proportion of their moisture and still recover when water becomes available. This ability to withstand water loss is paralleled by a variety of ways of obtaining and retaining water. Deep root systems are one method: those of mesquite (*Prosopis juliflora*) may reach down more than 50 m. Spreading root systems are another: although plants like the saguaro cactus (*Carnegiea gigantea*) appear to be widely scattered, the lateral roots may be in contact with each other. Succulent plants such as cacti store water in roots, stems or leaves, to the point where some may be used as emergency water sources by desert dwellers and travellers. To prevent water loss by transpiration, plants show a variety of adaptations like thick waxy **cuticles**, downy hair coverings, small fleshy leaves or rolled-up leaves reduced to virtual spikes, and the ability to shed leaves, twigs, whole branches or even to stop growing altogether when water is particularly scarce (Fig. 3.6). There is also the ephemeral habit in which annual plants remain dormant as seeds until rain comes. The water dissolves an inhibiting chemical in the seed coat and germination takes place followed by the entire life-cycle of the plant through to setting seed within a few weeks. Most desert shrubs possess features which appear to be defensive adaptations against grazing animals. Thorns and spikes are com-

Plate 16 Desert plants in the southwest USA: pictured are a spiny shrub with a thick cuticle; sage brushes with deep rooting systems and the facility of shedding branches; and a water-storing cactus.

mon; some species of *Euphorbia* contain a poisonous or irritant latex, some other shrubs have high contents of tannin in their leaves and the creosote bush has a pungent smell. The effects, on man at least, of the extract of the pods of *Cassia senna* are well known if not indeed proverbial.

The outcome in terms of plant communities is a vegetation closely related to water; as more water becomes available, the density of plants increases. Relatively moist areas such as the Mojave (shown as semi-desert or arid steppe in some classifications) have a vegetation of low (0·5–1·0 m) shrubs such as creosote bush set 2–3 m apart, as the 'dominants', and other shrubs interspersed. At intervals, clumps of low (0·2–1·0 m) succulents such as *Opuntia* are found. When the rains come (regularly in spring in this case), the ephemerals grow and the desert blooms. At the other extreme, the dune lands of the Sahara have sparse populations of only a few plants, such as the drinn grass (*Aristida pungens*), and in the rocky and stony deserts there may be large areas with no

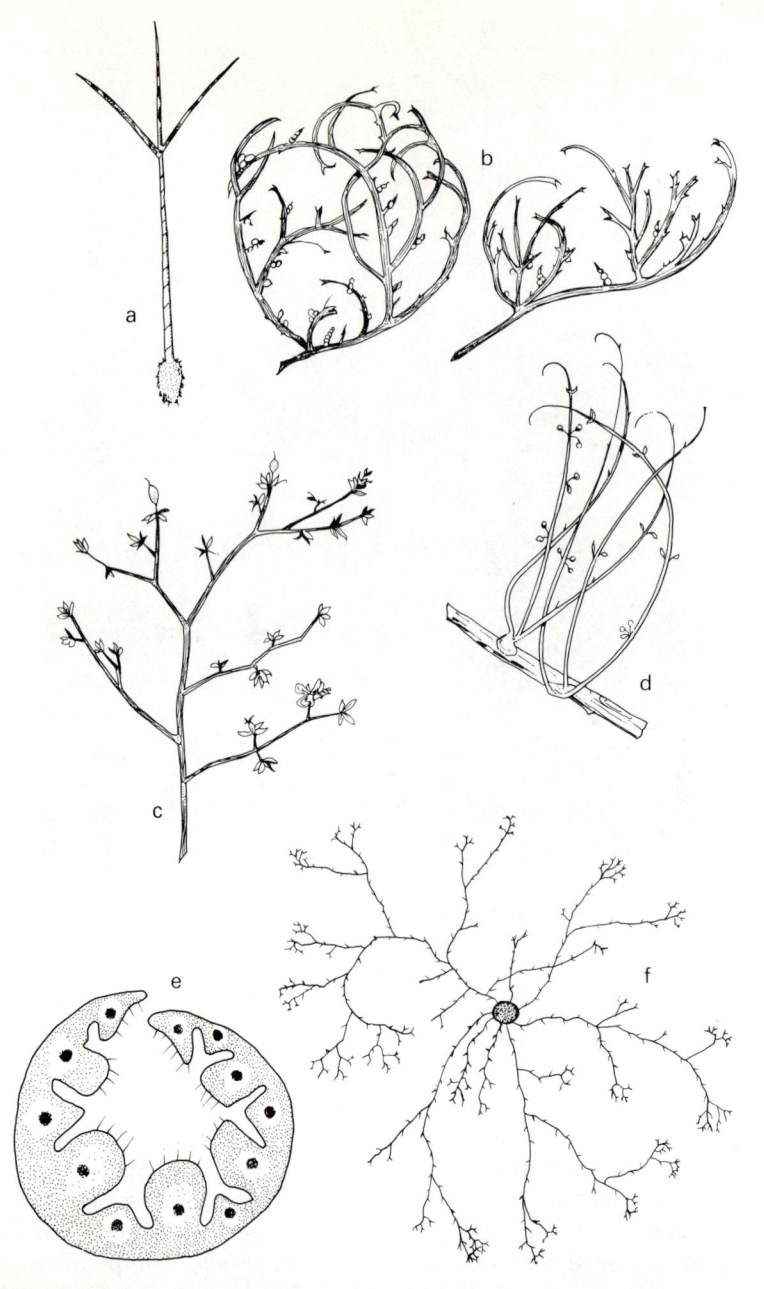

Fig. 3.6 Adaptations of desert plants: (a) Dispersal unit of grass (*Aristida*) (2 cm); (b) 'Rose of Jericho' (*Anastatica*), dry (8 cm) and wet (10 cm); Creosote bush (*Larrea divaricata*) (30 cm); (d) Aphyllous plant (*Calligonum comosum*) (25 cm); (e) Rolled xerophytic grass leaf in section (2 cm); (f) Plan of rooting system of a cactus (50 cm). (Drawings not to scale.)
Source: Cloudsley-Thompson, 1975.

vegetation at all except where a little finer soil material can accumulate and support a grass or a small shrub. So a great variety of vegetation density can occur: the well-known organ pipe or saguaro cactus deserts of Arizona look almost like arid forests with their layers of high and low cacti and shrubs, whereas a salt lake may be apparently entirely devoid of life. In between, most intermediate forms can be found with shrub deserts being perhaps the commonest.

To a casual observer, animal life in deserts is very sparse. But it is by no means absent even though NPP is very low: as with plants, a number of adaptations have evolved which facilitate the survival of the creatures in the prevailing conditions of heat and drought. Most species, therefore, are nocturnal, since they minimize waterloss by being active at night and probably avoid some potential predators as well. Any animal which has a moist or porous skin, such as the desert forms of worms and slugs, are always nocturnal. Some animals are active in the day but even they mostly leave the ground surface by burrowing, climbing plant stems or flying when the temperature reaches 50°C; only a few grasshoppers, beetles and spiders are active during the very hottest conditions.

Physiological adaptations are numerous and mostly centre around a lack of need for free-water intake and mechanisms for avoiding its loss. Many birds, insects and rodents, for example, gain their water in their food rather than from free water; when water becomes available, the African and Asian wild asses (*Equus asinus* and *E. hemionus*) can drink it in quantities amounting to a quarter of their body weight. The camel can lose 30 per cent of its body weight but can then drink up to 120 litres of water at a time. Water loss is minimized by a variety of means, of which perhaps one of the most common in vertebrates is an excretory system which produces concentrated urine and relatively dry faeces, and may lack sweat glands almost together. The antelope ground squirrel (*Citellus leucurus*) can drink salty water more concentrated than that of the sea, and the sand rat (*Psammomys obesus*) secretes a urine up to 4 times as concentrated as sea water. Kangaroo rats (*Dipodomys* spp) and other small desert mammals have a nasal mechanism for reducing water loss while in their burrows: the walls of their nasal passages are cooled to 28°C by the burrow air but when saturated air from their lungs at 38°C passes over the passages, condensation occurs and this water is swallowed. Coping with heat also elicits unusual phenomena: the camel's body temperature can vary over a wide range and it does not start to sweat until its body temperature is 40·7°C. Thus it stores heat by day and can lose it at night when the associated water loss is lower; the same phenomenon is shown by the ostrich, gazelles and the oryx. The North American jackrabbit does not burrow and ingests no free water and it is suggested that its very large ears (a characteristic of many desert animals and presumably also a defence aid against predators) radiate heat to the air when the animal is resting in the shade.

A spectrum of animal life is achieved in spite of the adverse conditions. Arthropods, lizards, tortoises, snakes, birds, rodents and larger mammals are all found, including a variety of carnivores which is surprising considering the low NPP. Most of them obtain water from their prey and the foxes are apparently the best able to cope with extreme aridity. Many of the predators take a wider range of prey than their relatives in lusher lands and some, such as the fennec fox, *Fennecus zerda*, eat not only lizards, insects and rodents, but dates as well. Desert coyotes will eat grass just as their urban cousins will strip citrus tree in Los Angeles (p. 227). As with plants a flush of rain will transform the apparent barrenness into an abundance of life. Insects move in rapidly to pollinate flowers, dung beetles appear to roll away the droppings of mammals, and the ants harvest grass seeds. Caterpillars, crickets, flies and wasps rarely seen at other times of year also descend upon the ephemeral plants. The rains may trigger off other processes: the maturation of the desert locust (*Schistocerca gregaria*) occurs in response to aromatic chemicals produced by desert shrubs at the time of the rains. Calving of mammals usually occurs at the time of the rains if these are regular. At the rainy season, the camel ruts and then has a 12-month gestation period so that young are produced at a time of maximum forage.

The importance of life below ground is emphasized in modern production studies: in many communities up to 80 per cent of NPP is underground and green material may only be 1 per cent of the total NPP of perhaps $90 \text{ g/m}^2/\text{yr}$. It follows that any human activities which affect the above-ground plant biomass have the potential to change the ecosystem radically. Grazing of domesticated animals is the obvious method and so it need not surprise us that extension of deserts is taking place where heavy grazing of animals like the goat occurs (see also p. 311). Nevertheless, pastoralism of animals like the camel can be a stable system in deserts provided that movement is frequent. In deserts used for recreation, severe damage to the ecosystem can result from vehicles which damage the plants, especially succulents, and where, for example, the use of dead cactus 'skeletons' for fire robs the soils of organic matter as well as destroying the scarce cover for arthropods. The natural rate of decomposition of leaves, animal shoots and seasonally dying parts appears to be fast, but woody material is much slower to decay. This means also that human-introduced debris remains virtually unaffected except by mechanical weathering for a very long time. So although the deserts scarcely look like a fragile biome they are certainly capable of modification by man. Of more current importance, perhaps, is the way in which some activities can extend them, in the phenomenon known as 'desertification', which is discussed on p. 311. The arid lands of the world, including the savannas and steppes prone to desertification, are estimated to cover 35 per cent of the land surface.

Deciduous forests of temperate climates

In most of western and central Europe, the northeastern USA and Japan south of Hokkaido are found forests which respond to a climate having marked hot and cold seasons, with winter temperatures which fall below the freezing point of water; the precipitation is of the order of 750–1500 mm/yr. The deciduous habit of the dominant trees can be seen as a form of dormancy which is a seasonal response to low ambient-energy levels and the unavailability of water. The distribution of this biome is notably confined to the northern hemisphere: areas of similar climate south of the equator are occupied by evergreen forest, a feature probably to be explained in terms of the evolutionary history of the Angiosperms during the Cretaceous rather than as a result of subtle differences of contemporary climate.

The biome is dominated by trees of 40–50 m height. Their leaves tend to be broad and thin, compared with the leathery but narrow leaves of tropical genera, and a large number of them produce nuts and winged seeds rather than pulpy fruits. Acorns, beech mast, chestnuts and the winged seed of the sycamore are examples of this habit. Dispersal by animals becomes very important if the seeds are too heavy for transport by wind, and indeed the density of herbivores (e.g. mice and voles) may affect the regenerative success of the trees. These trees are occasionally the habitats of climbers such as ivy (*Hedera helix*) and wild vine (*Vitis* sp), and epiphytes like mosses, lichens and algae grow on the trunks. Dominance by 2–3 species is common and single-species stands are often found (Plate 17). This contrast with tropical forests is mirrored in the wind pollination habit of most of the trees. A dense canopy is usually formed and the amount of light percolating through determines the character of the lower layers of vegetation, as does the leaf-mosaic density of the individual species of dominant tree. The dominant trees are typically oaks (*Quercus* spp), beeches (*Fagus* spp), elms (*Ulmus* spp), limes,

Plate 17 Deciduous forest. A temperate zone deciduous forest in England, dominated by oak (*Quercus robur*). A ground layer of grasses and fern can be seen, but there is only very slight evidence of a shrub layer.

lindens or basswoods (*Tilia* spp), tulip trees (*Liriodendron* spp), chestnuts (*Castanea* spp), maples (*Acer* spp) and hickories (*Carya* spp), though the species variety per unit area never approaches that of the tropics. In North America, the maximum is about 40 spp/ha, in Europe 8 spp/ha. Soil parent material may affect the distribution of forest-tree species; in Britain freely drained gravels are associated with woodlands of *Quercus petraea*, *Fagus sylvatica* (beech) and *Carpinus betulus* (hornbeam), whereas loamy soils derived, for example, from clays favour *Quercus robur*, elms (*Ulmus* spp) and lime (*Tilia* spp). Thin soils over chalk are also associated with *Fagus sylvatica*, while hard limestones often carry woodland fragments of ash (*Fraxinus excelsior*). Wet soils are tolerated by willows (*Salix* spp) and alder (*Alnus glutinosa*).

Below the trees a shrub layer may form, especially in gaps in the forest canopy, and a great variety of species may be involved depending on the continent. Birches (*Betula* spp), ash (*Fraxinus* spp), hazel (*Corylus* spp), members of the Rosaceae (e.g. the genera *Prunus*, *Rosa*, *Rubus*), grow in the gaps and form part of the succession back to mature forest. Some of these shrubs are spiny or have prickly leaves (e.g. holly, *Ilex aquifolium*) and may thus protect the seedlings of forest trees from browsing mammals. The seasonality of the forest is often reflected in the ground flora which exhibits two assemblages: a pre-vernal group which puts out leaves, flowers, and sets seed before the dominant trees have come into leaf; and an aestival group which can tolerate the lower light levels of the canopy in full leaf. In Europe, the wood anenome (*Anemone nemorosa*) is an example of the former, and dog's mercury (*Mercurialis perennis*) of the latter. Bracken fern (*Pteridium aquilinum*) is also characteristic of gaps in the canopy. A lower ground layer of mosses and, on very dry sites, lichens may also be found, and the presence of this stratum is less obviously dependent upon the summer light intensity.

The animal communities too are responsive to the climatic regime. Migration to warmer climates for the winter is common among insect-eating birds such as the warblers (Sylviidae); hibernation is found among those less capable of long-distance travel, like the black bear. Others, like the deer of various species which are chiefly browsers, remain active all year, digging through the snow for food when the above-snow browse is exhausted. A typical food web for a deciduous woodland is shown in Fig. 3.7. The importance of small mammals and insects as herbivores can be seen, as can a top carnivore level. (The complexity of this web shows also the general point that an organism may not fit easily into a single trophic level.) In this diagram, deer are omitted but various species are present in many deciduous woods of Europe and North America and, where not persecuted by man, their main predator the wolf is present as well.

The litter layer is the site of a diverse flora and fauna which are responsible for the release of inorganic mineral nutrients back to the soil. There are two main stages in this process. Firstly, the primary decomposers (millipedes, woodlice, beetles and earthworms) attack the litter and break it down into smaller particles. In the second phase, these fragments together with the faeces of the litter animals form the food for the secondary decomposers (mites and springtails) which further comminute organic material. Wet material is broken down by bacteria, fungi and protozoa as well and so more or less complete mineralization is achieved: there is little long-term accumulation on the surface, although some centimetres of humic material are nearly always present. In

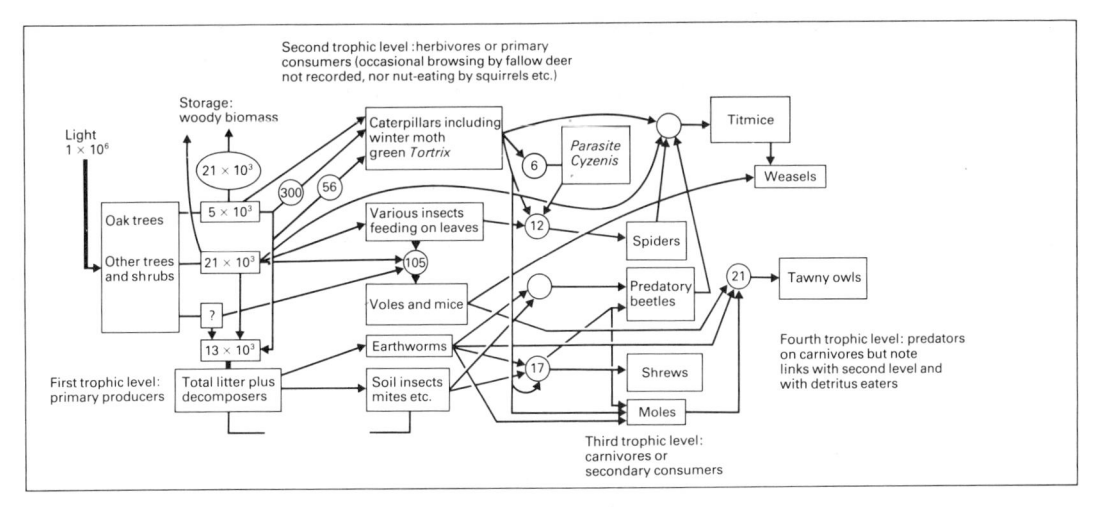

Fig. 3.7 Levels of biomass and energy flow in a deciduous woodland in midland England. Biomass (in rectangles) and transfer values (in circles) are in kJ/m² (biomass) or kJ/m²/yr (transfer).
Note—1 joule=4·2 calories.
Source: Learmonth and Simmons, 1977, after Varley 1974.

poorly drained soils, the breakdown is much slower and the build-up of humic material is greater. Under normal circumstances the soil fauna helps to mix the organic matter with the mineral components of the solum, so that horizonation is not distinct; the end result is a brown soil with a gradually diminishing content of organic material below the surface: the brown earth soil typical of this biome. It is not universal, though, for acid leachates from the litter may mobilize iron and other elements and redistribute them lower down the soil profile, producing a podsol.

The functional dynamics of deciduous forests have been more intensively studied than many other biomes. We know, for example, that the accumulation of biomass in North American forests seems to level off at *ca* 400 t/ha when the forest is 200 years old; European forests of comparable age seem to have less biomass. Mineral cycling has been intensively studied, notably in Belgium and in the Hubbard Brook study, New Hampshire. The minerals in the shed leaves are clearly important in the supply of nutrients to soil fauna and flora and hence to the trees again. As might be expected, though, the rate of uptake into the trees is much lower than in tropical forests: at 200–500 kg/ha, about 25 per cent of the value of tropical rain forest. Hence the soil is a proportionately much larger repository of minerals than in tropical forests. Losses to the runoff are therefore high by tropical standards but compared with the quantities in the intrasystem cycle (see p. 70), the nutrient cycle can still be described as 'tight'. In temperate climates losses to runoff are made good from inputs in precipitation and by the relatively fast weathering of the rocks, and the minerals are lifted into the biosphere by trees with deep taproots (uncommon in the tropics), soil creep on slopes, and soil animals. Weir (1969) has suggested that in England, significant quantities of nutrients may be imported into some woodlands by rooks (*Corvus frugilegus*).

No account of this biome can underemphasize the effect of man, for it is in this zone that Western industrial civilization grew up, preceded by a long period of agriculture. There have been two main effects: the first of these is obviously the clearance of forest and its replacement by agriculture. Notably, the resulting ecosystem had a greater proportion of its nutrients in the soil (as distinct from the vegetation) than in the tropics, so that fertility remained high, particularly where the annual leaf-fall was mimicked by annual inputs of animal manure from mixed farming; the dung helped to maintain the crumb structure as well as the nutrient levels. Secondly, the remaining forests have been heavily managed for an immense variety of purposes, from timber production through forage for domesticated herbivores to amenity in terms of visual pleasure or hunting. The effects on forest ecosystems are still being unravelled by historical ecologists but it is clear, for example, that the demand of shipbuilders for timber in Europe in the sixteenth to eighteenth centuries encouraged a move towards single-species woodlands of oak, just as demand from furniture-makers in the eighteenth and nineteenth centuries encouraged single-species forests of beech. Earlier still, common-rights legislation enabling villagers to take wood 'by hook or by crook' may have brought about pollarding; and a rural society's demands for pliable but straight poles produced many an area of coppice. Species like hazel (*Corylus avellana*) were often used because of the concomitant nut yield, but oak coppice and lime coppice were also found. The upshot is that it is unlikely that any deciduous woodland in Europe is 'natural' (i.e. unmanipulated by man), although it may be 'primary' in the sense that woodland has grown on that site from time immemorial (i.e. at least since the Dark Ages). In North America, the shorter period of European occupation makes more likely the preservation of some areas of virgin deciduous forest. But the more we know about the Indians, the more likely it appears that they affected the dynamics of the forests. Because of the longer drier summers, they were able to set fire in the deciduous forests, something relatively rare in Europe, and so this aid to hunting doubtless affected the plant and animal communities of the forests. So if a forest is to be called 'natural' it needs to have a certificate of its pedigree which shows that during 'prehistory' and 'historical times' all the techniques of the ecological historian, from pollen analysis to lichen-diversity counts, have failed to reveal any trace of human effect.

In Britain, it is now certain that many of the areas of semi-natural vegetation have had a long history of alteration by human activity: woods, heaths, moors and wetlands alike. The numerous excellent accounts of their plant and animal ecology (e.g. Tansley 1968; Burnett 1964; Pears 1977) should be read with such an historical perspective in mind: the study of a single wood in Cambridgeshire by O. Rackham (1975) is an outstanding example of such a viewpoint.

Boreal forests

Vegetation maps invariably show strikingly this great unbroken belt of forest across northern Eurasia and North America. To the north, the boundary is the Arctic treeline (sharper on world-scale maps than in reality) and to the south there is mixture with a variety of communities, mostly deciduous forests; in the Western Cordillera of

North America, however, a variety of the coniferous forests stretches southwards down the continent towards Mexico. The physiognomy of all these forests is dominated by the needle-leaved evergreen coniferous tree, with the genera *Picea* (spruces), *Pinus* (pines), *Abies* (firs) predominating. *Larix* (the larches) is also found; some of its species are deciduous.

The formation is found in a characteristic climate of long and cold winters with considerable snow-fall forming part of the precipitation of 250–1000 mm/yr. In winter the soils are frozen to a depth of about 2 m but if there is a thick snow cover then its insulating nature may keep the soil temperature as high as $-7°C$. The summer, although short, has a long daylength and the average temperature of the warmest month exceeds $10°C$.

In such a physical environment, the evergreen conifer has particular adaptations suited for survival: provided water is available it can photosynthesize all year round, and the needle-leaves help it resist drought when water is locked up as ice and when strong winds increase transpiration rates; the shape of the crown enables it to shed snow so that branches are not broken off by the weight. Large trees up to 40 m high are common and these dominate the structure of the biome: in a continuous stand relatively little light penetrates the canopy so that another layer of trees is uncommon (Plate 18). There is usually a continuous ground cover whose components vary according to local conditions of drainage and light: at the dry end of the spectrum a lower cover of lichens and mosses may be found; where there is more ground-water the vegetation will include low heath shrubs such as *Vaccinium* and *Empetrum*; and where it is very wet bog-mosses such as *Sphagnum* will cover the ground, often forming open bogs in very wet hollows.

The great dominant genus of the biome is *Picea*, the spruces. The forest tends towards single-species stands and so great swathes of spruce are found, with more diversity near breaks in the forest such as streamsides, areas of windthrow and burned areas. The firs, pines and spruces tend to be more restricted in their occurrence in the main sweep of the biome, although the larches for example are very important in Siberia, and *Larix dahurica* is said to be the northernmost former of true forest in the world and certainly reaches farther north than any other coniferous species. Its limit is closely paralleled by several broad-leaved species which reach more or less tree-size and whose deciduous habit is probably an aid to survival so far north. Thus birch, willows, alders and aspens are often found, particularly in two contexts. The first of these, as just hinted above, is at the northern or altitudinal limit of the coniferous forest: in Lapland, for example, a broad belt of birch scrub separates the pines from the tundra-type vegetation and in the Rocky Mountains the aspen (*Populus tremuloides*) occurs right at the tree-line before the alpine communities take over. The second role of the deciduous species is as pioneers in deforested areas as when windthrow has opened up the coniferous blanket, or when fire has removed the dominants, or where succession is kept at an early stage along rivers subject to flooding and the shifting of sand and shingle banks.

The size of the dominants and their ability to maintain growth during apparently unfavourable conditions means that a biomass is achieved not far short of some more southerly communities: in Hokkaido (the northern island of Japan) for example, the boreal stands of *Abies* and *Picea* averaged a biomass of 26 kg/m^2, whereas the broadleaved evergreen forests further south averaged 43·2 kg/m^2, NPP being 2 kg/m^2/yr in the first

Plate 18 Coniferous forest in North America. There is only one tree layer and the shrub layer is mostly young trees. This forest is on a dry site and the accumulation of litter is not very deep. Note the large amounts of dead wood, characteristic of natural and near-natural forests.

community and $2.16 \, kg/m^2$ in the second; the world average NPP for the boreal forest biome is $0.8 \, kg/m^2/yr$ and a biomass of $20 \, kg/m^2$ (Whittaker and Likens 1975). The annual litter fall is about 1–2 per cent of the above-ground plant biomass (i.e. in the region of $0.04 \, kg/m^2/yr$). The seasonality of the climate is strongly reflected in the breakdown of the litter, which is slow compared with other biomes and often results in a deep accumulation of organic matter on the forest floor, especially in wetter areas. Up to ten times the annual litter fall may accumulate on the forest floor, so that a litter biomass of 100–500 kg/ha is found. (In deciduous forests the equivalent measurements are five times the annual fall and 100–150 kg/ha. In both types of forest these figures are exceeded in wet habitats (Rodin and Bazilevich 1967).) The slow mineralization of this material is short-circuited by fire which under natural conditions is probably quite frequent and runs along the forest floor consuming the relatively low amounts of litter accumulated since the last fire. Fire-protection policies allow debris to accumulate and so a big fire may then result with the fire running up the trunks (aided by resinous drips) and then igniting the crowns. Such a fire is disastrous in commercial terms but probably infrequent under natural conditions. A shallow burn has another role of reducing humus thickness so that seedling trees can more easily root in the mineral soil, and to that extent it may aid forest regeneration. Given the low levels of soil fauna and flora compared with other forest biomes, especially the larger soil animals, a low-rate mixing of mineral and organic matter is to be expected. A clear break between the litter layer and the top mineral horizon is often seen in the soils, and the **chelating** effect of the leachates from the acid litter encourages pan formation. The podsol is therefore the characteristic soil of the boreal forests.

Periodicity is also found in the animals of the forest ecosystems. Seasonal change brings in some animals from further north in winter, such as the barren-ground caribou (*Rangifer tarandus*) and sends rather more to winter further south: birds are especially common summer visitors to the biome, including the numerous insect-eaters. Another aspect of periodicity can be seen in sudden population 'explosions' of some animals, notably insects such as the sawfly and spruce budworm which are important pests in commercially exploited forests.

Some animals are year-round residents. Rodents are a characteristic group and generally survive the winter under the insulating cover of the snow blanket. The beaver (*Castor fiber*) is also typical of this biome and is perhaps analogous to the elephant in its alteration of the local habitat. Of the larger mammals, various sorts of deer are characteristic including the moose (*Alces alces*) which is largely an animal of the secondary vegetation produced by fire. The diversivores are typified by bears, including the common brown bear of North America (*Ursus arctos arctos*) as well as the remnant populations of the grizzly (*Ursus arctos horribilis*). Small carnivores such as lynx (*Felis lynx*) and wolverine (*Gulo gulo*) prey chiefly on the rodents, as do the owls and hawks; the characteristic large carnivore is the wolf (*Canis lupus*) which is an important influence upon the populations of deer, caribou and moose. Studies of wolf–moose interaction on Isle Royale, Michigan, for example, have shown that in the absence of wolves, the moose breed to the point where they over-use the browse and then die of starvation and disease. Kept to lower numbers by wolves, the moose survive in better condition, and the vegetation is also kept more stable. Even the most casual observation of the forests in the low-lying areas of protected ecosystems such as National Parks will often show a 'browse-line' where the herbivores have eaten virtually all the young branches and the bark during winter. This will prevent regeneration and may even cause death of the mature trees and is symptomatic of too little predation: the ungulates may be protected but their wider-ranging predators may well stray outside the Park boundaries and be shot or poisoned.

Of the high biomass of these forests, only a small part forms food for animals at any one time, so that the grazing-chain secondary production is never very high: the biomass of vertebrates in one forest noted by Sukachev and Dylis (1964) totalled only 2·5 kg/

ha and they ate only 344·9 kg/ha/yr of food, both herbivore and carnivore. The gallinaceous birds of a Lapland forest consumed only 0·1 per cent of the green parts and 1·2 per cent of the berries of their habitat. Invertebrates, however, probably have a greater capacity to affect above-ground vegetation, and their effect on the soils and the mineral cycles is perhaps the biggest single contribution of animals to the ecology of these forests.

Man has used this biome ever since it was formed. It was the basis for many stable hunting and gathering cultures based on deer, moose and fish. Now it is the focus of the world's softwood lumber industries whose activities alter the ecology considerably: sometimes temporarily and sometimes over a long period, especially when the frequency of fire is increased beyond that to which the forests are undoubtedly adapted. Also, much of the nutrient supply of the ecosystems is tied up in the biomass and so rapid-cycle exploitation and whole-tree harvesting may reduce the productivity of this biome permanently (Tamm 1976). Recreation may also exert effects on the ecosystems: trails used by people and horses in the forests consist of a narrow (4–5 m wide) band of altered vegetation. As with recreational pressure in other biomes, some species decrease in these conditions and others increase, but a bare strip along the trail corridor is usual above a certain level of use (Dale and Weaver 1974).

The variants of this biome alluded to in the opening paragraph are the forests of the Western Cordillera of North America. These reach their maximum biomass on the coastal ranges where there is a high precipitation including moisture from the fogs which drift inland from the Pacific, although at high altitudes most of the precipitation falls as snow. The grand fir (*Abies grandis*), Douglas fir (*Pseudotsuga menziesii*), western hemlock (*Tsuga heterophylla*) and Sitka spruce (*Picea sitchensis*) are typical of this facies and usually have a low understorey of heathy shrubs and lichens whose biomass is about 3 per cent of the trees (Turner and Singer 1976). In northern California are found the remnant forests of the coast redwood, *Sequoia sempervirens*, specimens of which comprise the tallest trees in the world (Plate 44). Its near relative, the Sierra redwood (*Sequoiadendron giganteum*), grows in drier coniferous forests on the western slope of the Sierra Nevada of California, where its size enables it to be labelled the plant with the highest biomass: the blue whale of the plant kingdom. All such forests are heavily used for timber, which makes very important the protection of the only remaining populations of these giant redwoods (Simmons and Vale 1975; Vale 1975). The ecology of the riverside stands of coast redwood where the largest trees grow is complex: the mineral nutrient supply needed for their growth is provided by silt from floods and by mineralization of litter by fire, both of which processes are inimical to the human use of the area for other purposes so that conflict exists not only over designation of reserves but their management, and the use of the watershed beyond the reserves. None of which problems should prevent the conservation effort continuing.

The tundra

In the northern hemisphere, the boreal forest gives way northwards to a zone of scattered clumps of trees in a low open vegetation and then finally to a broad belt of treeless vegetation in which busy willows, birches and alders are the highest vegetation. The transition appears to coincide with a mean daily temperature of at least 10°C in the warmest month and a growing season of less than 3 months and 10°C of accumulated month degrees over a threshold value of 6°C. On both continents, the boundary between forest and tundra lies further north in the west where climate is moderated by warm westerly winds. The ground however remains frozen all year ('permafrost') except for the few top centimetres which thaws during the summer. Water which is frozen for the rest of the year is then made available to the plants. Precipitation is low (250 mm/yr) but is accompanied by slow rates of evaporation even in summer. Low-lying areas tend therefore to be waterlogged whereas ridges are very dry because even snow is blown off them by the winds and so the soil freezes to a lower temperature than where an insulating cover of snow is found. In central Alaska, an air temperature above the snow of $-50°C$ was measured at the same time as $-11°C$ at the surface of the soil beneath the snow. A similar set of measurements is depicted in Fig. 3.8.

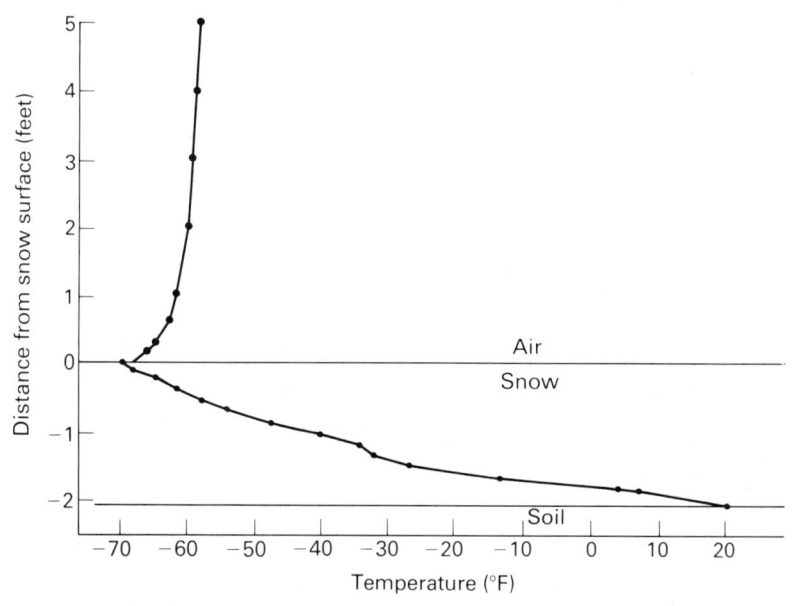

Fig. 3.8 Vertical temperature gradients one day in Alaska during a cold spell. Note measurements are in Imperial units: −70°F=approx. −57°C, 1 ft=0·3048 m.
Source: Odum, 1971.

The growing season is not so short that higher plants cannot grow, and is made longer by the long hours of daylight during the summer. A NPP average of 140 g/m²/yr is achieved, which is low, but higher than deserts and the open ocean (Table 2.4). With higher plants, much of this productivity is below ground—for example the ratio of root: shoot NPP at Point Barrow, Alaska, was 1 : 1.2—and the litter layer acts as a considerable store of nutrients because of the very slow rates of decomposition. Areas of relatively high vegetation ('low tundra') are characterized therefore by a thick spongy mat of living and partly decayed vegetation overlying a peaty horizon; near Hudson's Bay, the railway rests on 6·5 m of frozen *Sphagnum* peat. In summer its surface is saturated with water and small ponds are scattered throughout (Plate 19). The plants tolerant of these conditions are predominantly grass and sedge—for example, near Point Barrow the coastal plain is dominated by the grass *Dupontia fischeri* and the sedges *Carex aquatilis* and *Eriophorum scheuzcheri*. In sheltered valleys, thickets of dwarf birches (*Betula nana*) and creeping arctic willows (e.g. *Salix arctica*) can be seen. In drier places, heathy plants like *Cassiope tetragona* are found. Most plants are less than 20 cm high and thus lie beneath the snow in winter. The higher plants are usually perennials and reproduce vegetatively by rhizomes, bulbs or runners. The less productive 'high tundra' is much lower in moisture and frequently lacks a permanent snow cover in winter. The soils are low in organic matter and are vegetated by only a sparse growth of mosses and lichens, with

Plate 19 Tundra. A tundra scene in Greenland. The rocks beyond the pool shelter a low scrub but the ridge beyond is devoid of woody vegetation. The pool is fringed with a cottongrass of the genus *Eriophorum*, tolerant of both water-logging and acid conditions.

only the latter in the most exposed places. Such vegetation is, however, permanently available for herbivores during the winter.

The plant life is limited, among other factors, by the low availability of nutrients. At Point Barrow, Alaska, less than 1 per cent of the nutrient capital of nitrogen and phosphorus is resident in the living components of the ecosystem. Most of the rest is in non-exchangeable pools locked up by the slow rate of decomposition and so the small quantities of available nutrients (Fig. 3.9) must 'turn over' many times during a growing

Nitrogen Budget

- Senesced Plant Material: 2.2
- Animals 2.8
- Shoots 1.8
- Stem Base 1.2
- Rhizome 0.9
- Roots: 3.0 | 2.0
- 2.2 g N/Season
- 0.1 Soluble N
- 0.1 Exchangeable N
- Non-Exchangeable Soil Nitrogen: 207 (0–5 cm), 253 (5–10 cm)

Phosphorus Budget

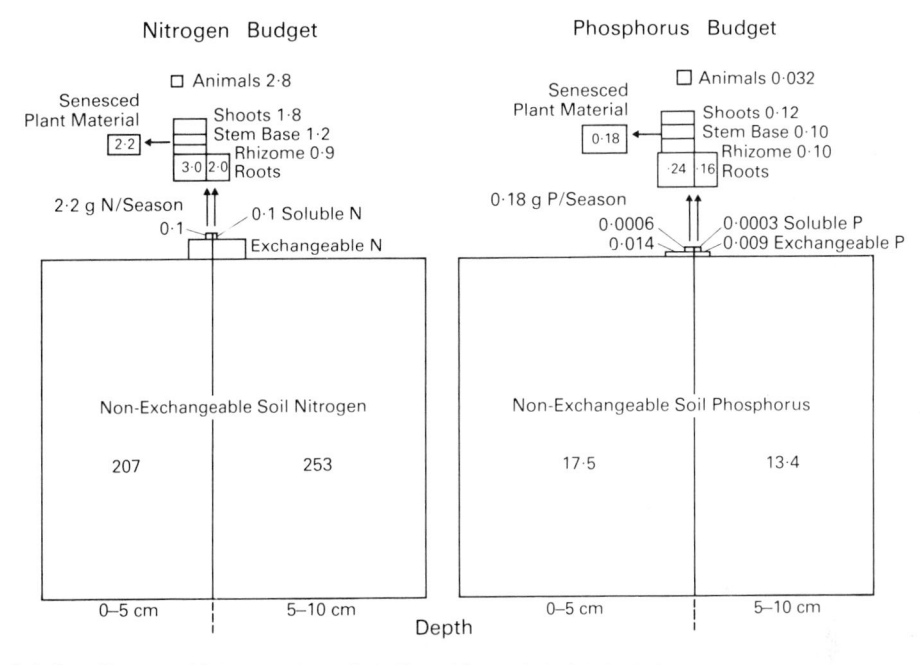

- Senesced Plant Material: 0.18
- Animals 0.032
- Shoots 0.12
- Stem Base 0.10
- Rhizome 0.10
- Roots: 24 | 16
- 0.18 g P/Season
- 0.0006
- 0.0003 Soluble P
- 0.014
- 0.009 Exchangeable P
- Non-Exchangeable Soil Phosphorus: 17.5 (0–5 cm), 13.4 (5–10 cm)

Depth

Fig. 3.9 Standing-crop biomass content (g/m²) and flows (g/m²/yr) of phosphorus in major components of tundra at Point Barrow, Alaska. The contrast between exchangeable and non-exchangeable components is especially striking.
Source: Rosswall and Heal, 1975.

season: soluble plus exchangeable nitrogen 11 times, and soluble plus exchangeable phosphorus 200 times or 3 times per day. In this respect, remark Bunnell *et al.* (1975), the tundra ecosystem is more like tropical than temperate systems.

The animal life is conspicuously adapted to the climate in one way or another. Warm-blooded animals which live above the snow are generally covered in warmth-retaining fur (e.g. the musk-ox, *Ovibos moschatus*; Arctic hare, *Lepus arcticus*; wolf, *Canis lupus*; caribou and reindeer, *Rangifer tarandus*). There are other anatomical and behavioural adaptations: the snow-shoe hare (*Lepus americanus*) has pads twice the size of an ordinary hare on its hind feet to enable it to cross snow; ungulates which sink into the snow will make sunken trails, just as they will clear away snow with their front feet to get at food beneath. They will then often be accompanied by ptarmigans which also feed on the exposed vegetation. Migration is also practised, with the yearly round of the caribou herds a much-studied feature. The caribou drop their fawns on the tundra during May, having wintered at the forest edge. They then range up to the shores of the Arctic Ocean, frequenting high ground to avoid the flies. They are constantly attended by wolves, who keep the herds moving and so avoid over-use of the forage. During the summer there is a brief burst of invertebrate life, in which insects are the most conspicious, especially those of the mosquito and blackfly kind. The tundra is also a favoured nesting place for many species of migratory birds, few of which winter in the north. They arrive in spring and carry out their reproductive cycle very rapidly, so that the young are ready for the southward migration at the end of the short summer. Animals which live beneath the snow, such as the lemming, have relatively thin fur and possess few of the modifications required by the animals of the open air of winter.

Some animal populations in the tundra are marked by oscillations of abundance. The snow-shoe hare, for example, fluctuates on a 9–10 year cycle, with a population peak being followed by a crash. Its chief predator, the lynx, follows the same cycle of abundance with a 1-year lag. Perhaps the most famous cycle of abundance is that of the lemming where the genera *Lemmus* and *Dicrostonyx* have species which in both Eurasia and North America become very abundant and then achieve spectacularly high mortality rates. At the especially high peaks which occur at 4-year intervals, the famous lemming migrations occur in Northern Europe, when the populations migrate southwards irrespective of any obstacles in their path. In North America, the ecological effect of the peaks on the vegetation was shown in experimental plots, where ungrazed areas had 36 per cent more grass in August than those utilized by lemmings. (By contrast, reindeer in Norway grazed and trampled only about 2·1 per cent of the above-ground NPP (Ostbye *et al.* 1975).) One effect, however, of heavy grazing by small mammals and invertebrates is that more solar radiation can reach the soil and so nutrient cycling can accelerate to some extent: a finely tuned mechanism is probably at work.

Much more constant are those animal populations which are linked to the food webs of the sea. Seals, and to a lesser extent walrus, are obvious examples, together with various species of sea-birds, many of which (like skuas and gulls) can feed off either sea or land. At the top of this chain is the polar bear (*Thalarctos maritimus*), now a threatened species and the subject of a circumpolar international convention. The linkage between land and sea ecosystems is illustrated in Fig. 3.10 from Pruitt's (1978) excellent short account of the ecology of boreal land.

At the southern pole of the globe, there is an equivalent zone which includes the fringes of the Antarctic continent together with some Antarctic islands. Even in summer, it is colder than the Arctic and so the vegetation of the continent is dominated by mosses and lichens, with bacteria, fungi and algae being found under very severe conditions. Antarctica possesses about 400 species of lichens, and in the coastal areas some 75 species of mosses and 20 liverworts grow on rocky flats and at the edges of streams. There are only two flowering plants, the grass *Deschampsia antarctica* and the herb *Colobanthus crassifolius* (Caryophyllaceae). On more northerly islands such as Kerguelen and Macquarie Island, the vegetation resembles a luxuriant arctic tundra but as there is unlikely to be permafrost, it is perhaps sub-arctic rather than true arctic vegetation. As might

The tundra 125

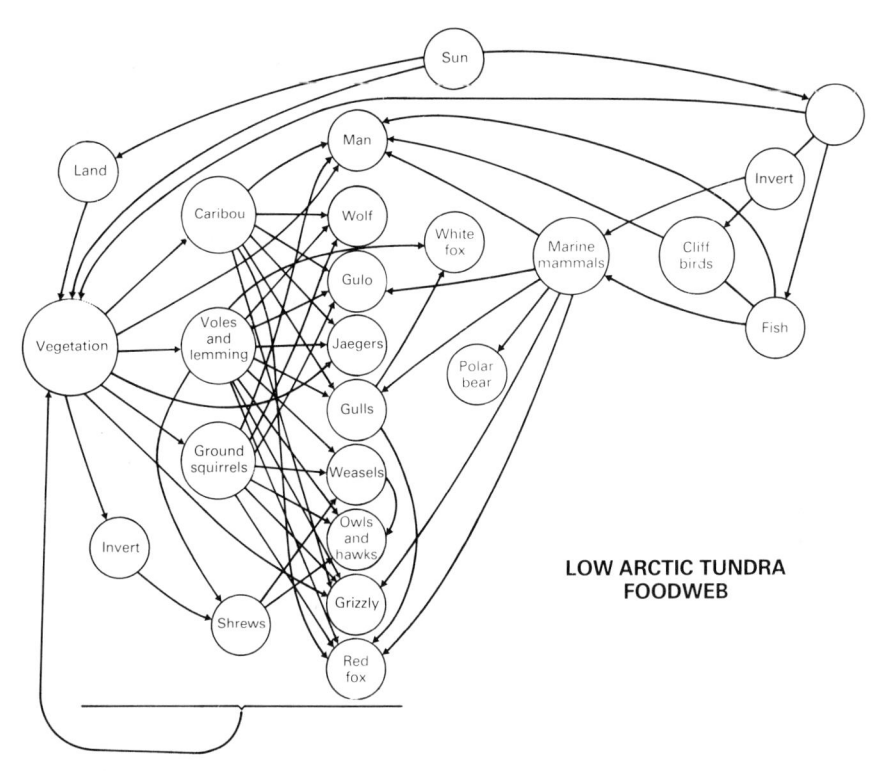

Fig. 3.10 A low Arctic tundra foodweb. Note that on land the herbivores are dominated by three groups and that these form the food supply of a number of carnivorous species.
Source: Pruitt, 1978.

be expected, the native island species tend to be endemics: for example the Kerguelen cabbage (*Pringlea antiscorbutica*) is the sole species in a genus confined to the one island.

Rather like the maritime fringes of the Arctic, animal life is quite abundant where the fauna is part of a food web which originates in the sea, for the NPP of these cold waters is high (see p. 86). Animals seen on land or just offshore are dependent on the sea for their food although ice-floes or the land may provide them with breeding space. Seals and the elephant seal are examples of such animals in the Antarctic: the Weddell seal, for example, is a fish and squid eater, the crab-eater seal feeds on krill (a small shrimp), and the leopard seal is a predator upon penguins. These birds are mostly confined to the Antarctic and sub-Antarctic islands and possess blubber, thick down, and oily feathers at the rate of $11/cm^2$ and so resist cold water and even colder winds. All the other birds are also sea-birds (e.g. petrels, skuas, gulls, cormorants and albatrosses) until the sub-Antarctic islands are reached: South Georgia is the home of one species of pipit and one pintail duck.

The impact of man in such environments is bound to be strong, for with such a short growing season, plants have low recuperative and colonizing powers. Antarctica is a

126   *The biome scale*

specially protected area (see p. 317) but the Arctic has a long history of human habitation by hunter-fishers, the Inuit or Eskimo in North America, and similar groups in Eurasia, and a shorter but much more manipulative occupance by Western cultures. The hunter-fisher cultures were always sparse in number and lived almost exclusively off animals such as seals, caribou and fish, supplemented seasonally with berries, birds' eggs and chicks and small whales. The social structure seems to have prevented population growth beyond these resources so that although starvation was by no means unknown, it was a stable adaptation to the resources of the environment. But the Arctic is now exploited for oil, gas, minerals, tourism and for military purposes and the impacts on the tundra can be profound, not only by industrially-based users, but modernized natives as well (Ives, 1974a, b; Osburn 1974a, b; Müller-Wille 1974). Road construction, fire, and oil spills all increase the depth of thaw and the amount of surface subsidence (Haag and Bliss 1974). (The Trans-Alaska pipeline is in places raised off the ground (Plate 20) to prevent the oil warming the ground surface.) Such effects in general become less severe in the High Arctic because there is so little vegetation that its damage exerts less influence on the soil (Babb and Bliss 1974). Smaller impacts include more dust from gravel working, incineration and transportation, and the effect of taking year-round water supplies from lakes below the ice layer. This has the effect of reducing the amount

Plate 20 The Trans-Alaska pipeline raised off the ground so as not to melt the permafrost. A caribou (*Rangifer tarandus*) is in the foreground.

of unfrozen water for fish and other aquatic animals; both primary and secondary productivity may be affected. Increased human presence has affected the population of land animals: it is said that seismic testing depletes a large area of Arctic foxes, and accidental fire on the dry lichens of exposed ridges leaves them unvegetated for decades and robs herbivores of a source of winter food. Many of man's influences have come together in their effect on the barren-ground caribou of Canada, which declined from several million before European contact to 672 000 in 1949 and 200 000 in 1958. Excessive hunting was one cause, especially in the years 1949–60 when there were very low calf crops. It is difficult to enforce catch limits because of the terrain and because many of the hunters are Treaty Indians who cannot legally be obliged to observe hunting regulations. Also, the exploitation of the Canadian North has caused many forest fires in the lichen-rich forests where the caribou winter and so their food supply has been diminished (Kelsall 1968). There is an analogous conflict between wildlife protectionists who wish to stop all hunting of the single-tusked narwhal (*Monodon monoceros*) and those preservers of Inuit culture, where its hunting is a traditional activity. Wastes are particularly slow to decay and some attract scavengers including the polar bear. A large oil spill into Arctic waters might perhaps never decay entirely. The tundra, too, is the recipient and concentrator of radioactive fallout from atmospheric weapons testing (p. 248). In the case of Amchitka, an island in the Aleutians, underground nuclear testing was carried out under the tundra. The monitoring report (Merritt and Fuller 1977) concludes, perhaps surprisingly, that the greatest damage in this case was the deterioration of aesthetic quality on the island caused by all the peripheral and preparatory operations although it has been argued that the general reputation of the Arctic as a 'fragile ecosystem' rests rather insecurely on the undoubted damage done to permafrost by vehicles and engineering operations. Even the distant islands of the sub-Antarctic have not escaped human influence. Dogs, cats and rats were introduced to Macquarie Island (54°S), and rabbits have drastically modified much of the island's vegetation. South Georgia (54°S) has rats and reindeer, and foodstuffs imported to whaling stations resulted in the establishment of aggresive alien plants. Even Signy Island at 60° south has experienced sporadic human presence, but the main contemporary threat is from tourism and agreement between the Antarctic Treaty managers and tour operators has excluded the island from itineraries. (Jenkin 1975; Lewis Smith and Walton 1975; Collins *et al.* 1975).

Beyond the tundra are permanent snow and ice, virtually lifeless deserts. For most discussions, life ends at places like northern Ellesmere Island, where the sun is above the horizon for only 143 days each year, where the air temperature is not above freezing until June, and a cold season which may bring temperatures of $-55°C$ is under way by September. It is a matter for wonder that wild life can penetrate so far into such extreme environments.

Mountains

A high mountain or mountain range produces progressively severe conditions for life with increasing altitude. In some ways, like the lowering of temperatures, the gradients are analogous to poleward latitudinal changes, but in others, such as photoperiod, the two are not directly comparable. The zonation of montane ecosystems depends largely on latitude and altitude and a high mountain in the tropics might have four distinct belts above the tropical forests: (a) A warm temperate zone with cool nights carrying evergreen forest. (b) A cool temperate zone with a lot of cloud and high humidity. Trees are deciduous and the tree-line comes at the top of this zone with a stunted woodland. (c) An alpine zone of grasses, sedges, shrubs and herbaceous plants. In Africa some normally herbaceous genera have tree-like forms in this zone, for example, *Senecio* (cf. the groundsel weed, *S. vulgaris*) and *Lobelia* (cf. the common ornamental flowers). At the top of this zone, mosses and lichens become most common. (d) An arctic belt with lichens and a permanent snowline at 4570 m. Superimposed on such belts may be north-south differences with south-facing slopes likely to be arid in character, and if the mountains are set in an arid region, xeric vegetation such as tussocky grassland may be found anywhere above the treeline.

The biota of mountains show the kind of adaptations that are encountered in latitudinal belts, with the added genetic effect that isolated populations are more common since mountain ranges are rarely continuous. By contrast, some organisms range over a wider set of habitats since the belts are more compressed. The spectrum of animal species shows some particular features: beyond the tree-line most mammals are herbivorous: the only exclusively high-level predator is the snow leopard (*Panthera uncia*) of central Asia. The herbivores include a variety of wild sheep, goats, antelopes, llamas and alpacas, with the yak (*Bos grunniens*) being the only bovid. Small rodents are found as high as 3000–5000 m, e.g. the European marmot (*Arctomys marmota*) and the Andean chinchillas (e.g. *Chinchilla laniger*). Ground-living birds (e.g. ptarmigans) are commonest, with insect eaters invading the higher zones only in summer; nevertheless the Andes possess a hummingbird which lives at 3000–4900 m but which has acknowledged its idiosyncrasy by eating insects rather than nectar. Scavenging and partially predatory birds such as the golden eagle and condor are commoner than true predators. Most of the animals have a higher density of haemoglobin in their blood to enable them to survive in the thin atmosphere of high altitudes, and melanism is common among invertebrates, presumably to help them absorb solar radiation: there are even black earthworms. Insects tend to be smaller than their low-altitude relatives and to have reduced wing areas. For some of them, freezing point is their optimal temperature and they are unable to withstand warmer conditions. For all biota there is a seasonal rhythm which bursts into life during the summer and which in this respect resembles the pulse of Arctic and Antarctic lands.

The impact of man on high mountains above the tree-line was for centuries that of man the transhumant grazier, who has now been replaced in many regions by man the recreationist, if we except the odd solar observatory, radar station and the like. Even so, it is possible to damage the vegetation, disturb wild animals and cause soil erosion by trampling and by the grazing of pack stock: plants have little power of recuperation given their short growing season (Willard and Marr 1970, 1971; UNESCO 1973a; Webber 1978). Winter sports may easily compact the snow to the point when it provides little insulation for the plants and animals beneath: this effect can be observed on mountains as small (on a world scale) as the Cairngorm Mountains of Scotland. As in high

Islands

The major features of the flora and fauna of islands have already been mentioned (p. 50). To recapitulate briefly, they are a low species diversity compared with the continental masses, on a gradient which is progressive away from the continents, especially down island chains; adaptive radiation, where a restricted taxonomic group evolves to fill a large number niches, as with the finches of the Galapagos, or the Lobelias of the Hawaiian islands; constant immigration and extinction as the biota proceeds towards an equilibrium; a tendency for early stages of succession to be unusually common compared with the mainland because of the incidence of catastrophes which may affect all or much of the island, like volcanic eruptions, hurricanes and, to a lesser extent, *tsunami*.

Island ecosystem will of course vary with the usual environmental factors, and the

Plate 21 The unusual ecology and biota of the Galapagos have made them into one of the most famous examples of island biogeography in the world.

relief and substrate are major determinants for the term island encompasses such variation as exists between islands like Hawaii with the active volcano of Mauna Loa at 4170 m, and low atolls never more than 15 m above sea level. In between, there are many geologically more 'normal' islands with a great variety of terrain. In all of them, the coastal ecosystems (which in the case of small coralline islands may be their entirety) are intimately linked with the sea, which may provide nutrients, for example, via bird guano.

The island ecosystems are not necessarily unique in terms of their structure. For example, the island of Hawaii (3200 km from the nearest continent and 720 km from the next island group) possesses natural formations of evergreen rain forest, evergreen seasonal forest, savanna, grassland, scrub, alpine tundra and near-desert. These are arrayed in a general altitudinal sequence similar to continental tropical mountains, and the plants often exhibit similar forms and in some cases similar species. The moss *Rhacomitrium lanuginosum* occurs in the alpine tundra on Hawaii as it does on the Cairngorm Mountains of Scotland, for example, and the tree-like *Senecio* of tropical mountains (p. 120) is paralleled in the same zone by the silverswords (*Argyroxiphium* spp). We may also note that Vogl (1969) considers that fire is a normal component of many Hawaiian ecosystems composed of native species.

As Mueller-Dombois (1975) points out, the island communities are likely to be different at the level of fine structure: the species assemblages will differ markedly from their continental analogues, and there may be different spatial relationships along environmental gradients, both being mostly due to the poorer diversity of fauna and flora available to comprise the ecosystems. For example on the island of Hawaii there is only one woody vine in the evergreen forests, together with a species of shrub which sometimes will grow as a vine but only up to 5–8 m. Similarly the pioneer stage of colonization of new volcanic surfaces in the rain-forest zone includes no native grasses, their place being apparently taken by ferns.

In this context, the impact of man has often been severe since he has brought in flora and fauna which may be able to occupy unfilled niches in the native ecosystems or be able to displace the native biota. The importation of pigs, goats, sheep and rats to oceanic islands was probably the first large-scale man-made influence on their ecosystems, to be followed by intensive agriculture and then development for urban-industrial and military purposes. The chronology of these changes, and the relative impacts of aboriginal inhabitants and European-culture incomers, varies greatly from island group to island group (Murdock 1965; Rappaport 1965).

Some native species have been easily displaced because they lacked any behavioural mechanisms for adaptation to the invaders: flightless land birds which evolved in the absence of predation are easy game for rats. If the rats make the rails extinct, they can easily turn to other foods since they are not linked indissolubly in the way the northern wolf is to deer. On the other hand, some island species can compete well against exotics unless the habitat is disturbed directly (e.g. where the soil surface is churned up by pigs). The grass *Holcus lanatus* (known in Britain as Yorkshire Fog) will outcompete native grasses in Hawaii only in the presence of pigs, for instance; and in the same place, vegetation dominated by an exotic pan-tropic grass *Eragrostis tenella* and grazed by goats, was 50 per cent covered by an endemic vine (*Canavalia kauensis*) three years after the exclusion of the goats. Many other effects of human presence, increasing in influence on the biota with the import of more technology, could be described, as has been done by Fosberg (1963), and UNESCO (1973b). What seems important is that a way of achieving a population–resource–environment balance should be found which does not diminish the already impoverished island biotas and in so doing reduce their interest and fascination for scientist and seeker of far-away places, and their viability as a habitat for their everyday populations.

The Seas

As an environment for life, the world's salt-water bodies, which comprise 70 per cent of its surface, are very different from the land surfaces which they surround. The ocean basins are deeper than the land is high and life extends to all depths although the biota are much more concentrated at the interfaces with the land. The oceans are all connected, forming one continuous water body so that temperature, depth and salinity form the main barriers to the movement of marine organisms. Another difference is that all terrestrial organisms have to cope with the strain of gravity whereas biota suspended in sea water are freed from the requirements of energy and nutrient investment in large bones and woody tissues.

The geological frame of the seas consists of an offshore continental shelf which drops gradually to about 200 m and which on a world scale comprises about 8 per cent of the ocean area. At its outer edge the continental slope drops off at an average gradient of 1 in 15 down to a depth of 3000–6000 m, where the deep-sea floor is reached. This floor may be more or less flat but is furrowed in places by trenches where depths of over 7000 m are reached: in the Marianas trench 11 000 m is reached. Conversely, undersea mountains and ridges, sometimes volcanic, also interrupt the basal plains. Like the atmosphere, the oceans are in a state of continuous movement, in which the major elements are the great currents set in motion by constant strong winds together with the forces resulting from the rotation of the earth. These run clockwise in the northern hemisphere and anticlockwise in the southern hemisphere and are so effective in preventing stagnation that oxygen depletion is rare even in the deeper parts of the ocean basins. The major currents have superimposed upon them a pattern of upwelling zones along the west coasts of continents and around Antarctica. Here the winds move the surface water away from the coast and it is replaced by cool nutrient-rich water from depth. High primary productivity results from this combination of characteristics and so the upwelling zones are very important in terms of the diversity and productivity of their organisms.

In terms of mineral nutrients, the ocean is a solution of virtually all the naturally occurring elements but most are in very low concentrations. However, the proportions of the major constituents (Table 3.1) are nearly constant throughout and in the open oceans the salinity is within the range of 32–38‰ (parts per thousand). In the Baltic, fresh-water inflow produces 5‰ in the Gulf of Bothnia and 29‰ in the Kattegat; at the other extreme rapid evaporation brings about salinities exceeding 40‰ in the Red Sea. In most parts of the oceans the concentration of nutrients is so low as to form

TABLE 3.1   The major constituents of sea water

Ion	g/kg of water of salinity 35‰
Chloride	19·353
Sodium	10·762
Sulphate	2·709
Magnesium	1·293
Calcium	0·411
Potassium	0·399
Bicarbonate	0·142
Bromide	0·0673
Strontium	0·0079
Boron	0·00445
Fluoride	0·00128

a limiting factor in the size of marine populations, and this is especially true of nitrogen and phosphorus. Not only are these nutrients relatively scarce (e.g. the average concentration of nitrogen is 0·5 mg/l, cf. chlorine at 19 000 mg/l), but they vary in concentration with depth and from place to place and season to season. The light necessary for photosynthesis is also limiting but in a different way: the ocean absorbs nearly all the light incident upon it (the energy input from solar radiation is about 1×10^4 times that of wind or tides) but it is then scattered and absorbed by the pure water and by both suspended and dissolved matter, in a manner that is strongly dependent upon wavelength. The result is that photosynthesis is restricted to a zone less than 200 m deep, called the **euphotic** zone. There is then a 100 m **dysphotic** zone when any remaining light is of the wrong wavelength for photosynthesis, and beyond that depth there is the **aphotic** zone of permanent darkness where animals often have luminous organs for luring or spotting prey. Other physical factors have an influence on marine life, as well. Temperature, for example, is variable from $-2°C$ on the sea floor near the Antarctic continent to almost 30°C in the surface waters of equatorial regions. In low and mid latitudes there is a temperature-depth profile which typically consists of (a) a mixed layer up to 200 m depth where the temperature is uniform and similar to the surface measurement; (b) a permanent *thermocline* down to a depth of 500–1000 m, which forms a transition from the mixed layer to the (c) deep layer of cold water below 1000 m. There may also be seasonal thermoclines at 20–50 m during the summer in middle latitudes under conditions of surface heating from the sun and light winds. Recent research has complicated this picture by the discovery that there is a temperature microstructure in the oceans in the form of layers of uniform temperature in the order of centimetres to metres thick separated by very sharp gradient layers of 1 cm or less thickness. Such a microstructure may play an important role in the diffusion of substances and the dissipation of energy in the seas. Pressure, too, is variable with depth, increasing by one atmosphere for every 10 m: animals living at considerable depths have to be adapted to withstand very high pressures.

The picture of the oceans as being in a state of constant movement is completed by the realization that the sea floor is moving due to tectonic processes, spreading out slowly from the mid-ocean ridges: even the very floor of the ocean basins is changing.

The biota of the oceans can conveniently be classified into functional groups according to their way of life. In this system, there is first the plankton: small floating organisms more or less dependent upon currents for their movement. This group comprises photosynthetic **phytoplankton** together with **zooplankton** of both herbivore and carnivore trophic levels; some zooplankton can swim actively to maintain their vertical position but they cannot move against appreciable currents. The biota also comprises the neuston, which are organisms resting or swimming on the surface; the nekton which are able to navigate at will, such as fish; and the benthos which are organisms attached to or resting on the sea floor or living in the bottom sediments. The term pelagic is also used to include the plankton, neuston and nekton of open waters.

The primary producers of the oceans are predominantly phytoplankton, although large algal seaweeds may have a very high NPP especially in middle latitudes, down to a depth of 60 m. The phytoplankton are often divided into net plankton, which can be caught in the finest silk net of aperture size 0·06 mm, and nanoplankton which pass through the smallest possible mesh. The nanoplankton, especially tiny green flagellates (with swimming organs like tails) from 2–25 microns in size, seem to be the most important photosynthetic organisms of the oceans, and as such form the basis for the food chains. The group also includes dinoflagellates, coccosphaeres, fungi and bacteria. The nanoplankton are also found in the aphotic zone at depths over 1000 m but here they are presumed to have heterotrophic nutrition, using dissolved organic matter formed in the euphotic zone. They are probably links between primary production and the consumer organisms of the benthos. The first consumer level is that of zooplankton, which feed either directly on phytoplankton or on detritus derived from them. There are then the carnivorous zooplankton, and thereafter the many trophic levels of the long food chains of the seas including fish, comprise carnivorous animals; there are few large animals which are strictly herbivorous. The Crustacea comprise about 70 per cent of the zooplankton, if the temporary contribution to the zooplankton of the larval phases of many sea animals is excluded, and include the important group of the Copepods, about 2000 species of mostly herbivorous zooplankton. Their role gains significance from the estimate that at least half the NPP of the oceans is converted for a time into wax and that the copepods are the main producers of wax in the oceanic ecosystem (Benson and Lee 1975). The sea also contains sessile animals, many of which have a plant-like appearance—for example, sea-anenomes. These also provide shelter for smaller animals as well as food for carnivores during their larval stages, which are usually pelagic.

In terms of biological productivity by phytoplankton (whose importance makes them an index for the whole system), the continental shelves and upwelling zones are by far the most important (see Tables 2.4 and 3.2 and Fig. 3.11). The producer trophic level is dominated by the phytoplanktonic diatoms, dinoflagellates and microflagellates, these last being mostly in the size range 1–20 microns. Diatoms seem to dominate in northern waters and dinoflagellates in warmer climes. Equally, there may be a seasonal succession of diatoms followed by dinoflagellates in temperate waters as they warm up. The productivity of the near-shore or neritic zone is also enhanced by the large green, brown and red algae (the familiar seaweeds of rocky coasts) mentioned above. The consumers largely comprise zooplankton of the copepod and crustacean groups, together with some Protozoa, tiny jellyfish, free-floating polychaete worms and the arrow-worms (Chaetognathae). To these are added a plankton fauna which consists of the temporary larval

134 *The biome scale*

TABLE 3.2   Global planktonic primary production

Type of water	Mean NPP (mgC/m²/day)	World area (×10³km²)	(%)	Annual production of world (10⁹ tC/yr)
Open oceans of the sub-tropical zone	70	148	40·4	3·79
Open oceans between sub-tropical and sub-polar Zones	140	83	22·6	4·22
Equatorial currents and sub-polar oceans	200	86	23·6	6·31
Inshore waters	340	39	10·6	4·80
Continental shelves	1000	11	2·9	3·90
AVERAGE/TOTAL	350	367	100	23×10⁹ tC/yr

NB: Column 2 is rounded off to the nearest thousand. To convert gC to dry matter, multiply ×2·2. World total then becomes 50·6×10⁹ t/yr of dry matter. This calculation is on a slightly different basis from Table 2.4 which gives a total of 55 t/yr specifically including estuaries.
Source: Modified from Bunt, 1975.

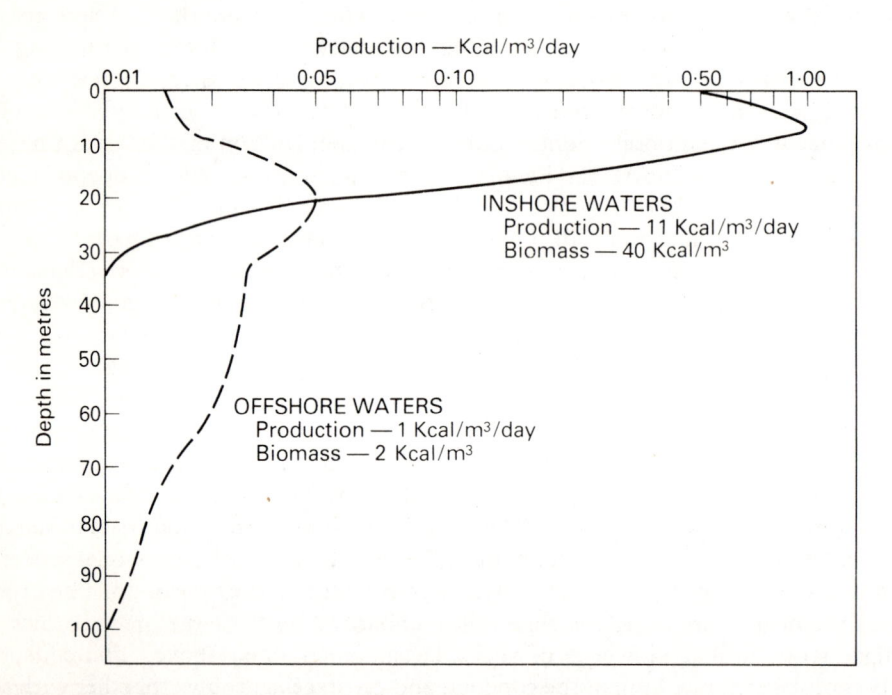

Fig. 3.11 Primary production and biomass in both inshore waters and the open seas. The off-shore productivity is much lower but takes place over a deeper section of the water column.
Source: Odum, 1971.

stages of animals which will later become benthic or free-swimming organisms, the meroplankton.

The larger consumers of the benthos may live on the bottom without being attached (as with some flatfish and lobsters), or may be sessile (sea-anemones, bivalve molluscs), or may burrow into the substrate if it is sand or mud (burrowing anemones, bivalve molluscs, gastropods, echinoderms and Crustacea). Bacteria are also found in large numbers in the surface sediments of the sea floor where they are thought to play a similar role in nutrient release to that in the litter layer of terrestrial ecosystems. The nature of the substrate exerts a strong effect on the taxonomic composition of the benthos of continental shelves, but on each type of substrate the secondary productivity declines with depth although the species diversity often increases even down to the abyssal zone.

The active swimmers of the shelf zones (nekton and neuston) comprise fish, the larger Crustacea, turtles, marine birds and mammals. Although individuals may range over a wide area, they are still limited in their distribution by the barriers of temperature, salinity and nutrients that affect the other components of their ecosystems. Even if not directly affected, they are tied down to their food sources. Because of the small size of plankton, fish which eat them are important links in the food chains over the continental shelves: the herring family (herring, menhaden, sardine, pilchard and anchovy), are very important in this, and in the Pacific some sardines are virtually herbivores. As the fish get larger their food sources may change to smaller fish, so that long food chains with tertiary and quaternary carnivores may be found. Most marine fish lay large numbers of eggs which receive no parental care and indeed are added to the food supply of the nekton. These fish also show a tendency to aggregate, or form 'schools', and to make seasonal migrations. This latter habit, along with selective feeding even of zooplankton, probably prevents 'over-grazing' of plankton as well as ensuring adequate nutrition of the fish. Birds are also important high-level predators of these seas; they spend some time on shore and thus return some of the nitrogen and phosphorus to the land from which it came, thus completing the nutrient cycle.

The productivity of offshore seas is highest in the upwelling zones where cold deep water brings up large quantities of available nutrients. Thus the fish productivity of the anchovies which dominate the fauna of the Peru upwelling zone has been estimated at 335 kcal/m^2/yr calorific content, as compared with 0·3 kcal/m^2/yr for the average of the world's marine fisheries, and 5·0 kcal/m^2/yr for the North Sea, an area of continental shelf. It is not surprising that the world's most productive natural fishery is found on the Peruvian upwelling, though in years when the upwelling fails to happen, the drop in catch is disastrous.

The open oceans beyond the continental shelves are populated entirely with pelagic and benthic organisms. The oceanic phytoplankton are predominantly very small and the zooplankton are the permanent kind (holoplankton), without the addition of larvae of other groups. Life in open waters is aided by flotation mechanisms such as spines, fat droplets, gelatinous capsules and air bladders, and the general paucity of nutrient is fought with recycling mechanisms which keep nitrogen and phosphorus, for example, closely bound into organisms rather than loose in the open water. Even though plankton productivity is low, they eventually support a characteristic fauna of oceanic birds such

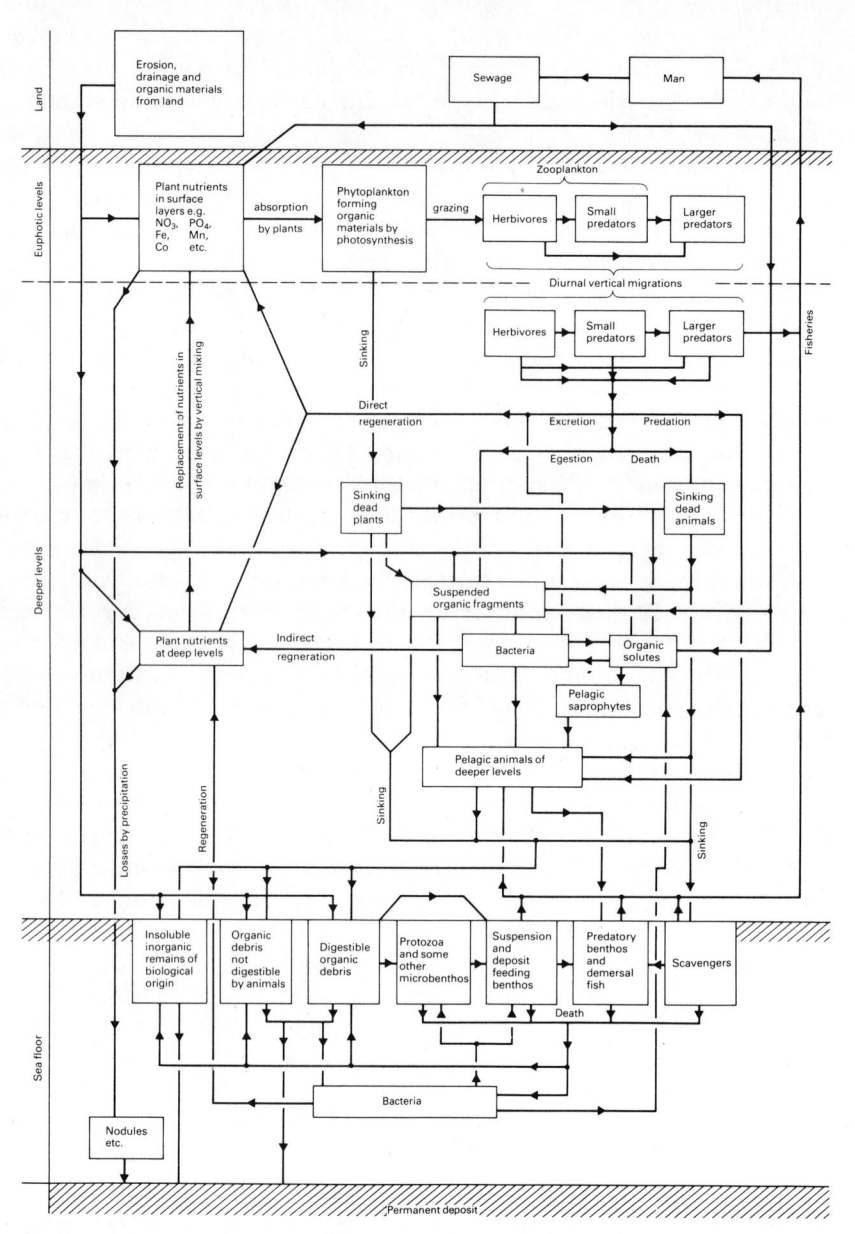

Fig. 3.12 A scheme for a marine food web arranged vertically. This diagram assumes that permanent long food chains are normally found rather than the shorter but shifting chains now hypothesized by some authors. Source: Tait, 1968.

as petrels, albatrosses, frigate birds and terns which only come to land in order to breed; most of these are bound to a particular type of surface water even though their movements appear to be unconstrained. Completely independent of the land as well are sea mammals like dolphin and whales. Zooplankton are the main food of the baleen whales and nekton animals such as squid are that of the toothed whales.

With increasing depth, the density of life gets lower, but is not absent in spite of the dark, the cold and the high pressure. There is a rain of detritus from the euphotic zone, although it sinks so slowly that much of it is decayed or dissolved before it reaches the abyssal depths; there is also transport of organic matter in the form of saprophytic plankton which are abundant between the euphotic zone and the bottom; and there is the formation of organic aggregates at depth, from dissolved organic matter. As the amount of dissolved organic matter at depth exceeds by a factor of 10 the quantity of particulate matter, the potential for conversion into ingestible food is quite large. The adaptations to life at depth are not all known or understood but among the obvious are the luminous organs, the enlarged eyes and the huge mouths of deep-sea fish like lantern fish, angler fish, hatchet fish and devil fish, all of which are less than 13 cm long. The benthic Crustacea and molluscs seem to be blessed with long appendages and abundant spines and stalks.

If we are to summarize the ecology of life in the sea, the conclusions of Wyatt (1976) are useful. He suggests that food chains are apparently complex (as in Fig. 3.12) but in reality are short and simple. But they are short-lived since as animals grow their prey changes and food chains continually form and disappear (Fig. 3.13). The summation of these simple units over longer periods of time gives rise to an apparent complexity of food webs. Thus the trophic level concept is of limited usefulness in terms of an individual species. Food chains may, however, have five or six stages, and both detritus and bacteria are important food sources in shallow marine environments, especially where they damp down fluctuations in seasonal food supply resulting from the pulsed

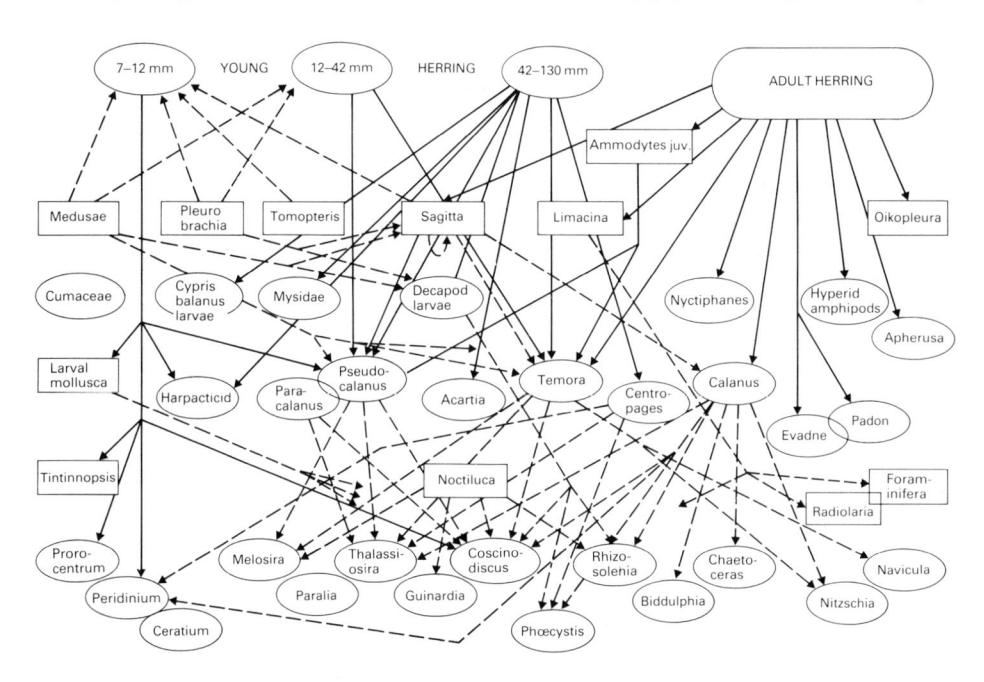

Fig. 3.13 Food webs for the North Sea herring (*Clupea harengus*) at various stages of its history. Source: Wyatt, 1976.

growth of phytoplankton. The efficiency of energy transfer between different stages in food chains seems to be very variable (Steel 1970).

The dynamics of marine population are, compared with many terrestrial ecosystems, so little known that the results of human interference are often hard to elucidate. The effect on the seas of eutrophication from sewage and fertilizer runoff, for instance, is complex and affects both species composition and productivity in a number of ways; analogous is the fate and effects of long-lived substances like chlorinated hydrocarbons such as DDT, and polychlorinated biphenyls (PCBs), although the toxic effect of these upon carnivores such as sea-birds is now well documented. Excessive growth of sessile green algae has underlain the depletions of the oxygen content of the bottom of shallow seas such as the Baltic and so a permanent anaerobic zone whose bacteria produce large quantities of hydrogen sulphide is established. The H_2S causes the release of phosphorus which adds to the growth of phytoplankton which increases the already heavy input of organic matter to the deeper waters, in a kind of vicious spiral (Jansson and Wulff 1977). Equally important, though not immediately quantifiable, are the effects of reclaiming estuaries, salt marshes and other intertidal communities which house the larval stages of nekton and benthic organisms. Best known of all is the ability of modern man to reduce populations of whales and fish to very low levels (see p. 300). Nobody now thinks that the oceans are so vast and so teeming with life that human effects are confined to a narrow zone round the shore, and that they are an inexhaustible supply of food or a bottomless sink for wastes. But putting such realizations into action is proving a slow process.

As we have seen, the marine ecosystems of the globe become more productive as the land masses are approached and it is not surprising that the most productive of all the salt-water systems are found at the land–water interface. None of them is very large in total area but extensive study in recent years has shown that ecosystems such as tidal marshes (found in all parts of the world except the very coldest), estuaries of all regions, and tropical formations like mangrove swamp and coral reefs are very high in biological productivity. Coral reefs, for example, average an NPP of 2500 g/m²/yr, which compares with the tropical rain forest at 2200 g/m²/yr; estuaries average at 1500 g/m²/yr, a figure close to that of the seasonal forests of the tropics. The following short accounts will show that all these very productive systems have natural subsidies of either energy or matter (or both): rivers flowing into estuaries, or the tides, or marine currents bring in organic matter or nutrients and remove wastes, so that the energy of the ecosystem can go into NPP rather than be 'spent' on maintenance and cycling processes. Other than this set of metabolic processes, the systems have some other features in common, such as adaptations of organisms to varying lengths of dessication as the tides rise and fall, and zonations resulting from various degrees of exposure of a coast to storms, but physiognomically they look very different.

Coral reefs are the most exotic of these ecosystems, especially to the inhabitants of cool temperate lands since they are not found where the water temperature is less than 21 °C (Plate 22). At first sight it appears to be a heterotrophic system because its structure is dominated by the calcareous exoskeletons of coral polyps, which are carnivorous animals, living mostly off zooplankton. Closer analysis shows that typically this biomass

Plate 22 A fringing reef of coral and algae, part of which is uncovered. The energy of the water brings in nutrients for the plants and zooplankton for the corals, and removes wastes. Hence such ecosystems are amongst the most productive in the world.

is exceeded by three times its weight of algae living in and around the corals. These algae not only provide the autotrophic fixation of energy but, some being calcareous, add to the structure of the reef as well, along with consumer organisms such as molluscs and Foraminifera. The proportion of algae and corals will vary within a reef according to the exposure to wave battering, but it is perhaps more accurate to refer to the whole ecosystem as a coral-algal reef. As every colour-TV viewer will know, the reef is the habitat for a considerable biomass of other animals, mostly belonging to the Foraminifera, Mollusca, sponges, worms, sea urchins and fish.

The high productivity is made possible firstly by the flowing of water round the reef which removes wastes and brings in plankton, leading to the second outstanding feature of its metabolism. This is its success in holding nutrients (especially phosphorus) tightly within the ecosystem and recycling them rapidly between algae and corals so that their growth is unimpaired by the general scarcity of phosphorus in the surroundings. The proximate source of phosphorus input appears to be the zooplankton, so that one role of the corals is of a nutrient trap for the algae. No organic matter is permanently locked up in the structure of the reef and so nearly all the phosphorus (and other mineral nutrients) are kept rapidly cycling within the reef ecosystem. If there are losses to mobile animals then they are presumably balanced by the gains from zooplankton brought in by water currents.

In spite of its rocky appearance, the coral reef is a relatively fragile ecosystem. Apart from direct damage from blasting, construction works or the removal of coral and shells to sell to tourists, certain land-use practices on nearby land masses will affect the reef. Engineering of river channels which flush a reef with large quantities of fresh or brackish water is one example but more common is the deposition of large quantities of silt which choke the reef organisms. Deforestation of a mountainous island will produce such an effect, as will dredging of a lagoonal harbour, and the 'reclamation' of coastal ecosystems which trap silt (of which the mangrove swamp is the most efficient) is usually sufficient to kill the reef. In recent years many coral reefs have been devastated by a population explosion of a predatory starfish, the crown of thorns (*Acanthaster planci*); the reasons for the outbreak are not known, but human activity in some form is strongly suspected.

The mangrove swamp is just one of a number of the ecosystems of the land–sea interface in the tropics. It is an inter-tidal community where organisms can withstand varying amounts of exposure to dessication and can tolerate varying degrees of salinity in the water. In spite of the apparent ecological stress of the fluctuating conditions, a vegetation dominated by trees develops, with such species as *Avicennia nitida*, *Rhizopora candelaria* and *R. mangle*. These form a scrub about 10m high and are zoned according to distance away from open water. Their most obvious characteristic is their tangled mass of strut-like and arching prop roots which raise the base of the tree above the silt and mud of the substrate. Branching from these roots, conical aerating roots grow vertically out of the mud. The evergreen shrubs form a dense community and few herbaceous plants are found; some red algae grow on the roots and a few epiphytic lichens on the trees. The animals are a mixture of land and sea creatures and in general the species diversity is low, though the biomass may be high. Insects are a notable group, often breeding in stagnant pools in rotten tree-stumps; barnacles encrust many roots; a complete zonation of crab species exists; and a bird fauna is found. Mammals are rather scarce although there may be otters and small cats together with crab-eating racoons and monkeys. Several mangrove fish ('mudskippers') of the lower zones have modified fins which enable them to 'walk' on the exposed mud and even to climb trees.

The primary productivity does not compare with terrestial tropical forests but the leaf-fall forms the basis of a set of food chains populated by a visible fauna with a high biomass. The fallen leaves are food for a host of crabs, shrimps, insect larvae, nematodes and molluscs which in turn are eaten by small carnivores that nourish many species of offshore fish and wildlife. The mangrove is therefore an exporter of organic matter into the sea where it is relatively scarce and so is a nursery for commercial fish. Its root system is an efficient silt trap and probably a mangrove or similar community on shore is the only way a coral reef can flourish offshore except in the context of an atoll or other small coralline island. These highly useful features have not prevented wholesale 'reclamation' of mangroves and other tropical coastal swamp ecosystems for agriculture, housing, airports, industrial development and similar features. The mangrove does not perhaps help itself for it is aesthetically unattractive: consider N. Polunin's summation in his *Introduction to Plant Geography* (1960):

> Even the trees are likely to be mis-shapen and lowly, while bubbles of stinking gas rise from the rotten mire, and a teeming population of crawling creatures adds to the atmosphere of gloomy squalor.

Back in the fresh air, the last of these productive inter-tidal environments is the estuary with its set of mud flats and salt marshes. With tidal effects, strong currents, high turbidity, variable salinity, the estuary imposes a high degree of stress upon its organisms: a wide tolerance of variability in salinity is necessary, for instance. Organisms tend to be endemic to the ecosystems, with a few from the open sea and even fewer from the fresh water upstream of the estuary. Plants are salt-tolerant (halophytic), grass or rush-like genera like *Spartina*, *Salicornia* and *Scirpus*, together with green algae like the sea-lettuce *Enteromorpha*, and the mud is covered at low tide with diatoms and blue-green algae. There are few phytoplankton because of the high turbidity.

Yet the estuary has ecologically redeeming features. Dissolved-oxygen levels are high. The intermixture of salt and freshwater acts as a nutrient trap, keeping river-borne nutrients in the estuary for a long time in spite of the river current. In the tropics at least, nutrients come from the sea also and especially from deep waters below the euphotic zone where they have not been depleted by the phytoplankton (Rodriguez 1975). There is a diversity of autotrophic organisms programmed for year-round photosynthesis even in seasonally pulsed climates. And the tide does a lot of work in removing wastes and transporting nutrients and organic matter so that the permanent biota can be sessile and do not expend energy on excretion and food gathering. The outcome is an average NPP of $1500 \text{ g/m}^2\text{/yr}$ (compared with continental shelf figures of $360 \text{ g/m}^2\text{/yr}$ and open ocean of $125 \text{ g/m}^2\text{/yr}$) a quantity similar to tropical seasonal forests. The productivity of the algae and higher plants goes mostly into a decomposer chain in which bacterial breakdown is an important stage, as is the activity of molluscs, worms and other detritus feeders. Bivalve molluscs (e.g. clams, cockles, mussels) in particular exhibit a high productivity: in northern Europe figures of 200 g/m^2 of dry meat biomass of mussels has been reported. A properly managed mussel bed should provide 2000 kg/ha/yr live weight, about 50–100 times the yield from beef cattle on grassland (Nelson-Smith 1977). The visible carnivore fauna, including the birds which find this ecosystem such an important feeding ground, live off these detritivores. Importantly for resource use, two-thirds of the commercial fish species of continental shelves spend their larval years in estuaries, and others must pass through them on their way from ocean to up-river spawning ground. The estuaries are therefore important nurseries of commercial fish. Estuaries are apparently also exporters of vitamin B_{12} for coastal waters as a whole (Stewart 1972).

Again, this ecological value has not prevented the widespread reclamation of apparently barren terrain (of interest only to the bird-watchers) for industry, housing, airports and agriculture. The level nature of the flats and marshes make it cheap to keep out the sea, and the upper parts are often well above high tide and so partly leached of their salt content, which gives them a good potential for agriculture. Estuaries are also prone to water contamination, with toxic chemicals, oil, sewage, heavy silt loads from poor land management, and calefaction. Governments in particular are very slow to realize the real and potentially lasting values of estuarine and tidal marsh ecosystems.

Fresh waters

Surface fresh water occupies an interesting position in the hydrological cycle, being as it were the 'bottle-neck' between oceans and atmosphere, and accounting for only 1·5 per cent of the unfrozen fresh water on the globe. Within this somewhat scattered biome, there are two main types of ecosystem. They are characterized respectively by standing water as in lakes, ponds, swamps and bogs (lentic habitats); and by running water as in springs, streams and rivers (lotic habitats). Sharp boundaries are unusual within each series and both of them are liable to change on relatively short time-scales: small lakes fill up, for example, and rivers change their characteristics as they approach base level.

As in the air, a number of environmental factors may be limiting for life. Temperature is less variable than in the atmosphere but exerts effects through the thermal stratification of lentic bodies as well as upon the rate of photosynthesis. Many aquatic animals have a narrow tolerance of temperature (i.e. are stenothermal) so that even moderate raising of the temperature by human activity (calefaction) can have widespread effects. The penetration of light is often limited by suspended material which restricts the photosynthetic zone. This is to some extent self-limiting as well, since the denser the plants the less light will penetrate beneath them. In large water bodies such as Lake Baikal or the Great Lakes, there is a zone without light, as in the seas. In the air, the concentration of oxygen is more or less uniform but in water its solubility is affected by temperature and by salinity. In nature, the maximum value for oxygen is 6 ml $O_2/1H_2O$, which is 3 per cent of the concentration in air, and the minimum value is nil, usually at the bottom of deep lakes with a heavy 'rain' of organic matter where the bacteria use up all the oxygen and are replaced by other (anerobic) bacteria which do not require it. Carbon dioxide is much more soluble in water and less limiting to plant growth than oxygen to animals, but artificial enrichment may accelerate photosynthesis although it is also likely to harm the animal life. In nearly all freshwater ecosystems, nitrogen and phosphorus seem to be limiting, and in soft water minerals like calcium may also be limiting: the hardest fresh waters have a salinity of 0·5‰ compared with 32·37‰ for sea water.

Spatially, individual freshwater ecosystems are often isolated from each other by the land and by the sea and so organisms without a suitable means of dispersal may fail to colonize otherwise suitable places: fish are a good example. But most of the small organisms have spread very widely so that algae, crustacea, protozoa and bacteria are very widely spread: at the level of family and genus the lower plants and the invertebrates of fresh water are highly cosmopolitan. E. P. Odum (1971) cites the example of the same water flea *Daphnia* being found in both the United States and England, rather like Baskin Robbins's ice-cream, we may suppose. Physiologically, all aquatic animals have to cope with a set of body fluids with a higher concentration of salts than in fresh water: if their membranes are permeable then water enters the body by **osmosis**; if their membranes are impermeable then they must have a salt concentration mechanism. Protozoa have vacuoles and fish have kidneys and both must excrete water otherwise the organisms would swell up and possibly burst. The difficulties of the regulation of osmosis may explain why many groups of marine animals have never invaded fresh waters.

In lentic waters (ponds and lakes) three main life zones are evident. The first of these is the littoral zone, with shallow water where the light penetrates to the bottom. The producers are visually dominated by rooted plants such as bullrushes (*Typha* spp) lilies (*Nymphaea* spp) and pondweeds (*Potamogeton* spp) together with small floating plants such as the duckweeds (*Lemna* spp). They form a series of concentric zones which, accumulating silt and organic matter, advance into the open water; at the end of such a succession under natural conditions, the lake becomes wet woodland. Productivity of open water is enhanced by the presence of a plankton consisting of diatoms, green algae, and blue-green algae, many of which are single-celled forms able to reproduce rapidly ('bloom') if nutrients cease to become limiting.

Consumers in the littoral zone are often vertically rather than horizontally distributed but in any event are denser here than in any other part of this ecosystem. Insects and their larvae inhabit the penthouse, crayfish and mayfly nymphs the ground floor, and freshwater mussels, worms and the larvae of midges are in the basement. Moving in and out of the littoral zone are fish, turtles, snakes and amphibians; the common frogs (*Rana* spp) for instance moving up the trophic levels from herbivorous grazers when tadpoles to predators when adult, as a baby moves up from creamed spinach to its first piece of roast beef.

In large lakes, phytoplankton dominate the production of the extensive open-water zone. The algae of the littoral zone are found as well as dinoflagellates, all of them having processes or bladders or other aids to floating so that they stay near the surface where the light intensity is greatest. It is these groups which so often colour lakes green. The zooplankton is composed of relatively few species but a large number of individuals, especially copepods and cladocerans. Most of them are permanently planktonic and do not have 'temporary' planktonic larval forms of other animal groups, as in the sea. The fish are the dominant larger animals, along with fish-eating birds. A profundal zone without light may exist where the food supply comes ultimately from the littoral zones or the upper layers of the open water. The bottom fauna includes consumers like blood worms and annelid worms, small bivalve molluscs and in some places the larval forms of midges which move up and down diurnally (Fig. 3.14).

In most temperate-zone lakes, there is a thermal stratification. In summer the top layer is warmer and gets mixed to a depth of about 5m; it is called the epilimnion. There is then a steep thermocline to a colder water mass below. If the thermocline is below the level of effective light penetration then the lower water mass (hypolimnion) is depleted of oxygen and 'summer stagnation' occurs. When the temperature falls the entire lake begins to circulate and oxygen is restored to the lower layers; at the same time, nutrients freed from decomposing organic matter by the bacteria of the lake bottom

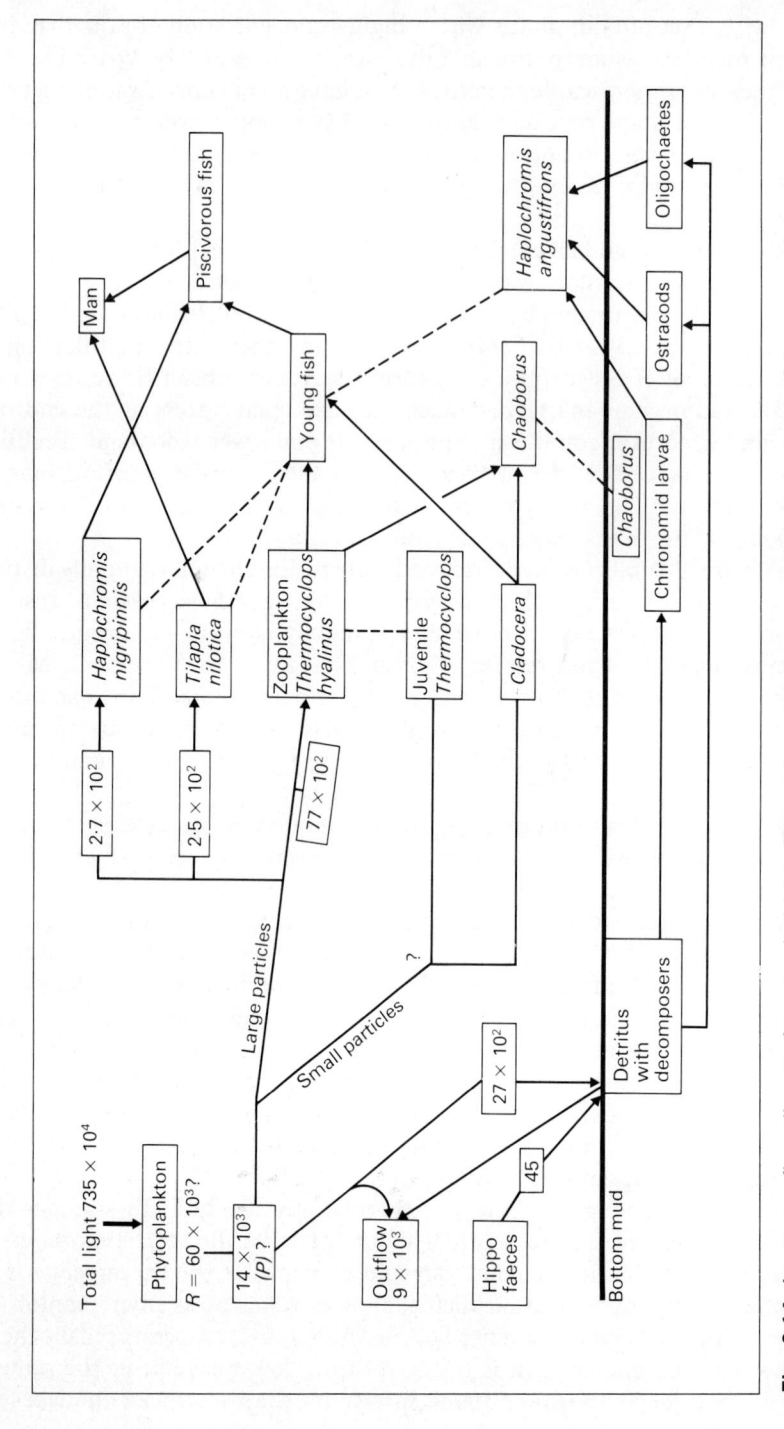

Fig. 3.14 An energy-flow diagram for a tropical lake. Pecked lines link the young and adult stages of the same species. Values in kJ/m²/yr.
Source: Varley, 1974.

are released upwards for use by the phytoplankton. Tropical lakes are weakly stratified and get mixed irregularly but usually during a cool season if there is one.

The ecology of lakes may be summarized according to their nutrient status. At one end of a spectrum are oligotrophic lakes with scarce nutrients, a deep hypolimnion not subject to oxygen depletion and a low density of plankton; at the other is the eutrophic lake with a higher NPP due to dense phytoplankton and marginal vegetation and suffering severe oxygen depletion in the hypolimnion during summer. In nature, succession gradually turns oligotrophic lakes into eutrophics but this process can be accelerated by human activities which result in enhanced levels of nitrogen and phosphorus, as with fertilizer runoff and sewage, for instance. The release of limiting nutrient levels encourages blooms of algae and these include blue-greens which can fix atmospheric nitrogen, so only need the one element to make them bloom. They and others are either resistant to grazing or so abundant that no consumers can eat them all. They may rot on the surface or may sink to the bottom where bacterial decomposition depletes the water of its oxygen and so anerobic bacteria (which often produces H_2S) take over the task. Deep lakes with a cold-tolerant population of fish may lose these if the rain of organic matter causes oxygen depletion at depth. Many lakes now have a permanent anerobic layer at the bottom and are subject to scums and smells; the process is reversible in small lakes but expensive and often politically fraught. Natural eutrophication involves a GPP of $>165 \, g/m^2/yr$, cultural eutrophication of over $770 \, g/m^2/yr$ and it is estimated that about one-third of the lakes in the USA have cultural eutrophication.

In running streams (lotic ecosystems), the strength of the current is a major control on life and one which is subject to great variations within short distances and on vertical gradients as well. A stream can be subdivided into rapid and pool stretches each with characteristic life-forms: in rapids, the consumers have adaptations such as hooks, sticky slime, and flat bodies, or continually move 'upstream' against the current. Oxygen and temperature distribution are much more uniform than in lakes, and springs tend to be at a constant year-round temperature. This can be very hot: bacteria in hot springs can survive and reproduce at $88°C$, fish and insects at $50°C$; blue-green algae tolerant of $80°C$ temperatures are the main producers.

Streams again differ from lakes in being the recipients of a higher proportion of organic matter from outside the immediate system, i.e. from upstream or from the banks. Fig. 3.15 shows some of the flows in the River Thames in England, to the point of including the lost baits of sport fishermen.

In world terms, fresh waters are very productive. Likens (1975) quotes NPP values for lake phytoplankton ('fertile sites') of $100-900 \, g/m^2/yr$, of $400-2000 \, g/m^2/yr$ for submerged higher plants, and of $3000-85\,000 \, g/m^2/yr$ for emergent vegetation. This puts them amongst the highest categories of primary producers in the world (Table 2.4). Within the biome, tropical lakes and tropical rivers can be the most productive (although some of the 'black water' tributaries of the Amazon are noticeably low) at $1000-2500 \, gC/m^2/yr$ (*ca* $2200-5500 \, g/m^2/yr$ dry matter) compared with a maximum of $650-950 \, gC/m^2/yr$ (*ca* $1430-2090 \, g/m^2/yr$ dry matter) for temperate rivers and lakes, and $35 \, gC/m^2/yr$ (*ca* $77 \, g/m^2/yr$ dry matter) for Arctic lakes. So their potential as a natural renewable resource is high and is only likely to be diminished by human action in treating fresh

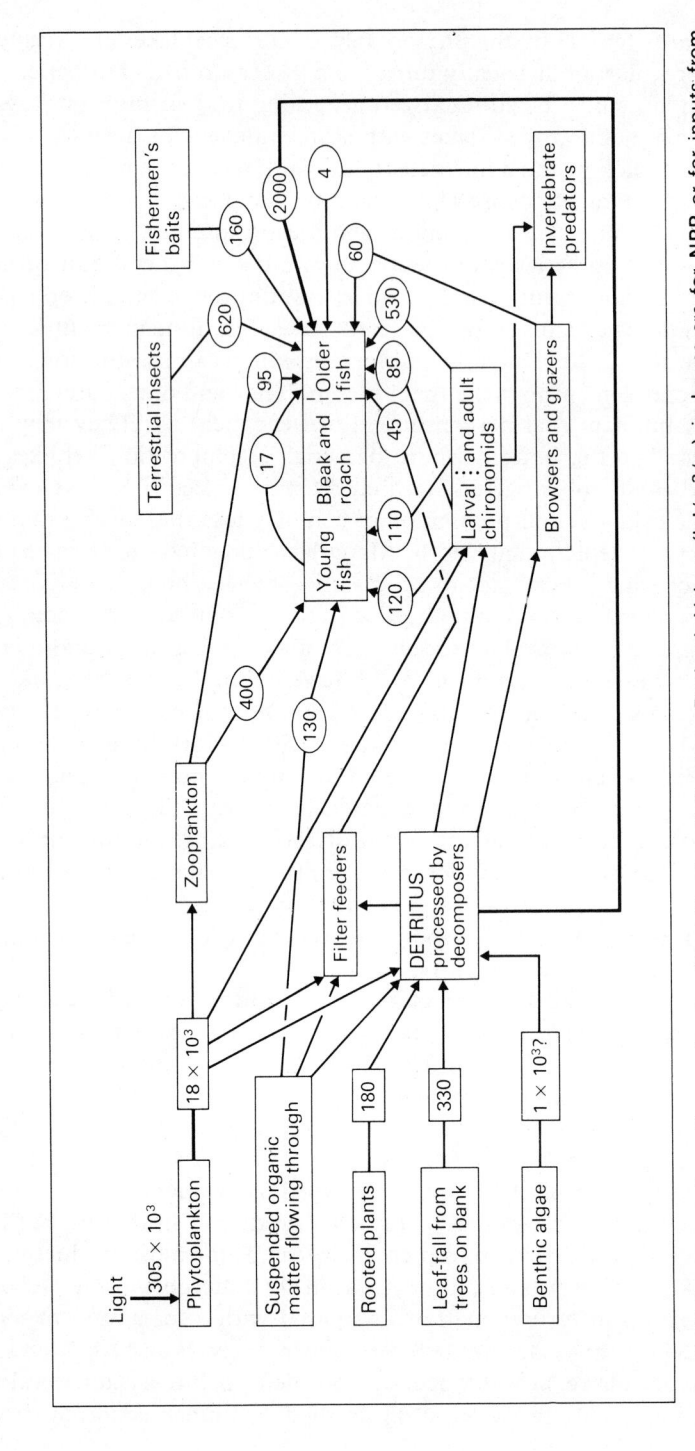

Fig. 3.15 An energy-flow diagram for the River Thames at Reading, England. Values (kJ/m²/yr) in boxes are for NPP or for inputs from outside the rivers system at that point. Values in circles represent the energy transfers to the various trophic levels.
Source: Varley, 1974.

water as a cheap source of water for urban, industrial and agricultural needs, and as a convenient waste pipe for the disposal of effluents of all kinds.

Further Reading

CLOUDSLEY-THOMPSON, J. L. 1975: *Terrestrial environments.* London: Croom Helm.
  1977: *Man and the biology of arid zones.* London: Edward Arnold.
CUSHING, D. H. and WALSH, J. J. (eds.) 1976: *The ecology of the seas.* Oxford: Blackwell.
DI CASTRI, F. and MOONEY, H. A. 1973: *Mediterranean-type ecosystems. Origin and structure.* Ecological Studies 7. Berlin, Heidelberg and New York: Springer-Verlag.
EYRE, S. R. 1968: *Vegetation and soils: a world picture.* 2nd edn. London: Edward Arnold.
FOSBERG, F. R. (ed.) 1963: *Man's role in the island ecosystem.* Honolulu: Bishop Museum Press.
GOLLEY, F. B. and MEDINA, F. (eds.) 1975: *Tropical ecological systems.* Ecological Studies 11. Berlin, Heidelberg and New York: Springer-Verlag.
IVES, J. D. and BARRY, R. G. (eds.) 1974: *Arctic and Alpine environments.* London: Methuen.
MYERS, N. 1972b: *The long African day.* New York: Macmillan.
OVINGTON, J. D., HEITKAMP, D. and LAWRENCE, D. B. 1963: Plant biomass and productivity of prairie, savanna, oakwood and maize-field ecosystems. *Ecology* **48,** 515–24.
REICHLE, D. E. (ed.) 1973: *Analysis of temperate forest ecosystems.* Ecological Studies 1. Berlin, Heidelberg and New York: Springer-Verlag.
RICHARDS, P. W. 1952: *The tropical rain forest.* London: Cambridge University Press.
ROSSWALL, T. and HEAL, O. W. 1974: *Structure and function of tundra ecosystems.* Stockholm: Swedish Natural Science Research Council Ecological Bulletins 20.
SUKACHEV, V. N. and DYLIS, N. V. 1964: *Fundamentals of forest biogeocoenology.* Edinburgh and London: Oliver and Boyd.
TAMM, C. O. (ed.) 1976: *Man and the boreal forest.* Stockholm: Swedish National Science Research Council Ecological Bulletin 21.
WEAVER, J. E. 1954: *North American prairie.* Lincoln, Nebr.: Johnsen.
WHITMORE, T. C. 1975: *Tropical rain forests of the Far East.* Oxford: Oxford University Press.

Part II
Cultural biogeography

The first part of this book tried to create a set of datum lines: the way the world of wild plants and animals functions and would be distributed if left to itself. The rest of the work is concerned with the effects of man upon these patterns: in changing the genetic make-up of plants and animals, of redistributing them about the world and in consequence altering the structure of many ecosystems. Human activities not directly concerned with biota also exert their effects: the gathering of resources, the building of cities and the dispersal of wastes, for example, all change biotic distributions and frequencies. A basically historical approach is adopted: the various cultural stages of human development, pre-agricultural and early agricultural, form the background to a consideration of the effects, both inadvertent and deliberate, of industrialization and the growth of science. The importance of biota as a resource is examined, and lastly the ecology of man in relation to the whole of his environment, plants and animals included, is outlined with an eye to both present and future.

4
Pre-agricultural man's effects on plants and animals

For much of man's tenure of the earth, he has been a hunter of animals of land and water, and a gatherer of plant material. Looked at in perspective, perhaps 90 per cent of his evolutionary span has been spent at this level, considering that agriculture began to emerge about 9000–7000 BC, and that industrialization on a large scale is a product of the nineteenth and twentieth centuries. Even though most of the hunters have gone, a cultural mode which lasted so long and which clearly had great survival value needs some examination in a book dealing with man's relationships to other living organisms. The different scale of this type of connexion can be indicated firstly by Fig. 4.1 which shows the diminution of the proportion of hunters in the human population and their shrinking distribution, and secondly by realizing that during the pre-agricultural period both absolute numbers and densities of people were low compared with the later magnitudes made possible by agriculture and by industrialization. Deevey (1960) suggests that in the lower Palaeolithic of one million years ago, the total human population was 0.125×10^6 and the density was $0.06425/km^2$. On the eve of agriculture, 10 000 years ago, the corresponding figures were 5.32×10^6 and $0.04/km^2$. The rise in absolute numbers represents a spread outwards from an African heartland of evolution and the increase in density reflects in part the biotic richness of some of the post-Pleistocene habitats occupied by Palaeolithic and Mesolithic cultures. So of the $80\,000 \times 10^6$ men who have ever lived on earth, about 90 per cent have lived as hunters, about 6 per cent by agriculture and the remainder in industrial societies. If the present relationships of man and environment prove to be non-viable, then a future archaeologist would classify our planet as one essentially occupied for a long period by hunters, followed by a rapid efflorescence of technology which led either to the extinction of *Homo sapiens* or back to the hunting mode (Lee and DeVore 1968). So the ecological relationships of these people, whether in the past and reconstructed by palaeoecology or in the present or recent times and recorded by ethnology, should not lightly be ignored. The ecological relationships, and in particular the possibility of habitat manipulation by evolving man in Africa during the Pleistocene and early Holocene is of considerable interest also, but the study is at too early and fast-changing a stage to be properly summarized in a book like this.

The ecological impact of pre-agricultural communities

The term 'hunter' conceals a variety of practice but essentially all pre-agricultural folk were food gatherers rather than food producers. They lived off the usufruct of natural

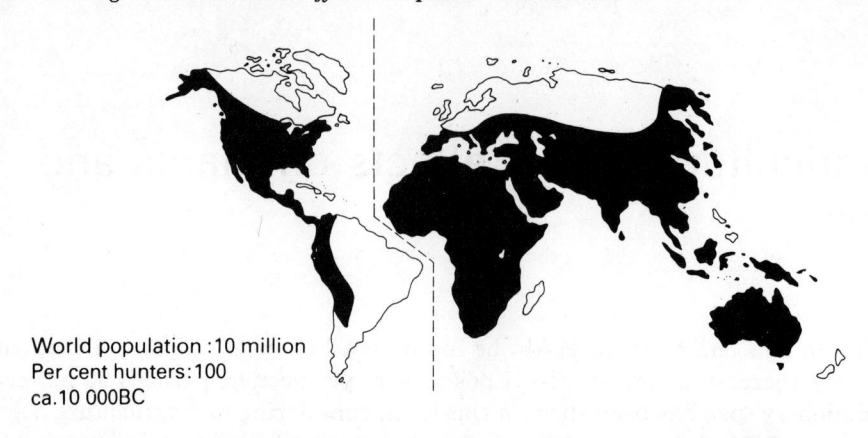

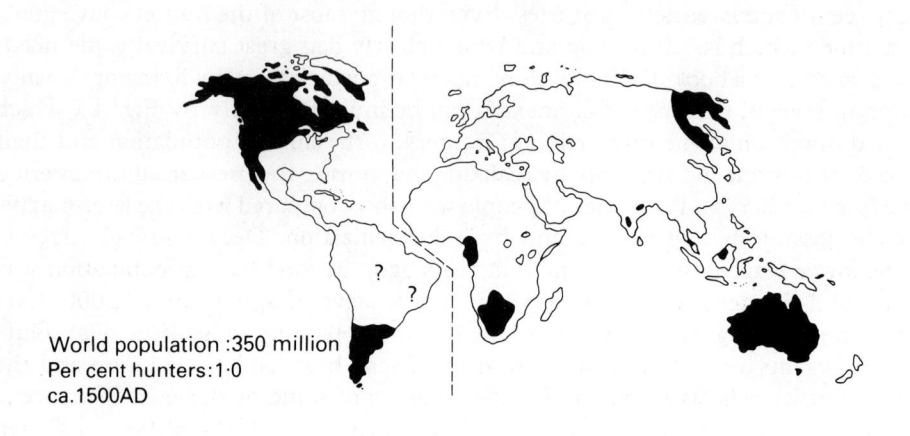

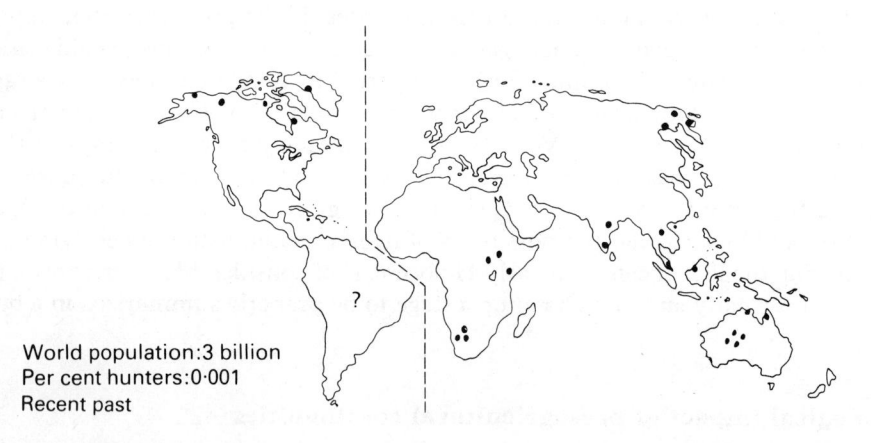

152 *Pre-agricultural man's effects on plants and animals*

World population : 10 million
Per cent hunters : 100
ca. 10 000 BC

World population : 350 million
Per cent hunters : 1·0
ca. 1500 AD

World population : 3 billion
Per cent hunters : 0·001
Recent past

Fig. 4.1 Hunters of the world at various periods showing their former importance.
Source: Lee and De Vore, 1968.

systems, although some evidence exists to show that they also manipulated whole systems in some places rather than merely cropping individual species of plant and animal. The likely impact upon the local ecosystems of groups of men with a simple technology depends upon variables such as the variety of food resources, the density of each and their temporal incidence; on the human side, the effectiveness of hunting and gathering techniques and the conscious conservation of resources are equally important. The mix of plant and animal foods appears to be related in post-glacial times to climate and biome type: no hunting group seems to derive less than 20 per cent of its diet from mammals, but that proportion increases in northerly latitudes. In the tropics, plant material constitutes the bulk of the diet (Table 4.1) but gathering of plant material is important

TABLE 4.1   Primary subsistence source by latitude (for contemporary and recent groups)

Degrees from the Equator	Primary subsistence source		
	Gathering	Hunting	Fishing
>60	—	6	2
50–59	—	1	9
40–49	4	3	5
30–39	9	—	—
20–29	7	—	1
10–19	5	—	1
0–9	4	1	—
WORLD	29	11	18

Source: Lee 1968.

in all but the most northerly cultures and so a closeness to plant resources as well as animal behaviour was doubtless characteristic of most prehistoric food-gathering groups, although it is hard to tell how accurate extrapolation backward in time can be when the base is a set of peoples living in the interstices between modern agricultural and industrial groups. A difficulty in discussing the effect of hunters upon plants is the lack of data, recent or historic, which chronicle the long-term effects of gathering upon the population dynamics of an individual taxon. Some instances show that even where a particular plant forms a staple, its life and regeneration are never at stake from its use as a primary food source. One obvious example is the nut of the Mongongo tree (*Ricinodendron rautanenii*) which accounts for half the weight of the vegetable diet of the !Kung Bushmen of the Kalahari. But many nuts are always left on the ground, and the Bushmen also have available 84 other species of edible food plant and, as Lee (1968) puts it, 'Many species, which are quite edible but less attractive, are by-passed, so that gathering mever exhausts *all* the available plant foods of an area.' Woodburn's (1968) account of the Hadza of Tanzania stresses that gathering activities allow little attention to the conservation of resources but no failure of them is mentioned. Overall, we are left with the impression that plant gathering in both recent and historic times did not greatly affect the population dynamics of selected species, but it must be remembered that an easily extirpated species would have disappeared very early in its exploitation history.

A great deal more work exists on the relations of hunters of both past and present with animals, particularly large mammals. Most herbivorous species seemed to have constituted a resource in some place at some time; fast carnivores such as lion, tiger and wolf may have been hunted for prestige or because they preyed on desired herbivores, but they were not an element of subsistence. The central role of herbivore mammals in cultures as spatially separate as the later Magdalenian (Pleistocene) cave-dwellers of the Pyrenees, and the Netsilik Inuit of the Canadian North in recent times is not however disputed, and in cold climates, virtual dependence upon animal resources, including fish, seems to be a long-lived feature of subsistence patterns. In lower latitudes and at periods of warmer climate, the animal contribution becomes nutritionally less important, although it may still be a preferred food and a source of personal prestige to the hunter (who is usually male), while the necessary but unglamorous plant food is gathered by the women. There is no lack of evidence to show that hunters were very keenly attuned to the habits of their prey, as any predator has to be, but the key question for the present discussion is whether any deliberate conservation measures were taken to ensure the future supply of a food species. Secondary activities which may impinge on the relationship are well attested, such as gambling for arrow-heads which leaves the majority of men without them during a hunting season, and propitiatory dancing and other ritual activities. But direct evidence of practices which directly affect the structure of animal populations is difficult to collect and interpret. At Star Carr, a Mesolithic site in northern England, the large numbers of red-deer bones were dominated by those from males, which led Clarke (1972) to suggest that selective culling of males was practised. Given the harem-like social structure of the red deer at rutting time, killing males would have the least effect on reproductive success since non-breeding males would simply move up in the social hierarchy to where they could acquire a group of hinds to serve. By contrast, evidence from a number of palaeo-Indian bison kills from the Great Plains suggests that herds were stampeded into traps such as parabolic sand dunes or over cliffs into arroyos without regard for age and sex of the beasts; remains of foetuses are often found to corroborate this interpretation (e.g. Wheat 1972; Frison 1974). From different sources of evidence, Ray (1975) has concluded that North American Indian groups at the time of contact with agents of the Hudson's Bay Company reacted very negatively to proposals to cull animal populations in a manner designed to eliminate or reduce the chances of over-exploitation, and that the Indians were normally indiscriminate in their culling practices.

In the present state of knowledge perhaps only a no-model model is valid: there existed a variety of practices, some of which were deliberately conservative, others deliberately not, others by chance one or the other. But all must be placed in the context of a low population density and usually access to alternative food resources, so that recovery of animal populations from heavy exploitation could have taken place if the hunting pressures were lifted, other factors being constant. When they were not, the type of 'overkill' discussed below (p. 158) might well occur.

One tool of use to hunting and gathering populations was fire, a source of concentrated energy to pre-agricultural peoples. By burning vegetation they might alter its composition in order to encourage desired species selectively and to eliminate useless ones. Fire might also be used as a direct aid to hunting by means of frightening animals to run in a particular direction. The association of man with fire appears to be of great antiquity, stretching back about 700 000 years ($\pm$ 10 000) in the Petralona caves of northern Greece and we may presume that knowledge of fire as a manipulative tool soon followed its possession for cooking, warmth and protection. Not all ecosystems will burn: the individual plants must be sufficiently close to each other to allow a conflagration and they must be dry enough. But as the accounts of biomes in Part I showed, many of the terrestrial vegetation types of the world have enough dry material above ground to be seasonally fired.

Ground temperatures in grassland fires reach about 100°C, whereas forest fires are hotter, 150°C being commonly reached. The effects of the heat upon individual plants are variable according to their state of growth and any adaptations to fire such as a thick bark, but in general the above-ground parts of all except mature trees are killed. Thus rapid shifts in the species composition of non-forest vegetation, and long-term changes in forest composition (due to the elimination of certain species at the seed or seedling stage) are to be expected. Animals show a more clear-cut response, being either killed or escaping to an unburned area, although the flying ability of some insects may be impaired, allowing easier predation (Daubenmire 1968).

The nutrient levels in the soil may also be affected. Nitrogen and sulphur are volatilized and lost from the vegetation stand which is burnt; other nutrients are changed to simple salts which may be water-soluble. Measurements on *Calluna* heath in Scotland, for example, showed that potassium salts were readily dissolved from the ash; calcium and magnesium were less soluble, and phosphorus was in a near-insoluble form (Miller and Watson 1973). Significant nitrogen losses are reported by some authors; others show that if leguminous plants are present in or invade the burn then nitrogen levels increase. In general, woody species that do not sprout from their roots can be either eliminated or reduced to a thin stand of small individuals by repeated burning. Herbs (forbs) are usually favoured over grasses in both annual and perennial grassland although in dry steppe, annuals may gain at the expense of perennials, a trend also brought about by heavy grazing.

The NPP of the community may well be increased by fire but in grazed ecosystems the gain in forage may be offset since grazing may have to be reduced before the burn in order to produce enough litter, and after it as well if the new forage is not to be so severely grazed as to weaken the plants' chances of survival. Post-burn vegetation is relatively rich in protein and minerals and so is subjected to particularly heavy selective grazing pressure, a process intensified by the local absence of less nutritive and less palatable old leaves and stems (De Vos 1969). Thus sheep in Scotland are particularly attracted to post-burn patches because they discriminate in favour of vegetation with high nitrogen and phosphorus content and against highly fibrous plants (Miller and Watson 1973). It is notable too that burning and grazing together in the Scottish island of Rhum preserved a floristic diversity. In experimental plots, the number of species fell by half over a twelve-year period of exclusion of both influences but only by a third where grazing was maintained in the absence of fire (Ball 1974). The new grass on recent burns is very good forage for ungulate grazers, as is new leafy shoot material for browsers, which after a few years is not only abundant compared with unburned areas, but higher in protein as well (Dills 1970).

Such is the antiquity and widespread incidence of fire in natural and semi-natural ecosystems that it is often possible to distinguish species of plants adapted to withstand fire: these are called **pyrophytes**. Adaptations include a thick bark such as is found in the California coast redwood (*Sequoia sempervirens*), and closed-cone pines which release the seed only when the cone is subjected to high temperatures as in the Bishop pine (*Pinus muricata*) of California.

The actual use of fire by hunting peoples is well chronicled from a number of environments. In late-glacial tundras of Europe, charcoal layers are found which can be interpreted as resulting from efforts to channel hunted animals. In upland Britain during the Mesolithic, fire was apparently used near the tree-line to open up the forests and enhance sands and grassland growth; with cessation of burning, trees grew again over some of the scrub-clad areas (Tallis 1975, Jacobi *et al*. 1976). In near-recent North America, the deliberate burning of vegetation in the course of hunting and gathering activities has been recorded for over 100 separate groups, including 35 in California alone. Similar accounts come from many other parts of the world, from Tierra del Fuego to Tasmania, and from China to Turkestan (Stewart 1956). The effects of such fires on plants and animals vary with the original habitat, but we may note two outstanding generalities: first that both the species diversity and net primary production of many communities (and especially of forest understorey vegetation) is increased by fire; hazel (*Corylus avellana*) is a European example in this context, and analogous examples to increase wild grass seeds, nuts, berries and sunflowers have been reported from North America. Fire was also used in California to burn off vegetation which obscured the fallen acorns that constituted the crop. The second generality is that in the case of large herbivorous ungulates, firing of the habitat significantly increased its capacity to support herbivore populations since both the quantity and quality of the forage improves, sometimes in the order of 300–700 per cent in woodland areas (Mellars 1976).

The antiquity and near-ubiquity of the use of fire by man as an agent of ecological manipulation is unchallenged, and it probably constitutes the greatest single force in changing biotic patterns which man was able to wield before onset of the Neolithic 'revolution'.

Hunters and hunted: stability and instability

One of the preconditions for the long-term success and stability of the hunting mode is a stable relationship in which the sources of plant and animal food are not exhausted by human use. The selective processes (natural and cultural) within such a system will not be for super-successful predators but for a relationship that can continue to maintain itself (Jarman 1972). During the Pleistocene period of Europe, for example, there are many examples of Palaeolithic hunters living off the animals of tundra and tundra–forest interface and for several hundreds of years maintaining an economy based largely upon Eurasian reindeer. In this context, heavy use of plant material would not be expected and so the centrality of the dependence upon the animals is corroborated. A herd-following economy was probably practised, and so migration of human groups between the winter and summer ranges of the animals would have been essential. Archaeological sites suggest locations near to migration routes and feeding grounds, but sufficiently far away to minimize unnecessary disturbance of the animals. Furthermore, selective killing was carried out (Sturdy 1975).

A similar relationship has been postulated for Europe during the Mesolithic of the middle post-glacial period when the vegetation had come to equilibrium after the rapid warming of the 8000–6000 BC period. As Jarman (1972) says, 'from Star Carr in Yorkshire to Sidari in Corfu a very few species provided the bulk of animal protein ... the red deer was commonly the most important of these species.' (These data come from bone representation at archaeological sites and the ubiquity of red deer in the faunal spectrum is emphasized in Fig. 4.2.) The sites cover a wide time-span and so it is difficult to escape the conclusion that deer were important features in the economic pattern for a long time and in most cases were superseded only by the advent of a cereal-based agriculture. (Their importance does not preclude the value of plant food in the diet (Clark 1976) but even if a preferred rather than a staple food, the relationship with deer could have been important in the lives of the Mesolithic folk.) The way in which this

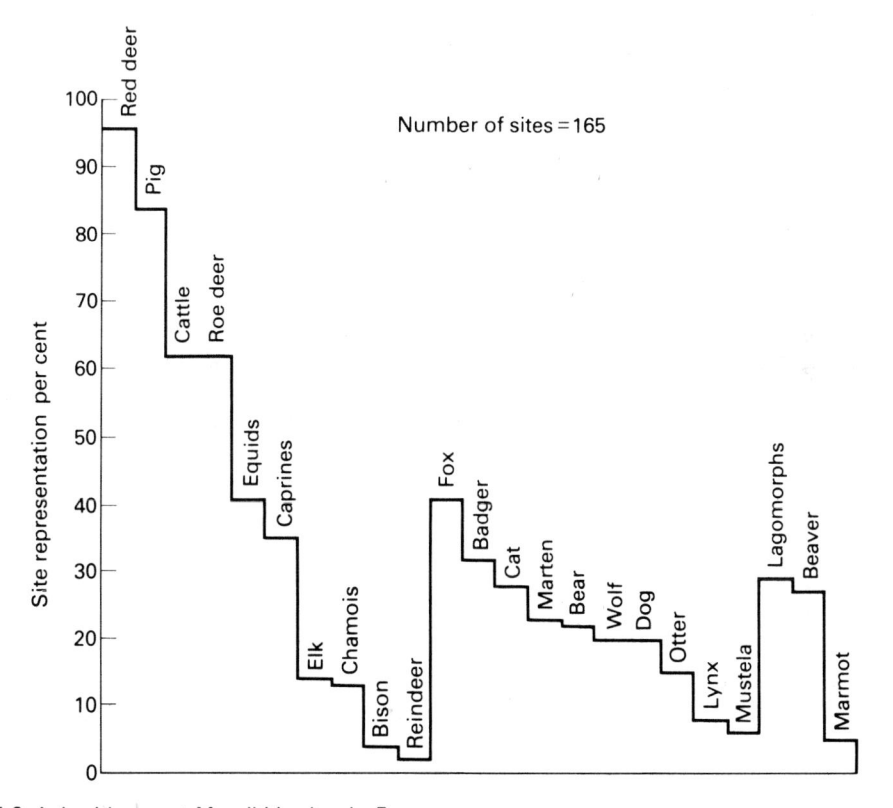

Fig. 4.2 Animal bones at Mesolithic sites in Europe.
Source: Jarman, 1972.

long-term relationship was probably maintained is discussed below (p. 160): here we merely emphasize its longevity in the period of a relatively stable climate and vegetation type, ca 7000–3500 BC in the British Isles, a little earlier in continental Europe.

A considerable contrast to these examples is provided by the history of large warm-blooded creatures (usually mammals but with some birds as well) at the end of the Pleistocene when some 200 genera of them became extinct. The phenomenon is perhaps most closely observable in North America: here, two-thirds of the large mammal fauna of the end of the Pleistocene (ca 13 000 BC) disappeared, including 3 genera of elephants, 6 genera of giant edentates (armadillos, ant-eaters, pangolins and sloths), 15 genera of ungulates, and various giant rodents and carnivores. There is no evidence of such extinctions in earlier periods which might have been expected if Pleistocene climatic changes had been the proximate cause, nor any firm evidence of survival of these genera past the 500-year period during which the extinctions appear to have taken place. These data have led to the hypothesis that it was the introduction of man as a predator against which these animals had no genetically implanted defence behaviour which was the cause of the extinctions, the idea of 'Pleistocene overkill' (Martin 1967). In North America, the date of introduction of man via the Bering Strait land bridge is usually taken as being about 12 000 years BP and a simulation of the extinction pattern supposes a wave-front advance by hunters whose population periodically exploded because of the new and favourable habitat (Martin 1973; Mosimann and Martin 1975). At the advancing front, a human population growth rate of 3·4 per cent p.a. may have given a density of 0·4 persons/km^2 and a forward movement (due to hunting-out of the large mammals) of 16 km/yr. Thus the front could sweep from Canada to Mexico in 350 years (Martin 1973; Fig. 4.3) and extirpate the large mammals through superior predation techniques.

Some support is given to this hypothesis by the effects of the arrival of man into other hitherto unpeopled places. The disappearance of the moa from New Zealand occurred within a few hundred years of the human occupance of the islands; the megafauna (species with adults weighing > 50 kg) of Madagascar disappeared within a similar period after the first human occupation in 1000 AD including a large terrestial bird *Aepyornis*, and a pygmy hippopotamus; in Java and Celebes, populations of dwarf elephants did not long survive the coming of man. It would appear, therefore, that when introduced into a new habitat man the hunter is capable of rapidly exterminating flightless birds and large mammals. The end of the Pleistocene in the Old World presents a somewhat different picture (Reed 1970). In Eurasia, only a total of 9 species were made extinct, compared with the New World total of at least 24 genera. The Eurasian losses comprised the woolly mammoth (*Mammuthus primigenius*), woolly rhinoceros (*Coelodonta antiquitatis*), giant Irish elk (*Megaloceros giganteus*), musk-ox (*Ovibos moschatus*), steppe-bison (*Bison priscus*), a buffalo of northern Africa (*Homoioceros antiquus*) and three species of associated carnivores. All except the buffalo are animals of the cold steepe which was widespread outside the ice-sheets of Europe during the last glacial phase and all the herbivores, especially the mammoth and bison, had been hunted for tens of thousands of years yet survived in large numbers until about 12 000–10 800 years BP. But this time-horizon does not mark the introduction of man to Eurasia, nor even a particularly great cultural change, so that it appears much more likely that climatic

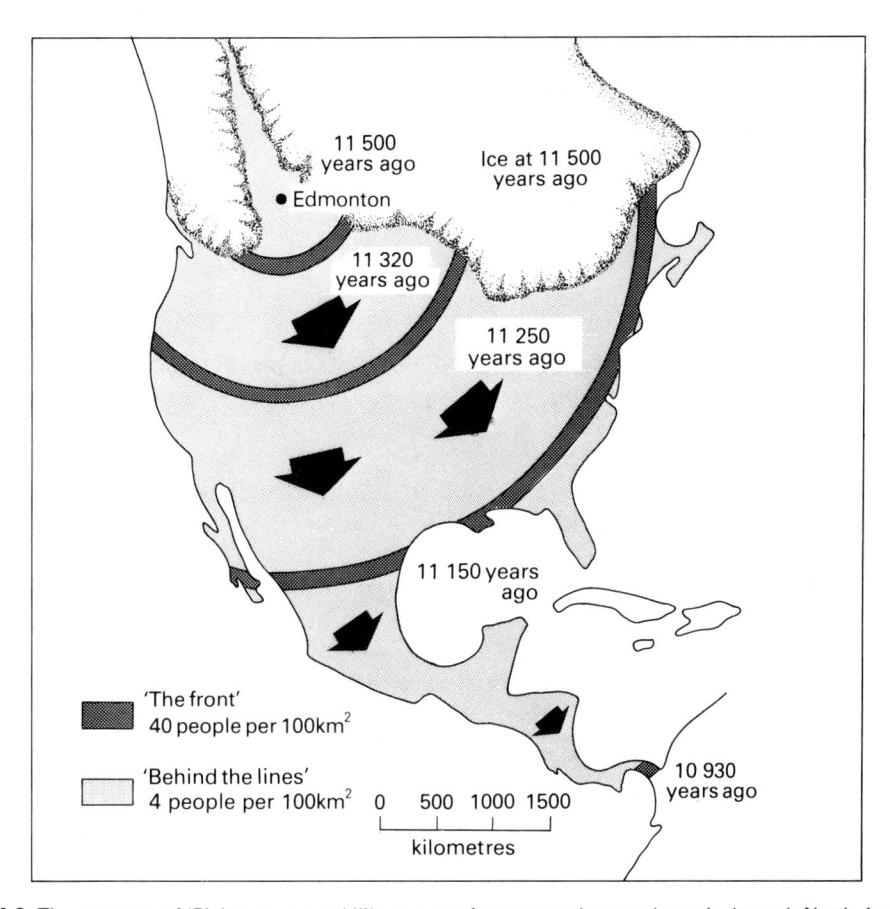

Fig. 4.3 The progress of 'Pleistocene overkill' seen as a front sweeping southwards through North America. Source: Martin, 1973.

change leading to forest growth of birch and pine was the main cause of the animals' disappearance. Of course, continued hunting when the population was under environmental stress would have hastened their demise but the change in habitat appears to be the primary reason. At any rate, when their habitats returned briefly in the cold period of 10 800–10 150 years BP, the animals did not. Some other species of the cold steppe proved more adaptable: the reindeer persisted on the tundra, and both the saiga antelope (*Saiga tartarica*) and wild horse (*Equus przewalskii*) took to the grassy steppes of central Eurasia. Comparable with Eurasia is a site in New South Wales, Australia, which appears to show 7000 years of coexistence of man and megafauna without any evidence that the human societies were the cause of the eventual extinction of the other animals (Gillespie *et al*. 1978).

So in different parts of the world, a similar phenomenon occurring at the same time appears to have had different primary causes. Not in doubt is the fact of faunal loss and it seems likely that in some places at least this phase marked the beginning of the man-related faunal extinction which was to continue through the rest of post-glacial history (see also p. 218). Whether man or climate was the cause of extinction and of the consequent shift of hunters to other species, the relationship between man and mammals at the end of the Pleistocene could scarcely be said to have been stable.

Red deer (*Cervus elephas*) in Mesolithic Europe

The relationship of man to red-deer populations as sources of meat has been mentioned briefly above; here we discuss how a stable relationship might have been attained in the cultural context of Mesolithic times. There is initially the direct evidence of selective culling of males, and further that young individuals feature more frequently in the osteological record than random killing would suggest. Yet in a red-deer herd there were probably more females than males. Explanations include the possibility that females are more wary than males but opinion favours the idea that exploitation of males and young individuals would have the least effect upon the continued success of a breeding herd. Therefore the majority of the kill is of males which might not be part of the breeding stock and of young animals not of breeding age (Jarman 1972).

Further evidence of management comes from an indirect source. In biostratigraphic investigations of upland peats in Britain, evidence of fire during the Mesolithic is frequent. This comes from charcoal layers in the peat and from pollen analytical evidence of forest recession. It is suggested (Jacobi *et al.* 1976) that burning of these uplands prevented trees from colonizing them, the actual form of the charcoal suggesting recurrent firing of the vegetation. The optimal strategy would probably be to burn the upper edge of the forest at intervals of 5–15 years. This would improve both the quantity and quality of the browse for red deer by encouraging the growth of shrubs such as hazel. Thus the maintenance of a healthy herd of deer would be facilitated with the happy spin-off of large crops of the nuts of *Corylus avellana*.

With the coming of agriculture, this practice ceased in many places. On some parts of the uplands it led to the colonization of trees over previously open terrain (Tallis 1975); elsewhere the practice of burning and the removal of forest seems to have induced a change in the water balance of the soils. Since the evapo-transpirative effect of the trees was no longer present, the soils became waterlogged in these areas of high (1500 mm) year-round precipitation, and where the slope was low peat formed (Moore 1975). So a 'blanket' of peat was initiated in some of the areas of human management of vegetation during the Mesolithic: the man–deer relationship may have been kept stable but there was, as so often occurs, an unforeseen repercussion elsewhere in the ecosystem.

Recent hunters and their relations with plants and animals

The literature on the subsistence patterns of recent hunting and gathering groups is not usually very helpful on the impact of exploitation on the animal and plant populations. What does perhaps emerge as one generalization is that the people are prepared to undertake a great deal of movement in order to get their foods and that they vary the size of exploiting groups according to the availability of resources. The G/Wi bushmen, for example, who live mostly off plants, split into smaller bands during the off-season (Silberbauer 1972) and this presumably diminishes the impact upon the plants at a time when they are least able to withstand heavy cropping. Another general observation is that the variety of flora and fauna cropped is generally wide and so there is usually an alternative should the preferred crop fail. The Mistassini Cree of central Labrador (Rogers 1972) eat every little plant material but a wide spectrum of animals: apart from fish and some eight species of birds, they killed and ate moose and caribou (2497 kg out of a total of 3814 kg of mammal food in one season) and also bear, beaver, hare, muskrat, porcupine, mink, squirrel, marten and otter. Even in very restricted environments like the borderlands of the Arctic Ocean a variety of animals are hunted, although one of them will be the most important. The Inuit groups west of Hudson's Bay were, in their traditional economy, dependent to a great extent on the ringed seal but they also hunted the bearded seal, walrus, narwhal and beluga whale, and valued the occasional polar bear and musk-ox. Some subgroups added to these the inland resources of fishing and caribou hunting (Damas 1972).

Not all relationships with wild organisms are economic: the Ainu of Hokkaido were participants in a bear cult which involved the capture of young cubs which were kept in the settlement for one or two years and then ritually killed. Adult bears were killed as well, for food, but none of the literature suggests that there was any evidence of a long-term effect upon the bear population (Watanabe 1966, 1972a, 1972b).

The studies of Heizer (1955) reveal numbers of cases of strictures on the harvesting of natural resources in simple cultures, some of which are relevant to the discussion of hunters, but they do not say except by implication whether they were successful in perpetuating the wild species' populations. Nevertheless, we may note that some Great Lakes Indians knew to strip bark from only a portion of basswood (*Tilia* spp) trees, and the same was true of the Vancouver Islanders' use of cedar bark. The Choctaw regulated the number of animals that might be killed by one family and the Kaska allowed the trapping of some species only at two- or three-year intervals; similarly the Iroquois spared the females of hunted species during the breeding season. Some groups even avoided a particular taxon altogether in the belief that they were descended from such species. If we accept that these tribes survived successfully until contact with Europeans, then perhaps it is legitimate to infer that their conservation practices were successful ways of managing wild populations. But as the work of Ray referred to on p. 154 shows, not all Indian groups appeared to carry out some form of conservation.

We can conclude in an interim fashion by suggesting that most hunting cultures have been too few numerically to exert long-term effects upon most animal populations, even where no conscious conservation policies were followed. The presence or absence of these latter seems very variable and no doubt some cultural and ritual practices had repercussions on the animal populations: a comprehensive study would be most useful. We need to re-emphasize as well the vegetarian emphasis of most hunting and gathering people: both in recent times and archaeologically, the hunting activities have a high profile but *de facto*, plants, and normally wide range of them from a wide spatial area, were the critical resource for pre-agricultural peoples except in very cold lands.

The decline of hunting and the transition to agriculture

The complex cultural reasons for the adoption of agriculture and the decline of hunting, especially in the hearthlands of agricultural innovation (p. 167), are not part of this book's task. But in the millennia preceding the emergence of fully-fledged agriculture, the acquaintance of hunting groups with their biotic resources must have been close. With plants, it is quite probable that some cultivation of wild plants took place in pre-agricultural groups in the form of a planned but small intervention in the life-cycle of a favoured plant, rather as in recent times such groups have replanted wild yams at a more convenient spot, or arranged for the diversion of a small stream to irrigate a wild grass which has nutritious seeds. The removal of 'weed' species from a bed of useful perennials is another form of cultivation which might have a long Pleistocene ancestry (Bronson 1975). Such rudimentary practices need not have any genetic effects upon the plants and so the threshold of domestication is not actually crossed; nevertheless the right preconditions for the commitment to agriculture as a means of subsistence are present, should the animals fail or diminish through over-hunting (Smith 1975) or if cultural reasons compel a shift in life-style.

Although the emergence of agriculture can be quite closely documented in terms of the recognition of genetically distinct strains of plants and animals, it is worth stressing that the roots of the practice must go much deeper into prehistory and that the division between hunter-gatherers and agriculturalists, useful though it is, must in reality have been blurred, both by the cultivation practices of hunter-gatherers and the hunting activities of agriculturalists.

Further Reading

CLARKE, J. G. D. 1972: *Star Carr: an essay in bioarchaeology*. Reading, Mass.: Addison-Wesley Modules in Anthropology.

DAUBENMIRE, R. 1968: Ecology of fire in grasslands. *Adv. Ecol. Res.* 5, 209–66.

HIGGS, E. S. and JARMAN, M. R. 1972: The origins of animal and plant husbandry. In E. S. Higgs (ed.), *Papers in Economic Prehistory*. London, Cambridge University Press.

KOZLOWSKI, T. T. and AHLGREN, C. E. (eds.) 1974: *Fire and ecosystems*. London and New York: Academic Press.

LEE, R. B. and DEVORE, I. (eds.) 1968: *Man the hunter*. Chicago: Aldine Press.

MARTIN, P. S. 1967: Prehistoric overkill. In P. S. Martin and H. E. Wright (eds.), *Pleistocene extinctions: the search for a cause*, New Haven: Yale University Press, 75–120.

STEWART, O. C. 1956: Fire as the first great force employed by man. In W. L. Thomas (ed.), *Man's role in changing the face of the earth*. Chicago: Chicago University Press, 115–33.

5
The domestication of plants and animals

The advent of domestication has meant a different type of man–environment relationship from that pertaining in hunting and gathering societies. In essence, domestication means changes in wild plants and animals: firstly, the alteration of their genetic structure by substituting human selection for natural selection. This process is still happening and although emphasis in this section will be on domestications in ancient times, a continuum view of domestication is desirable (see also p. 185). Secondly, the process has meant the alteration of ecosystems. This may have been deliberately engineered in order to accommodate the tolerances of the genetically novel biota, as in the practice of weeding in order to eliminate the competitors of chosen crops. At other times it has been accidental, as with the gradual shifts in vegetation caused by pastoralist economies (p. 305) or as with the heavier silt loads contributed to streams after the clearance of forest for agriculture.

The exercise of control over the genetics of plants and animals is usually called a revolution, often the Agricultural revolution, or the Neolithic revolution. But although the new relations between man and nature may have been revolutionary, the phenomenon itself does not derive solely from one time or place. Domestication in ancient times probably took place in many locations and was probably the culmination of a long period of association of man with particular species of plants and animals. Many biota were repeatedly domesticated at different times and places or may have been adapted in several regions simultaneously. Only where the spread of a domesticate from a centre of domestication has been particularly confined, as with the Yak (*Bos grunniens*) for example, can we be confident in locating its centre of origin; elsewhere we must recognize that a complex history is more likely than a simple one: that for example one species of rice was domesticated in Asia, but another in Africa, or that the wild pig was so widely distributed that parallel domestications in much of the Old World must have been very likely. The ideas of N. I. Vavilov, that there were a limited number of hearths of domestication which could be identified by the centres of diversity of wild relatives of important domesticates, must now be replaced by a more complex picture which recognizes that the antiquity of man's close association with plants and animals must go back beyond culturally 'Neolithic' phases and that wherever such closeness occurred there was a potential for domestication, some but not all of which came to fruition. But even though the details of the process are inevitably complex (and labyrinthine in their reconstruction), one result stands out starkly: the average population density of hunter-gatherers in recent times suggests that $0.1/km^2$ was for millennia the typical carrying capacity of the Earth for that type of economy. Early (*ca* 2000 BC) dry farming in Southwest Asia supported 1–2/km², and with irrigation this could be multiplied sixfold. Farming of domesticated crops thus represented an intensification of the use of resources, which enabled the human population to increase markedly.

Definitions of domestication and the nature of the evidence

Domestication is not necessarily to be equated with cultivation, for it is possible to cultivate wild plants or to husband wild animals in a purposeful way: wild grasses can be cropped year after year and protected from grazing animals but left to seed themselves; wild animals can be fed supplementary food in severe periods. Domestication means the bringing about of genetic changes to plants and animals which make them better suited to conditions provided by human societies than to those of their natural environment and when the process is complete the biota are fitted exclusively to a man-made environment and would not survive in the wild. Some domesticates become feral and may survive: horses, cattle and camels have been known to succeed thus, but many breeds of goat and sheep would only survive in the absence of any effective predation. Among plants, most cereals, for example, would be unable to survive the competition offered by plants which are kept down by the processes of cultivation. As an extreme example there is the contemporary tom turkey which has been bred for so much breast meat that it cannot effectively mount the female.

Proceeding from an early stage in most domestication when there was much interbreeding between the selected domesticate and its wild ancestors, there has been a general tendency to genetic isolation in which the domesticate is no longer interfertile with its wild relatives, even if they still exist. But in early times the distinction between wild, husbanded and domesticated biota must have been a tenuous one for many centuries at least, a fact to be remembered when evaluating evidence from the far end of agricultural time.

The evidence for the nature and progress of domestication in ancient times comes from a number of diverse sources. Archaeological sites have probably contributed most in the way of animal bones, carbonized seeds, impressions of grains on pottery, pollen, and the contents of coprolites, which may all point to the separation of cultivars from wild ancestors; the remains of molluscs, insects and pollen grains may point to habitat changes from which agriculture or pastoralism can be inferred. All such evidence has been culturally sieved, and the reconstruction of the numbers, morphology and role of a domesticate from the pieces of bone recovered during a dig under unfavourable conditions may often present difficulties and engender controversy. Direct representations of crops and domestic beasts are rare, so that those from ancient Egypt form a particularly valuable source of evidence (Plate 23). Even indirect cultural evidence may have to be used, for instance to infer the existence of agriculture from digging sticks, irrigation channels, field systems, and storage pits, even where no remains of plants have survived. Purely biological evidence may come from the study of the contemporary distribution of the supposed wild ancestors of the domesticates (Fig. 5.1) or from the cultivar itself by observation of its genetic structure through a controlled breeding programme designed to uncover recessive genes which may hold clues to its former state,

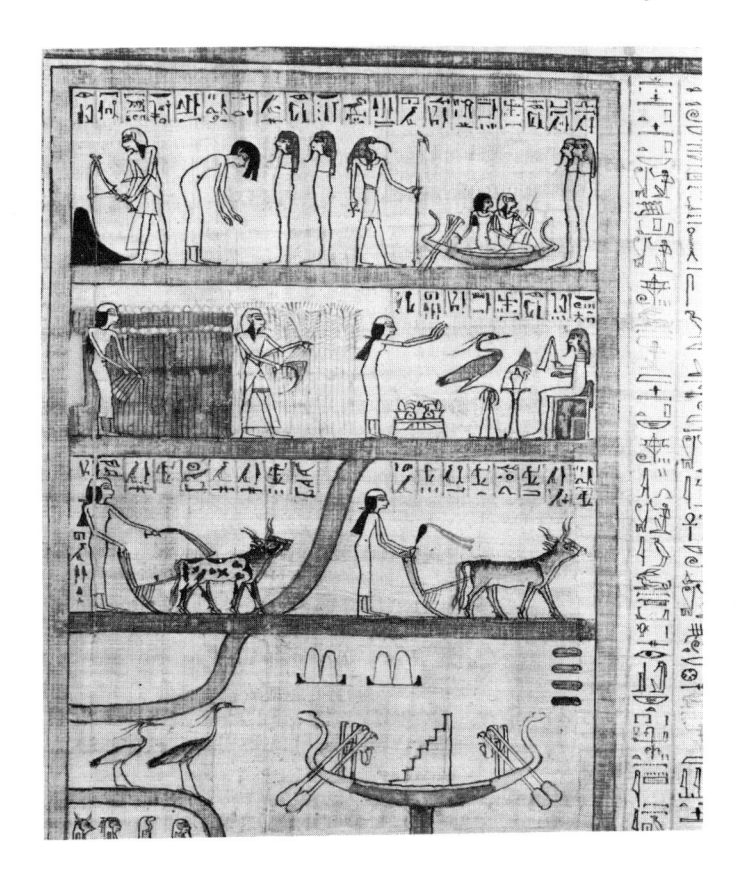

Plate 23 Agricultural scenes from the *Book of the Dead* of the priestess Anhai of ancient Egypt.

or by laboratory investigation of genetically controlled facets of its biochemistry. The outcome of investigations of all these types of evidence has been to produce a generally agreed set of histories of most of the major crop plants and animals, though there is still room for much more evidence.

Pre-domesticatory relationships

Much of the evidence for domesticated animals and plants comes from a stage when domestication appears to be fairly complete, and authorities suggest that there must have been a long period of association and exploitation of the species before it could be said to be truly domesticated. In the case of plants, an example that might be cited is the harvesting of wild grasses which were the ancestors of what became cultivated cereals. There must have been a period, as Zohary (1969) suggests, when grasses were

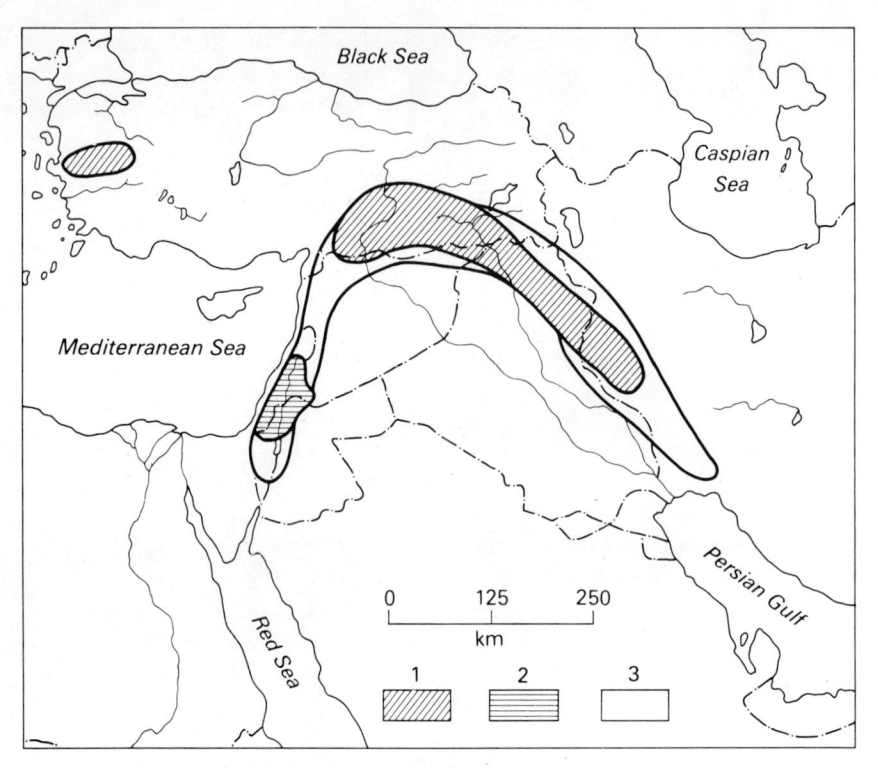

Fig. 5.1 The natural distribution of the wild ancestors of wheat and barley: 1, Einkorn (*Triticum monococcum*); 2, Emmer (*T. dicoccum*); 3, Barley (*Hordeum* spp.)
Source: Isaac, 1970.

harvested and all the grain obtained used as food. This is a form of gathering rather than domestication when the wild strains would maintain themselves spontaneously. Man's harvest of the wild plants would select differentially in favour of the less brittle (and hence most easily gathered) plants and so the quick-shattering heads would remain to form the next generation. Under conditions of true domestication, planting by man would ensure a reverse orientation: non-shattering heads would still be selected but now for planting as well as ease of harvesting.

As an example of an animal relationship, we may amplify the evidence accumulating about the interaction of men and red-deer (*Cervus elephas*) populations in Britain during the late Mesolithic (*ca* 7000–3500 BC), mentioned on p. 160 where human groups were burning the upper edge of the forests on the uplands in order to encourage shrubs and herbs at the expense of the mature forest. Thus a management of deer herds might have occurred without any attempt to control their genetics (Simmons 1975; Mellars 1976; Jacobi *et al.* 1976). But the relationship could have been even closer: Simmons and Dimbleby (1974) note the evidence for very high levels of ivy (*Hedera helix*) pollen at some Mesolithic sites and suggest that the explanation is to be found in the practice of gathering ivy in late autumn (at which time it flowers) and using it as deer feed. Whether this was to attract wild animals or to feed to tethered beasts is not known but again it suggests that a close relationship was achieved (which we might perhaps call 'Men and the Art of Deer-Cycle Maintenance') but not apparently one of proper domestication. A parallel course in Southwest Asia is suggested by D. R. Harris (1978) who argues for a long phase in which populations of sheep, goats and cattle became semi-domesticated as free-ranging animals breeding in the wild but loosely controlled through the provision of salt: recent examples of taming of wild sheep, the mithan (*Bos frontalis*), and mountain goats give added value to this hypothesis.

The domestication of plants in ancient times

Preconditions

To some extent, the existence of the propitious circumstances for the occurrence of domestication is a primarily cultural matter and hence not a primary concern here, but the availability of plants that were in a sense pre-adapted for cultivation cannot be ignored since the majority of plant species have not in fact become cultivated. To Carl O. Sauer, a leading scholar of the domestication of plants and animals, the preconditions included the existence of an area of diversity of plants and animals which gave varied materials with which to experiment. This in turn might mean a well diversified terrain and probably woodlands, which were more easily dug over and cleared than the tight sod of natural grasslands. To Sauer (1969), all this suggested a well situated fishing folk living in a mild climate alongside fresh waters with the leisure to experiment, and he proposed, on the basis of intuition rather than hard evidence, that Southeast Asia was the first home of plant domestication. Although seed agriculture tends to dominate discussions of domestication, D. R. Harris (1969) has drawn attention to the role of vegeculture as an early form of agriculture, especially for root crops where a piece can be put in the ground and it will most likely grow. Midden heaps would be excellent habitats for such crops, and Harris wonders if vegeculture may have preceded seed culture.

Scientific work on the preconditions for and origins of domestication was for long dominated by the work of N. I. Vavilov (1951). On the basis of the patterns of variation of both domesticates and their wild relatives, he identified geographic 'centres of origin' of domesticated plants: six of these hearths were set out for the Old World, and two for the New World (China; Central Asia; Abyssinia; India; Near East; Mediterranean; Indo-Malaya; South America; South Mexico–Central America). The paradigm behind this work was that the domestication of a particular taxon occurred only once or twice and thereafter spread by diffusion amongst cultural groups. This set of views has been transformed by more recent work by archaeologists, ethnobotanists and geneticists, and the current picture is more complex. Some crops did not originate in the centres to which Vavilov has assigned them, some in more than one centre, and others over vast regions rather than one locality. Likewise wild progenitors which were widely distributed may have been manipulated by various peoples over their entire range, and may have been repeatedly domesticated at different places at different times or at several

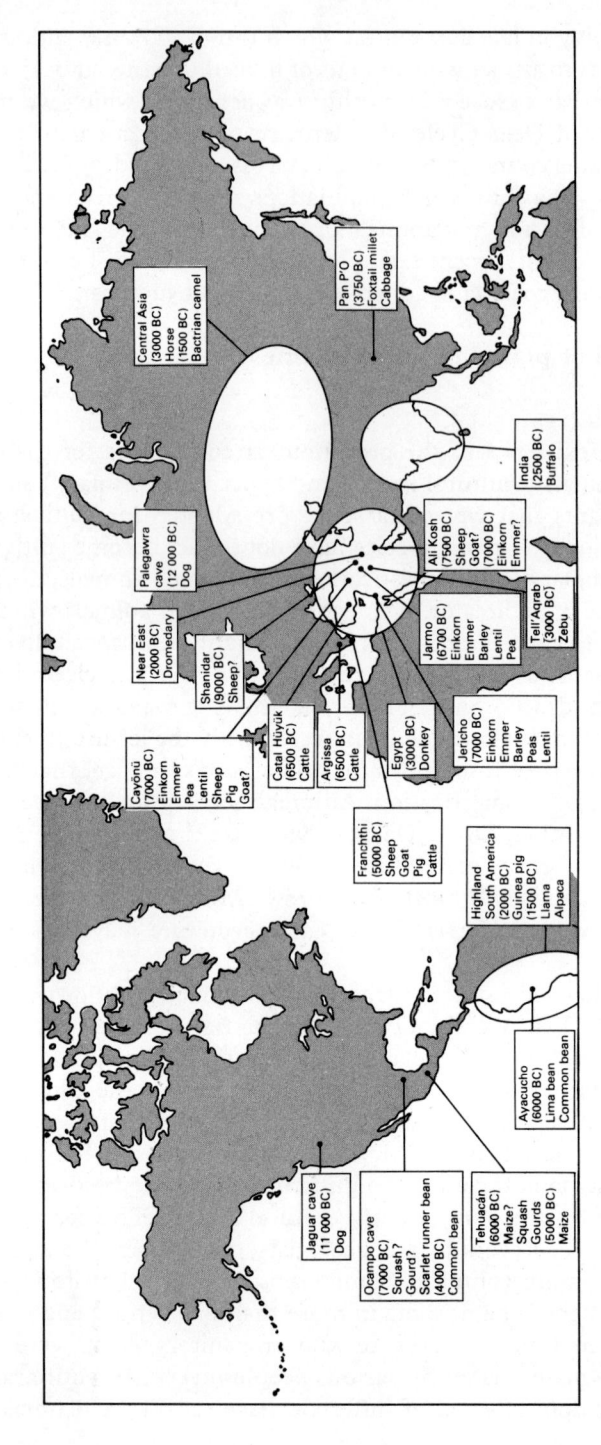

Fig. 5.2 The dates and locations of the earliest domestications. A remarkable congruity seems to appear, suggesting independent 'inventions'. Source: Harlan, 1976a.

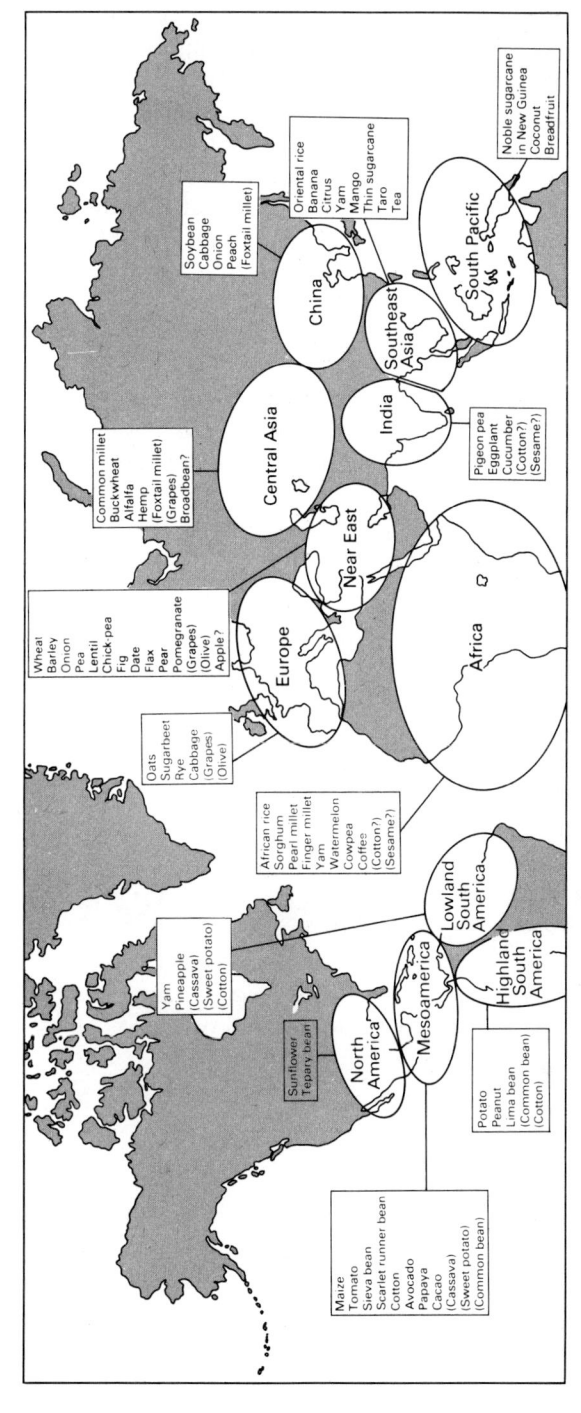

Fig. 5.3 Areas of major plant domestication. Bracketed names are of crops domesticated independently in more than one major focus (e.g. sweet potato, grapes, common bean). 'Near East' is equivalent to 'Southwest Asia' in the text. Source: Harlan, 1976a.

The domestication of plants and animals

places simultaneously. Thus we think of the origin of crop agriculture as the end point of a long process of interaction of men and plants (Harlan 1976a).

If the idea of a centre of origin in Vavilov's sense is now insufficient, the notion of regions where the earliest evidence of domestication of several plants is found has some validity, provided we accept that such a focus may not be the sole origin of a given cultivar. Figure 5.2 shows the dates and places of the earliest domestication of a number of important domesticates, from which it can be seen that Southwest Asia was a zone of such importance for plants and gave rise to many staple crops both of historic times and the present day that it deserves fuller treatment.

Cereals of the Southwest Asia focus

The distinctive mark of this area's contribution to agriculture is its concentration on seed agriculture, mostly of cereals but including pulses and fruits as well (Fig. 5.3; Table 5.1). It is supposed (Hawkes 1969) that the domestication of cereals from wild grasses

TABLE 5.1   Cultivated plants originating in Southwest Asia–Mediterranean basin

CEREALS
Avena sativa	Oats; secondary crop, N. Europe
A. strigosa	Fodder oats; addition crop, Mediterranean
Hordeum vulgare	Barley; primary crop, N.E.
Secale cereale	Rye; secondary crop, Anatolian plateau–N. Europe
Triticum aestivum	Bread wheat; addition crop, Transcaucasia–Caspian
T. diococcum	Emmer; primary crop, N.E.
T. monococcum	Einkorn; primary crop, Turkey
T. timopheevii	Very minor wheat; Soviet Georgia
T. turgidum	Tetraploid wheat; derived from emmer, N.E.

PULSES
Cicer arietinum	Chickpea; primary crop, N.E.
Lathyrus sativus	Grasspea; N.E. crop
Lens esculenta	Lentil; primary crop, N.E.
Lupinus albus	Lupin; N.E.
Pisum sativum	Gardenpea; primary crop with addition from Mediterranean
Vicia ervilia	Bittervetch; N.E.
V. faba	Broadbean, fava; N.E. or Mediterranean

ROOT AND TUBER CROPS
Beta vulgaris	Beet, mangel, chard; Mediterranean, W. Europe
Brassica rapa	Turnip; Mediterranean (also maybe China)
Daucus carota	Carrot; Mediterranean, widespread
Raphanus sativus	Radish; wild and weed races widespread

OIL CROPS
Brassica campestris	Rapeseed; E. Mediterranean
B. nigra	Mustard, mustard oil; E. Mediterranean
Carthamus tinctorius	Safflower; N.E.
Linum usitatissimum	Flax, linseed; primary crop, N.E.
Olea europaea	Olive; Mediterranean
Papaver somniferum	Poppy; possibly primary crop, N.E.

FRUITS AND NUTS
Corylus spp.	Hazelnut, filbert; Balkans to Caspian
Cucumis melo	Melon; N.E.
Cydonia oblonga	Quince; Balkans to Caspian
Ficus carica	Fig; Turkey–Iraq–Iran

Juglans regia   English walnut; Balkans to Pakistan
Phoenix dactylifera   Date palm; Lowland steppes of Near East
Pistacea vera   Pistachio; Turkey–Iran
Prunus amygdalus   Almond; Turkey to Pakistan
P. armeniaca   Apricot; Turkey–Iran
P. avium   Cherry; Balkans to Caspian
P. domestica   Plum; Balkans and E. Europe
Punica granatum   Pomegranate; Transcaucasia–Caspian
Pyrus communis   Pear; Turkey–Iran
P. malus   Apple; Balkans–Transcaucasia–Caspian
Vitis vinifera   Grape; Mediterranean

VEGETABLES AND SPICES

Allium cepa   Onion; Mediterranean
A. sativum   Garlic; Mediterranean
A. porrum   Leek; E. Mediterranean
Anethum graveolens   Dill; Mediterranean
Brassica oleracea   Cabbage, cauliflower, brussels sprouts, kale, kohlrabi, broccoli; addition from W. Europe
Carum carvi   Caraway; N.E.
Coriandrum sativum   Coriander; N.E.
Cucumis sativus   Cucumber; N.E.? India? (possible domestication in both areas)
Cuminum cyminum   Cumin; N.E.
Foeniculum vulgare   Fennel; Mediterranean (also widespread)
Lactuca sativa   Lettuce; Mediterranean
Lepidium sativum   Gardencress; Mediterranean
Petroselinum sativum   Parsley; Mediterranean
Pimpinella anisum   Anise; Mediterranean
Portulaca oleracea   Purslane; Mediterranean
Trigonella foenum-graecum   Fenugreek; Turkey

FIBER PLANTS

Cannabis sativa   Hemp; widespread, Eurasian
Linum usitatissimum   Flax; primary crop, N.E.

STARCH AND SUGAR PLANT (NOT ROOT)

Ceratonia siliqua   Carob, tree with sweet pods; E. Mediterranean

FORAGE CROPS

Agropyron spp.   The wheatgrasses; Eurasian, useful types from Turkey and USSR.
Agrostis spp.   The bentgrasses; W. Europe
Bromus inermis   Smooth bromergrass; Turkey to Central Europe
Dactylis glomerata   Orchardgrass, cocksfoot; Europe, Mediterranean
Festuca arundinacea   Tall fescue; Europe, N. Africa, N.E.
Lolium spp.   The ryegrasses; Europe–Mediterranean
Medicago sativa   Alfalfa; Central Asia, Turkey–Iran.
Medicago spp.   The medic clovers; mostly Mediterranean
Melilotus spp.   The sweet clovers; widespread Europe and N.E.
Onobrychis viciifolia   Sainfoin; Turkey
Phalaris arundinacea   Reed canarygrass; widespread Europe
P. tuberosa   Hardinggrass; Mediterranean
Phleum pratense   Timothy; widespread Europe
Sorghum halepense   Johnsongrass; Mediterranean, N.E.
Trifolium spp.   The true clovers; Europe, N.E.
Vicia spp.   The vetches; Mediterranean

DRUGS, NARCOTICS, FATIGUE PLANTS

Atropa belladonna   Belladonna; Mediterranean
Digitalis purpourea   Digitalis; Europe
Glycyrrhiza glabra   Licorice; Mediterranean, N.E.
Hyoscyamus muticus   Henbane; Mediterranean, N.E.
Papaver somniferum   Codeine, morphine, opium; Mediterranean
Plantago psyllium   Psyllium; Mediterranean

N.E.=Near East.
Source: Harlan 1975.

went through several stages: gathering first, and then more regular harvesting of a wild grass crop with some selection towards a non-brittle head. When sowing occurred, with careful retention of seed selected for a non-brittle head, then domestication can be considered complete, though further genetic change is not precluded.

In this region, two types of domesticates can be inferred: primary and secondary. The primary crops were domesticated directly from wild progenitors and were plants adapted to survive a long hot dry season, having thus large food reserves of obvious interest to human groups. In general they were plants intolerant of competition by large perennials, i.e. species from open, disturbed or unstable habitats: wheat, barley and flax, for example, are of this type. Secondary crops are thought to have started as weeds of the primary crop which in time were recognized as capable of useful domestication: oats and rye are examples.

The kind of detailed histories collected by Simmonds (1976) cannot be given here of all the plants which became significant crops but one or two species deserve a fuller treatment and the outstanding candidate is wheat (*Triticum* spp). The large seed size of wild wheats made them attractive to the collector, and their annual habit made them amenable to the dry farming in which they were probably first used. Their self-pollination helped the fixing of desirable characteristics and although their natural habitat of poor, thin, rocky soils made them suitable plants for early cultivators, they also responded well to richer habitats (Feldman 1976). The various species of wheat form a **ploidy** series with **diploids** ($2 \times = 14$), tetraploids ($4 \times = 28$) and hexaploids ($6 \times = 42$), with wild species being of both $2 \times$ (diploid) and $4 \times$ (tetraploid) types. Cultivars are of all three ploidy types. A simplified evolutionary history of wheats is given in Fig. 5.4. Placing this history in actual space and time is complex but a basic outline begins with the wild species in park-forests and steppes on the mountain chains of the fertile crescent and also in Anatolia and the Balkans. Cultivated wheats of diploid and tetraploid types occur at sites in north Syria, Iraq, Iran, Israel, Jordan and Anatolia in the period 7500–6750 BC, and a hexaploid wheat which was free-threshing in Iran–Anatolia by 6000 BC. It was the hexaploid form which proved most amenable to transport outside the Near East since *T. aestivum* grows better in continental climates: it was spread into central Asia and appears in the Indus valley by 4000 BC and China by 3500 BC. *T. aestivum.* var *aestivum* has become the most important wheat of the modern world, though other varieties of *T. aestivum* are grown, as well as relic populations of, for example, *T. monococcum* cultivated for fodder in mountainous parts of Yugoslavia and Turkey.

Barley (*Hordeum* spp) is another crop associated with ancient Southwest Asia (Harlan 1976b), and its domestication possibly pre-dates wheat: forms which could be either domesticated or wild have been found from Syria dated at *ca* 8000 BC. Definite evidence of cultivation occurs between 6000–7000 BC, usually along with remains of emmer and einkorn wheats, flax, peas, vetches and lentils. It became a crop of irrigation by 6000 BC and in ancient Mesopotamia it was nearly a monoculture by 1800 BC because it could tolerate the salt in hydraulic systems. Its spread then took in Spain by the fifth millennium BC and the lower Rhine a little later. Eastwards it reached India by 3000 BC and China by 2000 BC.

Rye (*Secale* spp) has been identified as a weed of other cereals. Its ultimate ancestor

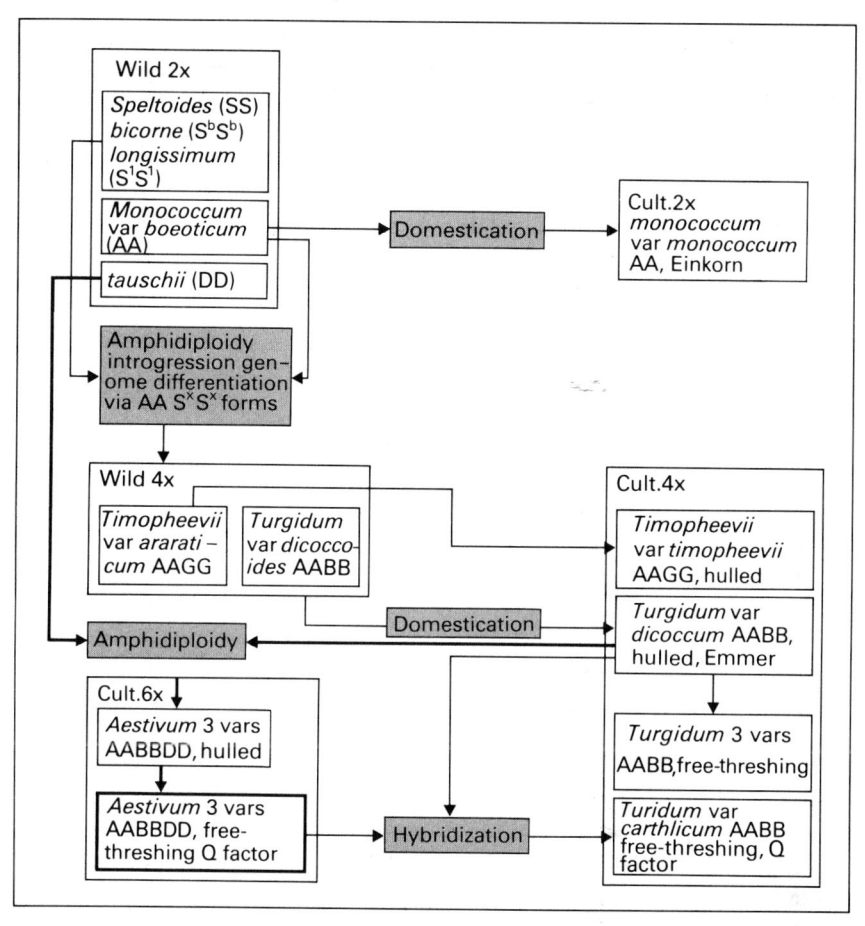

Fig. 5.4 Evolutionary relationships of the wheats, *Triticum* spp. 2×, 4×, 6× are ploidy series, ending with the hexaploid wheats of which *T. aestivum* is now the most important. Source: Simmonds, 1976.

seems to have been a perennial, *S. montanum*, which then gave rise to annual, weedy species of crops, a group of varieties known as the *S. cereale* group. From this set the cultivated *S. cereale* was developed. This probably took place in central-eastern Turkey–northwest Iran–Armenia and the most likely time about 3000 BC (i.e. much later than the primary crops of wheat and barley); it spread to Europe by about 2500–2000 BC but apparently not eastwards (G. M. Evans 1976).

Other plants of the Southwest Asia focus

As Fig. 5.3 shows, several other plants originated in this region, and together they have formed not only a disposal centre for crops of world significance but a characteristic type of Mediterranean and southwest Asian agriculture. The olive (*Olea europaea*) is a staple of this region: its wild ancestors are thought to be extinct, with the cultivated

174 The domestication of plants and animals

form having evolved as a hybrid swarm in the mountains of the eastern Mediterranean. It then moved westwards with a secondary centre in the Aegean and in southern Italy–Tunisia a tertiary centre evolved. A frequent companion of the olive is the grape which if not an actual staple ought not to be overlooked. Its evolutionary history is summarized in Fig. 5.5 where it will be seen that its centre of domestication is ancient Armenia,

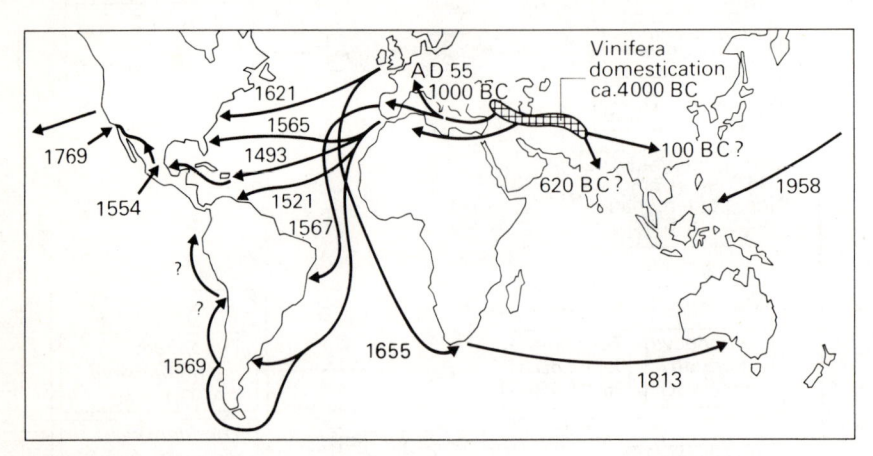

Fig. 5.5 The transference of the grape (*Vitis vinifera*) from its region of domestication to other parts of the world.
Source: Simmonds, 1976.

where the wild *Vitis vinifera* was first used *in situ* before the fourth millennium BC, after which time cultivation of the wine got under way in the Near East. By 55 AD vineyards were being established along the Mosel and its spread has continued practically up to the present day, being brought for example to the Philippines in 1958 from California (Olmo 1976).

Plants from other major foci of domestication

Probably at about the same time as cereal growing in Southwest Asia, agriculture was started independently in the Southeast Asian tropics. Here, the husbandry centred on bananas, sugar cane, tubers and probably legumes (Steensberg 1976), though it was also to develop the rice plant into a major cereal. Since not all other major domesticates can be discussed, two cereals will form the bread of this discussion, sandwiching a brief account of the tomato and the avocado pear. Regrettably, omission of important crops like cotton, the potato, sugar cane and many others, is inevitable: excellent concise accounts are given in Simmonds (1976); their original foci are shown in Fig. 5.3.

Maize (*Zea mays*) has had a relatively complicated genetic history, summarized in Fig. 5.6 where it can be seen that interbreeding with near-relatives such as *Tripsacum* and *Euchlaena mexicana* plays an important role. The geography is equally complex but some cultivars are probably firmly dated to 5000 BC at Tehuacan, Mexico. Many other races of *Zea mays* occur and the intricacies of their dispersal and breeding are dealt with

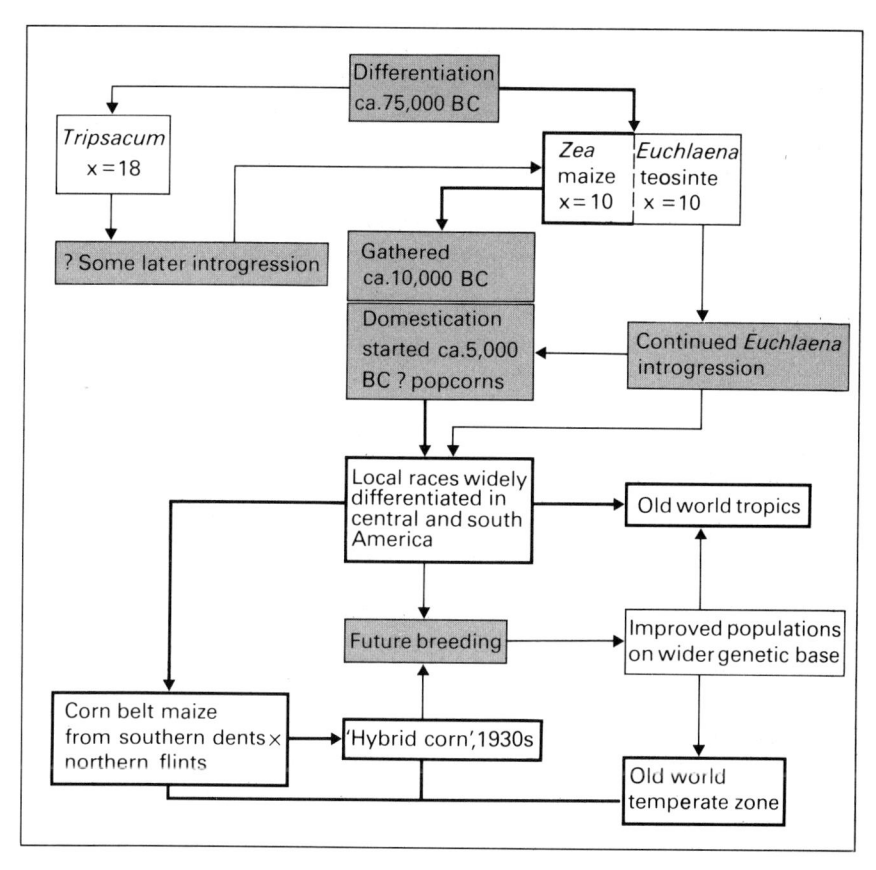

Fig. 5.6 Evolutionary relationships of maize, *Zea mays*.
Source: Simmonds, 1976.

by Goodman (1976); here it suffices to remind us that maize has become one of the world's great cereals, especially as a livestock food in individual nations. South and Central America were also the centres of origin of the tomato and avocado. The former (*Lycopersicum esculentum*) originates from wild ancestors in western South America and domestication probably took place in Mexico where the progenitor was possibly a weed of maize and potato crops. At any rate, it was well developed as a cultigen by the time of the conquest of the New World. Likewise, the avocado (*Persea americana*) developed in Central America where it is still known as the 'butter of the poor'. It was probably taken into cultivation about 5000 BC in southern Mexico and the same cultivars are the basis of modern production.

Unlike maize, rice (*Oryza sativa* and *O. glaberrima*) is usually fed directly to humans, for a large number of whom it is a staple food. The antiquity of domestication of *O. sativa* is not in doubt: remains from India dated to 4530 BC, and China from 3280–2750 BC have been found, but there is as yet not enough evidence to draw firm conclusions about the first place of its domestication. (*Oryza glaberrima* is confined to West Africa

176 *The domestication of plants and animals*

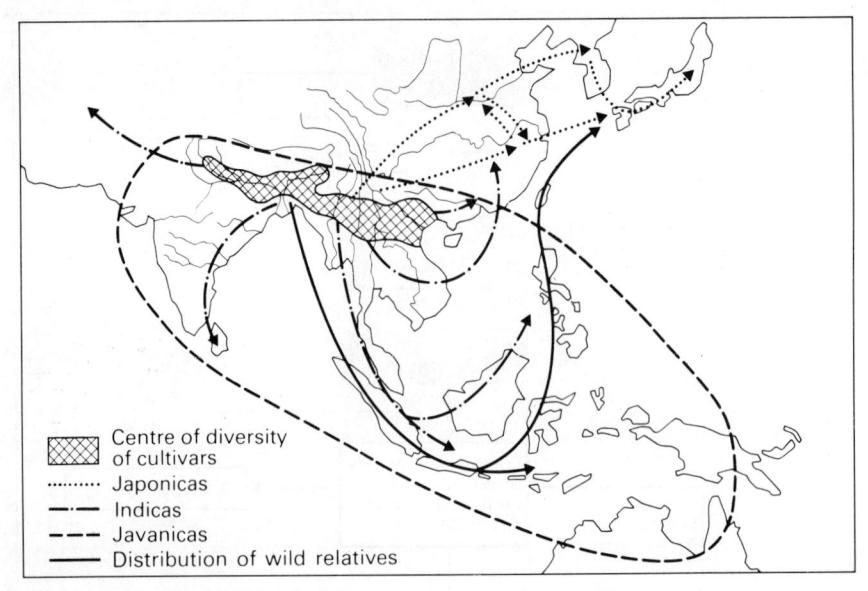

Fig. 5.7 The evolution and spread of races of the Asian rice *Oryza sativa*. It is presumed that domestication first occurred in the area of greatest diversity of cultivated varieties, cross-hatched on the map.
Source: Simmonds, 1976.

and represents an independent domestication, probably dating from around 1500 BC or later, in the upper Niger.) The routes of the spread of Asian rice are summarized in Fig. 5.7 but no chronology can be attached to most of it although its arrival in Japan (300 BC) and the Yangtze valley (200 AD) are known (Chang 1976). The dry-land rices appear to be secondary and developed as highly cultivated varieties out of the primary semi-aquatic line of evolution.

One general conclusion may perhaps be drawn from an examination of the history of most cultivated plants. This is the importance of weedy races as an intermediate stage of domestication: i.e. plants which readily propagate themselves but which require habitats disturbed by man; such races can be found in the history of most domesticated plants.

The domestication of animals

Preconditions

As with plants, there must have been a long period of association between animals and human groups before the somatic changes which can be identified archaeologically began to appear. With animals, behavioural characteristics must have been important in deciding whether a species became part of an early domesticatory relationship (Table 5.2). F. E. Zeuner, in his review of the subject published in 1963, focused attention on the practice of bringing up young animals as pets and thereby taming them. The domestica-

TABLE 5.2 Animal behavioural characteristics favourable and unfavourable for domestication

Favourable characteristics	Unfavourable characteristics

1 Group structure
  (a) Large social groups (flock, herd, pack), true leadership.
  (b) Hierarchical group structure.
  (c) Males affiliated with female group.
	(a) Family groupings.
	(b) Territorial structure.
	(c) Males in separate groups.

2 Sexual behaviour
  (a) Promiscuous matings.
  (b) Males dominant over females.
  (c) Sexual signals provided by movement or posture.
	(a) Pair-bond matings.
	(b) Males must establish dominance over or appease female.
	(c) Sexual signals provided by colour markings or morphological structures.

3 Parent-young interactions
  (a) Critical period in development of species-bond (imprinting, etc.).
  (b) Female accepts other young soon after parturition or hatching.
  (c) Precocial young.
	(a) Species-bond established on basis of species characteristics.
	(b) Young accepted on basis of species characteristics (e.g. colour patterns).

4 Responses to man
  (a) Short flight distance to man.
  (b) Little disturbed by man or sudden changes in environment.
	(a) Extreme wariness and long flight distance.
	(b) Easily disturbed by man or sudden changes in environment.

5 Other behavioural characteristics
  (a) Omnivorous.
  (b) Adapt to a wide range of environmental conditions.
  (c) Limited agility.
	(a) Specialized dietary habits.
	(b) Require a specific habitat
	(c) Extreme agility.

Source: Bowman 1977.

tion of scavengers may have started this way. These included the wild ancestors of the dog and pig which might have found sustenance in the rejected food wastes of settlements of all kinds. After agriculture had become established, then crop-robbing animals would no doubt impinge upon people: cattle and buffalo are thought to be examples of this group. Whatever the focus of initial contact, a process of feeding either adults or young, followed by taming, might lead to true domestication in the case of suitable species. In some cases the wild ancestral species might mix with the group in the process of becoming domesticated; in others the course of domestication, even though protracted, might result in only marginal changes in the genetics of the animal, as has probably been the case with reindeer.

As with plants, a spectrum of domesticates was established, most of which owe their continued existence to man. Some might survive now under feral conditions, depending on the environmental circumstances (some breeds of dog, probably cattle, and specialists like camels) but others very likely would not: sheep have apparently lost the considerable 'intelligence' possessed by their wild ancestors. Again as with plants, the total variety of ancient domestications is wide, but the number of economically important species is limited: the dog, goat, sheep, cattle, Asian buffalo, pig, horse and cat are overwhelmingly the most important domesticated animals on a world scale, followed by a number of regionally important beasts like reindeer, banteng, yak, elephant, camel, ass and llama.

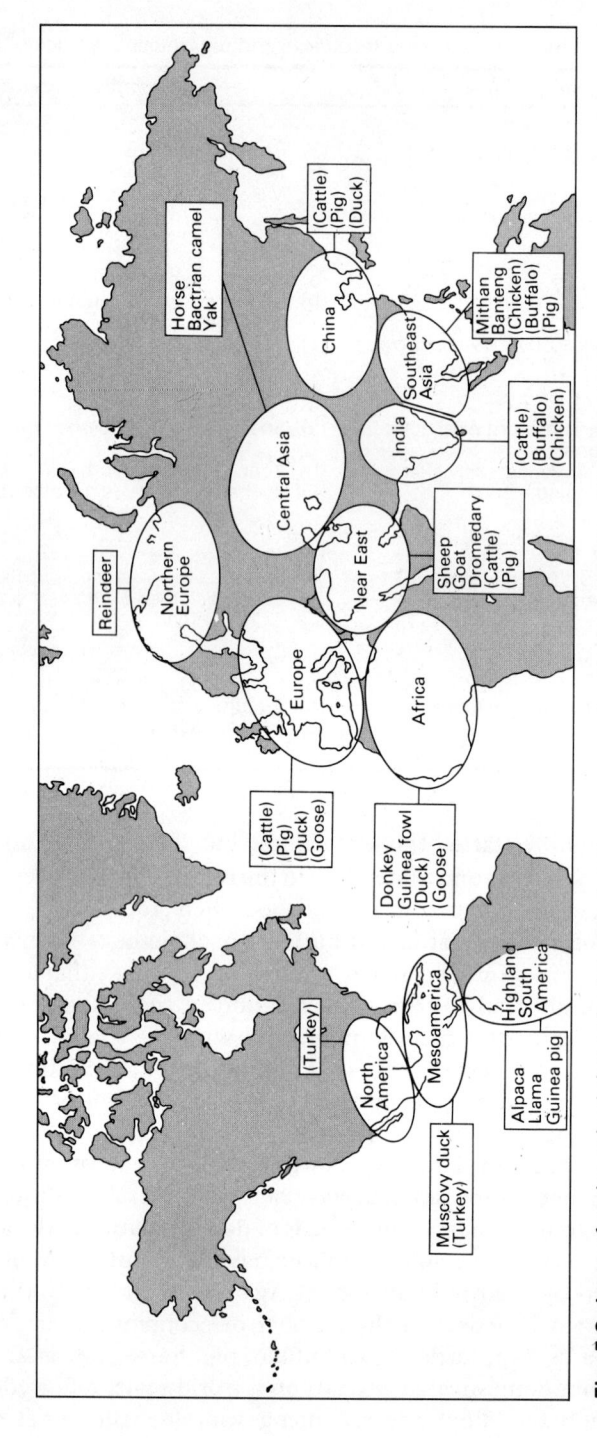

Fig. 5.8 Areas of major animal domestication. Bracketed names are of animals domesticated independently in more than one major focus, e.g. the pig. 'Near East' is equivalent to 'Southwest Asia' in the text.
Source: Harlan 1976a

Nor is manipulation finished: breeding improvements continue and new domestications are tried (p. 262).

Ancient domestications in Southwest Asia

As Fig. 5.8 shows, there is some parallel between plants and animals to the extent that Southwest Asia seems to have been the focus of domestication for species which today are of considerable importance. This is especially so in the case of goats and sheep; cattle were domesticated in this focus but probably not only here, as Fig. 5.8 suggests. The ancestry of goat and sheep is summarized by Bender (1975) and centres around animals of the varied terrain of Southwest Asia. A major problem of the earliest remains is distinguishing sheep and cattle bones, and the term 'ovicaproid' is sometimes used for the complex of the two species. The most likely ancestor of the domesticated sheep (*Ovis aries*) is the Asiatic mouflon (*O. orientalis*) which is found in a discontinuous distribution from Cyprus to the Punjab and Kashmir. The wild bezoar (*Capra hirens aegagrus*) is found continuously over much the same area, although extending further eastwards, and has strong claims to be the progenitor of the domestic goat, *Capra hircus* (Harris 1962a). Just as sheep and goats are difficult to tell apart osteologically, so are the early domesticates from their wild forbears, and hence the details of initial tamings are confused. At any rate, domesticated goats seem to be separable from wild ones by the middle of the seventeenth millennium BC at Jericho and Jarmo (Iraq). A roughly parallel chronology is suggested for sheep, with the most probable locus of initial domestication on the eastern and central Asian periphery of the Near East (Isaac 1970). The acquisition of tame versions of these two beasts conferred two advantages on their early possessors. Firstly, the wide environmental tolerance of sheep and goats allowed the penetration and utilization of marginal grasslands and eventually the development of pastoral nomadism, and secondly, unpredicted surpluses of domestic plant crops could be 'banked' in this form of live storage.

Wild cattle, *Bos primigenius*, were found throughout Europe, Asia and North Africa (though not in Ireland), and were most likely domesticated severally as groups of people came into contact with local populations. However the evidence for domestication is usually clear since domestic animals are smaller than the wild beasts, which weighed about one tonne and had a horn span of a metre. During the process of domestication, calves would have been easier to handle, but the problem of milk supply would require captivity of the mother also. Early domesticates appear at sites alongside sheep and goats, in the hilly, grassy and open-forested flanks of the Zagros, Lebanese and Palestinian mountains, with definite proof appearing about a thousand years later than sheep and goats. Unlike wild sheep and goats, the wild *Bos primigenius* or aurochs has become extinct, the last survivor dying in Poland in 1627 (Plate 24).

Other animal domestications

No account of animal domestication ought to omit that most useful of animals, the pig. The initial domestication seems to have been in Anatolia and Kurdistan and was

Plate 24 A picture of the last living specimen of the aurochs (*Bos primigenius*): seventeenth century AD.

under way by the middle of the seventh millennium BC. The wild pig, *Sus scrofa*, was distributed from the Atlantic fringe of mainland Europe all the way through Eurasia and so as domesticates spread, interbreeding with wild boars took place and was even encouraged. One reason for this would be to restore the endocrine balance which is sometimes disturbed in domesticates and which interferes with successful breeding.

In one way the case of the dog (*Canis familiaris*) is analogous to the pig, since its ancestor the wolf (*Canis lupus*) was widely distributed over Eurasia and North America. (Older views identifying the jackal or a now extinct wild dog as ancestors have now largely been abandoned.) Isaac (1970) suggests that in Southeast Asia there were a number of small species of wolves and that one of these might have provided the first truly domesticated dog in the Old World. Isaac argues in parallel that North American dogs (one is found from Idaho, *ca* 8000 BC) are the result of introductions from Eurasia rather than an independent domestication; recent work on the Chinese wolf (*Canis lupus chanco*) suggests that it may well have been the ancestor of the dogs of North America (Olsen and Olsen 1977). As with the pig, wild creatures would have attached themselves to habitation sites and the young are eminently tamable. Breeding with wild stock would also have continued for many millennia.

These views make any case for a single area of domestication rather less strong. Whatever the precise details of the initial domestication, the usefulness of the dog, as auxiliary herdsman, guard, pet and occasional meal, is abundantly clear.

Most of the animals so far mentioned have become more or less universally distributed, wherever domesticated animals are used by human societies. There are some animals, however, which might be called specialities; those which have spread little from their original centres of domestication, either because their environmental tolerances are narrow or because no avenue of wider cultural acceptance has opened up. An example would be the South American Camelidae: the guanaco (*Llama guanacöe*), vicuña (*L. vicugna*), llama (*L. glama*) and alpaca (*L. pacos*). The domestication of this group within the puna of Junin in the Central Peruvian Andes has been postulated by Pires-Ferreira *et al.* (1976). They suggest a progression from generalized hunting through specialized hunting and the control of semi-domesticated Camelids, to the herding of a domesticated variety, after which time the appearance of distinct breeds of domestic llamas took place. (In that region the period of domestication was between 4200–2500 BC.) The Spanish conquest began a process of replacement of the llamas by sheep, and probably because of Spanish indifference to the llama group as well as environmental factors, it has never been exported to other regions of the world.

The Old World camel itself (i.e. the single-humped dromedary, *Camelus dromedarius*) is also restricted in its post-domesticatory spread, largely because away from desert-like conditions the horse is a better animal. The tame camel is very similar to the few wild camels that survive and has apparently altered little under domestication. This probably took place in Arabia later than the other large mammals but was accomplished by about 1800 BC, although the animal was probably not introduced into Egypt until the Graeco-Roman period. Important though the camel has been in the arid zones of the Old World, it has been little introduced beyond them, although extravagant plans to colonize places like Arizona and Australia with it have existed in modern times.

There is also quite a long list of minor domestications (Table 5.3), some of which might be considered a little freakish (e.g. the dormouse, fattened in clay pots by the Romans), whereas others have given rise to considerable industries, such as the silkworm. There are yet others which might be termed status domestications, where the animals are perhaps partly domesticated but could easily go back to the wild and survive. In ancient times in the Mediterranean Basin, the gazelle, oryx and addax were herded

TABLE 5.3　　Minor animal domestications

Animal	First domesticated	Time
Ferret	Palestine	Before 1000 BC
Mongoose	Egypt	New Kingdom
Rabbit	Mediterranean	After 4th C. BC
Dormouse	Italy	Roman Empire
Peacock	India–Mesopotamia	?
Guinea fowl	Greece	? 5th C. BC
Pigeon	Mesopotamia	4500 BC
Goose	Mediterranean	Ancient Greece
Duck	Mesopotamia	Ancient cities
	China	Unknown
Cormorant	Japan	1st C. AD
Cranes	Egypt	*ca* 1450 BC
Silk-moths	China	*ca* 3000 BC
Honey-bee	Egypt	before 600 BC

Source: Zeuner 1963.

by the temples, and predators like the cheetah and various falcons are kept by royalty and aristocracy to this day in order to maintain the association between the upper ranks of a social hierarchy and the desire to kill for sport.

Generalities of ancient domestication

Genetic and somatic change

In the case of plants, the changes in gene structure that accompany a long period of domestication (and eventually produce a genetically isolated population) led to certain persistent morphological and physiological alterations in the new plant (Table 5.4). The

TABLE 5.4 Adaptation syndromes resulting from automatic selection due to planting harvested seed of cereals

I. Selection pressures associated with harvesting result in:
  A. Increase in per cent seed recovered
    Adaptations: (1) Nonshattering
                 (2) More determinate growth
                     (a) Growth Habit I: Cereals whose wild races have lateral seed-bearing branches, e.g. maize, sorghum, pearl millet. There is a trend toward apical dominance resulting in fewer inflorescences, larger inflorescences, larger seed, greater daylength sensitivity, and more uniform ripening.
                     (b) Growth Habit II: Cereals with unbranched culms e.g. barley, emmer, rye, einkorn. There is a trend toward more synchronous tillering and uniform whole plant maturation.
  B. Increase in seed production
    Adaptations: (1) Increase in per cent seed set.
                 (2) Reduced or sterile flowers become fertile.
                 (3) Increase in inflorescence size, especially in maize, sorghum, pearl millet.
                 (4) Increase in number of inflorescences especially wheat, barley, rice, etc.
II. Selection pressures associated with seedling competition result in:
  A. Increase in seedling vigor
    Adaptations: (1) Greater seed size
                 (2) Lower protein, higher carbohydrate
  B. More rapid germination
    Adaptations: (1) Loss or reduction of germination inhibitors
               (2) Reduction in glumes and other appendages
III. Selection pressures associated with tillage and other disturbances result in the production of weed races
    Adaptations: (1) Plants competitive with cultivated races, but
               (2) Retain the shatter habit of wild races

Source: Harlan 1975.

doubling or multiplication of chromosome numbers (*polyploidy*), as described above in the history of wheat, usually increases size and robustness of the plants; this is particularly marked in the size of individual fruits although on the other hand, fruit-producing ability may decrease. Nevertheless, the fruit of the domestic red pepper is now about 500 times the size of that of its ancestor, and the maize cob of today is at least 7 times the length of early domesticates. Those parts of the plant which protect it against undue herbivory tend to be lost under domestication: bitterness and toxicity are obviously selected against, as are thorniness or the presence of mechanical methods of protection such as awns and glumes in cereals. Cultivation and selection work towards producing a plant whose edible parts ripen simultaneously, whereas in the wild a more continuously fruiting habit is commoner; where this persists under domestication it is as a result of deliberate husbandry as in the case of some citrus fruit. Lastly, the cultivated plant loses the ability to disseminate itself and acquires the tendency to become an annual, even though its progenitors were perennials; wild rye for example is a perennial, cultivated rye an annual. A summary of the adaptations resulting from man-directed selection in cereals is given in Table 5.4.

With animals, analogous morphological changes can be inferred from the archaeological record. Many early types of domesticated animals are smaller than their ancestors: dog, cat, cattle, sheep, goat and pig are examples. But some (llama, camel) are about the same size, and the domesticated horse is considerably larger than the wild equines that preceded it. Another general statement sometimes made (e.g. by Zeuner) is that domestication selects for juvenile characteristics but preserves them in the adult, a process known as neotony. It is apparent in the skull shapes of domesticated dogs, in the short face and small horns of some domestic cattle, and in the skin folds of adult dogs. Hair tends to become curly rather than straight and baby hair is retained; pied coats replace monotone colours; and the size differences between male and female are reduced. Selection based on non-utilitarian human values may also change the morphology, for example, in breeding goats or sheep with especially flamboyant horns, or in producing some of the more egregious breeds of dog. Castration may aid selection for certain types of behaviour, especially docility, and fattiness, but is obviously avoided where fierceness and agility are prized, as in the 'wild' bulls of southern Europe and Latin America.

In modern times, the changes wrought by scientific breeding (see p. 262) have been immense, but the differences brought about by many millennia of gradual change by subsistence societies must not be underestimated for it was then that the foundations of all our modern agriculture was laid.

Effects of domestication on local biogeography

Considered in total, the ecological consequences of domestication have been very great indeed but even in ancient times the local environment of foci like Southwest Asia must have undergone considerable changes. Many of the wild plants of the southwestern Iranian steppe, for example, have the same general growing season as early wheats and barleys, and compete for the same alluvial soils with low salinity as the cereals require. Flannery (1965, 1969) suggests that as cultivation increased in area and intensity, these plants (such as the legumes *Medicago* and *Astragalus* spp., and grasses of the genera *Avena*, *Cynodon* and *Phalaris*) became labelled as weeds. If they were deliberately removed, then other plants took their place, probably brought in as seeds along with imperfectly cleaned cereal seedcorn. Grasses like *Aegilops* and *Lolium* are among this group, and the woody legume *Prosopis* appears to increase along with the cultivated area; fortunately its seeds are edible. Early agriculture would have needed fallow periods and this practice would have provided a habitat for plants different from those of permanent steppe or grassland: species of bedstraw (*Galium*) and plantain (*Plantago*) may be cited. Irrigation often brings with it salinity and a shift not only to more salt-tolerant crops but to salt-tolerant weeds as well. The canal bank became a niche for crop plants such as the date palm and the onion in early Mesopotamia and also as a retreat for crop pests such as the bandicoot-rat, *Nesokia indica*. An interaction between domesticates and a wild grass is shown by the case of *Stipa*, a feathergrass of the Zagros and Taurus ranges. Its seed has a sharp callus which penetrates the soil and is driven by the hygroscopic and twisted awn which rotates according to its state of humidity. This was of little concern to wild sheep with a very hairy coat but domesticated varieties with woolly coats caught the *Stipa* seeds which then 'burrowed' into the animals' skin and allowed infection. In broader terms, we must also think in terms of gradual shifts in the flora of areas grazed by flocks of domesticates since their food preferences are often narrower than their wild ancestors. Many areas of apparently wild vegetation are doubtless areas of cultural vegetation altered by the relentless selectivity of thousands of years of pastoralism.

In general, the effect of early food-producing groups was to extend their impact upon the environment to abiotic components of the ecosystem (Butzer 1974). Conscious large-scale destruction of native vegetation now became characteristic, as did the deliberate eradication of plants and animals identified by farmers and pastoralists as non-useful. In addition to the introduction of domesticates, selective favouring of wild plants useful to the economy (oak and elm as animal feeds, for example) would also have taken place. In the abiotic sphere, the agricultural use of the soil led to changes in soil humus, structure and chemistry. Soil erosion became a potential danger wherever the native vegetation was cleared from hillsides and uplands, and salinification became more widespread in areas of irrigation (Dimbleby 1976).

The suggestion has been made (Lewis 1972) that the early Neolithic occupants of the Near East burned woodland areas persistently and that the area of grassland was increased at the expense of the oak-pistachio ecosystem. This would have produced an increase in the number and range of sheep and goats, and probably the numbers of wild gazelle and wild cattle as well. Both sheep and goats could be hunted or herded as semi-domesticates by the seasonally transhumant hunter-gatherers of the region and this closer association might well have been an early step in domestication. If so, then it is notable as an example of domestication taking place along with major biogeographic changes.

Why domestication?

The discussion of the several conflicting hypotheses about the reasons for the inception of domestication (Cohen 1977; Orme 1977), is not germane to the major themes of this book, which is more concerned with the results. Thus the arguments over the role of climate in making hunting and gathering less secure, the effect of population growth in forcing a shift from hunting large mammals, to using a wide spectrum of wild foods, to cultivation and domestication are not of primary importance here. Neither are the importance of religious factors in using plants and animals to demonstrate the cycle of death and rebirth, or the role of harvesting methods in favouring the emergence of cereal cultivation, rather than the continuance of wild-seed gathering. But what most scholars emphasize is the long period of association between biota and men before recognizable domestication becomes detectable, and that modes of husbandry which alter neither genetics nor morphology may have had a long history which preceded true domestication (Harris 1978). Rather than thinking of domestication therefore as a 'Neolithic revolution', we might prefer to see it as one stage in a process of evolution which is still continuing today, with further refinements in breeding of old domesticates, and some addition of new organisms to the spectrum, an attitude well illustrated by Higgs and Jarman's (1972) diagram (Fig. 5.9) of the continuing development of the domestication of species.

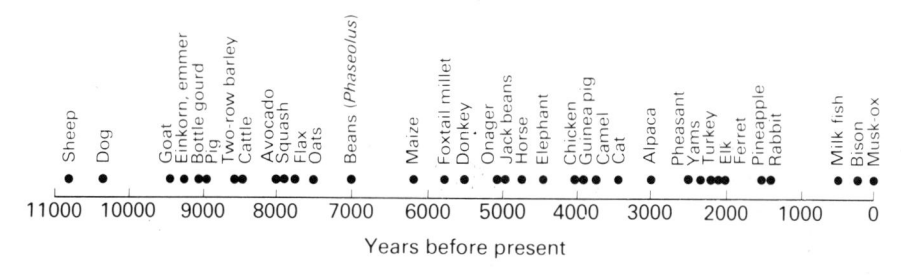

Fig. 5.9 The continuing process of plant and animal domestication with approximate dates for each group or species.
Source: Higgs and Jarman. 1972.

The establishment of agricultural complexes

The primary domestications were followed in many cases by diffusion, and independent domestications also took place in suitable regions. Coupled with differences in climates, cultures and availability of wild biota, this means a differentiation of types of agriculture in the world. By about 500 BC, therefore, a number of distinctive agricultural complexes were to be found (Grigg 1974). The basic Southwest Asian type of agriculture was dependent upon the major cereals wheat and barley, legumes, and the important animals were sheep, goats, cattle and pigs. A number of variants of this basic type were found, including the irrigated systems of the Nile, Indus, Turkestan and Tigris–Euphrates; some of these areas had added cotton to the basic set of crops. Other variants existed in the Mediterranean (with the notable addition of tree crops such as vine, figs and olive) and Northern Europe where oats and rye assumed a more important role than further south.

In Southwest Asia, tropical vegeculture had developed, often on a shifting basis and using as staples crops such as taro, yam and banana. These crops were often replaced by wet-rice which became the typical agriculture of this region (along with water buffalo and zebu cattle) until colonization by Europeans. This system was different from that of northern China, where the foxtail millet and soybean combined with the pig to offer a viable form of subsistence, based on conserving moisture rather than irrigation.

In Africa, shifting cultivation was the norm at this period, with tropical vegeculture based on yams forming one distinctive mode, and millets and sorghum another. Shifting cultivation was also characteristic of the Americas. Here one system was based on the cultivation of root crops, including the potato as well as yams, and adding animals like the llama group of Camelidae. The other grouping which spread northwards to occupy much of North America by contact time, was the maize–beans–squash complex, in which domesticated animals were not important.

This cross-section must not be taken to imply a static situation: expansion of cultures was a continual process and crops spread outwards in many ways, not the least as a result of the great explorations of the fifteenth century AD as well as the earlier establishments of great empires. So even before the coming of industry and the ease of transport afforded by steam, the exchange of cultivated plants and animals became a normal feature of human rearrangements of the Earth's biogeography.

The diffusion of domesticates in later prehistory and early historic times

Not all domesticates remained confined to their original hearths: some were thus restricted but others moved to other places. Their translocation took place at varying rates and on widely differing scales, for some moved little at all outside their hearths, whereas others became of worldwide significance even in pre-industrial times. Bearing in mind that data for the early post-domestication movements of many plants and animals are scanty, and that any subdivision by scale is bound to contain an arbitrary element, we can distinguish three types of diffusion.

1. Plants and animals that originated in a limited area, which achieved only local significance, and which spread little beyond the ranges of their immediate wild ancestors. Reasons for such limitation include the failure of the species to adapt to environmental conditions elsewhere due to a lack of genetic plasticity, and the lack of opportunity to be transported to a place where a suitable biological and cultural niche might be present. The multi-purpose animals of high mountains in both Old and New Worlds provide examples: the Yak (*Bos grunniens*), for example, was (and is) found both wild and tame in Tibet and adjacent provinces of China (Plate 25) but elsewhere is rare, since most of its practical uses can be supplied by other animals. Only in its native habitat does its ability to survive and to provide skins, meat, milk, dung, traction, and burden on narrow mountain tracks become essential. A similar role was performed by the llama (*Lama glama*), alpaca (*L. pacos*) and vicuña (*L. vicugna*) in Andean America. The llama was primarily a beast of burden and the alpaca's fine wool was an important textile fibre in the highlands of South America. Neither spread beyond the range of their probable wild ancestor, the guanaco (*Lama guanacöe*).

Analogous examples of plants include species such as the marsh elder (*Iva annua*) which Yarnell (1971) considers to have been cultivated by Early Woodland cultures in the lower Mississippi valley of North America but which did not spread far beyond that centre either in time or space. The 'golden fruit of the Andes' (*Solanum quitoense*) was for long little known outside its native Colombia and Ecuador, in spite of its potential uses as a fruit and a sweet-sour drink (NAS 1975). Lesser millets of West Africa such as *Brachiaria deflexa* and *Digitaria iburua* are similar examples (Harlan 1975).

Plate 25 A yak (*Bos grunniens*) bull in Nepal. Although ideally suited as a domesticate to this environment, it has never been taken elsewhere for use.

2. Plants and animals which underwent diffusion from their original hearths to the point where they became of continental significance. This was possible because adaptation to different natural environments could be achieved by selective breeding or by the creation of favourable niches by the domesticators or indeed because environments were relatively uniform across long distances. Further, the cultural channels for transmission and acceptability were present, for example through the spread of a religion which favoured the domestication of a particular animal for sacrificial purposes (Isaac 1970), or through the migration and land colonization of a culture-group bringing their domesticates, both plant and animal, with them.

Examples of this group are numerous. The maize-beans-squash complex of Meso-America had spread from the southwest of the USA to New Zealand between 700 and

1400 AD and later expanded further westwards. From centres in the Americas and Asia, the grain amaranths (*Amaranthus* spp) spread widely through those continents until relegated to the margins of cultivation by the advent of maize or rice (Sauer 1950).

The dromedary (*Camelus dromedarius*) forms an example of an animal which spread quickly beyond its original focus of domestication (inspired by the horse-borne nomads of central Asia sometime before 2000 BC) to Syria, Palestine and Anatolia and then by 1000 BC to Mesopotamia, to Egypt by 300 BC and Libya by the second century AD. Spread thereafter was conditioned strongly by environmental factors since other animals performed better outside the rigorous environments where the camel's virtues were most valuable. The diffusion of the Asian buffalo and the ass were analogous in character. The persistence of a contemporary domesticate related to a prehistoric culture is possible in the case of the goose (*Anser anser*), which was strongly favoured in the culture of Hallstatt (Iron Age) Europe and is still important in the economy of eastern Europe from the Balkans to Poland: an example, suggests Zeuner (1963), of the survival of a goose economy from Iron Age times to the present day.

3. Even in pre-industrial times there were plants and animals which had achieved extra-continental significance. Such biota had to possess wide adaptability, including the genetic plasticity needed for behavioural changes in animals, be capable of easy cultural acceptance, and be suitable for transmission by trade, migration or cultural change. Any list of examples must include the pig (*Sus* spp) where diffusion was probably aided by independent domestications outside the earliest hearths. Coupled with its fecundity, diversivorousness, and portability, this meant the transport of the animal practically wherever men went including some remote islands in the Pacific where it incidentally effected some very far-reaching changes in flora and fauna.

Numerous plants achieved wide transmission in the pre-industrial era. The sweet potato (*Ipomoea batatas*), for example, has wild varieties in tropical central America and northern South America and had spread to Polynesia in pre-Columbian times, having the same name in coastal Peru as in Maori (Baker 1970). Columbus recorded its widespread adoption in the Americas and then a rapid burst of dispersal occurred in the sixteenth century following the tracks eastwards of Portuguese voyagers and westwards (e.g. to Guam and the Philippines) with the Spaniards (Yen 1976). The peanut (*Arachis hypogaea*) has its origin in the Matto Grosso of Brazil but primitive domesticated forms are found also in China where the Portuguese voyagers had arrived by about 1500 AD. Cotton existed in the wild state in diploid forms in Africa, Australia and from East Africa to Pakistan. One of these became cultivated in Southwest Asia and thence became the cultigen *Gossypium arboreum* in India by 5000 BP. Wild tetraploids existed in western South America and Middle America and these were domesticated by at least 5500 BP. Introgression with one of the American diploid species as well as inter-tetraploid introgression produced the progenitors of Sea Island, Egyptian and upland varieties of cotton which dominated its commercial rise (Phillips 1976). The Irish potato (*Solanum tuberosum*) seems to be first recorded in Europe in 1596 but only later did it achieve the status of a staple food in cold climates (Klages 1942; Salaman 1949; Dodds 1965).

The spread of these domesticates was not, of course, uniform and lacunae in the distri-

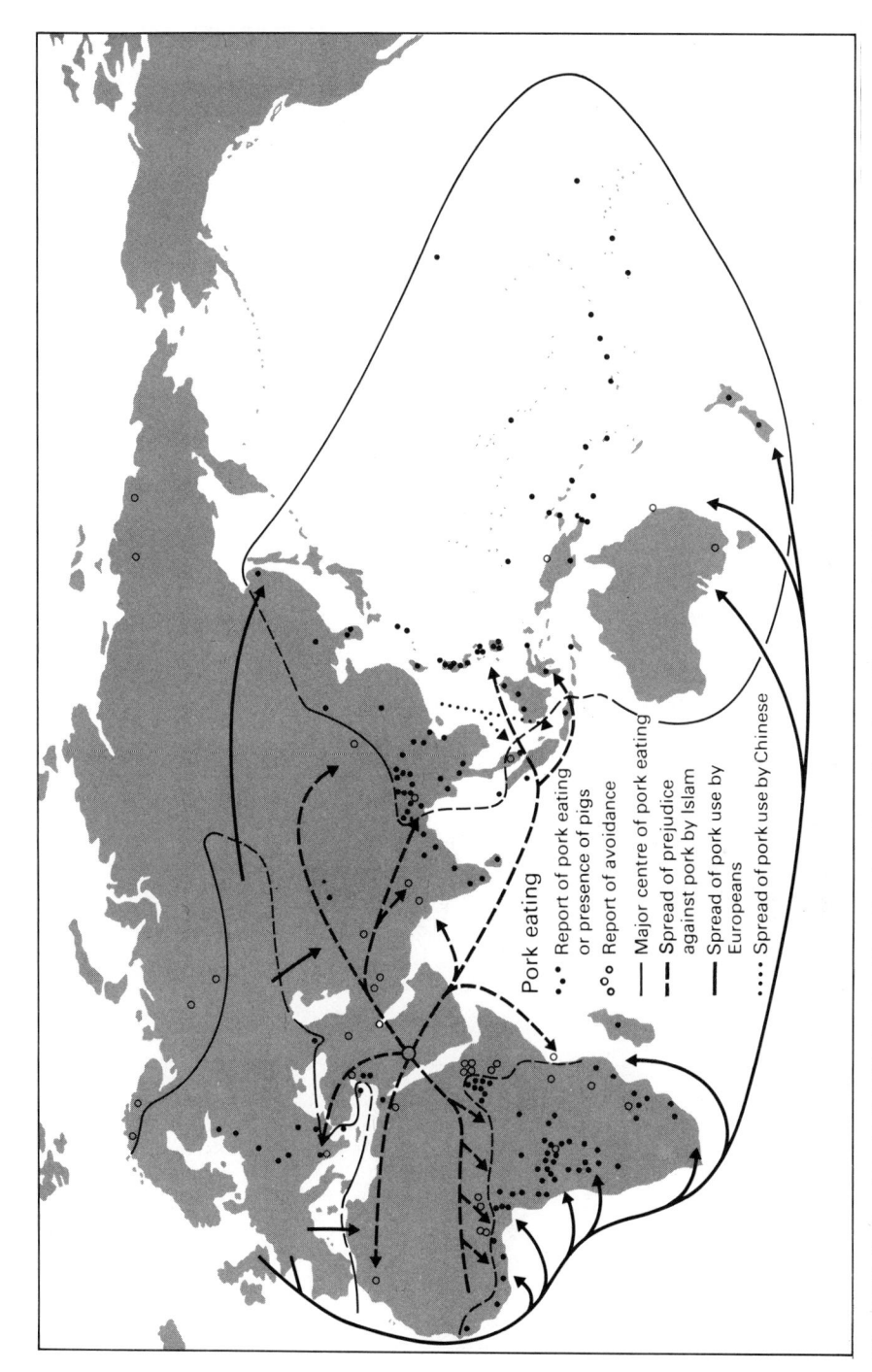

Fig. 5.10 Pork eating in the Old World: Negative (avoidance) and positive (emphasis).
Source: Simoons, 1961.

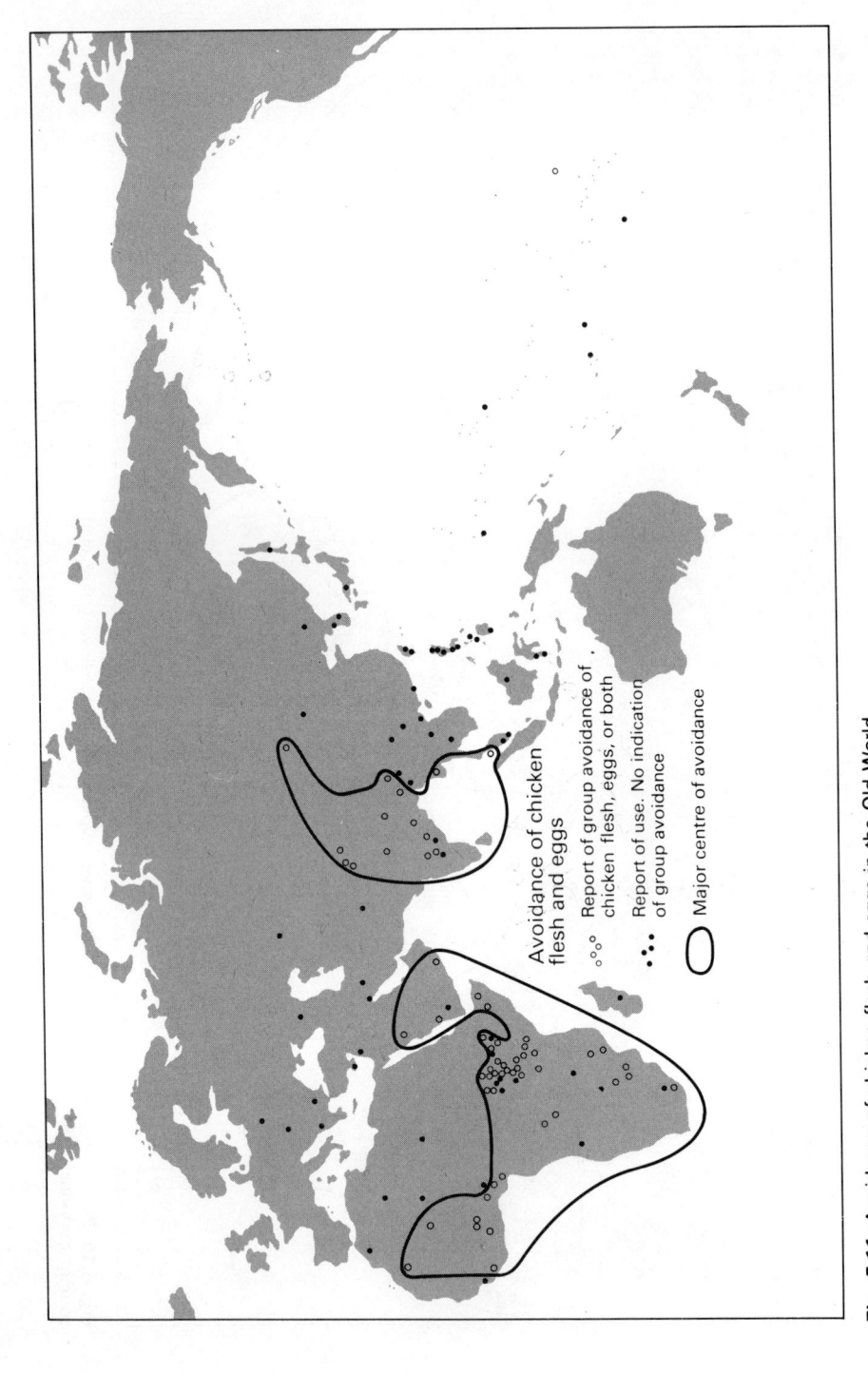

Fig. 5.11 Avoidance of chicken flesh and eggs in the Old World.
Source: Simoons, 1961.

butions of the domesticates often occurred. Most interesting of these are the areas where cultural rejection of a domesticate has occurred in places to which it is suited environmentally. The Middle East, for example, became a centre of pork avoidance as the pig became an unacceptable item of food among Muslims, Orthodox Jews, Yezidi, the Mandaeans of Iraq and Iran, and both Christians and Pagans in Ethiopia (Fig. 5.10). This avoidance was apparent by the fifth century BC, largely because the pig was becoming a sacrificial animal. The reasons for the prohibition are uncertain but Simoons (1961) argues that the 'rational' explanations will not stand up to close examination, and that conflicts between settled pig-rearing peoples and pastoralists (for whom the pig was unsuitable) are at the root of the growth of an out-group aversion to the animal which became religiously codified at the eastern end of the Mediterranean in the years after 1400 BC. Other food avoidances in the Old World include the rejection of beef in India due to the sacredness of cattle in the Hindu religion; the prohibition of chickens and eggs in sub-Saharan Africa and parts of South Central and Central Asia (Fig. 5.11); and an avoidance of horse-flesh, which was spread to Europe by Christians, and which is breached from time to time when food shortages occur.

In general it seems that animal foods excite disgust more readily than plants, and that 'rational' explanations pale in the face of tokenism, magic and religion, and the fear of ritual contamination, as reasons for not eating animals which exhibit the dual roles of being both sacred and unclean.

Further Reading

BENDER, B. 1975: *Farming in prehistory*. London: John Baker.
CLARKE, J. G. D. and HUTCHINSON, J. (eds.) 1976: The early history of agriculture. *Phil. Trans. Roy. Soc. Lond.* **B** 1–213.
COHEN, M. N. 1977: *The food crisis in prehistory*. New Haven and London: Yale University Press.
HARLAN, J. R. 1975: *Crops and man*. Madison, Wisconsin: Crop Science Society of America.
HARRIS, D. R. 1978: Alternative pathways towards agriculture. In C. A. Reed (ed.), *Origins of agriculture*. The Hague: Mouton, 179–243.
ISAAC, E. 1970: *Geography of domestication*. Englewood Cliffs, New Jersey: Prentice-Hall.
REED, C. A. 1969: The pattern of animal domestication in the prehistoric Near East. In P. J. Ucko and G. W. Dimbleby (eds.), *The domestication and exploitation of plants and animals*. London: Duckworth, 361–81.
SAUER, C. O. 1969: *Agricultural origins and dispersal*, 2nd edn. Cambridge, Mass.: MIT Press.
SIMMONDS, N. W. (ed.) 1976: *Evolution of crop plants*. London: Longmans.
SIMOONS, F. G. 1961: *Eat not this flesh*. Madison, Wisconsin: University of Wisconsin Press.
UCKO, P. J. and DIMBLEBY, G. W. (eds.) 1969: *The domestication and exploitation of plants and animals*. London: Duckworth.
ZEUNER, F. E. 1963: *A history of domesticated animals*. London: Hutchinson.

6
Man and biota in pre-industrial times

The advent of domesticates defines a new type of man–nature relationship, in which human societies garnered food and other biological materials from essentially 'tame' plants, animals and ecosystems, as distinct from the 'wild' which was the habitat of hunters and gatherers, even allowing for a small amount of environmental manipulation by the pre-agricultural peoples. With the acquisition of domesticates, the wild systems (including the seas) continued to be a major supplier of biological materials, such as plants impossible to cultivate or perhaps culturally undesirable to have ubiquitously available near a settlement: drug plants are one example. More common was the practice of using the wild lands as sources of animal flesh: in medieval Europe domesticated swine were depastured through 'the waste', which was mostly woodland, and wildfowl were taken from unreclaimed wetlands; certain social classes were allowed to hunt for venison and wild pig. Even in densely populated China, most villages has a sacred grove of trees which housed, besides the spirits of departed ancestors, trees and shrubs with products of medicinal or culinary value.

The very ubiquity of the wild may well have provided a lodestar for man: nature to a great extent provided the norms, although the balance inherent in this idea was eventually to swing to the human end of the beam. On a world scale, much of the land surface in 1500 AD must have still been clad in natural or near-natural ecosystems (see the continental maps in Bennett (1975) for estimates), and man's activities in creating early stages of succession were confined to a number of large areas (China and Southeast Asia, Western and Central Europe, parts of Meso-America) with small patches elsewhere. Lee and De Vore (1968) suggest that in 1500 AD, one per cent of the world's population were still hunter-gatherers and so presumably still living off largely wild ecosystems.

However, farming was the new relationship between man and biota which spread out from its hearths to occupy the surface of the land where conditions were suitable. In the most general terms, agriculture represents a replacement, either temporarily (as with shifting cultivation) or permanently, of the natural ecosystem with a man-made ecosystem containing crop biota whose genetic structure is much affected by human purposes. So the very complex food web of a natural ecosystem may be replaced by a simple food chain PLANTS–MAN, or with a domesticated animal interposed, PLANTS–ANIMALS–MAN, and from these sources, plus the wildland's products if available, sufficient calorie, protein and mineral nutrition has to be derived for the human population. The efficiency and transfer of energy and protein in a simple plant–man chain can be quite high: 30–40 per cent of the net primary production of the plant can be harvested as food, and 70–80 per cent of the harvest can be digested by humans. The protein content of various plant foods varies from *ca* 6–20 per cent of dry matter; in general the production of protein by a plant involves energy-expensive metabolic steps which use up carbohydrates which would otherwise accumulate as energy-rich tissue. The total organic yield of a crop is often therefore inversely related to its protein content, which may cause malnutrition difficulties in places where the human populations are very dense or rapidly growing. Supplementation of protein may come from fresh or dried fish, from wild animals, or domesticated animals. In the latter case a lot of energy is traded for the protein because the animals are usually inefficient converters of plant energy into protein, losing energy via respiration, selective intake, differential digestion and faeces. The most efficient way of using animals is either in a detritus chain, especially if they will eat domestic refuse and parts of plants inedible by man (hence one reason for the success of the pig and fowl in many pre-industrial societies) or as croppers of wild plants inedible to man.

Pre-industrial agriculture: the dissemination of biota and of knowledge during historic times

The folk-movements of early times were responsible for much dissemination of agricultural practices from their hearths. To this was added the development of trade and exploration in medieval and later historical times which provided an important route for diffusion of cultivates. In particular, it allowed people to import and experiment with new crops and to test their genetic plasticity against the different environments to which they were imported.

After the fall of Rome, the major empires had relatively little contact with each other and only the Arabs were significant in the transmission of crops over long distances, for example in bringing citrus fruits, rice, cotton, and sugar-cane into the Western Mediterranean. By contrast, after the fifteenth century, European colonists took their crops with them, especially those which would sell at home, like sugar-cane. In terms of the introduction to the continents of diffused domesticates before 1750 AD, we may note that Europe was dominated by crops from Southwest and Southeast Asia until the sixteenth century. After *ca* 1500 AD potatoes and maize came from the Americas (although Sauer (1960) has argued for a pre-Columbian introduction of maize into Europe via Asia Minor), and then grasses were developed from indigenous strains. Timothy went from England to New England where it was popularized by one Timothy Hansen, whose name it brought back to England. The leguminous forage crop lucerne or alfalfa, for example, was brought in classical times from Iran but was rediscovered during the sixteenth century in Spain, from where it spread quickly to Italy, France and Germany. By the eighteenth century, the food plants now important in Europe had all been introduced, even if not widely cultivated.

America likewise contributed important crops to Africa. Manioc and maize were taken there in the sixteenth century as was the sweet potato; from the east, the Portuguese brought the Asian species of rice (*Oryza sativa*) in the sixteenth century where it replaced the native rice *O. glaberrima* in areas of large-scale cultivation. Asia was also the recipient of American crops such as groundnuts, potatoes, and manioc in the eighteenth century but none became significant with the exception of groundnuts after 1918. In cultural terms, the predominance of rice assured that 70 per cent of Asia's arable land is still planted to rice, although the introduction of plantation crops in colonial times (rubber, tobacco, oil palm, coffee) significantly altered economies. The least affected of any major region by crops from outside has been China.

The Americas themselves absorbed numbers of Old World crops as well as providing extended environments for indigenes. The Spanish and Portuguese took the Mediterranean fruits and the vine (not really a success until the nineteenth century in North America) and the Portuguese also took sugar-cane and bananas from West Africa; in 1718 the Dutch brought coffee to Surinam from Indonesia; Britain acted as an intermediate host for the potato before it was introduced into South Carolina in 1674. Of the animals, sheep, cattle, and goats were part of Columbus's cargo and spread rapidly through Central America and southern North America where ranching was established with the use of Spanish longhorn cattle. The merino sheep was confined to Spain until the sixteenth century, until it was taken to Latin America. The Guinea sheep, which is hairy rather than woolly, was taken to Brazil and the West Indies in the sixteenth century from the savanna of West Africa.

In Oceania, the Maori brought taro and sweet potatoes to New Zealand and the Spanish introduced American plants into island groups such as the Marquesas and Guam. Quite early in the eighteenth century, the merino sheep was taken to New South Wales via changes of boat in England (George III's farm at Windsor) and Cape Colony. Most of the introduced fauna of New Zealand however dates from after 1750.

It is clear therefore that the sixteenth century represents a watershed, and indeed it is common to think of 1500 AD as a date when the mainly self-contained but developing regions of the world coalesced so that the ecumene became to all intents and purposes continuous. The perfection of the oceangoing ship was to a large extent responsible for this and the ecological repercussions turned out so far beyond the plants and animals that the vessels carried.

Types of agricultural ecosystems

The alliance of pre-existing ecosystem types to the diversity of introduced biota and cultures produced a great variety of patterns of use of biota for food and fibre by the middle of the eighteenth century, before the impact of an industrialization based on fossil fuels. Three general types delineated by Harris (1969) can be discussed, understanding that there existed a great deal of variation within each of them:

1. The most important, both for that time and our own, is seed cultivation, i.e. the husbandry of a plant's seeds as a source of carbohydrates and of protein. In some variants, the seeds are then fed to animals to allow the production of animal flesh. The relatively few cultivars from the three main centres of domestication were the mainstay of this economy: rice, maize and wheat were the leading crops, together with beans, millets, sorghum, rye, barley and oats. With these plants both temperate and tropical

areas were colonized since between them most environments to the very limits of cultivated cereals could be covered, although by the mid-eighteenth century the great diversification of varieties brought about by scientific breeding was yet to come. Even so, the history of cereal yields in southern and northwest Europe is one of steady increase of output, even though the years of the 'Little Ice Age' (fifteenth to nineteenth centuries) and especially after 1650 AD (Van Bath 1963, Fig. 6.1).

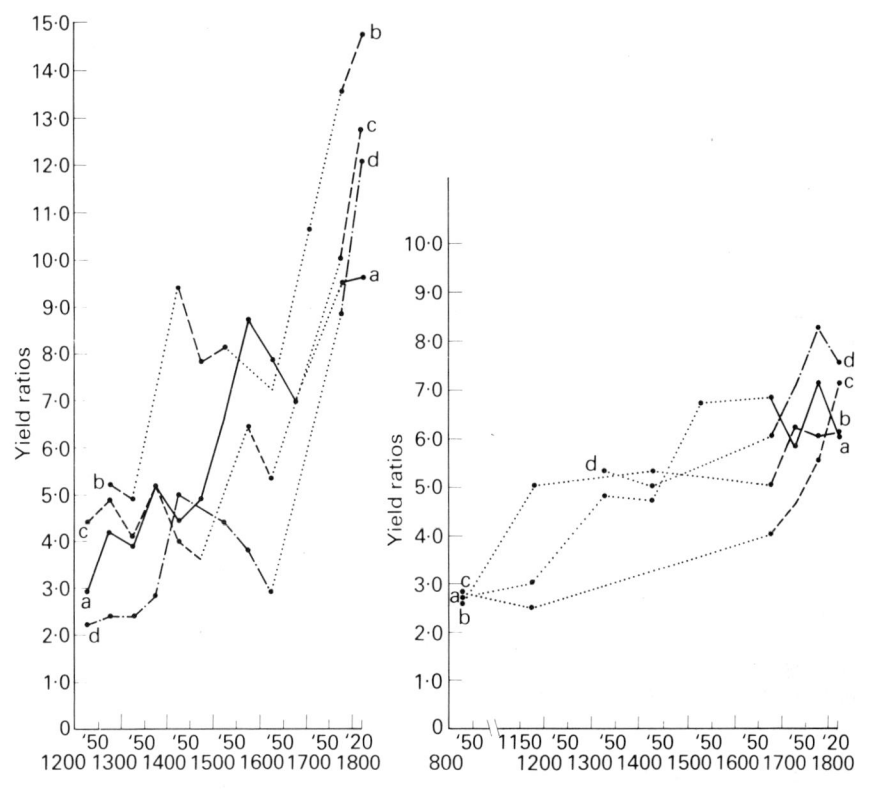

Fig. 6.1 Average yield ratios (i.e. seed corn to harvest) of (a) wheat, (b) rye, (c) barley, and (d) oats in 50-year periods. Left: England, Ireland, Belgium and the Netherlands; right: France, Spain and Italy. The overall upward trend in spite of the 'Little Ice Age' in northern Europe is marked.
Source: Van Bath, 1963.

2. Especially in the tropics vegeculture was a means of subsistence. In the tropics it has particularly been associated with the use of starchy roots such as sweet potato, manioc, taro and yams, often in the context of shifting agriculture, a resource base nowadays often apparently low in protein unless supplemented by gathering and hunting (Harris 1972). In temperate latitudes, the potato provides an outstanding example, although in the eighteenth century it had not reached the status as a staple which it was to achieve in the next century. Crops such as turnips, however, were becoming important features of European agricultural systems at about the same time.

3. Arboriculture also began to take its place among the important food producers. The specialized cultivation of a single species as a subsistence crop was especially marked in the case of the olive in Mediterranean lands, and the date in the arid areas of North Africa and the Middle East. Although often important as one element of the diet, and for trade, other fruit crops never attained such importance.

4. Pastoralism achieved a widespread distribution, mainly in marginal environments where particular animal species or specialized varieties could usually convert unpalatable or indigestible cellulose to meat, milk, blood, hides, bones, dung, traction or draught. But pastoralism also gained utility in less demanding regions where seasonal use could be made of an otherwise untapped resource, of which mountain transhumance of cattle and sheep in Europe and Southwest Asia is the most well known example. In fragile environments, pastoralism was often nomadic but if a halt could be made for a season, then it might be allied to a form of agriculture, in which the ability of the animals to manure the ground and pull the implements was of value (Barnard and Frankel 1964).

Until the advent of industrialization, all these types of subsistence had in common their reliance upon solar energy. Any other energy imports to the agricultural system like falling water or wind were also derived more or less directly from solar power, and so no 'energy subsidy' to the crops from fossil fuels such as oil and coal was introduced. The results were relatively stable systems many of which, given populations that were static in number or could hive off excess people to uncolonized areas, were stable ecologically for long periods of time. Diets were usually varied and adequate, although protection against lean years was often insufficient. Disruption of some of these systems by pre-industrial intensification (i.e. a shorter cycle for shifting agriculturalists) meant the conversion of forest to savanna and open woodland to savanna grass, and a general lowering of productivity and hence increased malnutrition. Nevertheless basic systems based on crops such as wet rice, maize, sweet potato, wheat and barley continued to provide subsistence for their people, though at a much lower level of output per unit area than later industrialized systems.

The ecology of grazing

The domestication of animals such as cattle and sheep allowed the adoption of nomadic pastoralism and the colonization of habitats hitherto reserved for hunters: after very early beginnings in Southwest Asia, cattle-sheep nomadism was firmly established between 4500–3500 BC in the western and central Sahara (Butzer 1974); in climatic contrast, domestic sheep became the mainstay of the economy of Iceland, although the similarity of desertlike terrain in both places is not without significance, and in both, long periods of grazing have had their ecological consequences in terms of altered vegetational cover and soil erosion. Some 60 per cent of grazing lands indeed are unsuitable for cultivation, and some animals can be used for draught or transport processes before ending up on the spit or in the stew. And although animal protein cannot be said to be absolutely necessary for human diets, the proper balance of amino acids is more easily achieved with meat than with plant material.

Each of the main pastoral animals have tolerances which fit them best for certain forage sources. The browsing goat, for example, can penetrate the furthest into arid lands since it will eat aromatic herbs and shrubs despised by other ruminants. It requires little free water and even living off thorny scrub will yield abundant milk and reasonable quality meat. Sheep are essentially grass feeders but at a subsistence level their meat is tastier and more tender than that of goat; the merino sheep especially will produce food on poor and dry pastures. Cattle, on the other hand, have the widest variety of uses (burden, traction, hides, blood, milk, meat, bone, dung), but are less adaptable than sheep and goats, having a higher water requirement and being more selective in their foraging. Many centuries of breeding have produced strains of all three animals which are tolerant of most of the tropical and temperate parts of the world. In more extreme environments, animals have been found that could be depended upon for subsistence: the camel is the outstanding example. Its ability to eat saline vegetation, to exist without free water for as long as thirty days provided the vegetation is reasonably abundant, and walk easily over desert terrains is well known; it lactates for 11–15 months, giving 1–7 litres/day of milk, and both hides and dung are useful for desert pastoralists, as is the meat if eating it becomes necessary. As is the camel to the deserts, the yak, llama and alpaca are to the high mountains of Old and New Worlds.

The tundra ranges of the world are of immense size (2·3 million km^2 in Canada, for example) but are probably best cropped via culling wild animals like the barren-ground caribou, *Rangifer tarandus groenlandicus* and the musk-ox (*Ovibos moschatus*) although a trial domestication programme for the latter has been introduced in northern Quebec. The Eurasian semi-domesticated reindeer (*R. tarandus fennicus*), was introduced to North America in 1891 and grew to 650 000 by 1932 but crashed to 25 000 by 1950 because of unsuitable herding practices; Scotter (1970) noted that the Inuit are bored by the monotony of reindeer herding. Nevertheless some form of cropping of terrestrial herbivores may be the optimum resource process both ecologically and economically for the native peoples of Alaska and the Canadian North. A small herd of reindeer was also introduced into the Cairngorm Mountains of Scotland but remains more of a curiosity than a resource; however the possibility of ranching red deer (*Cervus elephas*) on a commercial basis in these uplands is raised from time to time as an alternative to sheep rearing (Beddington 1975).

The use of animals in pastoral systems by no means exhausts their potential. Some are used in association with more intensive agricultural systems and others like the dog, in hunting and herding. Yet others are valuable for traction or transport, and others for skins or ivory. To the domestic animals of the pre-industrial period a few others have been added in recent times as food sources (p. 185) together with rats, mice and primates used in laboratory work, and the use of blood from, *inter alia*, cattle and horses for serum.

Pre-industrial use of aquatic animals

Most aquatic animals have been used by man as a relatively concentrated source of animal protein rather than for calories, although in some societies animals like salmon and seals were a staple source of energy. In pre-agricultural times, animals of rivers and marine littoral zones often formed highly reliable sources of subsistence, as witnessed by the high incidence of harpoons and leisters in the Mesolithic cultures of Europe or the wide variety of fish taken by all methods in the rivers and seas of North America, or the dependence of near-recent Inuit groups on the seal for winter food. The majority of fishing was still essentially Mesolithic in its ecology: men fished from banks, shores or small boats with lines or nets, or constructed fish weirs. They therefore hunted the fish: herding and farming of aquatic animals has only just begun, although fossil-fuel powered hunting is common.

Even though the outreach for fish and other aquatic resources is small in scale when not powered by fuel subsidies, populations of animals can still be strongly affected. This is particularly so where an animal population is relatively sparse like Steller's sea cow (*Hydrodamalis stelleri*), which was made extinct between 1741, the date of its discovery by Europeans in the Bering Straits, and 1830. The northern fur seal (*Callorhinus ursinus*) nearly came to the same end but international control beginning about 1900 seems to have ensured the survival of a stable population (Bennett 1975). Where an animal population is spatially confined at a stage of its life-span as with diadromous fish like salmon coming up a narrow river to spawn then the population may be threatened, though absolute numbers here are usually sufficient to bear exploitation. Where an animal is relatively immobile and can be easily garnered as with sedentary molluscs like the abalone (*Haliotis* spp); or where the animals are confined as in fish ponds, over-exploitation is easy. To what extent pre-industrial populations may have permanently affected fishery populations is not known: Rostlund's (1952) survey of native fishing in North America does not give any evidence for collapse of any fishery. On the other hand, the coming of missionaries in the eighteenth century to the Amazon and Orinoco caused immense inroads to be made into populations of the river turtle (*Podocnemis expansa*) since its meat was regarded as fish for fasting purposes and the eggs yielded oil for cooking and lighting (Smith 1974).

Fish ponds are perhaps one of the closest approaches to fish farming in pre-industrial conditions, for they are often fertilized by human sewage or water birds like ducks. In such cases, protein yields per unit area can be higher than for terrestrial animals, provided that the right type of fish are kept: carp has been a well-known temperate family, and the milkfish a tropical equivalent.

Other crops: an overview

Although food has been stressed, many biota have been used by man for other purposes, and most of them came into human economies in pre-industrial times, with exceptions like the great development of rubber (*Hevea brasiliensis*) in the nineteenth century. Fibres of different plant parts (Table 6.1) have been both gathered and cultivated, and a great number of tree species have formed the basis of the present-day timber industry. A rather smaller number were in use during pre-industrial times, with the emphasis on use of local species, although it is not suggested that there was no trade.

Beverage plants have also been domesticated (Table 6.2) though not in any great

TABLE 6.1 Fibres

Soft	Mainly from phloem	
	Flax	(*Linum usitatissimum*)
	jute	(*Abutilon avicennae*)
	hemp	(*Cannabis sativa*)
	ramie	(*Boehmeria nivea*)
Hard	From entire vascular bundles	
	Manila hemp	(*Musa textilis*)
	Sisal	(*Agave sisalina*)
Surface fibre	Cotton	(*Gossypium* spp)
	Kapok	(*Bombacaceae*)
	Coir	(*Cocos nucifera*)

Source: Spedding 1975a.

TABLE 6.2 Beverage plants

Coffee	*Coffea arabica*
Tea	*Thea sinensis*
Mate	*Ilex paraguariensis*
Chocolate and cocoa	*Theobroma cacao*
Alcohol	Fermented grains, roots, tree sap and fruits
Wines	Fermented grapes

Source: Spedding 1975a.

TABLE 6.3 Perfumes and flavours

Plant name	Product
Mentha piperata	peppermint
Melissa officinalis	balm
Salvia officinalis	sage
Prunus amygdalus	oil of almond
Artemisia dracunculus	tarragon
Vanilla spp.	vanilla
Zingiber officinale	ginger
Allium sativum	garlic
Allium cepa	onion
Smilax spp.	sarsaparilla
Rorippa armoracia	horseradish
Brassica alba	yellow mustard
Brassica nigra	black mustard
Cinnamomum zeylanicum	cinnamon
Myristica fragrans	nutmeg
Eugenia caryophyllata	cloves
Pimenta officinalis	allspice
Eucalyptus spp.	eucalyptus oil
Humulus lupulus	hops
Apium graveolens	celery
Coriandrum sativum	coriander
Angelica archangelica	angelica oil
Foeniculum vulgare	fennel
Anethum graveolens	dill
Petroselinum sativum	parsley
Pimpinella anisum	anise
Carum carvi	caraway

Source: Spedding 1975a.

number; nevertheless the variety of flavours of, for example, wine and coffee is remarkable. On the other hand, there is a long and doubtless incomplete list (Table 6.3) of plants valued for flavouring and perfumes. A miscellaneous group of plants yields useful materials, such as waxes, oils, gums and resins (Table 6.4).

TABLE 6.4   Vegetable oils and other constituents

Vegetable oils come from	cottonseed
	maize
	oil palm
	olive
	peanut
	safflower
	soybean
	(also castor bean, almond, sunflower seed, linseed).
Waxes	from whole plants, e.g. jojoba (*Simmondsia chinensis*)
	exudate from leaves of Brazilian palm (*Copernicia cerifera*)
Gums	many from legumes
Resins	mostly from Pinaceae

Source: Spedding 1975a.

Many other plants are used to produce drugs, some for social and some for medical purposes, if the two can be separated. Resins from the leaves of the hemp *Cannabis sativa* formed a social drug long before its discovery by the twentieth-century youth of affluent countries. More potent psychotropic drugs are also gained from plants: opium, morphine and heroin are alkaloids derived from the opium poppy (*Papaver somniferum*), and peyote is also an alkaloid, extracted from the button cactus *Lophophora williamsii*. The hallucinogenic properties of some fungi, notably *Amanita muscaria* have been used as the basis for shamanistic religions of both northern and Mediterranean regions and indeed been elevated by one scholar to be the origin of Christianity (Allegro 1970).

Many other plants have yielded drugs, such as curare, belladonna, quinine (whose anti-malarial properties were discovered in the West in 1638) and perhaps most widespread, tobacco. In its case there was conscious selection for particular qualities during the pre-industrial period, and it was especially needed in *Nicotiana tabacum* since some of the wild species which might be its possible ancestors contain unacceptably high quantities of nicotine in their leaves when mature. Breeding to develop strains in which the low-nicotine properties of immature leaves were retained must have been important.

Plants also provided materials for tanning, usually from forest trees, and dyestuffs. Indigo is a glycoside obtained from *Indigofera* species found in Asia but it was supplanted during the nineteenth century by the industrially produced aniline dyes; one of its precursors was woad (*Isatis tinctoria*), famous as the trendy gear of the Ancient Britons encountered by the Romans. Animals have little place in the drugs and dyes market, so the last words will be on cochineal, a red dye obtained from an insect (*Dactylopius* sp.) parasitic on cacti, the use of which originated with the Aztecs; and on musk, the basis for many a winning perfume, obtained from the scent glands of the Asian muskdeer (*Moschus moschiferus*).

Pre-industrial agriculture as an energy trap

Before the advent of fossil fuels, whether in past periods of currently industrial societies or in the poorer LDCs of today, agriculture is dependent entirely upon primary solar energy. Some stored energy in the form of wood or peat was available, and the power of wind and falling water could be harnessed, but the main inputs were the solar radiation which became trapped in photosynthesis, some of which might be fed to draught animals, and the energy of the people working the land, itself derived from recent plant and animal food energy.

Pre-industrial agriculture can nevertheless be a successful mode of subsistence. Spedding (1975a) estimates that whereas subhuman primates use 70 per cent of their energy in searching for food, in simple agriculture only about 30–40 per cent of the total energy intake is 'ploughed back' into the garnering of more food (Plate 26). The general efficiency of plant agriculture is confirmed by Rappaport (1971) for the Tsembaga of New Guinea who cultivate taro-yam gardens and the sweet potato. The input-output ratio as measured by the energy spent in agricultural tasks against the food yield in calories

Plate 26 A simple farming system in Ethiopia: the energy inputs come from the sun directly and indirectly via human energy and that of the domesticated beasts. These latter also help keep up nutrient levels via their dung. The nature of the vegetation of the background suggests that grazing animals are also part of the agricultural system of the area.

was 16·5:1 for the taro-yam plots and 15·9:1 for sweet potatoes, and would be as high as 20:1 if the gardens were close to the dwellings. (These figures should perhaps be lowered to take into account the loss of stored energy when the forest is burnt to create the swidden patch.) By contrast, 2:1 was the best ratio achieved for swine husbandry, but the pigs were not a major source of calories, being rather yielders of animal protein and of ritual satisfaction. The overall efficiency of such agricultural systems is lowered when work animals are used: a 652 kg horse working 12 h/day performs 8800 cal of work for 46 600 cal expended, so its all-day energetic efficiency is 16·5 per cent (Spedding 1975a). As late as 1901 in the UK, 30 per cent of the area of a lowland farm would be devoted to the upkeep of the horses. Nevertheless the animal may contribute to a subsistence economy by providing excreta for manure or fuel, and eventually meat, hides and other products. Thus in China during the 1930s a rice–beans–peanut economy using pigs and buffaloes, produced an energy ratio of 41:1. An Indian subsistence farm producing rice, cattle meat and milk had a ratio of 14·8:1, and simple farming of maize in Mexico, using only axe and hoe as tools, 30·6:1. Notably, the Dodo of Uganda who grow grains but whose culture centres around cattle from which they consume meat, milk and blood, produce a ratio of 5:1 (Leach 1976).

Energy use in cooking is also very important: wood, crop wastes and dung are used and these represent important components of ecological systems. Dung is fertilizer, and crop wastes can be similarly used, or fed to animals. Wood use normally means deforestation at a scale of 1–1·5 ha/cap/yr with the possibility of consequent erosion and desertification processes. At the end of it all, the conversion of fuels like wood to useful heat in cooking is about 5 per cent, compared to 25 per cent in a modern gas or electric stove (Leach 1976).

The human populations which can be supported by such cropping systems are variable according to the natural environment and are often difficult to estimate from contemporary statistics since an energy subsidy from fossil fuels is now so often found. Pastoralism requires a lot of land if overcropping is not to occur: 60–130 ha/cap is a normal figure, and is exceeded only by hunter-gatherers who may require 20–150 km^2/cap for survival, depending on the environment. Shifting agriculture based on roots in Asia can support 40 people per square kilometre, according to Grigg (1974), and the Tsembaga achieve 25/km^2 in New Guinea. Hunter (1966) reports that in the savanna belt of Ghana, the maximum supportable density under shifting agriculture is 54–59/km^2, after which emigration occurs. One of the most productive of systems is wet rice since double-cropping and potential for intensification exist even without energy subsidies, so that in Asia wet rice will support populations of the order of 400/km^2 and individual countries yield higher figures, for example Japan in 1970 had a population density of 1869/km^2 of arable land, and South Korea 1401/km^2 of arable land. Here there is extra energy input from fossil fuels, and the comparable figure for China is 770 and for North Vietnam is 906. Such densities appear to be stable in ecological terms, and have not passed the point of critical population density beyond which progressive deterioration of the land is inevitable. Demand for higher yield however produces phenomena such as emigration, soil erosion and the sort of technological change which involves large subsidies of fossil fuel. This theme is expanded on pp. 276–283.

Further Reading

BEDDINGTON, J. R. 1975: Economic and ecological analysis of red deer harvesting in Scotland. *J. Environmental Management* **3,** 91–103.

BENNETT, C. F. 1975: *Man and earth's ecosystems.* New York and London: Wiley.

GRIGG, D. B. 1974: *The agricultural systems of the world: an evolutionary approach.* London: Cambridge University Press.

LEACH, G. 1976: *Energy and food production.* London: IPC Press.

LEEDS, A. and VAYDA, A. P. (eds.) 1976: *Man, culture, and animals.* Washington, DC: AAAS Publication 78.

SPEDDING, C. R. W. 1975b: Grazing systems. In R. L. Reid (ed.), *Proceedings of the III World Conference on animal production.* Sydney: Sydney University Press.

7
Industrialization and biota

Perhaps ironically, the fossil fuels which man has harnessed to initiate the greatest alterations of all time in biota are themselves of organic origin. When oxidized, they provide large quantities of usable energy in relation to their bulk, in contrast to solar energy which is relatively diffuse. Before the widespread knowledge of the potential of coal, oil and natural gas which arose in the late eighteenth century and came to fruition in the nineteenth and twentieth centuries, only solar energy in its directly fixed form as organic material and indirectly as wind and falling water was utilized by human societies, together with the spatially sporadic use of sources such as geothermal energy. The steam engine conveniently marks the turn towards the new industrial order but more emphatically the railway locomotive and the steamship symbolize a new relationship between man and nature, for they carry their own fuel and are hence highly mobile. It becomes difficult to overemphasize the role of communications in altering biotic patterns, for not only do roads, railways, aircraft and ships provide actual niches for the dispersal of plants and animals but they act as rapid carriers of large quantities of materials and energy which can be used to manipulate ecosystems.

Industry has meant the sterilization of ecological systems under buildings and under mounds of stored and waste materials, as well as the destruction of habitats in their entirety. It has also produced concentrated wastes which are usually dispersed into water or the air and created substances, especially organic chemicals, which are unknown in nature and for which there are no biological degradation pathways. The effects of these changes are not confined to those areas where their results are most readily apparent. Cheap bulk transport has meant the outreach of societies for raw materials to all parts of the globe, with consequent effects upon plants and animals, not the least being the inadvertent intercontinental transfer of species. People and ideas have travelled faster and into more remote places, so that often fragile ecosystems have received considerable impact.

All this change, of a magnitude never before wrought by human societies, is underlain by three processes. The first is the rapid growth of the human population since the mid-seventeenth century with all that is implied for its impact upon plants, animals and ecosystems. The second is the accompaniment of intellectual developments which made the wholesale transformation and extirpation of nature culturally acceptable. The third is the growth and transmission of knowledge, and the rise of a class of people devoted entirely to that endeavour, sustained by the use of fossil fuels in freeing them from work in factory or farm. One result has been the application of a basically Western economy to many parts of the globe as part of the process of 'development'. Fossil-fuel-powered machines have been used to manipulate or destroy many different types of ecosystems in the cause of resource use and management: in deep-sea fishing, in forestry operations, and for outdoor recreation, for example. But perhaps the area of biotic change where the impact of fossil fuel use has been most felt is that of agriculture. The next section, therefore, examines the growth, and some of the consequences of, the use of machines, fertilizers and **biocides**.

Machines in agriculture

The introduction of machinery into agriculture marks the replacement of animal energy by power derived from fossil fuels. Steam power proved to be less easily applied in agriculture than in industry: it was used for drainage engines in Dutch polders and the English Fens at the end of the eighteenth century, for threshing by the 1820s, and in some steam ploughs by the mid-nineteenth century, but the major change of the nineteenth century in Europe was in fact the replacement of the ox by the horse. The horse was itself replaced by the tractor early in the twentieth century, with the machine achieving dominance after the 1920s. But in Europe, industrialization of agriculture became complete only after World War II. Between 1950–68 the number of tractors in West Germany increased seven times, and other machines have followed a similar trajectory: in Denmark in 1940 there were eight combine harvesters, in 1960 there were 11 570, and by 1968 some 40 000 (Grigg 1974). In Britain there were 3500 combines in 1946 and 66 000 in 1971; 1000 grain driers in 1946 had risen to 63 500 by the latter date (Blaxter 1973). The export of such machines to LDCs is part of the accepted process of development, and is justified on the grounds that it permits higher yields through the intensification of land use and the cultivation of hitherto marginal land. (Table 7.1 shows some statistics for recent growth in tractor numbers.)

Certainly, the involvement of fossil fuels has meant the cultivation of crops in areas

TABLE 7.1 Tractors in use

	Average 1969–71	1974	1976
All Developed Countries (DCs)	14 167 693	15 613 432	16 269 136
All Developing Countries (LDCs)	1 339 992	1 787 165	2 034 007

Source: FAO *Production Yearbook* 31, 1977.

otherwise impossible to clear of scrub or to plough, for example. Such intensification of land use has on occasion resulted in disaster: since some tropical soils when deprived of their litter layer and broken up at the surface either erode away very quickly, or turn to laterite. Even in temperate zones, it has been suggested that machinery may affect soil structure by compression, shear, slip, bounce and vibration. Particularly when soils are wet and plastic, considerable damage to their structures can be easily brought about since the bottom of a plough furrow may be spread out and smeared by both the tractor wheel and the plough, and the furrow slice may have its underside smeared. Exacerbation of these conditions happens in some farming systems and under certain weather conditions when organic manures have been replaced by chemical fertilizers so that the soil is low in the organic matter which contributes to the stability of soil aggregates (Agricultural Advisory Council 1970).

The tractor is only one part of the complex network of involvement of fossil fuels in agriculture. But mechanization is a major element in the conclusion that the output from fuel-subsidized systems is much higher per unit area: farms in the USA of 1880 had an average net plant productivity of $1.28 \, kcal/m^2/day$ whereas rice in the USA of 1964 was cropped at $10.0 \, kcal/m^2/day$ (Table 7.2). Similarly, grain in the Africa of 1936

TABLE 7.2   USA: Efficiency of utilization of energy: harvest of organic matter on farms

Year	%
1880 (solar energy only)	0.03
1964 All grains	0.12
1964 Rice	0.25

Source: Wagner et al. 1973.

averaged at $0.72 \, kcal/m^2/day$ compared with grains in North America (1960) at $5.0 \, kcal/m^2/day$ (Odum 1971).

Other on-farm processes such as grain-drying and harvesting are also powered by fossil fuels, and off the farm the transport, processing, storage and preparation of food in DCs (and to a lesser extent LDCs) are all underlain by conventional sources of fuel. Steinhart and Steinhart (1974) calculated that 13 per cent of total energy use (excluding solar energy) in the USA in 1970 was connected with the food industry in all its stages. In the UK, agriculture itself accounts for about 4 per cent of all consumption of fossil fuels (Blaxter 1975a). The use of fossil fuels has also increased agricultural output by irrigation and in industrial chemicals such as fertilizers and biocides. (The latter require large quantities of 'upstream' energy in their manufacture and transport as well as on-farm application.) The outcome in terms of crop is a rapid rise: Blaxter (1973) suggests that the increase in agricultural productivity per unit area in the UK since 1945 is almost entirely due to fuel imports and associated fuel-based processes. Further discussion of the role of fossil-fuel subsidies in food productions can be found on p. 276.

Fertilizers

The search for an adequate scientific knowledge of the mineral basis of plant nutrition began in the period 1750–1800 when pot cultures were used to investigate the best conditions for plant growth, although the discovery that plants gave off a gas led to the erroneous conclusion that plants generated mineral elements in the soil. A firmer foundation came in the years 1800–60 when, for instance, Dr Saussure demonstrated the role of plant roots in the uptake of nutrients, Liebig's work on the role of nitrogen appeared, and in 1843 the field experiments at Rothamstead (Plate 33) were begun. Systematic knowledge began to accumulate about the phosphate and alkali salt requirements of crops, the need of non-leguminous crops for nitrogen and the role of fallowing in N_2 accumulation. The founder of modern agricultural chemistry was probably Boussingault who, working in Alsace in 1840–70, carried out quantitative investigations of nutrient amounts in agricultural crops.

The offspring of these beginnings is the large-scale fertilizer industry of today which produces a great variety of chemicals for most agricultural systems. Fertilizer use, for example, is at the heart of the use of the improved varieties of rice and wheat which have underlain the 'Green Revolution', and on a world scale the use of fertilizers continues to grow. The average yearly world consumption of potash fertilizers 1961–65 was 10·2 million tonnes, and in 1971 it was 17·4 million tonnes. The regional breakdown (Table 7.3) emphasizes that the DCs which produce most of the chemical fertilizers

TABLE 7.3   Regional consumption of fertilizers
Total NPK fertilizers per ha of arable and permanently cultivated land (×100 g)

	Average 1961–65	1976
WORLD	272	636
Africa	47	128
North and Central America	405	864
South America	78	313
Asia	115	366
Europe	1037	2045
Oceania	338	346
USSR	161	727

Source: FAO *Annual Fertilizer Review 1977*, Rome: FAO, 1978.

also consume them, and that the areas of low agricultural productivity are also those receiving least material to upgrade their output. In 1970, Europe and North America used almost two-thirds of the world's total of NPK fertilizer and more than half of the nitrogen fertilizer. Developed regions applied an average of more than 90 kg/ha under cultivation, compared with 18 kg/ha in developing regions. West Germany, for example, consumed more N_2 fertilizer than the whole of Africa. Some changes in these trends may be seen in new methods of application such as the nitrogen fertilizer placed in a small mudball and inserted 10 m below the soil in rice paddy, and which actually increases the crop yield while reducing the quantity of fertilizer needed (IRRI 1974).

The use of fertilizers essentially imitates a natural process but with certain important differences. The spatial linkages of the cycle are more open: consider the movement of a molecule of phosphorus from a mine in Canada or Spanish Sahara through a fertilizer plant to a wheatfield then to a consumer and finally into an ocean as sewage. Again, the quantity per unit area on-farm is much more concentrated than in nature, and the whole cycle is faster. Thus ecological side-effects of fertilizers may be expected, especially where nitrogen and phosphorus, often limiting in natural systems, are used (see pp. 71–3).

Biocides

The very act of defining a plant or animal as a crop designates unwanted biota as weeds and pests. The methods of eliminating them have contained constant elements down the centuries (trapping, shooting, and scaring of larger animals, for example), prominent among which has always been the application of chemicals to kill both weeds and pests. Before the nineteenth century, several substances were used to kill insects: these included lye (a strongly alkaline solution of plant ashes), soap, turpentine, nicotine, pyrethrum, and mineral oil.

The development of the chemical industries of the West in the nineteenth century brought new weapons, based in particular upon arsenic. After 1868 Paris Green (copper acetoarsenite) was especially valued in the USA for use against the Colorado potato beetle (*Leptinotarsa decemlineata*), for instance, and on into the early twentieth century lead and calcium arsenates were much used, although increasingly regulated because of their potentially toxic effects upon human consumers. At the same time, sulphur, oil, rotenone (from derris root), phenols, pyrethrum and Bordeaux mixture (copper sulphate, quicklime and water) were widely used against insects and fungi. The transition to contemporary biocides came in 1939 with the first usage of an organic chemical (DDT: dichloro-diphenyl-trichloroethane) to control typhus-carrying fleas. Its value was widely recognized during World War II as an aid to the control of malaria mosquitoes and thereafter as a general medical and agricultural insecticide. Its chemical group, the chlorinated hydrocarbons (including such other synthetics as dieldrin, aldrin, endrin, heptachlor, DDD, TDE, DDE, methoxychlor, BHC-lindane, toxaphene, tetradifon and mirex), became by the 1960s the mainstay of the chemical pesticides. These were backed up by the organophosphorus insecticides (e.g. fenthion, diazinon, malathion and parathion); herbicides like 2,4-D and 2,4,5-T, fenoprop, and paraquat; and fungicides based on organo-mercury compounds. In the 1960s some 3,000 organic chemicals in over 10,000 formulations were in use as biocides in the USA, and certain herbicides like 2,4-D and 2,4,5-T were employed as agents of warfare in Vietnam.

World-wide documentations of the quantities used is very spotty, but Table 7.4 gives some idea of the quantities of two insecticides in common use. DDT is still very popular, although its ecological side effects (see p. 247) are increasingly recognized, leading to a diminution of the use of chlorinated hydrocarbons in some DCs. The natural product pyrethrum, derived originally from the plant *Chrysanthemum cinerariaefolium* and now

TABLE 7.4 Use of selected pesticides ($\times 10^2$ kg)

	Average 1969–71	1974	1976
(A) DDT			
India	28 981	40 000	65 500
Italy	17 808	20 192	18 537*
Austria	211	175	85
El Salvador	5 552	10 000	11 326*
(B) Pyrethrum			
USA	959	1 696	1 823
Austria	12	54	91
India	1 059	n.a.	n.a.

*1975.
Source: FAO *Yearbook* 31, 1977.

capable of industrial synthesis, is growing in use since it is less generally toxic and breaks down more quickly than the chlorinated hydrocarbons.

The effects of these biocides upon plant and animal target populations have naturally been variable. Any biological population will exhibit variable resistance to toxic substances, depending upon the dose, so that 100 per cent mortality of the target organism is rarely achieved (Brown 1971). Nevertheless, two major advantages have emerged from their use:

1. Increases in the health of human populations. The insect vectors of numerous diseases, especially typhus and malaria, have been brought to low levels in many places. People have therefore been freed of the source of the worst ravages of these diseases; one result has often been increased rates of population growth.

2. Greater absolute output of agricultural crops and, almost equally important, greater stability of output. Farmers have thus been able to plan ahead and invest capital with greater confidence. Lowering of costs was at first effected as a result of chemical pesticide use but ensuing energy costs associated with the manufacture and application have made them expensive to use.

Certain disadvantages are coming to light. The residues of some biocides produce side-effects (see pp. 241 and 247), and so considerable development of new compounds is now taking place, with a particular view to making them selective and non-persistent: the insecticide Sevin, for example, is claimed to control 160 different insects without residues in feed, forage, food or soil. New pesticides which do not easily produce resistant populations of target organisms are also now being marketed, such as the foramidines being used against cabbage looper, cabbage worm and diamond-back moth in the USA. Analogues of parathion such as fenthion, fenitrothion, ronnel, bromophos and iodofenphos, are all effective insecticides which are rapidly biodegradable, as are substitutes for DDT such as methoxychlor and the carbamate group (Metcalf 1972). Selective herbicides which combine the inhibition of germination in their targets with little or no effect on selected crop species (an example is penoxalin, used in Europe and New Zealand) are now in use (Moore 1975). Thus although modern biocides have been very successful in many places, the true costs of their use ought to incorporate an assessment of their effects as residual wastes in ecosystems peripheral to their targets. They ought also to include the costs of evaluating and monitoring the use of such poisons in the surroundings of plants and animals (von Rumker et al. 1974).

Introductions and extinctions

The Industrial Revolution accelerated the pace at which biota could be transferred by human activity from their native ecosystems to new locations. The scales of displacement by man of the world's estimated 10 million species of plants and animals had long varied from the local to the intercontinental but the nineteenth century brought a greater ease of transport of viable seeds, living plants and live animals over long distances. So while local and regional introductions are quantitatively important, some emphasis will be given here to long-range carriage. The means involved were all those methods of transport capable of taking organic material from one place to another, deliberately or accidentally, and range from specially constructed salt-proof cases for live plants on steamships to aircraft-tyre treads housing insects. The motives behind deliberate introductions were likewise diverse: ornament and pleasure, economic value, counter-predator are all uses to which exotic fauna and flora have been put. Some introductions, both conscious and unconscious, have taken hold in a moderate fashion, others not at all, and there are those which have been so successful in their new home that they have constituted what Elton (1958) called 'invasions'.

Even though the industrial revolution was an accelerator of such processes, the transfer of plants and animals from their native habitats has had a long history. One category of introduced plants and animals comprises those which were brought in for pleasure purposes such as hunting, or for scientific purposes, or for ornamentation, and which remained under close control, and have not spread into near-natural ecosystems, nor can be said to have become naturalized. The peacock in Europe may be cited as an example, as can many of the exotic plants brought to Europe by the highly motivated plant hunters from the Renaissance onwards. Some such plants were for taxonomic investigations at places like the Royal Botanical Gardens at Kew (e.g. their collections of Chinese plants), or for economic improvement (e.g. the germination of the seeds of the rubber plant *Hevea brasiliensis* smuggled out from Brazil, or the introduction of several species of fast-growing conifers from the west coast of North America, such as Sitka Spruce *Picea sitchensis* and the Lodgepole Pine *Pinus contorta*). The English and Continental gentry often took an interest in exotic tropical plants: the Sixth Duke of Devonshire's agent searched much of India for *Amherstia nobilis* and as many other orchids as he could find, suspending the epiphytes from his cabin roof on the way home: over a hundred species of orchid triumphally embellished the hothouses at Chatsworth. Some collectors were responsible for considerable depredations: a Prussian plant-hunter bought an Indian slipper orchid, nurtured a few specimens and then systematically destroyed the rest of the limited natural population so that he might have the monopoly on its introduction to Europe (Whittle 1975).

Naturally enough, many importations were unsuccessful except in very special conditions such as hothouses or zoo cages. A common feature of imported animals is that they may survive well enough as individuals but fail to breed for reasons which might include the lack of the correct behavioural signals in the absence of sufficient numbers of the appropriate species. Others may remain rare until some particular change creates favourable habitats or provides a method of dispersal: *Senecio viscosus* (sticky groundsel), a native of southern and central Europe, was first noticed in 1666 in the Isle of Ely but as late as 1915 it was found only in a small number of scattered localities. By 1940 it had spread to much of Britain and it is presumed that the growth of road transport provided a means of dissemination. Like several other introduced plants which grew well in waste places, the sticky groundsel found a very suitable niche in the bombed sites of British cities in World War II. *Senecio squalidus*, the Oxford ragwort, an escape from the Oxford Botanic Gardens and *Galinsoga parviflora*, Gallant Soldiers or Kew Weed, brought from South America to Kew Gardens from whence it became naturalized as a garden weed are similar examples.

Some introductions have proved to be successful in terms of the purposes for which they were imported but have remained in controlled populations. Several domesticated animals might be cited as examples but European cattle, taken to many places by colonists, are perhaps outstanding, by contrast with pigs and goats which have often gone wild in new habitats. However the transference of a species will usually involve its parasites as well and if these are conveyed to a widespread native host then their control may become difficult. A taxon may be introduced for some purpose or accidentally and then become uncontrolled to the point where it begins to disrupt the native ecosystems. Usually this happens because there is a vacant niche in a native biota of low diversity or because the newcomer has some competitive advantage against an indigene and can take over its place in energy flows and nutrient cycles, or simply because there are no native predators. Island biota have been particularly susceptible to invasions of a more or less inadvertent nature: the Brown or Norway rat (*Rattus norvegicus*) has left many ships for life ashore and has caused faunal shifts on many volcanic and coralline islands which lacked an aggressive and fast-breeding predatory rodent and which were also without a predator effective against the rat itself. Heavy grazing by introduced domesticated mammals has often reduced the native forests of islands to a xeric scrub of introduced species: this has been noticed on the drier parts of Antigua (Harris 1962b) and Oahu in the Hawaiian islands (Egler 1942). Deliberate introductions to florally and faunally depauperate islands usually result in large-scale ecosystem changes: the goats and pigs set ashore on Caribbean islands in the sixteenth and seventeenth centuries to breed and thus provide fresh meat for mariners in the age of sail frequently stripped the vegetation and disrupted the island ecosystem altogether. The presence of dogs, cats, pigs, goats, donkeys and rats on the Galapagos Islands is threatening the perpetuation of some of the special species of that place: pigs root out the eggs of the giant tortoises, for example; the Hawaiian petrel is threatened by cats and rats because it is a ground-nesting species; and the land iguana populations suffer from predation by feral dogs (Lewin 1978). In the Caribbean, the mongoose (*Herpestes auropunctatus*) was introduced from Asia into Jamaica in 1870 to control rats. Its voracity, however, meant that it has been responsible for the extinction of several reptiles and the decimation of some bird populations. In Australia, the chance introduction of North American succulents of the genus *Opuntia* rendered useless many grazing areas, interacting with drought and overgrazing. The latter process is also laid at the numerous feet of the European rabbit whose Australian history is well known. A controlling virus disease of the rabbit, myxomatosis, was also deliberately introduced. Australia has also been the scene of another and apparently successful introduction attempt for when domesticated cattle were introduced from Europe after 1788, none of the appropriate dung beetles accompanied them with the result that many pastures were becoming less productive since the effective area of pasture was being reduced by 20 per cent per year. Thirty million cattle thus sterilize $2 \cdot 4 \times 10^6$ ha/yr, some of it cumulatively since the dung-beetles associated with the native wild herbivores of the grasslands (e.g. the kangaroo) have never adapted to the altogether different conditions of the European cow-pat. Using strict conditions of quarantine to avoid the concomitant introduction of beetle-borne cattle diseases like rinderpest, other species of dung-beetle were imported from Africa and Asia, and two

212 Industrialization and biota

of these, *Onothophagus gazella* and *Euoniticellus intermedius*, have become established and are functioning well, eliminating most cow-pats within 24 hours. One problem however has been that the toad *Bufo marinus* (itself introduced from Hawaii to control sugarcane root beetles) has moved onto *O. gazella* as a food, seeking out fresh cow dung at night and swallowing, as they land, as many as 80 beetles a night. To counter this, both diurnal beetles and large heavily armoured beetles are being sought, such as those of the African genus *Heliocopris* which is about the size of a golf ball and which would eat its way through the body wall of any toad that ingested it (Waterhouse 1974).

New Zealand is another good example of the vulnerability of islands to ecological changes caused by introductions. At the time of European discovery it was about 68 per cent forest, and the animals consisted of birds, insects, a few lizards, the Maori dog and one species of rat. Captain Cook is thought to have set ashore pigs and goats,

Plate 27 Severe defoliation of native tree species in New Zealand by the opossum, introduced from Australia.

both of which have become feral and the goat especially is a pest. Forest, which now covers 14 per cent of the islands, is under severe attack from the most serious pest of all, the opossum (*Trichosurus vulpecula*) introduced from Australia in 1858 to be the basis of a fur industry. The opossum not only defoliates native forest trees but is moving onto the large plantations of introduced conifers such as *Pinus radiata* (Plate 27). Furthermore, the opening of the forest allows intrusions by various species of introduced deer which then browse out the lower layers of the vegetation. Regeneration becomes rare and so forests are replaced with fern or even sufficient bare ground to permit accelerated erosion of soil. In this case, therefore, introduced species have been responsible for wholesale changes in landscape and ecological dynamics (Salmon 1975).

As an example at a different scale, the example of the import of exotic trees into Britain after 1700 may be given (Fig. 7.1). The outstanding feature of the distribution is the

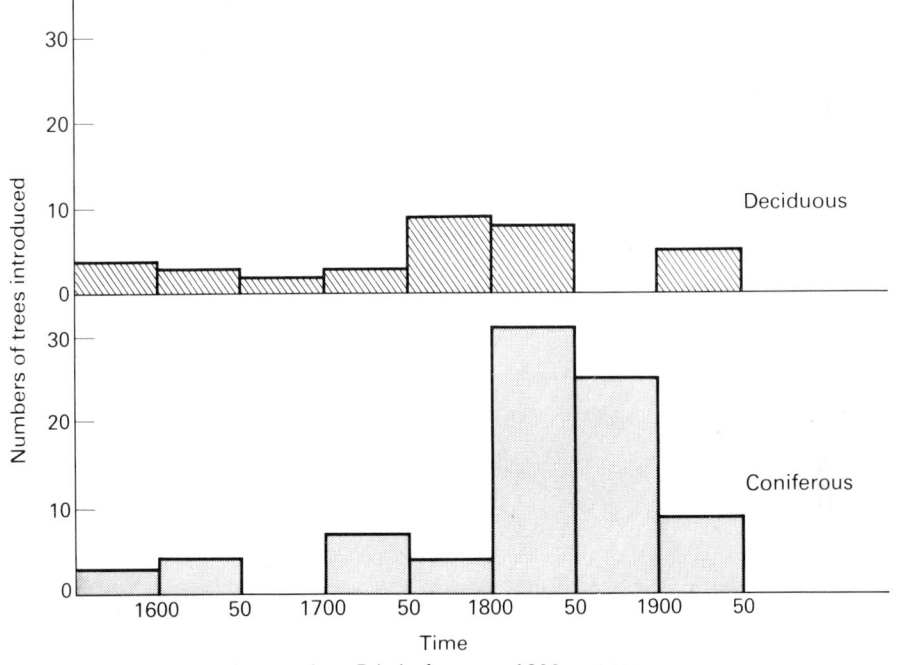

Fig. 7.1 Introduction of exotic trees into Britain from pre-1600 to 1950.
Source: Forestry Commission.

dominance of conifers which reflects among other things the paucity of softwood trees in the native British flora. The bulk of them were brought in between 1800–1900, which reflected the plant-hunting exploration of the time, the relative ease of carriage by steamship and an increased scientific knowledge about silviculture. In this crop, many economically important trees, including the Sitka Spruce (*Picea sitchensis*), the present mainstay of the Forestry Commission's plantations, were introduced to Britain. Many of the group survived only as estate or botanical-garden curiosities and have had impact on neither landscape nor economy.

214   *Industrialization and biota*

Some accidental introductions, on the other hand, may reach epidemic proportions, as with the chestnut and elm blights around the Atlantic, or the nuisance value of the European starling (*Sturnus vulgaris*) in North American cities which spread from Central Park, New York City after 1891 (Fig. 7.2). Similarly, the whole European population of muskrats (*Ondatra zibethica*) spread from 5 individuals kept in 1905 by a landowner near Prague.

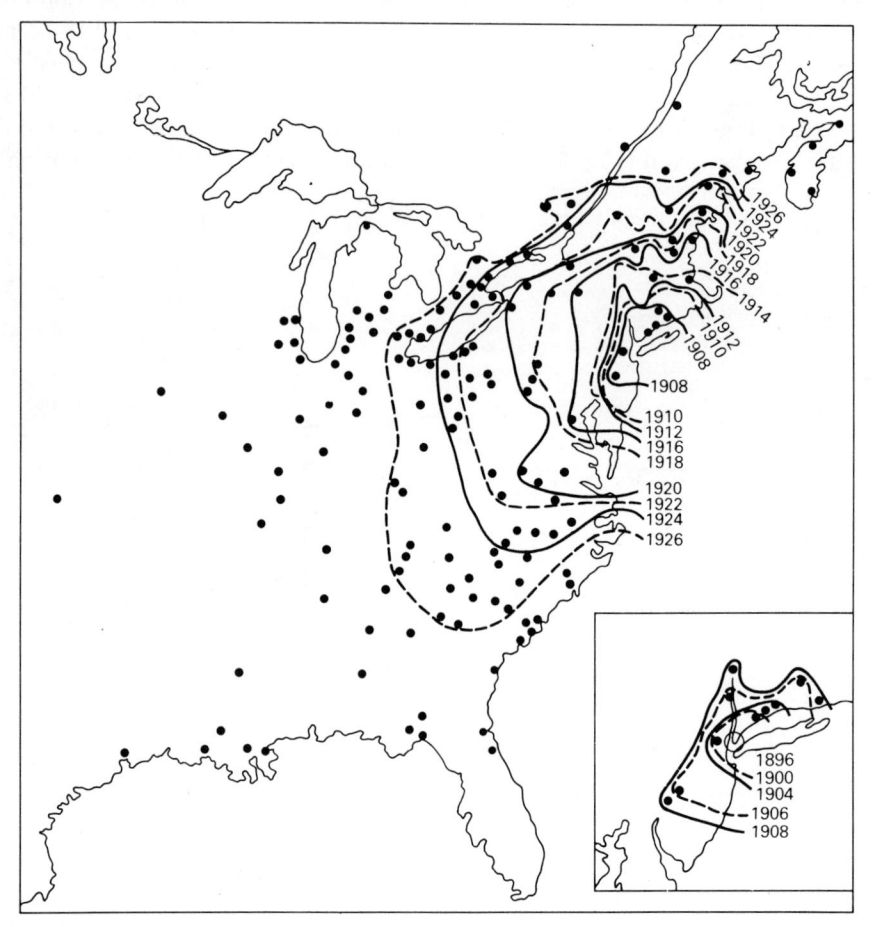

Fig. 7.2 Isochrones of the spread of the breeding range of the European starling (*Sturnus vulgaris*) in North America 1891–1826.
Source: Elton, 1958.

The last hundred years have seen the effects of comparable introductions in the oceans (Elton 1958). The digging of the great isthmian canals, for example, might be expected to have breached hitherto impenetrable floral and faunal barriers. The Suez Canal (opened 1869) provides a saline passageway, aided by currents and tides, for animals from the Red Sea into the faunally depauperate Mediterranean. The Red Sea crab (*Nep-*

tunus pelagicus) became common in the canal in 1889–93 and has now reached Cyprus. However the Panama Canal has not so functioned, because of a freshwater zone which includes the Gatun Lake. Only one species of fish is known to have made the crossing of the canal: the tarpon (*Tarpon atlanticus*).

Accidental carriage by shipping, in ballast or water tanks or on hulls, has steadily dispersed marine fauna from ocean to ocean. The Red Sea prawn (*Processa aequimana*) has been found with increasing frequency in the North Sea, and the Chinese Mitten Crab (*Eriocheir sinensis*) from the rivers of North China established itself in the River Weser in 1912 and has since colonized rivers from the Seine to the Baltic. Among plants the red algae *Asparagopsis armata* has spread from its homeland along the south coast of Australia, and New Zealand, to France, Algeria, Ireland and Great Britain.

The deliberate transmittal of marine animals is epitomized by the transplantation of oysters. *Ostrea edulis* was taken to America from Europe, *Ostrea gigas* to America from Japan. In most cases, predators came along too and very often spread to native mollusc populations as well as their original prey: the American oyster drill, *Urosalpinx cinerea* reached Europe from eastern North America in the late nineteenth century and is well established, and has gained a hold on the Pacific Coast of the USA as well.

Increased urbanization, more travel, and greater affluence have all contributed to growth in the number and diversity of biota kept for pets. The pet trade has transported great numbers of fish, especially tropical taxa, around the world and escapees have sometimes become established: 42 exotic species have been found in Florida waters between 1971–75, and 24 of them (together with 5 hybrids from native species) have become established, and at least four species have caused ecological disruption. Comparable releases of birds and mammals have also been made in Florida: some of these pose threats to agriculture (e.g. the gerbil, *Meriones unguiculatus* and the hamster *Cricetus cricetus*) as well as to native ecosystems. As with all introductions, these animals bring their parasites with them, and these can spread from their original hosts: a tapeworm of the Chinese grass carp spread from an introduction in the USSR to most European cyprinid fish. Micro-organism transport is even more likely and its transmittal to humans very probable at some stage (Courtney and Robins 1975).

A more or less complete picture of animals introduced into the British Isles and now wild, from the late Neolithic/early Bronze Age (Orkney and Guernsey Voles) to 1973 (Mongolian gerbils in the Isle of Wight) has been compiled by Lever (1977). Their origins and the time-pattern of the introductions are shown in Figs. 7.3 and 7.4; it must not be imagined that all of them are evenly spread over their new country (Plate 28). The pheasant, *Phasianus colchicus*, originally from Georgia and Armenia but probably first brought to Britain in the time of Edward the Confessor, is more or less ubiquitous in agricultural and wooded land throughout the British Isles; whereas the coypu (*Myocaster coypus*) escaped from a fur farm in the valley of the Yare in 1937 but have remained confined to the low-lying parts of eastern Norfolk and Suffolk ever since, with some isolated sightings outside this range which suggest that governmental control programmes have had some success in preventing its spread; the total population was probably about 7000 in 1976. Coypus undermine river and dyke banks and raid farmlands, especially for sugar-beet, and hence are much less beneficent an introduction than the beautiful

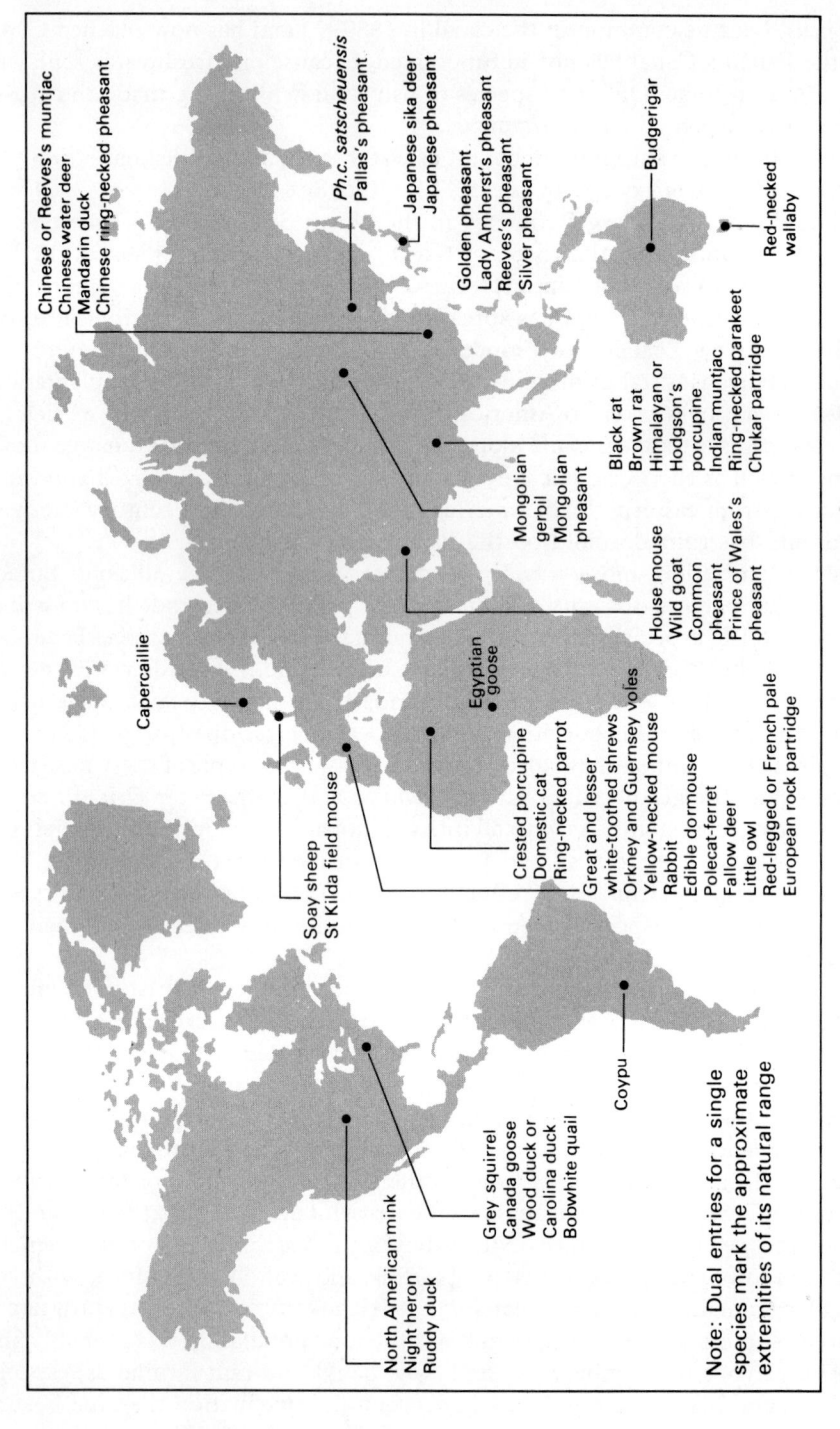

Fig. 7.3 The provenance of some of the animal species that have become naturalized in the British Isles from prehistoric times to the present. Source: Lever, 1977.

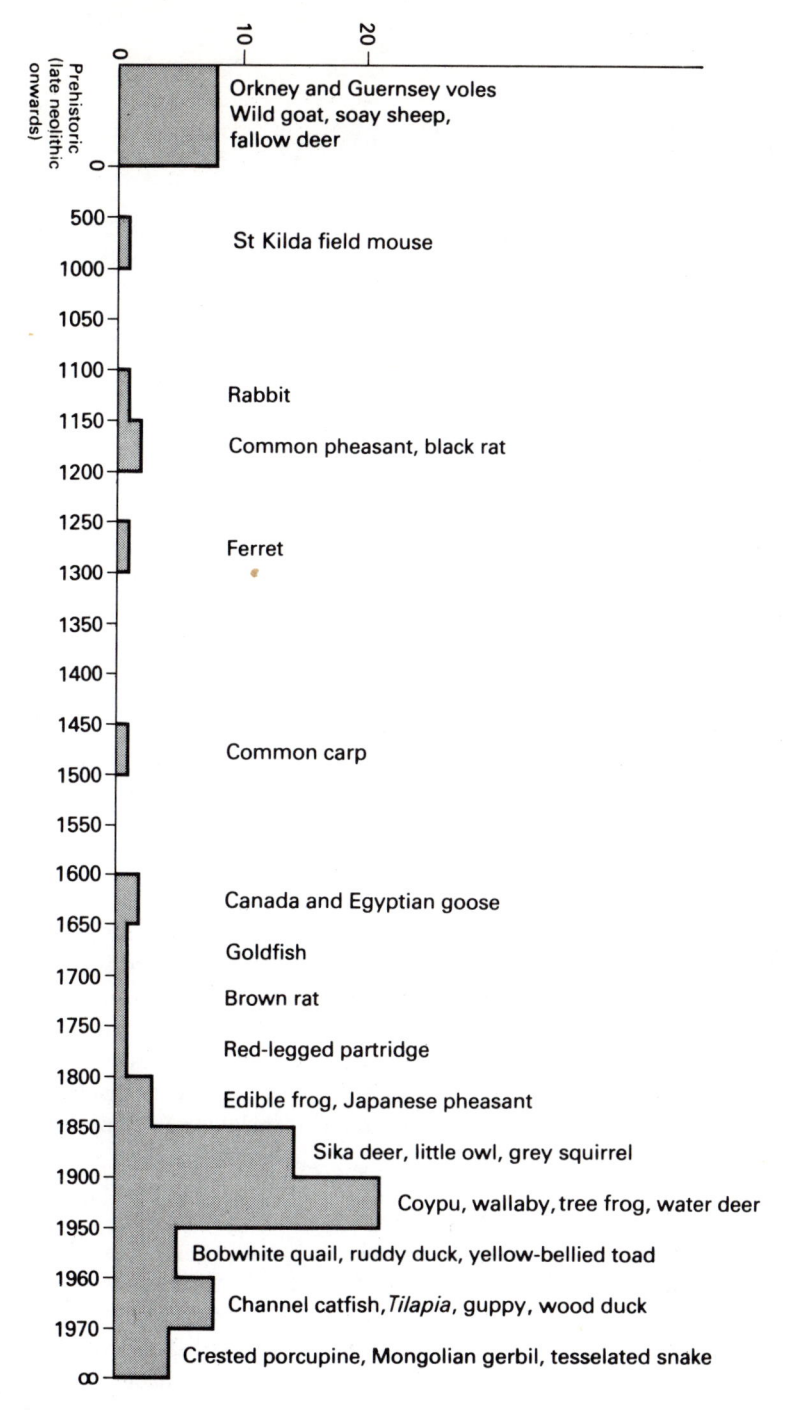

Fig. 7.4 The chronology of animal species now naturalized in Britain (vertical axis is number of species). The names are examples only. Compare the distribution with Fig. 7.1.
Source: Lever, 1977.

Plate 28 A small-scale re-introduction: a herd of reindeer in the Cairngorm mountains of Scotland, an area from which they had been extinct presumably for over a millennium.

and edible pheasant. Some introductions are little less than devastating: although Dutch elm disease has been endemic in the UK since the 1920s, the introduction after 1965 of a more virulent strain (a fungus transmitted by a bark beetle) from North America in imported logs has caused the death of enormous numbers of elm trees. Not only was the disease spread by the dispersal of the beetle, but elm trees with their bark on were shipped from infected to uninfected areas of the country as the price fell in places with a high mortality. A government ban on such translocation came too late (Sarre 1978).

Knowledge of the history of extinctions is in general less reliable than that of introductions, until very recent times. Nevertheless, a reasonably accurate estimate (Fig. 7.5) can be given for mammals and birds (Ziswiler 1967). The reasons why these warm-blooded animals became extinct in the period from 1600 AD onwards are both natural and man-made, but the overwhelming majority of extinctions (76 per cent of mammals and 67 per cent of birds according to Fisher *et al* 1969) fall into the latter category. The proximal causes are predominated by hunting (42 per cent of birds, 33 per cent of mammals) and thereafter introduced predators, other introductions, and habitat disruption share the responsibility, with the last category being marginally the most impor-

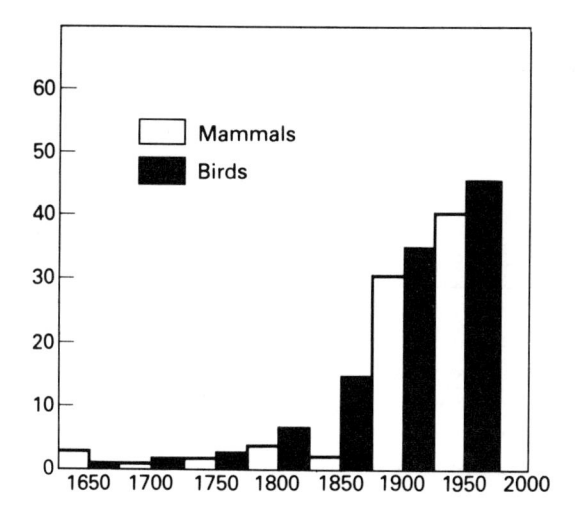

Fig. 7.5 Extinction of mammals and birds since 1600 AD.
Source: Nobile and Deedy, 1972.

tant (Guggisberg 1970). Looked at temporally (Fig. 7.5) the record of extinctions rises markedly after 1800 AD. Before then, habitat alteration and the indirect effects of maritime traffic had begun to add to the species finished off by natural means, but the nineteenth century saw a rapid increase in all the causes to which species extinction has been attributed. Table 7.5 shows the actual numbers of animals which have been extirpated. To these may be added the species considered now as very rare: at least 120 species of animals and 187 of birds. If it is accepted that the number of bird species in the world totals 8684 (Nobile and Deedy 1972) then 1·09 per cent have become extinct since 1600; for mammals, 0·85 per cent of a fauna of 4226 species have suffered the same fate. So about one-hundredth of the warm-blooded fauna has vanished since 600 AD; at current rates of extinction about one-fortieth may become so by 2000 AD, since the current rate of extinction among most groups of mammals is a thousand times greater than the high extinction rates that occurred at the end of the Pleistocene (Miller and Botkin 1974). In terms of all animals, the extinction rate of species and subspecies appears to be about one per year, compared with one every 10 years from 1600–1950 and perhaps one every 1000 years during the great dying of the dinosaurs (Myers 1976).

The spatial pattern of extinction is set out in Table 7.6 in which certain features can

TABLE 7.5   Extinctions since 1600

	Mammals, birds and marsupials
1600–1699	12
1700–1700	27
1800–1899	91
1900–	81

Source: Nobile and Deedy 1972.

TABLE 7.6   Extinctions since 1600

	Marsupials	Mammals	Birds
Asia	—	4	5
Africa	—	11	2
Central & South America	—	1	—
North America	—	10	7
Europe incl. Iceland	—	6	1
Antarctica incl. islands	—	1	68
West Indies	—	3	22
Hawaiian birds	—	—	24

Source: Nobile and Deedy (1972).

be distinguished. The relatively low figures for Europe and Asia are somewhat surprising, considering the population density compared with North America. But the most obvious fact is the vulnerability of oceanic islands, though this figure is exaggerated by the separate status accorded to some members of a subspecies: the figure for species in Oceania is probably comparable with those for the West Indies and Hawaii. Within the Pacific, habitat alteration is considered to have caused more extinctions than hunting overkill, and new animal hazards (predators, competitors, diseases and genetic swamping from compatible species) is third. The greatest spatial concentrations of endangered species within a single political jurisdiction are in Hawaii, New Zealand and the Galapagos (McTaggart Cowan 1976).

Even where extinction has not occurred, the use of animal parts for decorative purposes has had adverse effects on certain populations. As an example, the trade in feathers during the nineteenth century may be cited: fashion in the industrial world dictated the slaughter of many millions of birds to provide feathers to adorn elegant ladies. Rheas, herons, pheasants, ostriches, owls, jays and birds of paradise found their way from South America (the chief source), Africa, India and New Guinea to the markets of New York, London and Paris (Doughty 1975). Illicit trade in wild animals was attacked by the Washington Convention, implemented in 1976. However the UK statistics (Table 7.7) show several species which according to the Convention should not be in trade. Clearly more careful licensing and enforcement is essential.

Such situations were and are not accepted with equanimity everywhere. Many groups work hard to save threatened species and international agencies such as IUCN and the World Wildlife Fund are leaders of this movement (Goodwin 1973). The scientific arguments for preservation centre around the irreplaceable loss of genetic material which extinction means, and around the loss of taxonomic diversity: some species could have played critical roles in the stability of their ecosystems. Beyond these arguments, the cultural valuation of many of the now threatened species (e.g. tiger, Javan rhino, jaguar) is strong. Many nations are now working towards a system of reserves (see also p. 314) in which those species most threatened with extinction can be protected. As far as animals are concerned, six types are identified as being in particular danger (Terborgh 1974):

1. Species on the top trophic level and which are the largest members of groups, such as mountain lion, bison, California condor, tiger, golden eagle. Many have low reproduc-

TABLE 7.7 Selected imports of wildlife into Britain (January–July 1976)

*Leopard (*Panthera pardus*)	661 skins
*Jaguar (*Panthera onca*)	279 skins
Margay (*Felis wiedii*)	9 277 skins
Geoffroy's cat (*Felis geoffroyi*)	14 169 skins
Ocelot (*Felis pardalis*)	16 619 skins
Tiger cat (*Felis tigrina*)	9 000 skins
*Polar bear (*Thalarctos maritimus*)	101 skins
Malayan bear (*Ursus malayanus*)	100 live
Elephant (*Loxondonta africana*)	42 871 kg ivory
Estuarine crocodile (*Crocodilus porosus*)	100 live
Estuarine crocodile (*Crocodilus porosus*)	523 skins
Caiman (*Caiman crocodilus*)	310 live
Caiman (*Caiman crocodilus*)	5 021 skins
Tortoises (*Testudo graeca*)	245 026 live
(*Testudo horsefieldi*)	25 000 live
*Hawksbill turtle (*Eretmochelys imbricata*)	322 kg tortoiseshell
Monitor lizards*(*Varanus flavescens*)	646 389 skins
*(*Varanus bengalensis*)	462 live
*(*Varanus griseus*)	245 254 skins
(*Varanus exanthematicus*)	25 000 skins
(*Varanus niloticus*)	25 002 skins
(*Varanus salvator*)	22 008 skins
Iguana (*Iguana* spp)	25 000 skins
Tegu (*Tupinambis* spp)	40 013 skins

* On Appendix 1 of the convention—i.e. should not be in trade.
Source: Burton 1976.

tion rates and need protection from hunting combined with large wild areas in order to survive. Species often occupy a broad spectrum of habitats and so the size of park is probably more important than its precise location.

2. Widespread species with poor dispersal and colonizing ability. Their vulnerability is due to habitat destruction and they need protection in a small number of medium to large parks with their habitats intact. The parks must provide a maximum diversity of habitat types.

3. Continental endemics. These are usually common in the limited places in which they occur but have rather rigid habitat requirements. Destruction is often due to lack of awareness of their restricted nature and they can often be protected in small but strategically located reserves.

4. Endemics of oceanic islands. These comprise a majority of the avian extinctions of historic times. Reserves become very important in any tract of undisturbed vegetation and the introduction of further exotics should be cautiously undertaken. Wherever possible, elimination of feral populations of cats, cattle, pigs, goats or mongoose should be undertaken, as is being done by the US National Park service on the island of Hawaii.

5. Species with colonial nesting habits. These comprise marine turtles, birds, and mammals that come ashore to lay eggs or rear young and are thus extremely vulnerable to economic development and to introduced predators such as rats. They can be protected by setting aside their colonial habitats, which are not generally large, and by eliminating exotic predators on offshore islets.

6. Migratory species. These are subject to the pressures of change at each end of their route and may encounter both altered habitats and heavy hunting pressure on the way: for instance, the annual toll of songbirds migrating over Italy is enormous and prohibitions against shooting cannot be enforced even in National Parks. Protection of these groups often requires international agreement to protect suitable places at either end of a migration route, to say nothing of the stopping places along the way.

Nor need the extinction of plants be forgotten. Habitat changes due to climatic change and human activity have ousted populations of plants at many scales; as might be expected, island floras have been especially vulnerable to introduced diversivores and herbivores like the pig, goat and sheep. In 1774, Phillip Island, 1000 miles east of Sydney, Australia, was thickly wooded but within 100 years grazing animals had helped to remove the forest and along with it at least two endemics, *Hibiscus insularis* and a wild pea *Streblorrhiza speciosa*. Similar histories of total extinction can be told for nearby Norfolk Island where only two specimens of Norfolk Island pine (*Araucaria heterophylla*) were left in 1971; and for St Helena and for Hawaii, together with thousands of other small islands. Co-evolution of plants and animals on islands may mean that if one partner becomes extinct then the other is threatened. An example is the tree *Calvaria major*, endemic to Mauritius. Although once common, by 1973 only 13 trees, all over 300 years old, were left. The tree has seed surrounded by an endocarp wall as thick as 15 cm and the coincidence of the last known germinations with the extinction of the fruit and seed-eating dodo (*Raphus cucullatus*) which took place by 1681 has led to the suggestion that passage through the gut of the dodo (and, in particular, crushing in that bird's gizzard) may have been the only way in which the dormancy of the seed was lifted and germination could take place (Temple 1977).

Extinctions on continents tend to leave some specimens of the plant in another part of its range so that their flora is not so easily thinned out as that of isolated islands. Nevertheless, one estimate (Tinker 1971) suggests that 20 000 plant species (about 10 per cent of the global flora) must now be considered to be threatened with extinction. This estimate is refined by Raven (1976) who estimates that the world flora comprises 240 000 species and that of the 85 000 species in the temperate zones, 4500 are threatened of which some 450 are endangered. The equivalent data for the tropics (150 000 spp of flowering plants in total) are virtually impossible to obtain, but there are forecasts that 50 000 tropical species may become extinct by the end of the century. In the continental USA, 750 plant species out of a flora of 20 000 species are threatened (Elias 1977). Doubtless among these plants there are many that might be useful or beautiful or stabilizing components of ecosystems, and such is the interdependence in ecosystems that a disappearing plant can take with it 10–30 dependent species such as insects, higher animals and even other plants.

Biogeography of the city

The construction of cities and industry has been one of the main direct purposes of the post-nineteenth century world economy. Such places are not usually biological deserts, even though their productivity is much lower than the ecosystems they have replaced or in the food-producing or forest systems which probably surround them. Built environments act as resource converters in bringing in large quantities of energy and materials (including living organisms) from great distances and then expelling wastes. They also transform ecosystems: on the one hand they obliterate nature and on the other they create new habitats, effects which are sometimes the consequences of deliberate actions and sometimes accidental. Because the city is an open system with inputs and outputs of energy and matter, and transformations of them within its boundaries, it can be viewed in the same ecosystem framework as natural systems. Although ecology has become mainly a rural science, its analytical approaches should be valid for urban areas as well, although few lineaments of an urban-based ecology have yet appeared (Egler 1970; Detwyler and Marcus 1972).

The urban ecosystem

The processes of the city differ from those of other terrestrial ecosystems in two respects (Hughes 1974). Firstly there is the massive energy flow, based mainly on fossil fuels, which may exceed that of the surrounding rural areas by a thousand times; secondly materials are garnered from huge distances, are concentrated, transformed and stored and then unwanted wastes are ejected (Fig. 7.6). All the replacements come from outside the city boundaries, unlike a natural ecosystem, and most of the wastes are dumped outside as well, usually in a very concentrated form. Inside the city, the producer trophic level is virtually insignificant and the consumers use the energy flow for the chemical transformation of materials, the erection of structures and for transport. Wastes include heat and materials like sewage and solid wastes. Thus the biological requirements of urban man (air, water, space, shelter, energy as food and heat) are met by inputs from

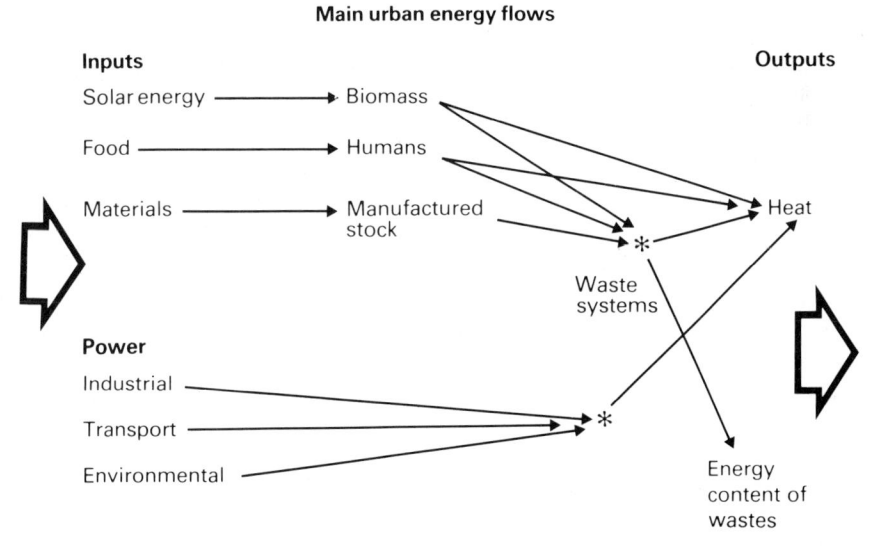

Fig. 7.6 Main energy flows of cities. The fuel-based inputs far exceed those based on solar power. Source: Hughes, 1974.

a wide variety of sources and places, but waste disposal tends to be concentrated in a very few and not-too-distant places. Wolman (1965) estimated some of the quantities per person per day for the USA as: 568 litres/day of water, 1·8 kg of food and 8·6 kg of fossil fuels. These were converted into 455 litres of sewage, 1·8 kg of solid wastes and 0·86 kg of air pollutants.

Urban biogeography: plants

In the course of transforming nature which the building of urban areas represents, many plants are killed and the habitats of the remainder much altered. Changes in conditions may mean that even plants deliberately preserved during construction may be unable to grow: some species of trees for example are unable to tolerate contaminated air or to withstand trampling of the soil above their roots. Severe air pollution by SO_2 may also mean the loss of practically all the lichen flora (see p. 233) creating a so-called 'lichen desert'.

On the other hand the city provides niches for 'new' or adventive plants of which the most obvious group are the ruderals: the flora of open spaces and other pieces of waste land. Such niches often offer freedom from competition and so plants of early successional stages are often successful. One of the outstanding examples is the colonization of bombed areas in European cities during and after World War II by fireweed (*Chamaenerion angustifolium*) which in natural conditions flourishes in open patches in woodland especially if burning has occurred. Such habitats may also provide suitable spots for the introduction of exotic flora (see p. 246).

In the same sense as the city itself is a conscious creation of man, so some of its plants reflect the overt desires of individuals and societies. Some trees, for example, are relics of pre-existing vegetation, whereas others are newly grown to improve the urban scene. In the more prosperous suburbs of North American cities, for example, street and garden trees may form virtually an open forest in which the houses and roads are set, although in a study of Chicago, it was particularly noted that the residents avoided planting the native forest trees and shrubs unless the garden was of a large size, commonly greater than 1 hectare (Schmid 1975). In semi-arid areas, the city may appear as a green oasis because of the trees in it. Secondary woodland may also close over abandoned agricultural land near urban agglomerations and so give a forested matrix to cities: about half the Boston-Washington megalopolis is wooded, for instance.

Even more deliberately managed are parks and green space in the cities. Many of these have to satisfy a number of uses and so vegetal variety helps to define areas for such uses: grass for open play areas, shade trees for summer picnics, shrubbery such as rhododendrons to provide shelter from winds for the less mobile, and aquatic vegetation such as pond lilies for visual appreciation. All these are part of the plant background to an urban park. Such parks vary in size greatly and may even be totally artificial to the point of being created on the roof-tops of large buildings (Plate 29).

More intensively managed are private gardens. These may start indoors with pots of tropical plants and move out to provide enclosures of private space, with evergreen hedging as one of the aids to seclusion. Other uses of the garden are to provide recreation

Plate 29 A microcosm of nature is created in the roof garden of the Kaiser Building in Oakland, California.

and aesthetic pleasure by gardening and the enjoyment of the results, and provide fresh fruit and vegetables; in suburban Greater London during the 1950s, Hyams (1970) estimated that 21 per cent of the gardens in working-class areas produced fruit and vegetables. We should note that the intensity of input into gardens, especially productive ones, raises their biological productivity well above that of most agricultural ecosystems. A study of 600 gardens in a London suburb (Best and Ward 1956) showed that the value of food output that of unit area was higher than average farmland and indeed close to that for the best farmland. Similar results were obtained from urban allotments, which produced 48 t/ha of vegetables (60 GJ and 788 kg protein per ha/yr) which is as high as any agricultural crop system and is achieved by virtue of labour intensity at a small scale, using the land and solar energy more fully than with single-crop farming. Seventy per cent of the energy costs were for fertilizers and so could be reduced by increased composting of organic wastes. The encouragment of urban wildlife, especially birds, through food plants, the direct provision of food, or breeding sites is another function of urban gardens. To these ends a wide variety of flora is employed: from wild plants brought directly from natural ecosystems to highly bred exotic food plants or flowers whose ancestors lived on other continents. Some gardens may be specialized, like those providing a pleasant variety of smells for the blind, or those designed to enhance meditation as in the stone and moss gardens of some Buddhist monasteries in Japan. In some cultures, however, the greatest unit inputs of energy, water and minerals are reserved for lawns, where the ideal is to have a bright green close-cropped sward without broad-leaved plants. Such lawns cover about 16.5×10^6 acres in the USA and accounted for an annual expenditure in the 1960s of $3 million (Detwyler 1972). An insight into their intensiveness can be gained from the study of Falk (1976) upon a suburban lawn in California. He calculated that energy inputs (labour, petrol, fertilizer) amounted to 578 kcal/m²/yr, which resulted in a primary productivity of 1020 g/m²/yr which compares with maize (1066 g/m²/yr) and tall grass prairie (1000 g/m²/yr). The input figure is roughly comparable to that for maize (715 kcal/m²/yr) when adjustments are made for the number of days of growth and irrigation need.

Needless to say, the city is not an ideal environment for some plants. Trees, for example, may suffer from reduced supplies of oxygen and water in paved-over or heavily trodden soils. Park and garden management may well interrupt nutrient cycles by removing litter not only in the shape of polythene bags and orange peel but as leaves, twigs and branches so that the plants become short of nitrogen, phosphorus and potassium. The applications of rock salt to roads may also cause damage, and spray drift may deposit insecticides on the wrong tree. Even if not lethal, such stresses may weaken an urban tree's susceptibility to disease, as may atmospheric contaminants. And finally there are the problems of mechanical damage and vandalism.

Urban biogeography: animals

As with plants, construction initially reduces the diversity of fauna. When the urban framework becomes stabilized, however, more animals are found within its confines, and like plants these have varying relationships with human presence. The deliberate introductions are dominated by pets of which the cat and the dog in all their manifold breeds are the most numerous. In 1973, dogs were possessed by 26 per cent of households in the UK and 46 per cent in the USA. The UK figure for cats was 16 per cent. In absolute terms, the UK population of dogs was about 3·2 million in 1956 and probably over 6 million in 1976; some estimates suggest that a million of these are strays (DoE 1976). Such a proportion exacerbates the problem of rabies which is liable to break out in animals like dogs and cats. Britain is currently (1979) rabies-free, having been so since 1902, and strict import controls aim to keep it thus in spite of the epizootic which has come near to the continental side of the English Channel. Pre-emptive vaccination could be possible but would likely be ineffective if nothing is done about fox populations, the main wild-animal vector of rabies virus. Reduction of stray-dog numbers would nevertheless be a useful first step (Taylor 1976) and any such measure which concomitantly reduced the incidence of dog excreta in towns would have considerable support from the non-owners in the population.

In addition, a wide variety of exotics are imported as pets to any country, with, in temperate lands, a possible preference for the colourful and obviously tropical such as budgerigars and some warm-water fish. (The residents of Washington DC are reported by Havlick (1974) to buy over 4500 t/yr of birdseed.) Zoos are another great repository of imported species though today there is a good likelihood that many of the animals on view are zoo-born. The purposes of zoological gardens are dominated by scientific endeavours, especially the breeding of rare animals and the public entertainment comes second (Tudge 1976), but in the 'safari park' type of zoo the reverse is usually true. Some animals may do very well in both, and the price of lions in Britain fell from £500 to £25 in the period 1971–76, and the country's lion population was estimated at *ca* 500. Both pets and parks suffer escapes and these are occasionally dangerous; more often they serve to fuel the stories of the enormous size to which escaped alligators have grown in the sewers of a city.

Of the animals not deliberately encouraged by man, there are two main groups: those which continue to exist in the city but at a relatively low level of population. Man-made deterrents to animal life in such places include the negative effects of roads, railways and power lines, nocturnal migrants in particular suffer heavy mortality from collisions with high towers and buildings, and in-flight collisions of birds and aircraft have meant considerable measures to make airports unattractive habitats for birds. There are also those whose populations expand because of some favourable feature of the habitat and thus become a feature of the city's biogeography, such as species of garden bird like thrushes and robins, and migratory water fowl may have no objection to ponds in city parks. Where suburbs are of a low density and other open spaces abound, like railway yards or rights of way under power-transmission lines, then larger mammals may persist, such as the fox, rabbit, racoon, chipmunk and skunk. As one instance, the islands of semi-wild landscape within the Los Angeles conurbation have provided a continuing habitat for the coyote. Not only does it predate upon cats and dogs but has become an urban scavenger and even a herbivore, stripping fruit from the lower branches of orange, avocado and plum trees. In spite of control programmes, the numbers are such that probably some 400 remain in the city (Gill and Bonnett 1973). Enough, apparently, to provide sufficient howling to halt concerts occasionally in the Hollywood Bowl, which is surrounded by chaparral and where no doubt the concert programme was punk rock rather than Palestrina. During winter, even larger animals are known to invade North American cities: white-tailed deer and black bear, for example. In leafy towns, certain aphids become pests because of their excretions ('honeydew') which coat parked vehicles. Usually, insecticides are used against them, although examples of biological control of the linden aphid (*Eucallipterus tiliae*) and the elm aphid (*Tinocallis platani*) by introduced parasites have been reported from Berkeley, California (Olkowski *et al.* 1976).

Bird populations in suburbs show, not unnaturally, that numbers of species and of individuals increase with time, especially in the first few years of domestic plantings. It also appears that the residential neighbourhoods of one study (Vale and Vale 1976) supported more species and individuals than the pre-suburban environments (Fig. 7.7.).

Some species have become characteristic features of the city in temperate latitudes: pigeon, starling, English sparrow, jackdaw and swift are examples, the latter two showing an ecumenical preference for towers of an ecclesiastical nature. But the numbers of pigeons and starlings may be such that they become labelled as pests along with the

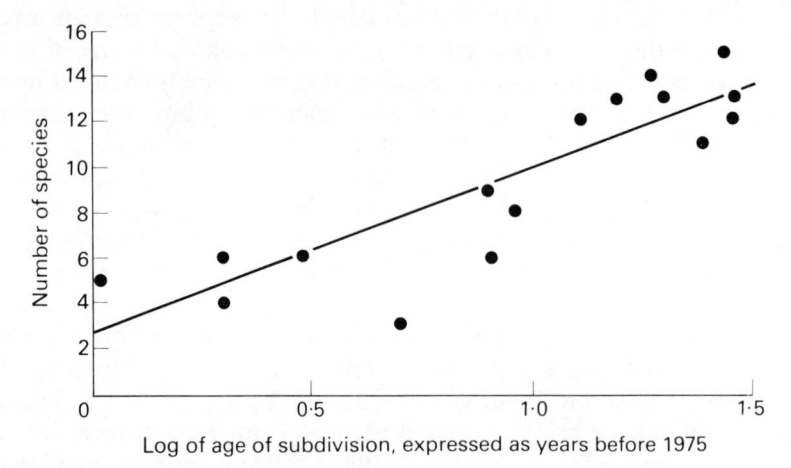

Fig. 7.7 Number of bird species against time in a suburbanizing area of California. Source: Vale and Vale, 1976.

less obtrusive but equally common (if largely nocturnal) rats and mice. In Britain, the herring gull (*Larus argentatus*) is currently proving an explosively successful colonist of the rooftops of coastal towns, where safe and secure nesting sites have helped in the 13 per cent per annum increase in its population since 1945 (Coulson and Monaghan 1978). The rhesus monkey of India fills an urban-scavenger sort of role and even prefers cooked to raw foods. These groups are viewed as problems for various reasons: rats and mice spoil food, eat through important timbers and sometimes electrical wires and above all are perceived as unattractive; in some countries, small urban mammals such as the squirrel may be a reservoir of rabies. The birds are generally credited with the crime of soiling buildings (starling roosts on ledges and their vicinities are apt to look like guano-rich islands), being noisy, and, most heinous of all, desecrating the statues of the famous. Minor problems include the stripping of bark from urban trees in winter by mice and rabbits, and the strewing of domestic garbage by racoons (Stearns 1972). To this must be added the anxieties caused in a suburban household by the defensive action of a skunk when attacked by the family dog taking its nocturnal exercise in the garden. When considering unwanted biota, we might note that a large department store contained a bacterial density of 4 million/m^3 of air and main streets 575 000/m^3, compared with 50/m^3 in a forest (Decourt 1974).

Man, city and nature

Although the biogeography of cities might at first seem marginal to any consideration of man and nature, we must not forget that for many people the plants and animals they see in the city are their main contact with non-human life. Exactly how this may affect their attitudes towards the protection of nature elsewhere, or towards their use of the resources provided by living systems, is unknown. But it seems likely that an outlook conditioned by a city in which plants and animals are reasonably common, even if in managed populations, is likely to be more conducive to an appreciation of what nature can provide than a totally sterile concrete karst. Management of cities in order to maximize the opportunities for a diverse wildlife is possible and there seems no reasons why planners should not take into account this facet of biogeography (Dansereau 1970; Gill and Bonnett 1973).

Further Reading

ECKHOLM, E. 1978: *Disappearing species: the social challenge.* Washington, DC: Worldwatch Paper 22.

EDWARDS, C. A. (ed.) 1973: *Environmental pollution by pesticides.* New York and London: Plenum Press.

ELTON, C. 1958: *Ecology of invasions by animals and plants.* London: Methuen.

GILL, D. and BONNETT, P. 1973: *Nature in the urban landscape: a study of city ecosystems.* Baltimore: York Press.

GUNN, D. L. and STEVENS, J. G. R. 1977: *Pesticides and human welfare.* Oxford: Oxford University Press.

LEACH, G. 1976: *Energy and food production.* London: IPC Press.

MYERS, N. 1976: An expanded approach to the problem of disappearing species. *Science* **193**, 197–202.

TERBORGH, J. 1974: Preservation of natural diversity: the problem of extinction-prone species. *BioScience* **24**, 715–22.

WHITTLE, T. 1975: *The plant hunters.* London: Picador Books.

ZISWEILER, V. 1967: *Extinct and vanishing animals.* New York: Springer-Verlag.

8
The inadvertent creation of new biotic patterns

The metabolism of animals and plants always produces substances which are released into their immediate environment and the evolution of ecosystems has usually ensured that such substances are taken up by another component of the system. In the case of man, his cultural wastes have always been added to metabolic excretions, and for as long as *Homo sapiens* has existed, these superfluous products have been emitted into natural systems. The capacity of the terrestrial, freshwater, marine or atmospheric systems to process these wastes was never in doubt while the density of man was relatively low and while his technology was pre-industrial. The deleterious effects to man himself of concentrations of waste such as were found even in pre-industrial cities were quickly realized, though largely because of features such as the smell—the role of micro-organ-

Plate 30 The red kite: once a common urban scavenger in Europe, now a rare bird of sparsely populated rural areas of Atlantic Europe.

isms in causing disease was not realized until the nineteenth century. Naturally, such substances attracted animals capable of utilizing the wastes: apart from micro-organisms, the rat, for example, profited from these conditions and in European cities the red kite (*Milvus milvus*), now a rare bird of remote areas (Plate 30), was a common scavenger. But as with so many aspects of the man–nature relationship, the industrial revolution of the West heralded a profound change. The linked processes of industrialization and urbanization have meant the concentration of wastes both from people and from manufacturing, and the extension of energy-intensive methods to farming has produced analogous effects in rural areas. Thus the smoke and sewage of a town of 1500 people in 1500 AD may have had relatively little effect beyond its boundaries, but a city of 1·5 million in 1950 AD is bound to have exerted a considerable effect on surrounding ecosystems by virtue of the quantity, concentration, and composition of its wastes. Not only are there now higher concentrations and waste flows of higher magnitude than ever before (Table 8.1), but science has produced many substances, especially complex

TABLE 8.1

(a) World production of industrial chemicals

	$\times 10^6$ t
1950	7
1970	63
1985 (projected)	250

(b) Direct release to the environment in 1970

	$\times 10^6$ t
Solvents	10
Detergents	1·5
Pesticides	1
Gases Chemicals	1
Lubricating and industrial oil	2·5
Miscellaneous	7

Source: Klein 1976.

organic chemicals, which are unknown in nature and for which no pathways (especially those which would degrade complex molecules into simple ones capable of uptake by plants as nutrition) exist in the ecosystems into which they are led. More than 1000 new substances per year now being produced (Biswas and Biswas 1976).

Many of the flows produced by human activity are intensifications of background flows, which already exist in nature; man accelerates the natural cycles and magnifies the flows within them and hence intensifies their impact upon ecosystems. Examples of substances which have natural pathways accelerated and intensified by man are those of mercury and phosphorus from the land to the oceans, of sulphur in the atmosphere, and oil to the oceans' surfaces; ionizing radiation is also continually present under natural conditions but is intensified at certain localities by man's manipulation and concentration of radioactive isotopes.

232 *The inadvertent creation of new biotic patterns*

Today there are numerous wastes which affect living organisms and they can be classified either by their own physiochemical characteristics or, more useful for the present purpose, by the type of biospheric system into which they are led by way of disposal. Not all of them affect biota directly: some only do so via secondary effects, and not all are directly lethal; some wastes exert only sub-lethal effects or may insidiously bring about long-term genetic change. But others are direct and obvious in the way in which they produce conditions inimical to some (and occasionally all) organisms. There exists also the chance of synergisms: the possibility that two or more wastes thought to be harmless may combine in the environment or in a receptor organism to produce a toxic or genetically manipulative effect. We may note too that many of the effects of wastes are side-effects in the sense that they take place after the substance has done its job: many of the deleterious effects of biocides, for example, have been caused by the action of the chemicals on animals which they were not intended to reach, let alone kill.

Put together with all the other effects of industrialization, some very far-reaching effects on the world's flora and fauna have been brought about, and some generalizations about these results can be found at the end of this chapter.

Wastes in the atmosphere

Many of the wastes led off into the air are waste substances from the combustion of fossil fuels for energy production; others arise as the result of chemical industries and also from the production of aerosols as propellants in spray-cans and in aerosol forms of metals, e.g. tetraethyl lead, which is an anti-knock additive in gasoline (Table 8.2). Particulate matter also arises from bare surfaces such as unplanted arable land, deserts and cities (American Chemical Society 1969; ten Houten 1969; Jacobson and Hill 1970; Mellanby 1972; Hodges 1973).

Most of these substances are gaseous or consist of very small particles and are thus quickly dispersed: if low concentrations affect plants and animals, they may do so over a large area when they fall out from the atmosphere, and in their residence time there chemical changes may have taken place. In this case they can be said to have direct effects; but indirect effects are also possible, if for example particulate emissions were to cause a global cooling then shifts in floral and faunal ranges would be expected.

TABLE 8.2   Incidence of effects of atmospheric wastes

Waste	By-product of energy production	Direct effect on biota	Indirect effect on biota
O_3	+	×	
CO_2	+	×	×
SO_2 & H_2S	+	× (becomes H_2SO_4)	×
NO_x	+	×	
Hydrocarbons	+	×	
Smoke	+	× (decreases radiation)	×
Aerosols			×
Fluorocarbons			×
Heat	+	× (but slight in atmosphere)	

Ozone, a by-product of photochemical air pollution, drifts out into rural areas from cities and is known to affect many plants: its mode of action seems principally to be to effect closure of stomata and to damage the palisade cells of the leaves (Howell and Kremer 1972).

Carbon monoxide is a product of fossil-fuel combustion and the man-caused emissions are probably several orders of magnitude higher than those from natural causes (80 per cent of the atmospheric load comes from automobiles). In the atmosphere it oxidizes slowly to CO_2, but close to emission sites its effects on warm-blooded animals is well known, since it combines with haemoglobin to lessen the oxygen-carrying capacity of the blood. Carbon monoxide levels of 10 ppm in cities are undesirably high but can exceed 100 ppm for short periods. Most observations have been carried out on man and we can assume that other mammals and birds suffer similar effects; no field results of effects on plants have provided any startling results.

A better known contaminant of the atmosphere is sulphur, mostly in the forms of sulphur dioxide (SO_2) but also found as hydrogen sulphide (H_2S). In nature it is emitted by volcanoes and by anerobic decomposition in mires. Man's emission of sulphur compounds is variously estimated at 30–70 per cent of all the atmospheric sulphur, depending on whether only SO_2 or total sulphur is measured. Much SO_2 is given off by the burning of coal and oil (ca 6 million t/yr in Great Britain during the 1960s) and it has a residence time of between 5 days and 2 weeks in the atmosphere. During this period, much of it becomes ammonium sulphate, which is not toxic to biota; some however oxidizes to SO_3 and reacts with water to form H_2SO_4 (sulphuric acid). Both the fallout of this acid and the direct effect of gaseous SO_2 are known to be damaging to plants and animals, although it appears that plants like rye grasses can evolve city forms which are resistant to contamination by sulphur. Effects on plants have been noted at sulphur-dioxide concentrations of 0·03 ppm, and they are highly damaging at 0·1–1·0 ppm, when leaf blotching due to cell damage, and reductions in crop yield are noticed. Weakening of a plant makes it more susceptible to injury by predators or parasites: the ponderosa pine in western North America succumbs more often to the western bark beetle when stressed by sulphur contamination. Lichens appear to be especially vulnerable, to the point where the presence or habit of certain species can be used as an index of the incidence of SO_2 contamination of the local atmosphere (Gilbert 1970; Hawksworth and Rose 1970; Fig. 8.1). A sequence of diminishing variety of species and luxuriance of individuals, depression of fruiting, and the trend of presenting the smallest possible surface cover at the lowest possible elevation, leads eventually to sterility and absence, and can be correlated with the everage annual quantity of SO_2 present in the air. A series of zones of SO_2 concentration, from 'pure' air to SO_2 levels of >170 $\mu g/m^3$ air can be designated upon their lichen floras: at the most contaminated end in Britain only *Pleurococcus viridis* is present on deciduous trees but is then confined to the base; in 'pure' air, species like *Lobaria amplissima* and *Usnea articulata* are found which cannot tolerate any SO_2. In recent years much of the earlier 'lichen desert' in cities has become colonized by a virtual monoculture of the lichen *Lecanora conizaeoides*. This was once a rare element in the British flora but was also known from the areas of high natural levels of sulphur near hot springs in Iceland and was clearly pre-adapted for a sulphur-

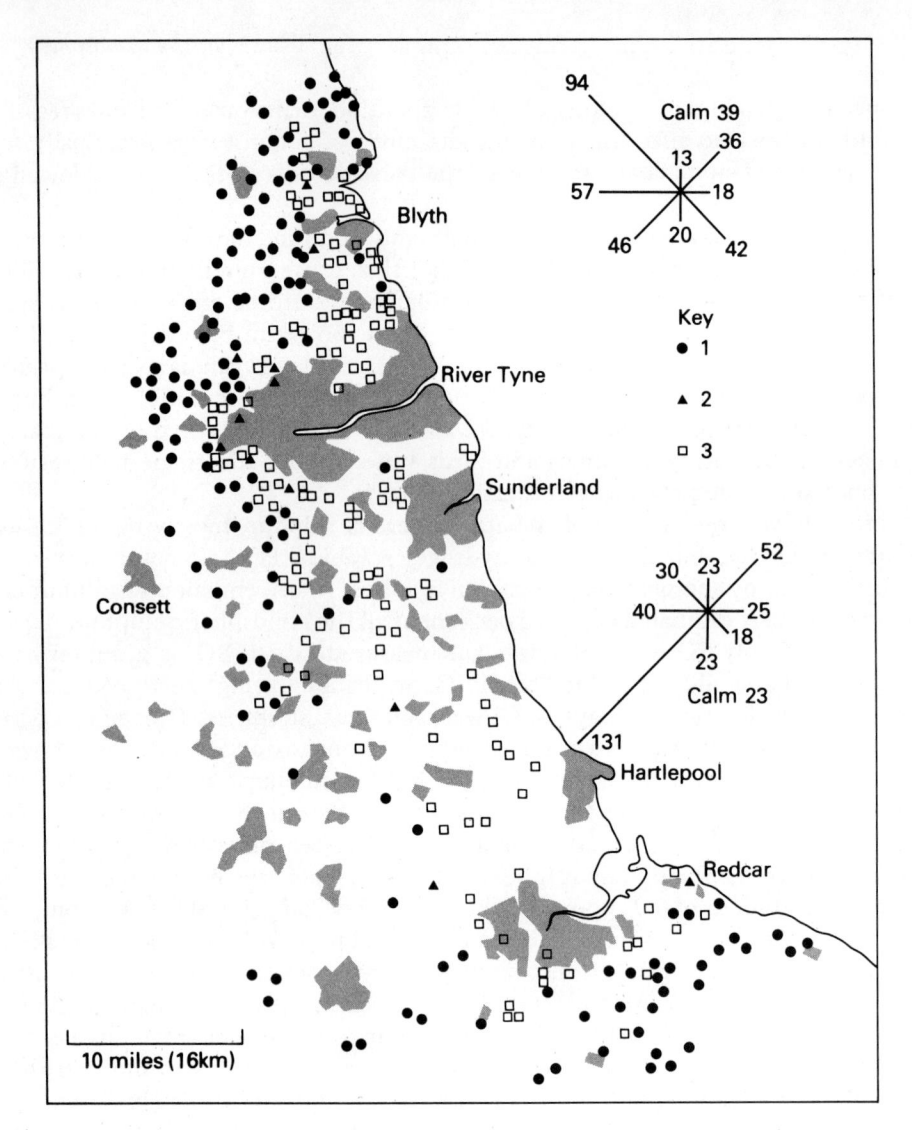

Fig. 8.1 Northeast England showing the limits of the lichen *Parmelia saxatilis*. Windroses for 1962-65.
1. Inner limit of *P. saxatilis* on exposed stone.
2. Inner limit on sheltered stone.
3. Absent from apparently favourable site.

Source: Gilbert, 1970.

tolerant role in cities (Cook and Wood 1976). One finding of such work is that plumes of SO_2 levels of 45 µg/m³ and above extend downwind from cities a great deal further than had been detected with conventional monitoring systems. The diminution of the green plant layer of the ecosystem means, as would be expected, a lessening in the biomass of herbivorous animals, e.g. psocids on ash (*Fraxinus excelsior*) bark in Britain. The numbers of predators in this system did not suffer much loss, however, because they feed on other organisms which fall out of the canopy (Gilbert 1971).

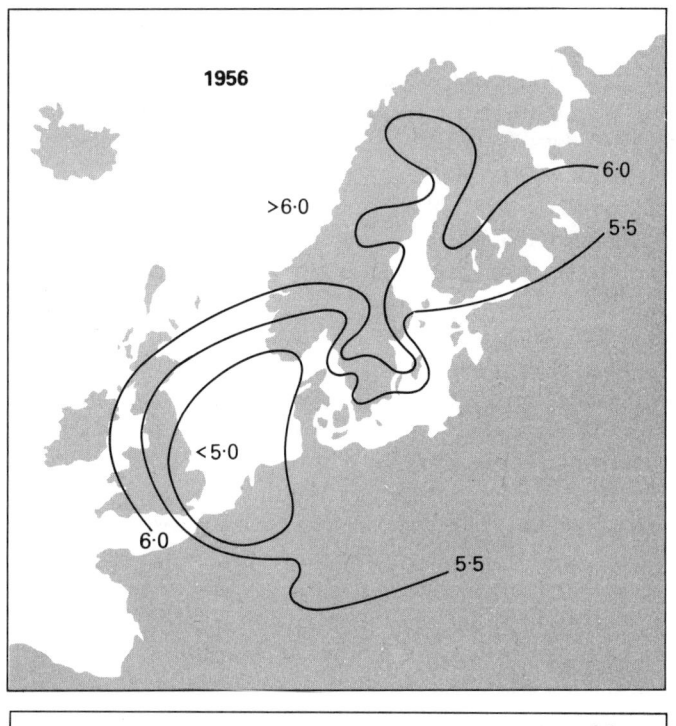

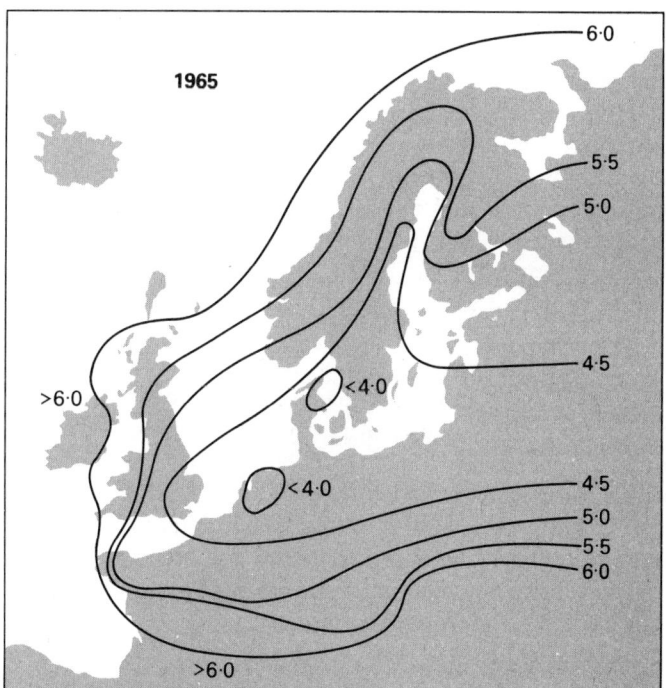

Fig. 8.2 pH in precipitation over Europe 1956 and 1965.
Source: Lundholm, 1975.

Knowledge of the effect of SO_2 on animals is mostly derived from medical studies: both H_2SO_4 and SO_2 irritate the respiratory system. Daily mean concentrations of 0·2 ppm SO_2 for 4–5 days have increased human mortality and so other animals are likely to suffer in some way.

A noticeable but indirect effect of SO_2 production is the acidification of precipitation which may reduce the pH of surface waters a considerable distance downwind from the point of emission. In 1958, precipitation was exhibiting values of less than pH 5·0 only in the Netherlands but by 1962 areas of less than 5·0 had extended over central Europe and the Netherlands was receiving precipitation of pH 4·0. In the late 1960s the 4·5 value had reached central Sweden (Fig. 8.2). The strong supposition is that SO_2 from Britain is the cause of the acidification and that it may cause shifts in flora and fauna of freshwater habitats: e.g. salmon do not persist below pH 5·5, and 4·0 is something of a lower threshold value for many forms of life (Bäckstrand and Stenram 1971). The picture is nevertheless somewhat more complex: the atmosphere above the Atlantic and Arctic oceans contains background sulphuric acid, for example, and the presence of acids from the breakdown of organic material and from soil erosion complicate the picture. The 'pure' water of Scandinavia is unusually sensitive to acid precipitation, so that the UK is perhaps not totally to blame, and anyway its SO_2 emissions are decreasing at present, whereas in the rest of Europe they are not (Yanchinski 1978).

Several industrial processes (e.g. phosphate-fertilizer manufacture from phosphate rock, aluminium processing, steelmaking, ceramic firing) cause the emission of fluorine into the air, in forms such as HF, H_2SiF_6, and SiF_4 which reacts with water to give HF and SO_2. Fluorides are readily absorbed by vegetation, and atmospheric concentrations as low as 0·001 ppm can, if persistent, cause leaf injury and the reduction of growth and yield. Within the leaves, selective concentration near the margins may turn at atmospheric level of 0·001 ppm into a cellular level of 150 ppm and cause necrosis. Animals which eat vegetation thus affected may suffer from fluorosis if their feed contains 40–60 ppm of fluoride or more. Slight poisoning shows as mottling of the teeth but more severe cases exhibit softening of the bones and eventual death.

Another set of gases which are produced in the natural world but which have their cycles magnified by human activity are the oxides of nitrogen (mostly NO and NO_2 but with some N_2O, and generally written as NO_x). The natural biological production is probably about 1×10^9 t/yr and forms part of the nitrogen cycle. The man-directed production (mostly from the burning of petroleum) of NO_x is about 48×10^6 t/yr and the gases have an atmospheric residence time of 3–4 days. The effect of NO_2 on plants is well known: leaf injury and retarded growth are observable at concentrations of 1 ppm/1 day or at 0·35 ppm for several months. NO_2 is fatal to humans and other warm-blooded creatures at a level of 100 ppm for a few minutes and a concentration of 15–50 ppm for 2 h damages key internal organs; 0·06 ppm over a long period produces an increase in respiratory diseases in man.

One of the major biotic effects of NO_x is caused when it interacts with hydrocarbon residues in the atmosphere to result in 'smog'. Hydrocarbons are naturally emitted into the atmosphere during the decomposition of organic matter, mostly as methane (CH_4) together with smaller quantities of terpenes and isoprenes, with a production of $1 \times$

10^9 t/yr of methane. Man-made emissions are of the order of 30×10^6 t/yr in the US and probably three times as great on a world-wide basis. But these emissions are usually hydrocarbons of higher molecular weight, which in contrast to methane are highly reactive. They are mostly products of petroleum combustion, especially in motor cars and so in city air they come into contact with the NO_x group described above. Nitrogen dioxide absorbs ultra-violet light and dissociates into NO and O; the oxygen combines with atmospheric oxygen to give ozone (O_3) which in turn reacts with NO to give NO_2 and O_2 so the reaction is cyclic. But the mixture of NO and O_3 reacts with the hydrocarbon products in the atmosphere to give aldehydes, ketones and peroxyacetyl nitrate (PAN). It happens mostly under atmospheric inversions in sunny climates (hence its 'type locality' in California) and it drifts out from towns to cause damage in rural areas. It seems to cause damage to plants at lower concentrations than primary pollutants (e.g. at levels of 10 $\mu g/m^3$) and millions of dollars' worth of damage to agricultural and forest crops are done every year in the USA. In the Los Angeles basin, photochemical smog has reduced the rate of water use, the photosynthesis rate, and hence the yield of lemons and oranges; the latter tended to drop off before they were properly ripe. In some cases the yield of mature fruit was reduced by as much as 50 per cent (Thompson et al. 1970). In animals the respiratory system is irritated; in man the initial symptom is eye irritation but most bronchial complaints are exacerbated.

Some of the hydrocarbons in smog are in aerosol form and this distribution is also characteristic of airborne lead. In the air, the natural concentration is about 5×10^{-4} $\mu g/m^3$, from silicate dusts and volcanic eruptions. However, lead is used as tetraethyl lead as an 'anti-knock' additive to petrol, and so US suburban areas have an annual average level of 1–3 $\mu g/m^3$, the Los Angeles freeways have 25 $\mu g/m^3$ and there are short-term concentrations in enclosed places such as tunnels of 44 $\mu g/m^3$. Lead falls out on roads, and presumably into fresh water also, at concentrations which in roadside plants may reach 500 ppm of lead, although such high concentrations are unusual: in central Europe roadside grasses range between 50–400 ppm. Plants are usually toxified by lead in soil at 10 ppm but it is noteworthy that some plants (e.g. the grass *Agrostis tenuis*) have evolved strains which appear resistant to lead and other metals (e.g. copper, zinc and nickel). These strains presumably evolved on soils above metalliferous veins and are particularly useful in re-vegetating old heaps of metal-ore spoil. Animals are highly sensitive to lead: 0·33 ppm will kill sticklebacks and 1·00 ppm/100 h is lethal to rainbow trout.

Above the size of the aerosol particle, larger fragments of a wide variety of materials find their way into the atmosphere and rain out onto ecosystems, normally within a few days; this is especially so of the larger particles. All sizes of particle occur as natural phenomena from volcanic eruptions and dust sources such as deserts, but receive additions from many human activities which create smoke or bare ground, and from urban areas. They may be composed, for example, of metal dusts, rubber particles, asbestos, and particles from smoke. Dust fall in cities may be of the order of 0·35–3·5 mg/cm²/month; in Britain large cities may get up to 525×10^6 kg/ha/yr but rural areas are more likely to receive 0.5×10^6 kg/ha/yr. The effect of particulates on biota may be direct if a toxic substance forms the fallout or if a poison has become adsorbed onto the particle, but in all cases, inhibition of photosynthesis is likely since leaves are prevented from absorbing light. This is in addition to any light-scattering effects of the particles while they are still suspended in the atmosphere.

Animals including man generally suffer from inflammation of the respiratory tract and cancer has been postulated as an effect of contaminated air of this type. In some lower animals, the phenomenon of industrial melanism has been marked out as an effect of particle deposition in urban areas: over 150 species of moths have darker (melanic) forms in industrial areas than in rural locations; the example of *Biston betularia* was mentioned on p. 25.

Another man-made source of particulate contaminants is the disturbance of habitats which are then colonized by ruderal plants whose pollen if emitted in very large quantities is an aeroallergen. This is especially true of the North American ragweed (*Ambrosia* spp), a plant of waysides and railroad tracks. Plant-control programmes are sometimes necessary to alleviate the suffering of asthmatics and rhinitics.

At an indirect level, one of the atmospheric contaminants which could affect biota considerably is CO_2. Produced in large quantities by the combustion of fossil fuels, the nineteenth-century concentration in the air of 290 ppm has increased now to 320 ppm. The fate of 'extra' CO_2 appears to be divided between remaining in the atmosphere (about $\frac{1}{6}$) and being absorbed into the oceans ($\frac{5}{6}$). The actual effect of this CO_2 in isolation is not accurately known; there is some further discussion on p. 25.

The heat which is produced in combustion processes is an analogous case. The rate of radiation out to space is finite and so at some stage the heat burden (currently at a very low proportion of the natural heat flux) might interfere with climatic patterns, as it does at present in some cities under certain meteorological circumstances. The same is true of the postulated effects upon the ozone layer caused by dissociation of ozone by emissions of NO_x from high flying supersonic aircraft and by the chlorofluoromethanes used as propellants in aerosol cans. Less ozone would mean much more UV radiation and hence such features as more skin cancer and global warming. Reliable empirical data are as yet in short supply.

Contamination of aquatic environments

Both fresh waters and the seas have obvious advantages as sinks for concentrated wastes: the rivers carry them away, and the oceans appear to be large enough to dilute even the most noxious substances to an innocuous level. It is not necessarily so simple, for rivers may have regimes which provide insufficient water at low-flow periods to dilute wastes significantly; and in the seas as in fresh water certain substances which are virtually insoluble in water may be highly soluble in fats and so migrate differentially to living organisms and there become concentrated. In spite of their size, therefore, the oceans may be especially vulnerable to the biotic effects of contamination by wastes, especially since most wastes led off into fresh water are likely to end up in the seas at some later time.

The contamination process with the longest history is probably the influx of soil into water courses. Some such erosion is natural but in most parts of the globe has been at times accelerated by human activities. Its effects on flora and fauna are quite obvious and where the problem is longstanding then of course ecosystems have developed adaptations to it such as faunal shifts towards fish species tolerant of heavy silt loads in rivers.

One of the most characteristic contaminative processes of aquatic systems in more recent years is the enhancement and acceleration of the nutrient status of the water by the introduction of concentrated volumes of plant nutrients, especially nitrogen and phosphorus; this process is called **eutrophication**. The man-made sources of N_2 and P include sewage, fertilizers, detergents and farm-animal wastes, so that both urban and rural areas contribute to the process. Table 8.3 compares some natural and man-made

TABLE 8.3 Flows of nitrogen and phosphorus in fresh waters of the USA ($\times 10^6$ t/yr)

	N_2	P
Natural flows	935–3800	222–643
Man-generated	3610	620–920

Source: Hodges 1973.

flows of these two elements. The human-generated concentrations may be very high: for example domestic sewage has ca 200 mg/l of N_2 and farm-animal wastes ca 700 mg/l; a large animal-raising operation may thus create as much waste as a small city. In the USA, for example, canneries, dairy plants, the potato-processing industry, meat slaughtering and packing, and sugar refining provide each day wastes equal in nutrient content to the sewage derived from 8·0, 11·3, 2·1, 13·0 and 4·8 million people respectively (Alexander 1974).

A major effect of these concentrations in fresh water appears to be to remove limiting effects upon the growth of algae, although the process is not completely understood. Algal blooms are the commonest result, especially in lakes, and these lead to lower oxygen levels because they inhibit photosynthesis in other plants and the later decay of the algae by bacteria also causes oxygen depletion of the water and may kill fish.

In the sea as well, blooms may result from such enrichments: those of dinoflagellates are best known since the proliferation of the organisms causes 'red tides' (a further discussion of the phenomenon and of other contaminants which directly affect fisheries is given on pp. 301–302). Secretions from these motile plankton may then cause toxic substances to build up in sessile marine organisms such as mussels (*Mytilus edulis*), and may cause paralysis if the shellfish are used for human food.

In addition to phosphorus and nitrogen, domestic sewage contributes large volumes of carbonaceous organic matter. This forms a substrate for bacteria which oxidize it to CO_2 and H_2O and hence bring about oxygen depletion. The intensity of the effect depends on the relative volumes of sewage and water and in rivers diminishes downstream from the point of input. If the sewage is concentrated in a small volume of water then the river will exhibit changes in colour and smell (H_2S will be given off) and will have a scum of suspended solids. In biological terms a badly contaminated river will have only bacteria, sewage fungus and *Tubifex* worms near the input. In temperate latitudes the most resistant fish appears to be the eel and the least the trout and they are

240 *The inadvertent creation of new biotic patterns*

Plate 31 Dead fish killed by pollution in an Ontario lake.

sometimes used as general indicators of water quality (Plate 31). A summary of downstream changes in water quality and biota is given as Fig. 8.3. Sewage and animal wastes also introduce bacteria and viruses into water but most of the significant types are infectious to man rather than other organisms.

Since the nineteenth century, the demands of industry have caused a great number of chemical substances to be released into water systems. Opencast mining may often allow the oxidation of pyrites (FeS_2) to sulphuric acid, which then acidifies the drainage. In the USA some 4×10^6 t/yr of H_2SO_4 is pumped into 17 000 km of streams. Another example is pulp mills where some 50 per cent of the wood input may appear as waste and this is released to rivers and lakes where it reduces photosynthesis and clogs the gills of fish. Toxic substances may also be present: methyl mercaptan, pentachlorophenyl, sodium pentachlorophenate, and sulphite liquor are toxic to most shellfish. In

Contamination of aquatic environments 241

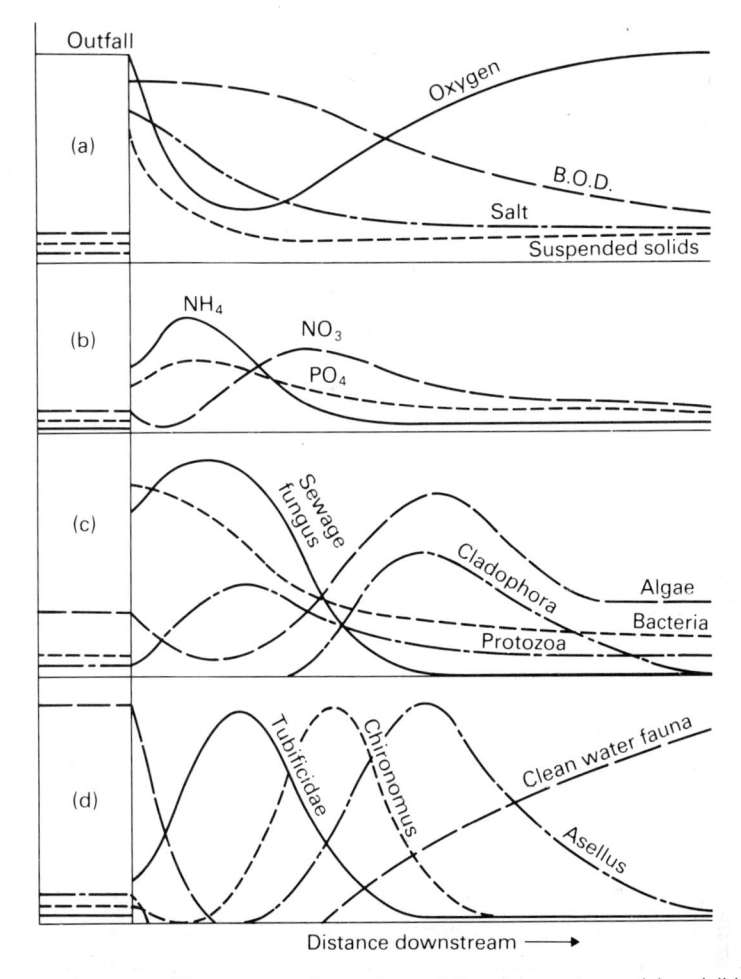

Fig. 8.3 The effects of organic effluent upon a river at the outfall and downstream. (a) and (b) are physical and chemical changes, (c) micro-organisms and (d) larger animals.
Source: Mellanby, 1972.

total, if all the organic chemicals produced in one year were evenly distributed over the land surface of the earth (without any degradation at all) then there would be a total of 7·47 kg/ha. Included in this total would be 0·07 kg/ha of pesticides and 1·12 kg/ha of fertilizers (Klein 1976). Given that such even concentrations are improbable, the biogeographic effects of such quantities of chemicals of the consequent concentrations are likely to be detectable.

Complicated organic compounds may often have no natural degradational pathways and so will decompose only gradually, displaying a tendency to accumulate in organisms which excrete them slowly (Rudd 1964). The **chlorinated hydrocarbon** pesticides (see p. 208) and **polychlorinated biphenyls** (PCBs) are examples of this class.

(Edwards 1973; Guenzi 1974). Their breakdown is via other similar substances (e.g. DDT yields DDD, DDE, DCB and DDA) which may be equally toxic or, occasionally, even more so. Most of them enter aquatic ecosystems via runoff or fallout and enter food chains differentially because they are more soluble in fats than in water, and display a tendency to concentrate at the higher trophic levels such as vertebrates. Fish-kills may be an immediate and direct result of high concentrations in runoff, but low initial levels may build up in, for instance, predatory birds to the point where eggshell thinning occurs and reproductive failure ensues. Via oceanic transport, DDT is now found in the fat of Antarctic crab-eater seals, and migratory animals may pick up the toxins at one stage of their journeys, even though the others are poison-free. Mobilization of body fat during migrations may then release the poison into the blood-stream and so dead animals may occur in relatively pesticide-free places. Such a sequence is thought to have happened with the Mexican free-tailed bat (*Tadaria brasiliensis*) which winters in Mexico and breeds in the southwest USA: DDT and DDE were implicated in this case (Geluso *et al.* 1976). The long-term effects of chlorinated hydrocarbons are not known: their effect on photosynthesis in marine phytoplankton is still disputed and in any case even if their use was stopped now, it would be 25 years before the substances came to equilibrium in the environment. However the world's biota probably contain in total less than $\frac{1}{30}$ of 1 year's production of DDT during the 1960s, which is probably a fortunate state of affairs (Woodwell *et al.* 1971). The distribution of residual DDT on a world scale (Fig. 8.4) shows heavy concentrations in places where it has been most used but with evidence for the world-wide transport which takes it to remote oceans.

PCBs (polychlorinated biphenyls) behave similarly (Peakall and Lincer 1970; Tatsukawa 1976). They are widely used in industry in large transformers and as plasticizers, hydraulic fluids and diffusion-pump oils. About 20 per cent of the total production may be released to the environment and it accumulates in food chains: in the Great Lakes, for example, 3 per cent of the fish catch were found to exceed the permitted level (5 ppm in the USA, 2 ppm in Canada), some having as high as 54 ppm of PCB. Regulations against discharge are tight nearly everywhere now, and Japan has banned PCBs altogether (ACS 1976), but the bulk of the environmental PCB load is now in the marine environment and will affect the biota for a long time to come; there is some evidence to suggest that PCBs reduce the size and production of phytoplankton and divert marine trophic pathways away from fish towards gelatinous predators like jellyfish (O'Connors *et al.* 1978).

Metals are the basis for many compounds used in all phases of human activity and certain of these have proved to be troublesome for biota. Mercury, for example, has been used as a fungicide as well as in industrial processes and mercuric wastes have accumulated in aquatic organisms which have then proved detrimental to man. Lead, cadmium, nickel, zinc, copper and selenium levels have also risen but in general these do not seem to have affected populations of organisms except where local concentrations are very high, e.g. downstream from smelting installations.

An immensely important power source for industrial societies is oil and although natural flows exist in nature, mainly through submarine seepage from the Earth's crust (*ca* 0.6×10^5 t/yr; Wilson *et al.* 1974), its main effect on ecosystems is through spillage

Contamination of aquatic environments 243

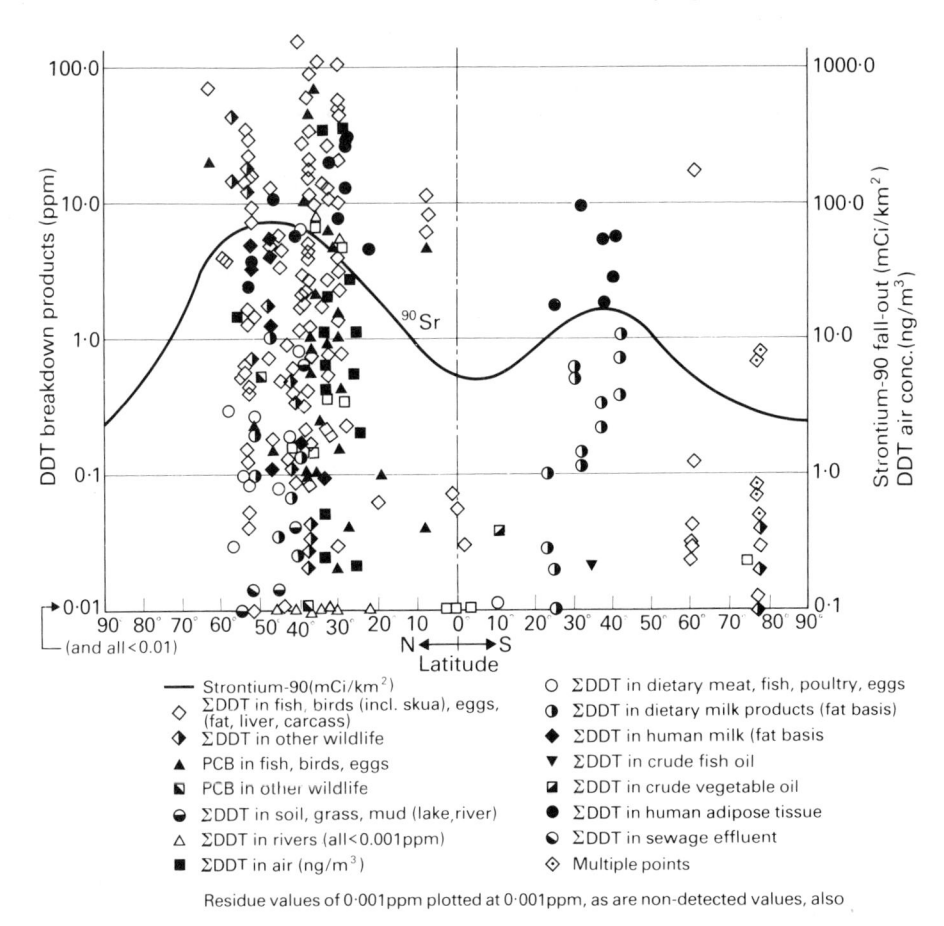

Fig. 8.4 The relationship of persistent DDT residues to global latitude. Strontium-90 from atomic weapons testing is also included in this diagram.

Legend:
— Strontium-90 (mCi/km²)
◇ ΣDDT in fish, birds (incl. skua), eggs, (fat, liver, carcass)
◆ ΣDDT in other wildlife
▲ PCB in fish, birds, eggs
▧ PCB in other wildlife
◐ ΣDDT in soil, grass, mud (lake, river)
△ ΣDDT in rivers (all < 0.001 ppm)
■ ΣDDT in air (ng/m³)
○ ΣDDT in dietary meat, fish, poultry, eggs
◉ ΣDDT in dietary milk products (fat basis)
◆ ΣDDT in human milk (fat basis)
▼ ΣDDT in crude fish oil
▣ ΣDDT in crude vegetable oil
● ΣDDT in human adipose tissue
◑ ΣDDT in sewage effluent
◇ Multiple points

Residue values of 0·001 ppm plotted at 0·001 ppm, as are non-detected values, also

estimated to be at least 1×10^6 t/yr (Hoult 1969). As an example of magnitude, the number of oil-pollution incidents reported off Britain's coasts in 1976 was 595, and in 1977 was 642. In 1978 large pollution incidents occurred in W. Europe with the sinking of two tankers, one off Brittany and the other off East Anglia. The processing of petroleum also produces wastes which constitute 3 per cent of the total volume treated and includes a wide range of complex organic compounds. Contamination of marine ecosystems from oil spills is well documented: the main large organisms to suffer are sea-birds since mollusca, for example, can often close their shells against the foreign substances, although they may not be able to do so if an emulsifier has been used to combat the oil. Indeed, detergents employed as clean-up agents probably exacerbate the effects of contamination by oil: limpets will normally survive the oil but show 100 per cent mortality if certain emulsifiers have been used (Boesch *et al.* 1974).

Radioactivity is also present naturally, but is increased from man-directed sources.

After nuclear explosions, some of the fallout comes down directly to the oceans, some indirectly as runoff from the land surface which however may take up a large proportion (90–99 per cent of strontium-90, for example) of the isotopes deposited on it. Nuclear-powered ships are at present a very minor source but nuclear power stations release low-level wastes via their cooling water including tritium and krypton. Production reactors such as Hanford and Windscale produce wastes which are led off into a river or the sea. Hanford releases 60 **radionuclides,** and Windscale produces a variety of long-lived radioactive wastes (Fig. 8.5). The fate of these wastes is very variable, depending upon local conditions: some are adsorbed onto sediments (especially muds rather than sands), others enter biological material and may pass along food chains and become concentrated. (Schultz and Klement 1963; Nelson and Evans 1969.) Table 8.4 shows some typical concentration factors for radionuclides released to the seas. On the Cumbrian coast of England people eating large quantities of local fish may be getting about 3 per cent of the IRPC maximum recommended dose, and a salmon fisherman was

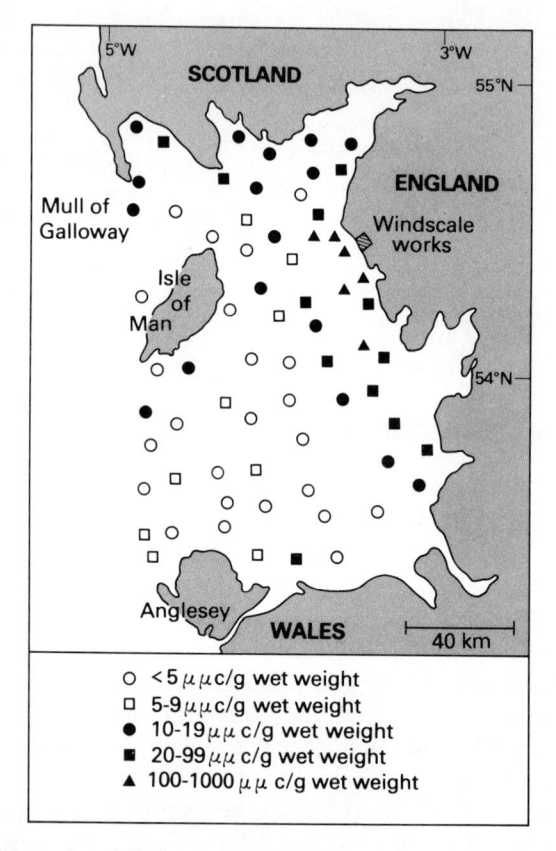

Fig. 8.5 Radioactive effluent from Windscale (man-made α-activity only) present in bottom invertebrates of the Irish Sea.
Source: Farvar and Milton, 1972.

TABLE 8.4   Concentration factors for radionuclides

Element	Algae	Crustacea	Molluscs	Fish
I-131	×5000	×30	×50	×10
Sr-90	50	2	1	0·2
C-14	4000	3600	4700	5400
Cs-137	15	20	10	10
Pu-239	1300	3	200	5

Source: Rice and Wolfe 1971.
Note: It is not implied that there is a food link along each row.

reported to have 7 per cent in 1976. The fate of released radionuclides is clearly of importance and the increase in numbers of fission reactors makes monitoring of levels in organic and inorganic receptors especially vital.

Power plants are an important source of alteration of biota in their local waters. A 1000 MW plant will require $4·9 \times 10^6$ l/min of water for cooling and will raise its temperature by 10°C. Organisms drawn into the plant will be subjected to thermal shock which will be lethal to many of them, especially plankton. Some of them may survive the thermal shock received in the condensers but be killed by continued immersion in the effluent plume of heated water. The photosynthetic ability of algae is reduced by up to 90 per cent in such conditions. Intake water is often chlorinated to prevent the growth of plants and animals in the intake structures and this may affect other biota, depending on its concentration; however, as little as 0·1 ppm of free chlorine may kill 95–100 per cent of the organisms, especially during winter months in temperate waters when many aquatic animals are attracted to the effluent plume. Alewives, menhaden, shrimp, small lobsters and flounders have all been killed in this manner in northeast North America. Simple passage through the generating station is also enough to kill a great diversity of organisms and is indeed probably the greatest source of damage to them. Bar screens and trash racks may inflict mechanical damage even to fish and crustacea which are deflected from the plant, but smaller organisms such as the larvae of fish, arthropods, molluscs and worms are subjected to shearing forces and turbulence and torn to pieces. Where a plant is situated on a river with distinct low-flow periods or on a shallow estuary, a very significant proportion of the total volume of water will pass through one or more power plants, so that much of the local fauna can be affected (Tarzwell 1972).

In general terms, the effect on lakes and rivers is mostly local or regional and removal of the contaminant allows the ecosystem to return to its former condition. In the oceans, the end-point for most inputs, the effects may be less obvious but possibly more far-reaching and less easily eradicated, particularly where the contaminants concerned are those which take some time to come into equilibrium with the receiving ecosystems.

Contamination of the land

The emissions into the atmosphere from industrial sources are dealt with under the appropriate heading but the acidification of rainfall may have indirect effects upon land biota, as distinct from any immediate toxification: Likens et al. (1972) thought that the acid rainfall accelerates the leaching of soils, so that elements such as calcium are lost to the runoff at a faster rate than normal. The subsequent reduction in pH of the soils thus decreases the competitive ability of the calcicole plants and increases the likelihood of an acidophilous flora. To this may be added that in Sweden during the last fifteen years there is thought to have been a reduction in forest growth rates because of acid rainfall (Engström 1971); and in the longer term, the prospect is opened up of greater inputs of agricultural fertilizers to replace mineral nutrients whose removal is occurring at an enhanced rate.

Terrestrial ecosystems are a common place for the disposal of solid wastes, which in North America are produced at a rate of 3·1 kg/day from family households. To these must be added the wastes of industry and agriculture, and in most countries disposal is usually in open dumps or in landfills where the rubbish is covered with soil or possibly an industrial waste such as slag or fly-ash. Open dumps obliterate pre-existing ecosystems (which may have been of low biological productivity like quarries, or very high like estuarine mudflats) and encourages the growth of populations of flies, cockroaches, mosquitoes and rats. All of these may be implicated in the spread of diseases to man and to domesticated animals. Controlled landfills are usually free of such problems although if the filling is not complete then rats may nest in hollow objects which are not totally filled up. Incineration is now held to be the most advanced method of disposal, but this transfers some contamination to the atmosphere.

Solid wastes are not solely an urban problem. The intensification of farming produces high quantities of wastes analogous to those of a city of which the most obviously land-based are the large numbers of dead animals (30×10^6 dead hens/year in the USA) whose carcasses must be buried or burned. In the same country some 250×10^6 t/yr of plant residues are burned, mostly to try to prevent the transmission of diseases from one generation of annual plants to another.

Open dumping of solid wastes is akin to and sometimes part of the process of the creation of the derelict land which is usually formed by the extraction of minerals and the dumping of mineral wastes from mines, refining and processing plants. The material is often too toxic or physically unstable to be vegetated for some time but eventually plants usually grow on the tips and in the holes. A ruderal fauna characteristic of open habitats normally develops; its composition may be modified by the presence of very high concentrations of toxic substances like metals. Succession may well ensue to the point where a scrub develops but spontaneous woodland is rare. The soil disturbance of waste areas and the lack of competitive pressure sometimes mean that introduced aliens find a suitable niche for growth and perhaps a nucleus from which they can spread out to become more established members of the biota. In Europe, the Cruciferae, Caryophyllaceae and Compositae are especially prolific in providing invaders: the Mediterranean *Gypsophila porrigens* is often found on rubbish dumps near British ports; *Centaurea diluta*, from North Africa, grows on dumps whither it comes from packets of budgerigar seed (Salisbury 1961). A number of higher plants show genetic adaptations to high concentrations of metal in areas where toxic wastes are exposed on the surface of the ground. Where several metals are present the plants are tolerant to them all but each tolerance appears to be genetically independent of the others. On non-toxic soils the tolerant individuals in a population are adversely affected by competition but sufficient persist to form a nucleus for the colonization of other contaminated places. In Britain, grasses of the genera *Agrostis* and *Festuca* have been particularly useful in providing tolerant races; world-wide some 21 species have been noted to exhibit this potential (Antonovics et al. 1971). The fauna of derelict land is most known for its dominance by fast-breeding rodents such as rats, though in later stages of colonization by plants, a diverse bird fauna may be present and mammals such as the fox may establish themselves.

Where a diversity of habitat and vegetation has built up, then an area of basically derelict land may have value for nature protection. In Britain, former gravel pits are sometimes set aside for this purpose: the combination of dry land and aquatic substrates, different plant species according to stage of colonization and the lower economic pressures to transform the land into something 'useful', may result in an 'island' of high biotic diversity and interest.

Biocides (see p. 207) may have effects on land fauna in their residual form, i.e. that part of the chemical which is surplus to the target organism and remains in the terrestrial ecosystem. Some is lost to the runoff, and some to the atmosphere in aerosol form whence it may rain out back to the land. Animal populations at high trophic levels have suffered most: the peregrine falcon and golden eagle in Britain are two examples. Diminution of the use of persistent chemicals (especially chlorinated hydrocarbons) seems to allow the populations of the affected species to recover to some extent.

The diminution in populations of target organisms may simply mean that another taxon becomes a pest. In Trinidad, the use of 2,4–D on sugar-cane decreased the number of broad-leaved weeds but these were replaced by a grass, *Paniculum fasciculatum*. Subsequently biocides were developed to control the latter, only to have it replaced by the sedge *Cyperus diffusus* (Fryer and Chancellor 1970). Spider mites, once a minor nuisance of agriculture in many parts of the world, have exploded into major pests as secondary-outbreak species which develop in the wake of insecticide usage. Similarly, the phenomenon of resistance is well known: selection acts upon natural variation and a strain of target organism arises which is resistant to the biocide. The spruce budworm of North America, once epidemic at 27-year intervals, is now perennially epidemic because spraying has induced resistance. Resistance to DDT and dieldrin has developed in some mosquitoes, including the carriers of yellow fever, *Aedes aegypti*, and malaria *Anopheles gambiae*. In the case of *Anopheles culicifacies*, a malaria-carrier in South Asia, DDT was used for 10–11 years before resistance developed but it then spread to nearly all the range of that mosquito within 6 years. The resurgence of resistance is unfortunate in many ways: it negates the benefits of the biocides and provokes the extra costs of developing a substitute. Total elimination of weeds and pests is probably never possible, and 'management' of them to an economically tolerable (and predictable) level is probably a more worthwhile aim, but not one usually acceptable in the case of human disease.

The application of herbicides to terrestrial ecosystems was a feature of US policy in the Vietnam war. In South Vietnam, an average of 21 kg/ha of active ingredients was applied to about 10 per cent of the country, and to 30 per cent of the zone between Saigon (Ho Chi Minh City) and the Cambodian border. Eighty-six per cent of the missions were against forest and mangrove, the rest against crops, and the effect in forests was usually defoliation followed by the invasion of grasses and bamboo. In the case of mangroves, very little revegetation has so far been reported (Westing 1976). Shrapnel has also reduced the value of the remaining forest because of its effect at the sawmills.

Concern over the effects of man-produced radioactivity on terrestrial systems centres on four phases: the possibility of a reactor accident, the storage of high-level wastes, fallout from the atmospheric testing of nuclear devices and accidents involving nuclear devices carried by aircraft or satellites. Neither of the first two has yet produced any documented effects on the populations of plants and animals. From the work of Woodwell (1962) we know that the type of plant community which would best survive heavy radiation (e.g. of the order of 100–300 **roentgens**/day) would be communities in the early stages of succession with a high proportion of ruderal plants. In forests, radiation of 20–60 roentgens/day causes damage to trees, especially conifers and it appears that

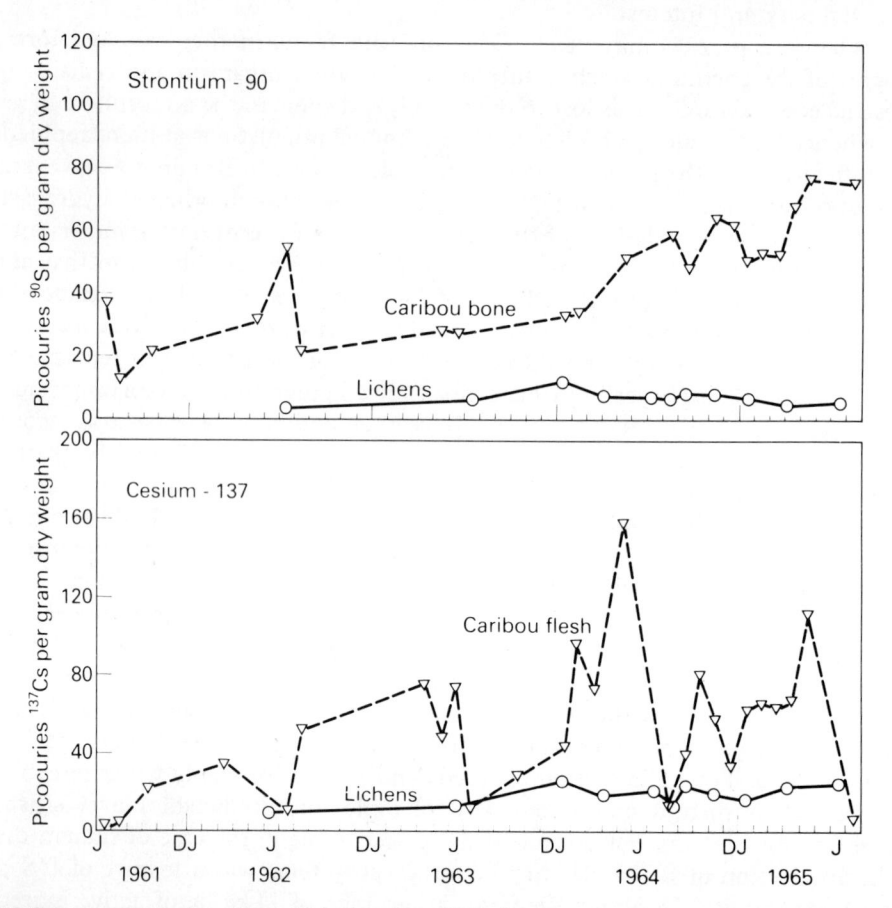

Fig. 8.6 Concentrations of strontium-90 and cesium-137 in lichens and caribou at Anaktuvuk Pass, Alaska, during 1961–65.
Source: Hanson, 1967.

of natural communities, needleleaf forests are the most sensitive to radiation and tundra the least. Complicated food-web effects may ensue: a leaf-eating insect not damaged by radiation but whose food species suffers defoliation may be expected to accelerate and intensify the effects of radiation as it feeds on any remaining leaves.

Atmospheric testing of nuclear weapons led to fears of the accumulation of radioactivity in humans and so monitoring of its pathways is well documented. Fallout was particularly heavy in the circumpolar regions of the northern hemisphere and studies (Hanson 1967) have shown that slow-growing foliose lichens are efficient accumulators of particles of, for example, strontium-90, cesium-137, and iron-55. This then passes on to herbivorous mammals such as reindeer and caribou and may then find its way to groups of Lapp and Eskimo folk who still live largely by hunting. Because of the long decay times of some of these elements (e.g. ^{90}Sr has a **half-life** of 29 years, ^{137}Cs half-life is 30 years) accumulation can occur, and Fig. 8.6 shows the levels of some of these elements in the Alaskan tundra ecosystem in the 1960s. ^{137}Cs and ^{55}Fe have also been shown to accumulate in aquatic food chains. Happily, fallout has diminished since the nuclear-test treaty ban came into effect in 1962 but continued tests by China and France have kept the levels above their 1950s intensities, and build-up in food chains is cumulative, although so far no obviously deleterious effects upon biota have been reported.

The same appears to be true of the accidents involving nuclear devices, where either clean-up operations or the low quantities of radionuclides involved have made it unlikely that large amounts of radiation have been added to ecosystems (Osburn 1974c). Examples have been the crashing of about 25 aircraft since 1945 while carrying nuclear bombs (e.g. in Greenland and Spain) and a Soviet satellite which crashed in northern Canada in 1978. By contrast, 23 nuclear tests on Bikini Atoll have left high radioactive levels in soil and water and differentially of ^{137}Cs and ^{90}Sr in the coconut palms, so that Bikini islanders allowed back after 1970 are being moved out again in 1978.

Overall effects of industrialization

If we look at man's impacts upon biota since the Industrial Revolution, then a number of trends can be discerned which encapsulate the information of the foregoing section of this book (Detwyler 1971).

There is, first of all, an increasing variety of impacts: in pre-industrial times there was hunting, fishing, gathering, agriculture and settlement; now, intensified agriculture, industry and urban spread are added and within each the diversity of impact is increasing. For example most of the persistent biocides have been developed since 1945 and new complex chemicals appear yearly. At any given place, too, the frequency and magnitude of impacts is likely to be increasing: the intensification of agriculture is one example; damage to biota from atmospheric contaminants another. Further, the spatial spread of processes which were once locally or regionally confined, appears to be inexorable. Contamination of the fauna of even remote places by persistent insecticides like DDT and its derivatives is one instance; suffusion of much of the atmosphere by increased levels of CO_2 and by particulates is another. Some of these impacts produce unforeseeable repercussions and synergistic effects: the biochemical transformation of mercury residues, for example, and the alteration of local climate and atmospheric constituents by urbanization and their effects on some plant groups. Even the course of evolution is likely to be altered by the extinction of some species and the favouring of others, e.g. organisms such as insects with short-life-cycles, which apart from man and perhaps some micro-organisms are the only large group currently increasing in number. Lastly, the relation to human populations is seen not only in the increasing magnitude of man's numbers but in his greater *per capita* demand for resources from the ecosphere and the consequent leading of wastes into living systems.

Diversity

One of the effects of these trends is loss of biotic diversity. This happens at two levels: there is of course the irreplaceable loss of a taxonomic unit when a species becomes extinct through human activity. Its place in the ecosystem becomes unoccupied and the energy and nutrient pathways become vacant: they may be taken up subsequently by another group or remain unused. An example of the reduction in diversity carried by contamination of a river is given in Fig. 8.7. At a second level, the controlled breeding of crop species and the replacement of a variety of cultivars by a few improved varieties means a great loss of genetic material, a condition which also applies to the total loss of taxa. This genetic pool may be needed in future as a breeding resource, e.g. for altered climatic conditions or for resistance to pests and so a concerted effort to preserve strains of cultivated plants must be made as a corollary to breeding programmes. This will necessitate the collection of germ-plasm resources in the field, its storage as seeds, tissue cultures or by other appropriate methods, its testing for useful characters and availability

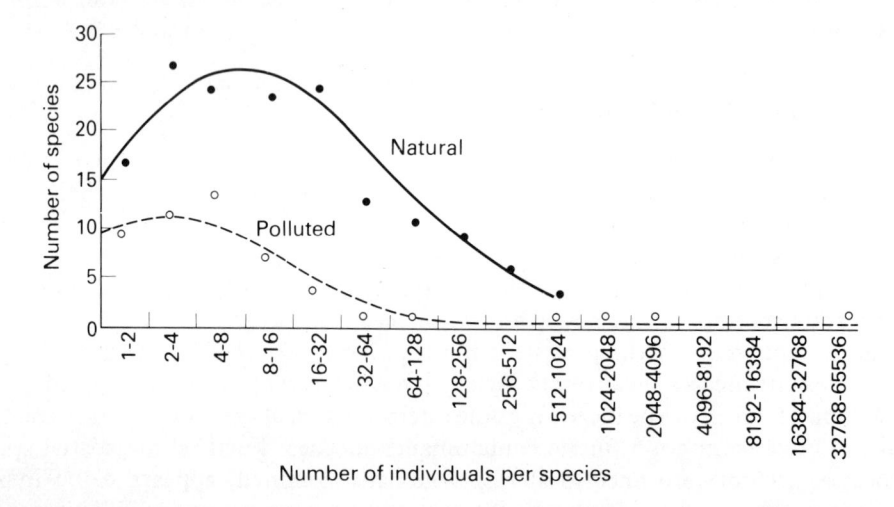

Fig. 8.7 A lognormal distribution of the relative abundance of diatom species in a natural stream and a polluted river.
Source: Ricklefs, 1973.

to breeders, the transmittal of information about stored genetic material, and the training of scientists and technicians, particularly from LDCs. A body called the International Board for Plant Genetic Resources (IBPGR) is starting this immense undertaking.

Anthropocentrically we must admit that the reduction of some elements of biological diversity has been a distinct short-term gain to people in a direct fashion. This is especially true of micro-organisms which cause disease and where combinations of better sanitation and diet, drugs and immunization have diminished the incidence of human diseases: the reduction of the incidence of the poliomyelitis virus in industrial nations is one example, and the virtual elimination of the smallpox virus on a world scale is another.

Contamination

There is no doubt that man's wastes since the Industrial Revolution have caused considerable biotic changes at all scales. The changes resulting from different contamination processes have enough in common for certain generalizations to be made. Woodwell (1970) summarizes them thus:

> Pollution operates on the time scale of succession, not of evolution to cure this set of problems. The loss of structure involves a shift away from complex arrangements of specialized species toward the generalists: away from forest, toward hardy shrubs and herbs; away from those phytoplankton of the open ocean that Wurster proved so very sensitive to DDT, towards those algae of the sewage plants that are unaffected by almost everything including DDT and most fish; away from diversity in birds, plants and fish toward monotony; away from tight nutrient cycles toward very loose ones with terrestrial systems becoming overloaded; away from stability toward instability especially with regard to sizes of populations of small rapidly producing organisms such as insects and rodents that compete with man; away from a world that runs itself through a self-augmentive, slowly moving evolution, to one that requires constant tinkering to patch it up, a tinkering that is malignant in that each act of repair generates a need for further repairs to avert problems generated at compound interest.

Little can usefully be added, except that most of this section of the book has provided evidence for the validity of Woodwell's statements, and that considerable research is now being pursued to try and convert wastes into a more benign form in which they can contribute to biotic productivity rather than the reduction of biotic diversity (Chemurgic Council 1972).

Substitution

In the Western world and some parts of the LDCs, one of the most noticeable effects since the nineteenth century has been the replacement of nature by man-made structures of an urban and industrial kind. Towns, factories, communication networks, waste heaps, mines and quarries have all increased in number and most of these are biologically inert for at least part of their existence and sometimes for all of it. Thus the actual biomass

Plate 32 The mosaic of contemporary ecosystem types:
(a) Mature, self-maintaining systems with natural components: an Australian rainforest.
(b) Highly productive but unstable systems with a low species diversity: agriculture in Thailand.
(c) A mixed landscape of productive and protective ecosystems: woodland and grassland in the English Lake District.
(d) A biologically inert system: near Chicago, Illinois, USA.

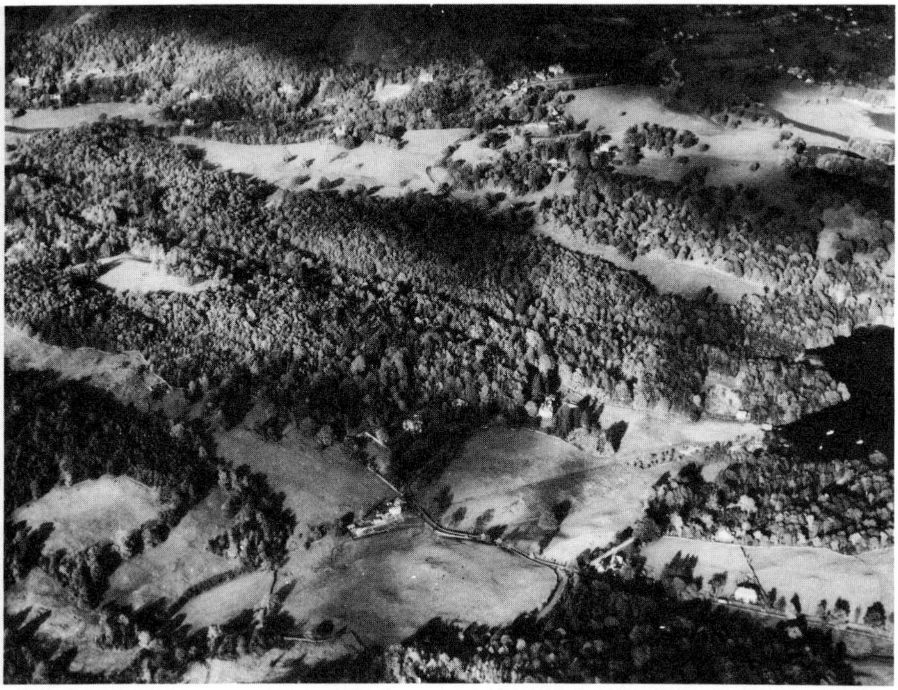

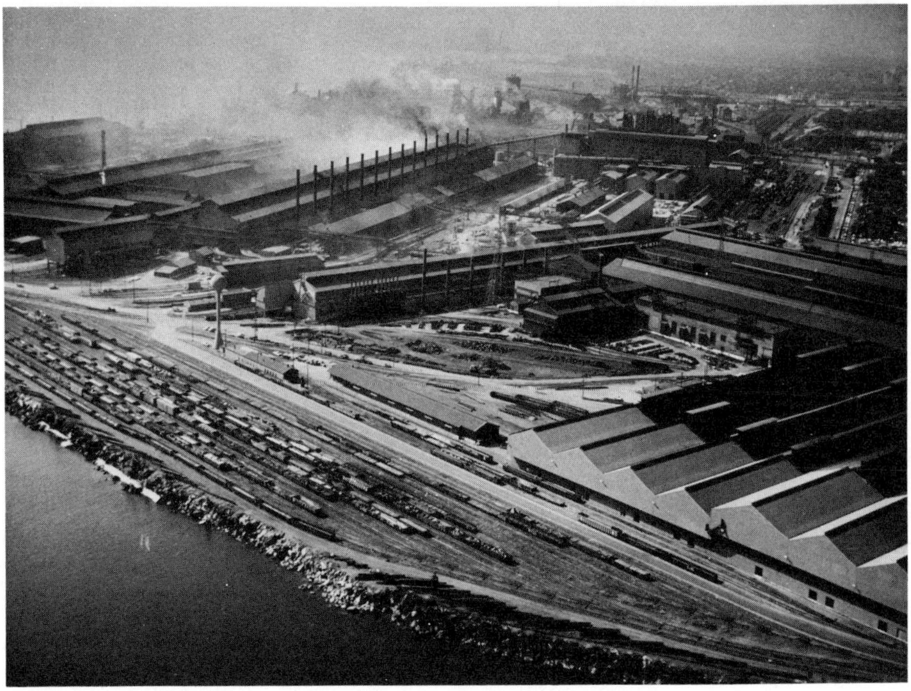

is lowered by industrialization: itself a consequence of the use of stored photosynthesis in the form of fossil fuels.

There is a tendency to exaggerate the quantity and rates of growth of these inert systems: in Great Britain the urban-industrial area increased by other 1 million acres between 1850 and 1900, and since 1900 to a further total of *ca* 4·5 million acres (Best 1965). This area is not all biologically unproductive, for it includes open space (20 per cent) and housing (nearly 50 per cent) which will often include gardens; perhaps 2·5 per cent of the surface of the country is biologically dead, if micro-organisms are excluded from the calculation. In cities, the ratio of inorganic mass to organic mass of all kinds is obviously very high but the consequences, if any, for man and for ecosystem development are difficult to understand. The lifeless proportion rises higher in industrial areas with a great deal of land dereliction resulting from mining spoil or processing wastes (Wallwork 1974). In the USA, perhaps 5–6 per cent of the land is biologically inert (estimated from Brubaker 1972).

The contemporary mosaic of ecological systems

There are different kinds of natural ecosystems, discussed on pp. 68–69. One type is the early successional stage of development which is characterized by such features as linear food chains dominated by grazing, a low species diversity, short and simple life-cycles of component organisms and very open nutrient cycles. Under natural conditions, these eventually become mature systems with characteristics such as weblike food chains dominated by detritus feeders, a high diversity of species, long and complex life-cycles and closed nutrient cycles with a slow exchange rate between biotic and abiotic parts of the system (Table 2.3). There are also inert systems in nature: completely unvegetated areas, for example, such as volcanoes, ice-caps and parts of deserts. In nature there is a mosaic of these types: successional phases lead to mature ecosystems which in turn have early phases within them caused, for example, by natural disasters like wind-throw in forests, or floods in riverine communities. In the long term the whole is stable and possesses feed-back mechanisms which ensure that no component increases out of proportion to the rest for an excessive length of time (Plate 32).

As Odum (1969) points out, ecosystems produced by man resemble those of nature in some ways (Fig. 8.8). In particular, the systems we depend upon for organic production on a short-cycle basis (i.e. most food crops), resemble those of early successional phases. Their biotic diversity is severely limited, nutrient cycles are open, and their resistance to external perturbations is low. Then again, man creates inert systems like cities (which resemble volcanoes in their emission of dust and SO_2) and render some land desert-like. The point is that the natural world has a balance of mature, inert and successional ecosystems but that man's impact is increasing the quantity of successional and inert systems at the expense of the mature or protective systems. The conclusion that Odum would have us draw is that the protection by legislation and management of the mature systems is vital to the continued functioning of the biosphere in a stable manner, a set of processes upon which the future of man may in many circumstances depend.

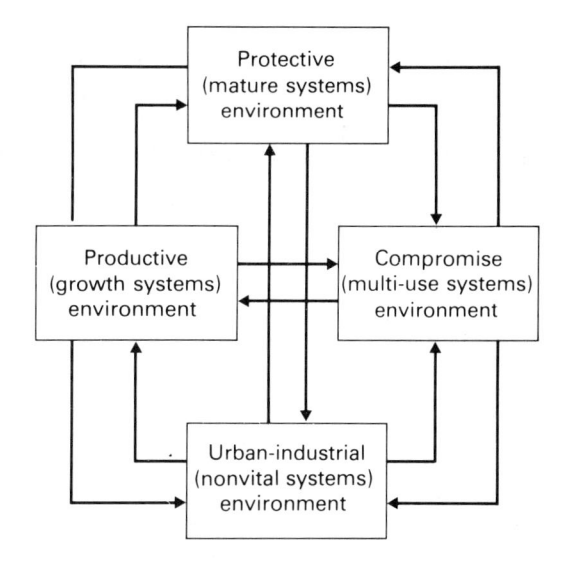

Fig. 8.8 A compartment model of the basic types of ecosystems produced by human activity. Source: Odum, 1969.

Further Reading

ALEXANDER, M. 1974: Environmental consequences of increasing food production. *Biological Conservation* **5**, 15–19.
ASKEW, R. R., COOK, L. M. and BISHOP, J. A. 1971: Atmospheric pollution and melanic moths in Manchester and its environs. *J. Applied Ecology* **8**, 247–56.
BÄCKSTRAND, G. and STENRAM, H. 1971: *Air pollution across national boundaries. The impact on the environment of sulphur in the air and precipitation.* Stockholm: Royal Ministry for Foreign Affairs/Royal Ministry of Agriculture.
MUDD, J. B. and KOZLOWSKI, T. T. (eds.) 1975: *Responses of plants to air pollution.* London and New York: Academic Press.
MURDOCH, W. W. (ed.) 1975a: *Environment, resources, pollution and society*, 2nd edn. Sunderland, Mass.: Sinauer, chs. 10, 11, 13.
ODUM, E. P. 1969: The strategy of ecosystem development. *Science* 164, 262–70.
PERRING, F. H. and MELLANBY, K. (eds.) 1978: *Ecological effects of pesticides.* Linnean Society Symposium 5. London: Academic Press.
WESTING, A. H. 1976: *Ecological consequences of the second Indochina War.* Stockholm: SIPRI/Almqvist and Wiksell.
WOODWELL, G. M. 1970: Effects of pollution on the structure and physiology of ecosystems. *Science* **168**, 429–33.

9
Man's deliberate creation of new biotic patterns

Even during prehistory, men created new genetic patterns and new ecosystems, as the history of domestication and, for example, pastoralism show. Throughout recorded history the scale of alteration has increased and it continues to increase, so that what is often called environmental impact is accelerating, having been given enormous impetus by the growth in human populations since the seventeenth century AD and by the Industrial Revolution. Again, the scale of urgency and need for changes appears to be increasing as the demands for biotic resources mount and as the environmental impact of other resource processes climbs apparently inexorably. All these processes have been considerably facilitated by the development of science and technology, which has made feasible one of the potentially most spectacular of all interventions in biotic processes: the designed creation of new combinations of genetic characteristics.

The methods employed rest on knowledge of the structure of the macro-molecules of the cell nucleus which carry genetic information. The key structure is a molecular chain in the form of a double helix of a nucleic acid called DNA (deoxyribonucleic acid). This is self-replicating during cell division, and different genetically transferred characteristics have specific locations along the molecular chain. Research has discovered bacterial enzymes which can sever and anneal molecules of DNA so that the genes of one organism can be linked to a segment of DNA from another to form a recombinant DNA molecule. The techniques are known as 'shotgun' experiments since the DNA chain is broken down into fragments a few genes in length. Each fragment is then linked to a vector molecule which carries its recombinant partner into a host bacterium, which is usually at present at strain of *Escherichia coli* used in laboratories (Fig. 9.1).

If developed to its obvious potentialities, the technique could produce all kinds of useful results. Bacteria, for example, might be programmed to produce useful products, especially of a pharmaceutical nature; growing crop plants might be given the genes for the fixation of nitrogen from the atmosphere and so fertilizers would become unnecessary. A brave new world of precise genetic engineering is opened up (Bulletin of the Atomic Scientists 1972; Cohen 1975).

On the other hand, some scientists argue that there are considerable dangers. A nitrogen-fixing bacterium such as *Rhizobium* could possibly be manipulated so as to destroy its nitrogen-fixing capabilities but retain the tumour-building characteristics which produce root nodules. A crop infected with such a strain might die immediately. Similarly, a harmless *E. coli* strain used for experimental work on tumour-causing viruses might escape from a laboratory or a shotgunned *E. coli* impregnated with the capacity to pro-

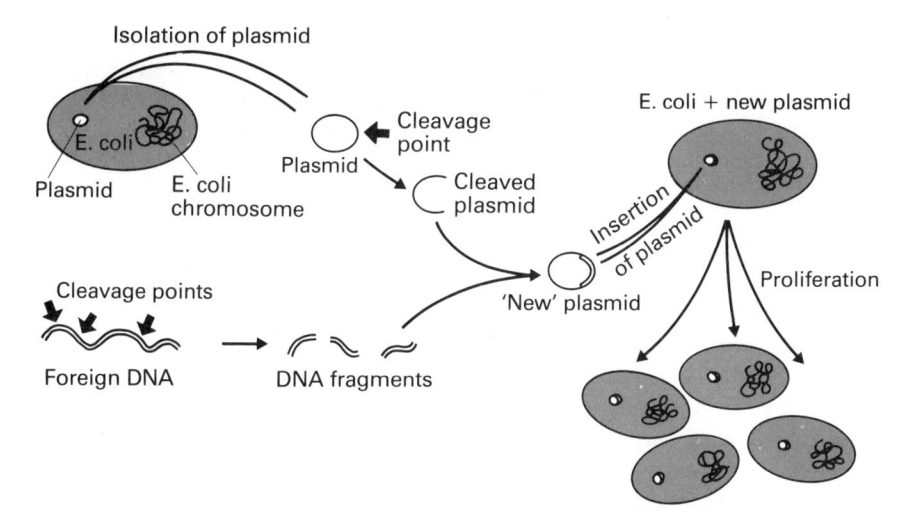

Fig. 9.1 The basis of genetic manipulation. A piece of circular DNA (a plasmid) is extracted from an *E. coli* organism, cleaved, and a fragment of foreign DNA inserted. This is reinserted into an *E. coli* bacterium where the foreign DNA will be expressed along with the carrier DNA from the plasmid.
Source: Freeman, 1978.

duce a male hormone might prove to be the ubiquitous (and inescapable) aphrodisiac that comic novelists have been dreaming about for years.

So experiments with recombinant DNA were under a virtual embargo until 1975 when guidelines were established in the USA, paralleled in the UK by those laid down by a Genetic Manipulation Advisory Group (GMAG), which specified the physical and biological levels of containment to be used in experimental work, such as not creating aerosols, the use of negative air pressure and air locks, and all the other methods normal when handling dangerous pathogens (Wade 1975). During 1976, a self-destructing *E. coli* was developed which is claimed will not survive out of its culture medium in the laboratory, partly because its growth needs a substance (diaminopimelic acid) not prevalent in nature.

Nevertheless, partly because it is difficult to screen for the unknown, human error cannot be entirely eliminated: even at the highest level of containment, Fort Detrick biological-warfare laboratory had 423 cases of infection and 3 deaths over a period of 25 years (Wade 1975). It is scarcely surprising that scientists have been divided about the desirability and safeguards of such a development (King 1977; Singer 1977), and that in 1976 the Cambridge (Mass.) city council became involved in the debate about the continuance of such work at MIT and Harvard University. The dangers and benefits seem analogous to harnessing the atom: potential plenty is a possible benefit and awesome destruction a distinct risk. As with atomic power, the benefits are relatively easy to see (e.g. new sources of hormones and new treatments for genetic diseases in this case) but the risks are mostly conjectural and therefore difficult to assess in terms of their probability and magnitude. Recombinant DNA, in effect, confers the power to intervene in evolution and create genetic combinations at a stroke instead of waiting

258 *Man's deliberate creation of new biotic patterns*

for them to be thrown up by the slow processes or organic evolution. At any rate, by 1979, the US Court of Customs and Patent Appeals had ruled that patent applications could be granted on two separate micro-organisms developed by industrial concerns. If upheld by the Supreme Court, this means that for the first time the whole of a living taxon is being claimed as the property of an individual or company.

Scientific manipulation of organisms

Plant and animal breeding

Of an estimated 200–250 000 species of flowering plants, about 3000 have attained a significant role in man's economies and most people receive their basic calorie nutrition from only twelve: rice, wheat, maize, potato and sweet potato, cassava, beans and soya beans, coconut, banana, sorghum and peanuts. Considerable attempts at crop selection have been made in order to increase the yields of basic plants: scientific breeding accelerated in the eighteenth and nineteenth centuries with the first attempts to hybridize wheat as a landmark at the end of the eighteenth century (Plate 33). From then onwards the aims of all plant-breeders have been roughly similar, beginning with increasing the size of the whole plant or the useful part of it. More recently attempts to breed for these qualities have been directed towards increasing the efficiency of carbon-dioxide assimilation by crops, or more simply the efficiency of photosynthesis. Experimentation

Plate 33 Agricultural experiment stations are the basis of scientific plant and animal breeding: Rothamsted, England. In the top right, a continuous barley experiment was started in 1852 and, top left, an 'exhaustion experiment' in the same year. The other plots are devoted to rotation experiments.

incorporating somatic cell genetics, use of tissue cultures, and an increased knowledge of the biochemistry of carbon compounds related to photosynthesis is at present under way. The potentialities of the plants possessing a particular parcel of photosynthetic pathways are being closely investigated. Such plants (called C_4 plants) appear to be 10–25 per cent more efficient in terms of dry weight gain/unit of solar radiation and also transpire less water per unit of dry matter produced. Ten families seem to possess the C_4 mechanism, including the Gramineae, so that cereal agriculture and grasslands are potentially within the practical application of discoveries about the C_4 syndrome (Zelitch 1975; Woolhouse 1978).

The aim of plant-breeders, however, may not always be directed towards productivity and size. New varieties of dwarf tropical rice are a case in point. The application of fertilizers to some tropical rice varieties simply induces tall leafy growth, and eventual collapse of the plant, so that a semi-dwarf variety IR-8 was created by manipulating a semi-dwarfing gene from a Taiwan variety of rice (Athwal 1975).

Breeding for yield has so far reached its apogee in the varieties of cereals produced by research bodies like the International Rice Research Institute in the Philippines; a similar institute for wheat (CIMMYT) exists in Mexico and a programme for crops of semi-arid areas such as sorghum and millet is now under way. The high-yielding varieties of wheat and rice now widespread in LDCs will produce crops up to 60 per cent heavier than traditional strains if conditions are propitious, which usually means a controlled water supply together with insecticides and nitrogenous fertilizer applied at just the right time. So these 'Green Revolution' crops increase output but tie the crop ecology to industrial-world conditions. One response has been an attempt to use some of the genetic qualities of traditional varieties of rice, for example, in the IRRI's variety IR-26 introduced in 1975 which crossed a new high-yielding variety with a traditional but low-yield rice which nevertheless had high disease resistance. The resulting strain should therefore yield well but be less dependent on imported pesticides. Coupled with these advances is a major concern to improve rice as a source of protein by breeding in genetic material that produces at least one fourth more of it.

Developments elsewhere include the breeding of cereals which will grow on saline soils (very useful for failed irrigation projects), and strains with higher protein content: most sorghums for example lack the amino-acid lysine. The achievement of higher protein content is best demonstrated by maize where a double cross of four inbred strains or mutants known as opaque-2 and floury-2 gave both high yields and a protein content of 10–12 per cent (compared with the parents' 8–9 per cent) with a high proportion of the hitherto deficient amino acids lysine and tryptophan. The benefits of the new high-lysine corn have already had significant effects in human health in LDCs (Harpstead 1971; Carlson and Polacco 1975).

Breeding may also be used successfully to extend the range of a crop plant. Research is currently under way in England on projects to adapt the structure of the maize plant to the British climate. Leaves on present varieties are carried horizontally, and encourage shading which in turn reduces photosynthesis. This is particularly crucial as a natural disadvantage of the climate is that light intensity is already much lower than for example parts of the USA where maize is grown successfully. Thus the Plant Breeding Institute is attempting to induce maize leaves to grow upright. An earlier example concerns the Australian wheat variety 'Federation' which was developed with narrow leaves and a short stem so that it lost less water than other strains and so could be used in drier places. Because of the introduction of 'Federation' in 1901 wheat production in Australia rose from 1 to 4 million tons between 1890–1920. In a different context, the breeding of plants to withstand high salt concentrations is now being undertaken both for potentially useful halophytes and for adaptable cereals, of which barley is one. The consequences of success here would mean the re-evaluation of salinized soils (whether naturally so or from failed irrigation projects), and the possibility of using seawater for irrigation without costly desalination.

Some breeding programmes are also aimed at utilizing all genetic material that could possibly be useful including that from other genera. Just as the history of, for instance, wheat and maize shows intergeneric transfer of genetic characteristics, so modern plant breeders use other genera. An intergeneric hybrid of wheat and rye called Triticale is replacing rye in some regions since it gives high yields but is disease resistant. It has more protein than wheat, a higher lysine content than rye, has potential for making pasta, pancakes, tortillas and chapatis, and has the additional, perhaps decisive, quality that it can be used to make whisky. The value of Triticale may be further enhanced if it extends cereals to places where rust normally limits wheat production, notably subtropical and tropical highlands. At present it is grown in fifty-two countries from the prairies of Canada to the foothills of the Himalayas. Research has proceeded still further with the production of a new strain X308 or 'Armadillo', achieved by crossing Mexican dwarf bread wheat with a Triticale plant. It combines a higher yield, better insensitivity to day-length and earlier maturing characteristics with better nutrient qualities than older varieties of Triticale (Hulse and Spurgeon 1974).

The back-crossing of some wheats with two species of wild grasses *Aegilops speltoides* and *Aegilops comosa* has also improved resistance to the yellow rust *Puccinia striiformis*. In a similar context breeding by crosses and chromosome engineering by irradiation transferred a gene with resistance to leaf rust from the wild grass *Aegilops umbellulata* to common wheat. A variety of mutant rice known as Novin 8 has been produced in Japan which reaches maturity two months earlier and has seeds with more protein. It has even been suggested that induced mutation could bring about the redistribution of protein in the rice kernel—normally in the outer layers and hence always destroyed when the rice is polished (Sigurbjornsson 1971). Breeding for pest resistance is often a critical part of a scheme of pest management (see p. 264): it was especially so before modern chemical biocides and may well achieve dominance again. The inroads made by pests may be illustrated by the losses of wheat on the Canadian prairies earlier this century: the variety 'Marquis' was much used but susceptible to black stem rust (*Puccinia graminis*) and losses of over 112 million tons were reported from Manitoba and Saskatchewan in 1925–35. In 1924 the stem-rust resistant 'Ceres' was brought in but both it and 'Thatcher' (introduced 1935) were both susceptible to leaf rust and only 'Regent', subsequently introduced, repelled both.

New plant architecture is also being tested in order to improve yields of crops such as durum wheat (*Triticum turgidum ssp durum*). Fewer leaves than normal, if not critically disadvantageous in terms of photosynthetic area, will reduce evapotranspiration and the number of damp spots for fungus diseases to flourish. Upright rather than droopy leaves allow more light to reach the lower leaves by way of compensation for fewer of them. A long peduncle to the grain head will reduce the frequency of rain splashes off the leaves and diminish the frequency of damage by scab (Breth 1975). Not all breeding programmes are successful: plant-breeding fertilizers and various management methods have all failed to raise soya-bean productivity in the USA by more than 1 per cent p.a. recently, compared with a 90 per cent increase in maize in the last decade.

Possibly the most important advance in plant breeding currently under investigation is the transfer of nitrogen-fixing properties to non-leguminous plants. Such a development would greatly reduce the need for nitrogenous fertilizers with their problems of costs and residual eutrophication (Safrany 1971). Research has included the finding of the 'switch' gene which allows a bacterium to fix nitrogen and why it is sometimes inhibited. Field-grown pearl millet (*Pennisetum americanum*), and guinea grass (*Panicum maximum*), lightly fertilized and inoculated with the bacterium *Spirillium lipoferum* produced significantly higher yields of dry matter than did the controls. Up to 42 kg/ha and 39 kg/ha of nitrogen were replaced by inoculation for each respective species. This confirmed that *Spirillum lipoferum* seemed to have an associative symbiosis with grasses, which had been suggested by work on rice rhizosphere associations at the IRRI and in Africa, and on maize, wheat and forage grasses in Brazil (Smith *et al.* 1976). Cell engineering, including the intake of foreign DNA by plant protoplasts and seeds, is presumably likely to undergo considerable advance in the next few years. By using cell cultures, generation time can be reduced from months or years to hours or days: resistant cells can be selected in a single generation and selected cells can be cloned to create large numbers of individuals in a few weeks. It is already possible to generate viable and reproductively fertile plants from single cells or small clumps of tissue in about a dozen species, including carrots, tobacco and maize. Within decades it may well be possible to introduce specific genetic materials into, for example, a cereal cell which would determine seed-protein content and composition, photosynthetic efficiency and the ability to fix nitrogen. Combinations of desirable plant features such as adaptations to lower temperatures or mechanisms to avoid water loss with high yield–high protein varieties of important crops may be made (Ledoux 1975).

All of these diverse advances in plant breeding and their future significance has been neatly summarized by Revelle (1976) who stated:

> By the year 2000 completely new crop species may have been created by applied research. An ideal new crop species would produce the edible portion of the plant with a photosynthetic efficiency two or three times higher than that of any existing food grain; it would fix its own nitrogen preferably in the leaves rather than the roots; the protein in the edible portion would have the balance of amino acids needed by human beings, and the plant would be water saving, that is it would evapotranspire much less water per unit of edible product than present day cereals. It would look and taste, however, like a present day cereal, and it would be equally capable of being made into bread or pasta or chapati.

The breeding of domestic animals has followed analogous lines. High yields of animal products at a high quality are sought, especially in the breeding of cattle, sheep, pigs and poultry. Genetic control methods are like those used for plants with similar emphasis on characteristics such as hybrid vigour (Lasley 1972). In pigs for instance, cross-bred swine farrow more piglets which are heavier at birth and more of which survive from birth to weaning. Attempts to concentrate the inheritance of a few outstanding ancestors (made easier by artificial insemination) may result in the uncovering of detrimental recessive genes and so defects like a twisted-skeleton deformity are transmitted by 25 per cent of AI Charolais bulls in Britain. Similarly, the central-nervous-system defect known as daft-lamb disease rose to virtually epidemic proportions in the UK in 1975. Some genetic manipulation is also attempted in the raising of animals suitable for intensive rearing systems: hornless cattle for example injure each other little and more can be fed at once. If genetic engineering becomes feasible then intensive-rearing agriculture will no doubt make use of **cloning** techniques for a totally uniform product under a standard set of conditions: real factory farming. Like all such monocultures, disease problems will be rife unless specific resistances can be genetically programmed: recently antibiotics were fed to pigs, poultry and cattle, and as a consequence resistant bacteria built up in the animals and in man also; accordingly their use is not now in favour. In general, breeding for disease-resistance is less advanced in animals than plants, and the biochemical genetics of disease resistance, as well as of metabolic efficiency, are likely to receive increased attention in the coming decade (Rendel 1975a). Control of diseases by drugs is more target-specific in animals and so residue problems are less intractable (though not with such examples as dips against insect larvae which get into the scavenger food chain via carcasses) thus in Europe, sheep-pox, rinderpest and cattle pleuro-pneumonia were all eliminated by 1900, horse glanders by 1928 and sheep scab by 1950. Rabies was also eliminated from the British Isles but is now back on the doorstep and its reintroduction seems inevitable.

As in the domestication of plants, the selection of domesticates by man leaves a lot of wild beasts untouched and new possibilities are sometimes considered (Jewell 1969). The British, for example, have always been regarded as being particularly conservative in their meat-eating habits. For centuries the Chinese have reared dogs, and the Somalis camels, for human consumption, yet the development of the broiler-rabbit industry in Britain was painfully slow. But in the developed world, tastes are in fact constantly changing: so much so that in the USA and Canada lamb and mutton have almost become a luxury (Blaxter 1975b). 'New animals' would need to present a large initial stock for the selection of desirable characteristics and preferably would harvest and convert vegetation otherwise inedible to humans. Most current domesticates are from the order Artiodactyla and this taxon probably contains the greatest potential; of the Perissodactyla only the zebra shows any promise. The Artiodactyla are mostly African, although outside that continent the moose (*Alces alces*) has some potential and is being managed in herds in the USSR. Inside Africa, the buffalo (*Syncerus caffer*) seems an obvious target for domestication attempts and experimental work with the eland (*Taurotragus oryx*) has been done (Plate 34). The eland is believed to have great potential for game-cropping or ranching purposes. It has a higher weight than the locally husbanded cattle

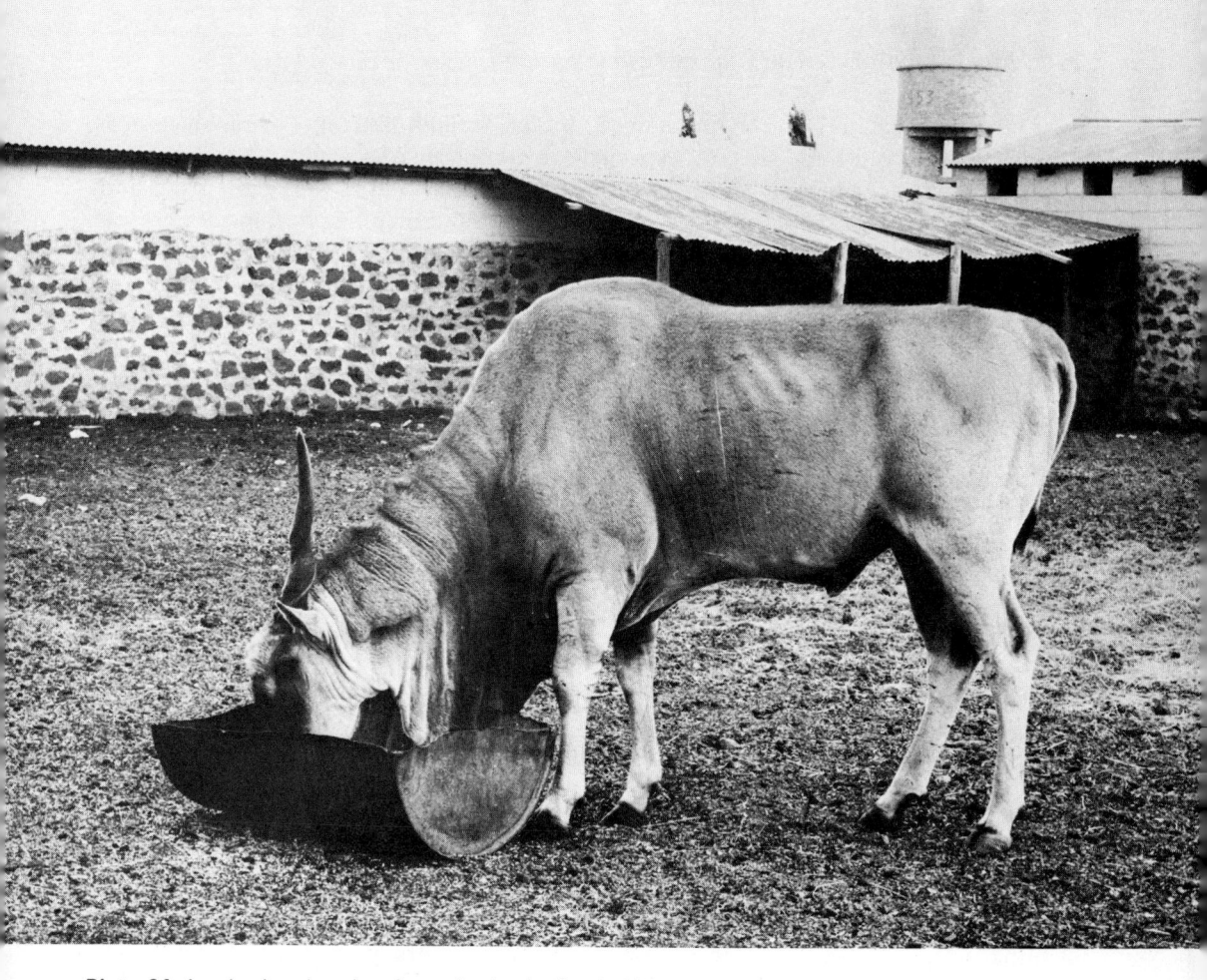

Plate 34 An eland undergoing domestication in South Africa.

and can be readily ranched. It is well adapted to hot, semi-arid environments, can utilize a wider spectrum of plants than domestic cattle, and its only drawback seems to be its ability to jump fences 2·5 metres high. Blesbok (*Damaliscus* spp) and springbok (*Antidorcas marsupialis*) have been the focus for study in the Transvaal: grazed at a capacity of one beast per 8·5 ha over a period of six months, they represented twice the live weight per km^2 of cattle (de Vos 1969). The kudu (*Strepsiceros strepsiceros*) is another promising species which has the great advantage of being a browser which will not compete for food with cattle or sheep, and requires little water. Warthog and bush pig have also been mentioned: the latter two, for example, are exempt from the swine fever that affects *Sus scrofa* in Africa, to which it is not native. The territorial characteristics of some of these animals might be useful in keeping them together in groups, and they are likely to keep moving over their range and cause less overgrazing than cattle. However, if domesticated, the spectacular weight gains and adaptation to environment shown by their wild relatives may not be maintained. So far, though, attempts to ranch African game have achieved some success, especially in Southern Africa.

Alongside game ranching, African experiments with harvesting wild game from both 'game farms' and 'game parks' have been tried. Although culling techniques are not yet properly attuned to the biology of the target species, yields of both meat and by-products can be very high, although some parasites can be transmitted to human consumers. Nevertheless, an adult elephant can yield up to 115 kg of dried meat and in Botswana, nearly 60 per cent of all animal protein is from wild animals, a figure said to rise to 80 per cent for meat in several West African countries (Young 1975).

Culling of such animals for food also encourages their conservation and so a two-fold benefit can be expected. But because of the difficulties of rationally culling such a variety of species, and because of population growth in Africa leading to demands for the replacement of wild lands by agriculture, not all commentators have forecast a bright future for African game, whether to be seen or eaten (Parker and Graham 1970).

In temperate zones, game cropping of the red deer and allied species may be potentially important. The red deer could produce four times as much meat per hectare as cattle. In addition, deer yield a greater amount of first quality meat in the total body than cattle (Blaxter 1975b). After Scotland, with an estimated red-deer population of 180 000, both West Germany (80 000), Austria (70 000) and Denmark (3000) have respectable numbers with a potential for controlled cropping; further east (Poland 32 000; USSR 200 000) the possibilities seem greater still (Bigalke 1975).

The future of animal breeding in the DCs is full of uncertainties: on the one hand there are intensively reared animals producing a uniform product (a chicken can be raised to 1·6 kg in 10 weeks on 4·1 kg of food, water, and a lot of energy) for which there is a high demand and these free land resources for other uses (Rendel 1975a). On the other hand, DCs feed these animals (and others like pets and domesticated fur-bearers such as mink) on grain and fish which could be used to feed the populations of LDCs. So animals which can live off the odd wild places and the scruffy bits of bush may be the ones the DCs will need in the future; as with the old varieties of crop plants, it is imperative that their genetic material is not lost (Frankel and Hawkes 1975, Rendel 1975b).

'Biological' control of crop pests

Attempts to achieve chemical control of crop pests by the use of biocides has been described above (pp. 207–209). There have been resulting problems of which the most important are the rapid development of resistant strains of the pests and the occurrence of unwanted side-effects like the poisoning of their natural enemies, as well as other biota which were not target organisms. Hence, interest has been revived in 'biological control' whose basis is the imitation of the feedback mechanisms of natural ecosystems in preventing or controlling population outbreaks of a particular species—i.e. the use of predators, parasites and pathogens. Included in the term are other non-chemical ways of control such as X-ray irradiation, plant-variety selection, hormones, and agro-ecosystem manipulation. All these techniques are directed not necessarily at totally eliminating target organisms but reducing their number to the point where their damage is below an economically acceptable level (van Emden 1974; Huffaker 1971; van den Bosch and Messenger 1973; Huffaker and Messenger 1976).

Control methods using predators and parasites may have involved their encouragement by providing off-season habitat or their attraction to crops before the main infestation of pests, a process in which chemicals may be used. Or, 'inoculation' may be attempted, when small numbers of the predators are placed on a crop in the hope that they will spread and become permanently established; this is normally used for an imported natural enemy left behind in another region or continent while the crop pest itself was brought in. A final technique is 'inundation' where the crop is flooded with laboratory-bred parasites or predators: this is especially used on annual crops. Examples of success include the parasitization of the coconut moth (*Levuana irridescens*) of Fiji by the fly *Ptychomyia remota* from Malaya in the 1920s: 32 570 parasitized larvae were released and the moth brought under control in two years. The woolly aphid (*Eriosoma lanigerum*) became a pest of apples in Britain after its introduction from North America in 1785, and a parasitic wasp (*Aphelinus mali*) was introduced; this could not survive the damp English winter and so apple shoots with parasitized aphids were held over winter in cold stores. (This pest is now chemically controlled with BHC.) In California, walnuts were attacked by an aphid *Chromaphis juglandicola* and so a wasp was introduced from France in 1961. It was unsuccessful but a different ecotype of the same species from central Iran was tried in 1968 and this was effective. Of passing interest is the fact that the English sparrow (*Passer domesticus*) was introduced into the USA in 1850 in the hope that it would control an infestation of a worm (*Eugonia subsignaria*) which was defoliating the shade trees of Brooklyn. Doughty (1978) does not record whether or not it succeeded, but does remark that the sparrow was itself regarded as a pest in most of North America by 1900.

There have been many instances of failure of this type of biological control, but the stated advantages are many (van den Bosch and Messenger 1973). They include the selectivity of the organisms involved; the absence of need for manufacturing since the materials are provided by nature; the ability of the beneficial organisms to find the pests for themselves and to multiply to the appropriate level; the inability of the target pest to develop resistance at all quickly; and the self-perpetuation of the entire process. Not all these advantages are always present in every attempt at regulation, and disadvantages have become apparent. These include the relative slowness of the process; the fact that such repression does not exterminate the pest; the unpredictability of the outcome of new control attempts; expense and difficulties in development; and the need for expert supervision (El Titi and Steiner 1975; Kiritani 1972).

A logical development from these forms of biological control is the use of microbial pesticides in which bacteria and viruses are disseminated as sprays or powders: a granulosis virus, for example, has been used against the pine sawfly in Canada. There are obvious dangers with such techniques and another idea is that of using chemical attractants and repellants to disturb the effectiveness of insect pests. Sexual attractants may gather together all the males into a lethal place or such chemical messengers (**pheromones**) may damp down reproductive behaviour. Irradiation of males may make them sterile, which is especially effective if the female mates only once. A more ecological method is system control in which the crop's environment is manipulated so as to minimize the success of pests, for example, by clearing up crop wastes, strip farming to control infestations, rotations to prevent the build-up of endemic pests, the growth of crops which trap the pests and which are then ploughed in or burned, and the timing of crop cycles to provide the least comfort for pest organisms. Lastly, the non-crop components of agricultural ecosystems may play an important role in reducing the impact of pests. Wild plants in waste places, on roadsides or in hedgerows may provide reservoirs not only of the pests themselves but of their controlling organisms (see Fig. 10.5, p. 284). A roadside may provide food for an insect which is then forced to feed on a nearby crop if the roadside is sprayed, for instance, so that it often seems prudent management to retain 'waste' land outside the crop area so as to maintain a reservoir of natural enemies of the crop pests.

An ecological approach to pest control involves both economic and environmental considerations, along with the whole spectrum of control methods including target-specific biocides where appropriate. The term 'pest management' is used for schemes which supplement natural controls and try to combine long-term reliability with low-cost protection of crops; in such a way some of the harmful side-effects of the blanket use of biocides may be avoided. Murdoch (1975b) has suggested that it should be possible to design agro-ecosystems to reduce the severity of insect-pest problems but one of the major problems is the lack of co-evolutionary links between interacting species, and the haphazardness of the collection of species in a crop field due to the frequency of human intervention. He notes that while a crop system should contain enough plant species to maintain continuity of any predator upon the pests, diversity *per se* is not a guarantee of agricultural stability. Mathematical models suggest that a patchy distribution in space enhances stability in predator-prey relationships.

An example of the unforeseen consequences of effective measures against pests can be seen in the effects of attempts to eradicate tse-tse fly in the Sahel zone of West Africa. The consequent control of trypanosomiasis in cattle was one of the factors which allowed a considerable increase in the number of cattle in this region in the 1960s; subsequent overgrazing (exacerbated by deep well construction) has enhanced the effects of and perhaps even helped to cause the tragic drought conditions of the early 1970s. Omerod (1976) remarks that extension of trypanosomiasis control could 'seem beneficial in one field of activity but cause disaster in another'. It is to combat narrow views of pest control that biological control and ecologically based pest-management strategies are seen as the best possibility for the future (Metcalf and Luckman 1975); however, less than 5 per cent of the Earth's 5000 pest insects are said to have had biological control projects initiated against them (De Bach 1974). The kind of classification which divides pests according to the organisms' 'ecological strategy' may well help to formulate rational pest management. The two basic types (although there are intermediates) are the r-pests such as locusts, aphids, mosquitoes, fleas and rats which share features such as high potential rates of increase, strong dispersal and host-finding ability and small size relative to other members of the same taxonomic group. By contrast K-pests such as the codling moth, the rhinoceros beetle and the tse-tse fly have low rates of potential increase, more specialized food preferences and relatively greater size. The first category often occur in the form of massive outbreaks attacking leaves and foliage; they have natural enemies but these seldom act before considerable damage has been done. *K*-pests are more of a constant problem but of lower numbers, and attack the fruit; they seldom have natural enemies. Clearly, different combinations of biocides and biological control methods are needed to manage these different types of pest (Conway 1976).

Varietal control looks at the problem from the plant-crop and attempts to make the plant less attractive to predation and pathogenic attack. Hairier plants, for example, may be more resistant to aphids, and rapid production of epidermal cork may inhibit some insects. Plants can sometimes be bred which flourish in spite of infestation, for example, by putting out new roots as soon as others are attacked. The IRRI in the Philippines has developed high-yielding rice varieties which have strong resistances: IR-30, for example, is resistant to bacterial blight, grassy stunt, greenleaf hopper and brown plant hopper and to zinc deficiency. It is also moderately resistant to blast, stem borer, alkali and salt injury and to phosphorus deficiency.

Nevertheless, epidemics of pests are everywhere likely: some may devastate a crop or other organism (e.g. chestnut blight, Dutch elm disease) and remain a persistent trouble, or may fall back to a non-epidemic level. An example of an eruption in progress which has not been either satisfactorily explained or controlled is that of the coral-eating crown-of-thorns starfish *Acanthaster planci*, which devastated many coral reefs in the Pacific after 1965; to follow both the course of its 'population explosion' and attitudes towards it continues to be most interesting (Branham 1973).

The scientific exploration of new foods for man

The existing food-supply systems like agriculture, grazing and fishing (discussed in Chapter 10) are plainly suffering from stress and so science is trying to extend the range of foods which can be incorporated in those systems or which come outside them. The quest for new sources of food begins with the use of existing biota or parts of them which are currently ignored or neglected (Siegler 1977). The immediate purpose often centres around the provision of more protein since until recently it was widely held that most malnourished people suffered from protein deficiency rather than a lack of carbohydrates (Crawford 1975; ACS 1969; Bender *et al.* 1970). One element in this search is the use of oil-seeds, especially the residue after extraction of the oil which is the current market crop. The residue of groundnuts after expression of the oil, for example, is 56 per cent protein; soya and cotton seed a little less. Similarly, sugar waste at one stage has a juice that contains 25 per cent protein; if not thrown away, this could yield 10 000 t of protein per million tonnes of sugar processed. An advantage of these crops is that they are often tropical genera and grow in those latitudes where the need is greatest; but they have to be stored and processed with great care since contamination can produce poisons such as aflatoxins.

One of the most ubiquitous sources of plant protein is leaves. Some are used for food by man but fruits, seeds and roots are normally more popular, often because of the fibre content of leaves. Given suitable techniques of extraction, processing and acceptability, there seems little reason why many kinds of plants, both cultivated and wild,

Plate 35 An irrigation canal in Florida clogged with water hyacinth (*Eichornia crassipes*). Normally regarded as an expensive weed, it might be cropped especially in LDCs.

should not yield plant proteins (Pirie 1971). Among the sources might be trees and bushes, prunings, seaweeds, water weeds and wetland plants, plants with slimy pulps, grasses, cover crops; sugar beet tops, pea and potato haulms, jute leaves and doubtless many others. If processing could be free of heavy technology, then LDCs would have a very useful protein source and since the production of leafy material reflects solar input, the tropics would gain most. Experimental work (Arkcoll 1971) suggests that in temperate zones, legumes might yield 1200 kg/ha of protein, and fertilized crops up to 2000 kg/ha; in the tropics, 3000–5000 kg/ha seems likely: 3500 kg/ha has been obtained from lucerne in India (Joshi 1971). Investigations into possible uses of the rapidly spreading tropical water hyacinth (*Eichornia crassipes*) are being made. These include composting, use in paper pulp, and as a component in cattle, poultry and fish diets (Plate

35). Such uses would save the enormous amounts of money spent in somewhat unavailing efforts at eradication (Bates and Hentges 1976).

Yet other apparently unpromising plants have been suggested by Pirie (1975). The agave catcus (*Agave* spp) of rough and arid land in Mexico might yield 6 t/ha/yr of mixed sugars with the potential for making molasses. Algae, while not the panacea imagined in past years, have still a useful role to play: *Spirogyra* is harvested as feedstock in Thailand, and the blue-green alga *Spirulina*, of saline and alkaline lakes, is harvested for direct consumption in Venezuela, Lake Chad and Mexico. The theoretical yields of protein from algae far exceed those for most other food products (Table 9.1). However,

TABLE 9.1 Theoretical protein productivity of various sources

Source	Protein, dry wt (kg/ha/yr)
Spirulina platensis	24 304
Chlorella pyrenoidosa	13 680
Fish	627
Peanuts	470
Milk	100
Eggs	60

Source: Harlan 1975.
Note: The level of energy subsidy to each system is not specified.

industrially-grown algae probably need too much care and constant inputs of carbon dioxide, nutrients and energy to make the major elements of feedstock or human diets. Since most areas of human dietary deficiency are in the tropics, the search for plants of promising economic value in those regions has had a particular point. One such study (NAS 1975) yielded 36 species which were little exploited but which had potential value as a source of food, forage or industrial raw material. They included cereals such as the channel millet (*Echinochloa turnerana*) of inland Australia, and the marine eelgrass *Zostera marina*, normally harvested for seed only by the Seri Indians of the West Coast of Mexico. The list, interestingly, includes the jojoba (*Simmondsia chinensis*), a shrub of arid lands in Mexico and the southwest USA which produces a liquid wax with considerable potential since it appears at present to be the sole substitute for the spermwhale oil which is used as a lubricant in mechanisms that most withstand extreme pressure. A plan for southern California and Arizona calls for the establishment of ten reserves of 4000 ha each over a five-year period, each hectare producing 2000 kg/yr of oil (Maugh 1977). The residual seed meal after processing also contains 35 per cent protein and might be made acceptable as livestock feed (Yermanos 1974). Another semi-arid plant of North America, the guayule (*Parthenium argentatum*) was once the source of 50 per cent of natural rubber consumed in the USA but its cultivation was finally abandoned after 1954. But its physical properties are similar to rubber from *Hevea brasiliensis* and so it could lessen the dependence of the USA and Mexico on imported rubber from politically unsettled areas like Southeast Asia and ecologically unstable places like South America, where leaf blight has all but destroyed the rubber industry.

Like jojoba, there is a major by-product: in this case a wood fibre which could be made into cardboard or paper. Minor by-products include a wax from the leaves and a pine-like resin (Maugh 1977).

In most places, macroscopic saprophytes are regarded as marginal items of diet but the role of mushrooms as protein concentrators should not be overlooked. Few inputs are needed to get a low-cost product high in protein, perhaps by a factor of ×1000 compared with the same area devoted to beef. Lysine levels are high, so mushrooms would be very suitable for rice-dependent regions; in India and Taiwan, the common mushroom is being supplemented by species tolerant of higher temperatures such as *Agaricus bitorquis* and the Japanese *shiitake* mushroom.

The energetics of animal husbandry have led to less emphasis on the animal kingdom as a new source of food, apart from developing ideas like game cropping (p. 264) and fish farming (p. 302). Nevertheless, meat, milk and eggs are concentrated sources of high quality protein, vitamin B_{12} and iron. Milk production if increased might help to reduce malnutrition in children: where lactose intolerance occurs in Africa and Asia it mostly affects adults. Further, if animals were fed on waste products such as cane molasses, cottonseed, rice bran and residues from breweries and distilleries then they would not be taking away plant foods from people, but functioning rather as members of a detritus chain (Loosli 1974). Technological processing has helped in some cases to preserve flesh which would otherwise deteriorate: in Canada, for example, a fish-protein concentrate is made by defatting and deboning fish and producing a tasteless white powder (FPC). This is suitable for animal fodder or for diet supplementation and it provides a livelihood for aboriginal fishermen as well. The detritus chain of aquatic environments is often little promoted except for sessile molluscs like oysters and mussels: controlled aquaculture (see p. 304) could probably increase the contribution of this energy-rich pathway by several orders of magnitude. Terrestrial invertebrates generally are reckoned to be unclean but animals like the giant African snail which can weigh up to 3·5 kg can scarcely be overlooked as a barbecue item. On the other hand, the search for protein in many places has eliminated practically all the visible wild fauna.

Beyond the harvesting and processing described above there are foods which can be produced only by the application of sophisticated technology. They are thus inextricably linked to industrial societies and at present are mostly used by them for animal feed; it is not yet apparent whether they will ever become significant sources of food in LDCs. One example is the rendering of soy beans to a flour which is then spun, shaped and flavoured before being presented as Textured Vegetable Protein (TVP), often in the form of a meat substitute. The soya isolate is 95·98 per cent protein and forms about 40 per cent of the final product. It is low in the amino acids cystine and methionine but supplementation is easy. TVP is not at present cheap and so is largely an 'extender' for meat products in DCs. Industrially-produced substances like soy flour and amino acids can be used to supplement foods, as in Brazil where a soft-drink is fortified with soy protein (Altschul and Rosenfield 1970).

The idea of converting many wastes to foodstuffs via industrial technology has mostly followed the path of employing micro-organisms, usually yeasts or bacteria, to grow on the waste and convert all or part of it to a protein-rich product known as single-cell protein (SCP). Examples of substrates used are:

1. Hydrocarbons from oil. For a number of years at Grangemouth in Scotland, BP produced 4000 t/yr of an animal feed by continuously culturing the yeast *Candida lipolytica*. In this process 1 kg of feedstock produces approximately 1 kg of yeast; this is centrifuged off into a cream which is 15 per cent solids and then spray-dried to a final product with only 5 per cent water. Such a product depends naturally upon oil supply and on energy costs generally, although if done 'in house' at a refinery, these are less important.

2. Methanol. This can come from naphtha, coal and natural gas, including otherwise wasted flare-off gases. The methanol is soluble in water and so is suitable as a substrate for SCP fermentation. The organism currently most favoured is the bacterium *Methylophilus methylotrophus* and a pilot plant in England has been producing 1000 t/yr since 1973 with a commercial (50–70 000 t/yr) operation planned for completion in 1979. Taylor and Senior (1978) estimate that world SCP production will be 760 000 t/yr in 1980 and 2 900 000 t/yr in 1985, with the bulk of this production being in the industrialized countries.

3. Organic wastes such as sewage, pulpmill whey, and sorted domestic garbage may also have potential for growth of micro-organisms: an analogue of the decomposer chain in natural ecosystems. Even plastics such as polyethylene can be oxidized with nitric acid and then used as foodstock for yeasts and bacteria. Such processes would have the other great utility of reducing environmental contamination by these wastes.

4. It may also become feasible to recover proteins from food-industry wastes by adsorption from the waste streams onto spherical particles of precisely controlled size. In laboratory trials, protein recoveries of 80 per cent from milk whey and 70 per cent from cereal-industry effluents were achieved.

But many of these methods tie the users to the industrial world and its costs and problems and in some ways many LDCs would be better off using sunlight more efficiently. This is especially true in the tropics where countries like Tanzania, Kenya, Mozambique and Zaire might become producers of a great deal of plant protein. One last difficulty of all these foods, as Pirie (1969) discusses, is acceptability: this may often mean some form of industrial processing although perhaps intermediate technology might provide this rather than the sophisticated plant used in the DCs. Unless dire circumstances arise, human acceptability thresholds are likely to be higher than those of the Australian sheep which in 1974 were said to be thriving on a diet of old government reports.

Further Reading

BIGALKE, R. C. 1975: Technological problems associated with the utilization of terrestrial wild animals. In R. L. Reid (ed.), *Proceedings of the III World Conference in animal production*. Sydney: Sydney University Press.

BLAXTER, K. L. 1975a: The energetics of British agriculture. *Biologist* **22**, 14–18.

CONWAY, G. 1976: Man versus pests. In R. H. May (ed.), *Theoretical ecology: principles and applications*. Oxford: Blackwell, 257–81.

HUFFAKER, C. B. and MESSENGER, P. S. (eds.) 1976: *Theory and practice of biological control*. London: Academic Press.

LOOMIS, R. S. and GERAKIS, P. A. 1975: Productivity of agricultural environments. In J. P. Cooper (ed.), *Photosynthesis and productivity in different environments*. IBP Studies 13, 145–72. London: Cambridge University Press.

PARKER, I. S. C. and GRAHAM, A. D. 1970: The ecological and economic basis for game ranching in Africa. In E. Duffey and A. S. Watts (eds.), *Scientific management of animal and plant communities for conservation*. Oxford: Blackwell, 393–404.

PIRIS, N. W. 1969: *Food resources conventional and novel*. Harmondsworth: Penguin.
— 1975: Using plants optimally. In J. Leniham and W. W. Fletcher (eds.), *Food, agriculture and the environment*. Glasgow and London: Blackie, 48–70.

REVELLE, R. 1976: The resources available for agriculture. *Scientific American* 235, 164–78.

SIEGLER, D. S. (ed.) 1977: *Crop resources*. London: Academic Press.

VAN DEN BOSCH, R. and MESSENGER, P. S. 1973: *Biological control*. New York: Intertext Publications.

YOUNG, E. 1975: Technological and economic aspects of game management and utilization in Africa. In R. L. Reid (ed.), *Proceedings of the III World Conference on animal production*. Sydney: Sydney University Press, 132–41.

10

Productive resource processes underlain by biota

At this point, it is the intention to consider living organisms used by man as components of ecosystems and to discuss their characteristics and productivity in that context. Additionally, if we consider biota as resources then certain ideas follow. Man becomes central to the flows of energy and matter because he directs them towards himself as efficiently as he thinks possible; his effects upon crop biota in that manner will have become apparent so far in this book. But human effects upon plants and animals extend to the entire process whereby a resource is taken from its state in nature, used by man, and then put back into ecological systems as a waste material. This flow is called a resource process and where men garner materials like food, fibre, wood or water from them, they can be termed productive resource processes. On the other hand, there are resource processes which are protective rather than productive: where, for instance, the biota form the basis of a natural or natural-seeming ecosystem which is valued as such as a wilderness or a nature reserve or an outdoor recreation area. Such places are just as much resources for man as are fields of wheat or water impoundments. In the following sections, therefore, some of these processes will be examined in order to evaluate the role of the key biota within the context of the ecosystems in which they occur, man's position in altering the pathways of the systems and where possible the limits to their development which seem feasible or desirable—i.e. their carrying capacity (House and Williams 1971).

Of all the man-centred resource processes of the present day, none attracts more attention than food supply. While agriculture is clearly the most important of the resource processes which yield food, other systems also feature: the grazing system using both domesticated and wild animals upon largely unimproved ranges (see pp. 196–7); the yield of fresh waters, which is mostly fish, and the seas: again mostly fish but with molluscs, crustacea, seaweeds and whales as further products. Lastly, there are industrial and other novel foods, the essence of which was discussed in the last part of Chapter 9.

Agriculture

The nature of agriculture as an ecological system has been under considerable scrutiny in recent years. Looked at thus, it appears as a series of displacements of natural ecosystems and their replacement by man-made and man-directed ecosystems. (Rarely, however, is the new system viewed *in toto* as an ecosystem, but attention is focused almost exclusively on the crop component (Potts and Vickerman 1974).) In Table 10.1

TABLE 10.1 The displacement of biological processes in agriculture

	Biological process eliminated or reduced	Non-biological method of control or substitution	
DEVELOPING AGRICULTURE	Predation	Fencing and housing	
	Removal of dead animals by scavengers, etc.	Burial or burning	
	Disease	Drugs, preventive medicine	
	Parasitsm		
	Crop pest damage (predation, parasitism of pests)	Pesticides (pesticides)	
	Damage by pests		
	Weeding	Herbicides or cultivation	
	Nitrogen fixation	Artificial nitrogenous fertilizers	
	Harvesting by man or by grazing animal	Machinery	Processes involving fossil fuels
	Ensilage by fermentation	Silage additives	
AGRICULTURE	Haymaking	Artificial drying	
	Rumination	Grinding	
		Barley instead of forage	
	Grazing	Zero-grazing	
	(geese)	(force feeding)	
	Foraging by pigs and poultry	Mechanical feed supply	
	Natural service	Artificial insemination	
	Hormonal control of reproduction	Exogenous hormones	
	Ovulation	Egg transplant	
	Milking (by progeny man)	Milking machinery and artificial rearing	
	Natural incubation	Electrical heating	
INDUSTRIAL SYSTEMS	Biodegradation of excreta in soil	Oxidation ditches, lagoons	
	Faecal deposition by the animal	Mechanical spreading on the land	

Source: Spedding 1976, with minor additions.

we can see the biological processes involved in much of agriculture and their replacement during its development from simpler forms to the highly mechanized and scientific processes in Western farms today. One obvious outcome is that rich people eat at the carnivore trophic level more often than do the poor, who are mostly vegetarian. In part this is due to diversion of foodstuffs from the LDCs (e.g. the Peruvian anchoveta catch is used mainly for animal fodder in DCs) and in part to the use of large quantities of fossil fuel in the food systems of the richer lands. Yet in spite of the energy subsidy and the technological developments in agriculture, the climatic component of the ecosystem remains very important. The effects of climatic change, however, are likely to be more complex than first reactions would indicate. The cooling of the northern hemisphere over the last 25 years has in fact produced some better crops: the cool, wet summer of 1958 produced record yields of grain in both the USSR and USA. The greatest hazard of climatic change seems to come from a heightened variability of weather: more droughts and more floods, more very hot summers and more very cold winters. Given such fluctuations, yields 10 per cent below 'normal' are likely to happen in some years, and prudence would seem to indicate the storage of buffer stocks from the better years (Gribbin 1976; Schneider and Temkin 1978).

The actual biota which form the basis of agricultural products, both in non-industrial and industrial systems, are relatively few compared with the diversity of the biotic world as a whole. Mangelsdorf (1961) points out that man has used at least 3000 species of plants for food, and about 150 of these have entered the world's commerce. Nevertheless the majority of the world's population are fed by the twelve species of plant given on page 258. In spite of considerable breeding programmes to extend the ranges of many of these crops, they are still grown in particular regions of the world as determined largely by climate and culture; maps may be found in sources such as the *Oxford Economic Atlas of the World* or the *Oxford World Atlas*. Their cultivation represents the replacement of natural ecosystems with high biomass and high nutrient storage by man-channelled systems with low biomass and low nutrient storage; in effect, systems formerly 'nutrient-tight' are 'leaked' away to the consumer. Nevertheless, some crops achieve quite high productivities under competent management; Table 10.2 gives some

TABLE 10.2   Productivities of major crops

Species	Lat	Location	Annual production (t/ha)	Notes
Maize (corn)	30°N	Egypt	29·1	About 40% of this
Zea mays	42°N	Italy	26·0	is dry grain
	52°N	England	17·0	
	38°N	USA	26·0	Stockton, Calif. A record crop for this latitude
Sugar-cane				
(*Saccarinum* sp)	21°N	Hawaii	67·3	
	27°S	Swaziland	63·0	About 38% of this is sucrose
Sugar-beet				
(*Beta vulgaris*)	21°N	Hawaii	39·7	About 45% is sucrose
	36°N	California	42·4	About 45% is sucrose
	52°N	Netherlands	22·0	
Soybeans				
(*Glycine max*)	33°N	Japan	6·3	
	42°N	Iowa, USA	10·4	
Wheat				
(*Triticum vulgare*)	27°N	Mexico	18·3	About 40% is dry grain
	46°N	Washington State, USA	29·8	
	65°N	Alaska, USA	4·5	
Manioc				
(*Manihot esculenta*)	8·5°S	Java	41·0	
	9°N	Sierra Leone	33·0	
Potato				
(*Solanum tuberosum*)	38°N	California	22·0	
Rice				
(*Oryza sativa*	38°N	California	22·4	About 50% is dry grain

Source: Loomis and Gerakis 1975.
Note: These values are for the annual production of the standing crop and not necessarily the portion (usually 50%, see last column) which enters trade and forms the basis of most production statistics.

details of biological productivities of major crops. These statistics, while incomplete for all major crops because of the difficulties of finding comparable data, nevertheless point to the biological conditions for high crop production: dependence upon latitude, for example, is clearly still a major factor with most crops. The graminaceous species of tropical and sub-tropical origin excel at the lower latitudes, with sugar-cane being the outstanding example, although the grass *Pennisetum purpureum* seems to have produced the largest standing crop of any species. At higher latitudes up to 30°, the Sorghums perform very well, and of the underground crops, manioc (where 95 per cent of the plant weight is storage roots and stems) is the best yielder of dry matter. Legumes, valued for their nitrogen-fixing properties, seem to be relatively low yielders at all latitudes. But even at periods of maximum growth, no plant seems to 'fix' more than 3–4 per cent of the total daily radiation of appropriate wavelength.

But, in general, agriculture is operated for economic return rather than biological productivity: this is nowhere more apparent than in the industrial nations where mechanization has forced its imperatives upon the agro-ecosystem in the name of cost-reduction. So selection by industrialization takes place: a good-tasting apple which bruises easily is clearly unsuitable for mass production and packing; potatoes are favoured which hold their tubers close to the stem for easy mechanical harvesting irrespective of their other qualities, and pigs are bred to conform with modern tastes in lean bacon.

Crop biota are the products of a system which is fuelled by energy, either solely from the sun or from stored photosynthesis as fossil fuels as well. Thus there are two types of agriculture: firstly, the non-industrial which is powered by solar energy and natural substrates together with recycled farm wastes; and secondly, industrial, underlain by solar energy to which is added fossil fuels and industrially-produced agricultural chemicals. So the crops in this sytem receive an 'energy subsidy' which appears in farms as mechanization, fertilizers and biocides, all of which have their 'upstream' energy uses at the places of manufacture and in their distribution. The process of applying these energy subsidies to an agricultural system so that there is a higher productivity is called intensification. To be complete, perhaps the energy converted into knowledge should be added, and certainly the energetics of the complete food system, including the energy expended in food distribution and processing as well as shopping, should be accounted for. It is clear that such lavish expenditure upon the food system is the prerogative of the richer industrial nations of the world. For instance, the USA applies about one horsepower of mechanical power/ha of arable land, whereas power from all sources (human, animal and mechanical) is estimated at 0·20 hp in Latin America (UCFTF 1974); one application of a herbicide incurs a total energy cost of 200 GJ/t (Green and McCulloch 1976).

The role of energy subsidy in food production can be illustrated by the balance sheet for Israel worked out by Stanhill (1974), Fig. 10.1. Here the solar-energy flow is converted to food with the assistance of large quantities of fossil-fuel energy, most of which must be imported from across Israel's borders. The diagram shows the heavy energy cost of irrigation, and that 70 per cent of the plant-crop energy is diverted into livestock production in Israel is negative to the extent of -1123 kal/m^2/yr: each kilocalorie of

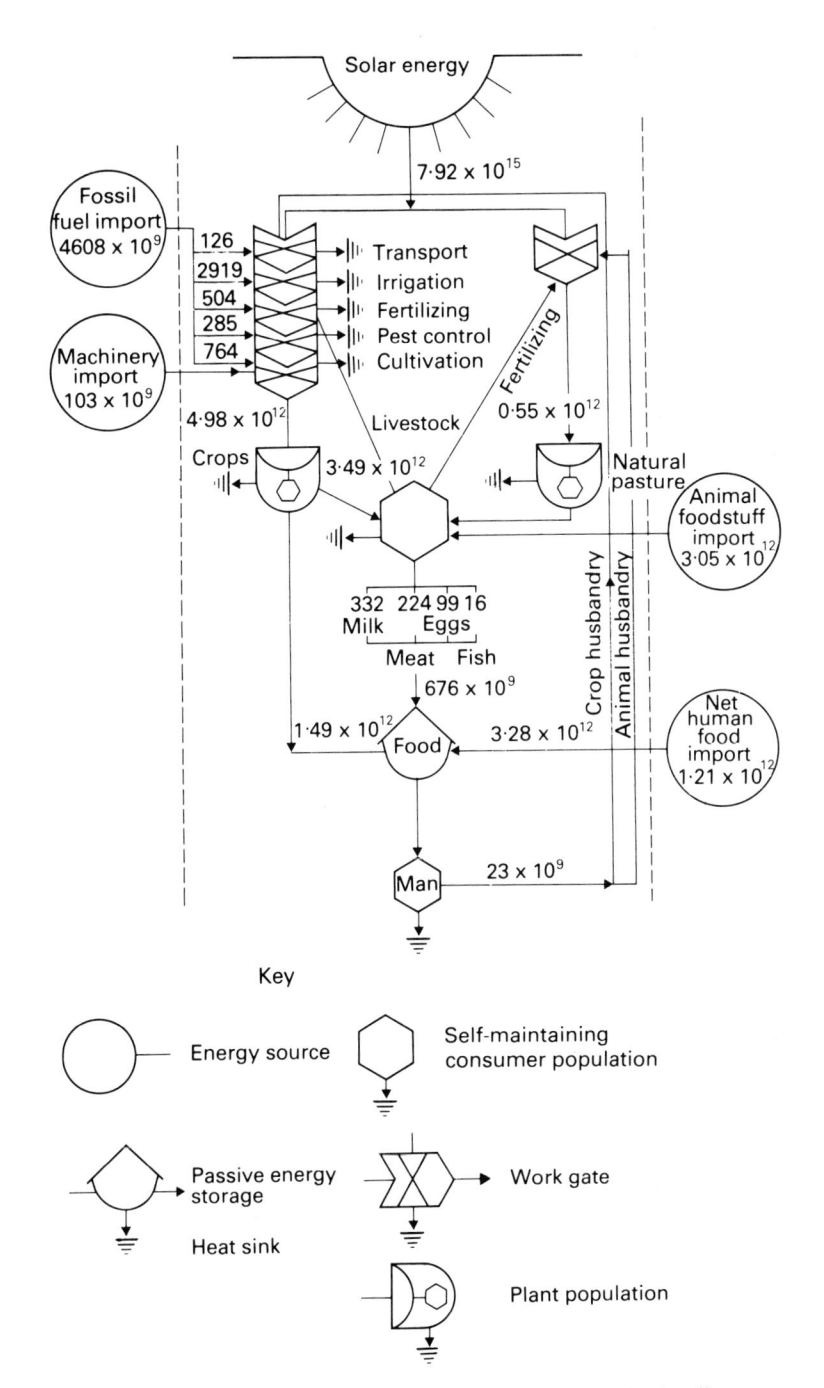

Fig. 10.1 Energy flow and food balance in Israeli agriculture 1969–70. The broken lines represent the pre-1967 borders containing 4105 km² of cultivated land; all quantities in kcal/yr. Large quantities of imported energy (as fossil fuel and machinery) make possible (i) irrigation, and (ii) a high production of animal protein.
Source: Stanhill, 1974.

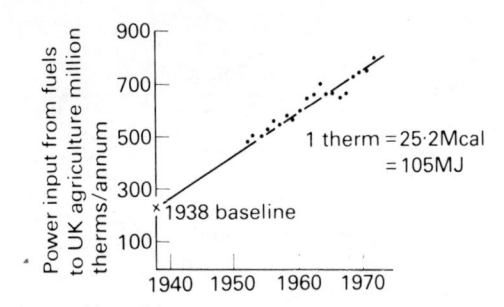

278 *Productive resource processes underlain by biota*

Fig. 10.2 Energy input as fuel into UK agriculture 1938–71.
Source: Blaxter, 1973.

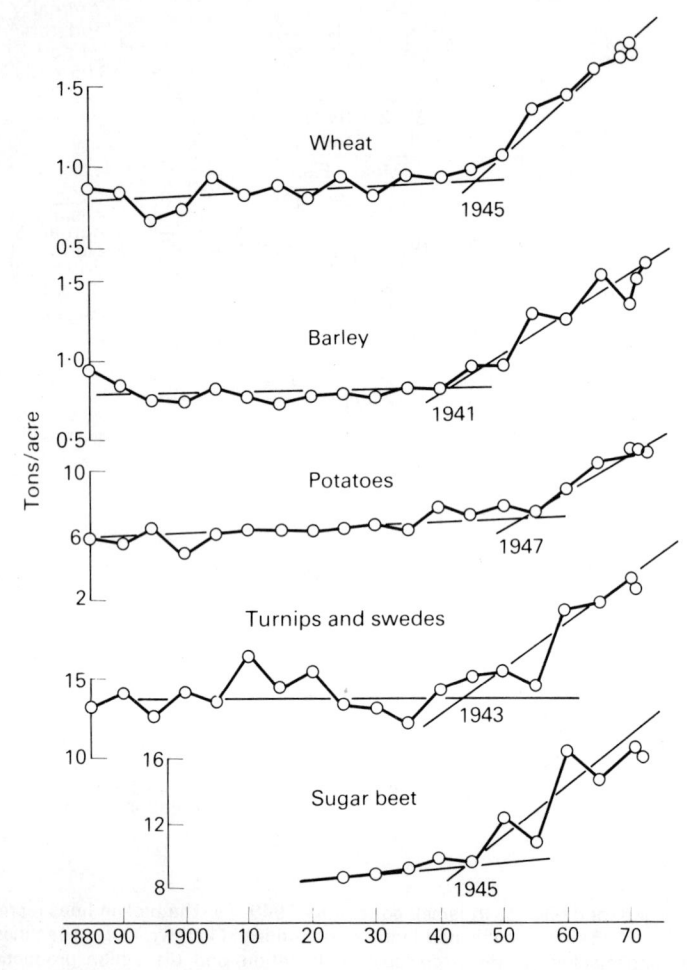

Fig. 10.3 Crop yields in the UK 1885–1973.
Source: Blaxter, 1973.

being imported. The overall conclusion is that the overall energy balance of human food production in Israel is negative to the extent of $-1123\,\text{kcla}/\text{m}^2/\text{yr}$: each kilocalorie of unprocessed human food required 2·44 kcal of fossil fuel. Allowing for the possibly atypical role of irrigation in Israel, the situation is an exemplar of many industrial countries, especially the small ones. (Irrigation is a high user of energy: if $6·17 \times 10^3\,\text{m}^3$ (5 acft) of water is lifted 76·20 m (250 ft), the energy required to supply the water will be 20 times the energy required for field operations in the crop (CAST 1973). This does not of course apply where gravity feed can be used.)

In the UK, similar studies (Blaxter 1975a) indicate that the ratio of edible food output at farm gate to fossil-energy use is 0·34:1, another negative balance. Blaxter regards all the post-1945 increases in UK agricultural production as the outcome of increased fossil-fuel use (Figs. 10.2 and 10.3). Even with these inputs, dissipation of the energy present in food production is at a high level: of the energy fixed in a potato crop just before harvest time in the UK, only 30 per cent becomes human food. The output:input ratio for a 1 kg white, sliced wrapped loaf at point of sale in the UK is 0·525:1 since a total of 20·7 MJ of energy has been expended in providing 10·6 MJ of food energy (Leach 1976). Analogous figures for an animal system show that only 4 per cent of the energy value of the dry matter of the primary productivity ends up as human food. Fossil-fuel inputs in the USA are very high: the intensity of fuel inputs means that every 1 kcal of fuel results in 2·5 kcal of maize at the farm gate (Pimentel *et al.* 1974), a positive energy balance. When the whole food system (on-farm, processing, commercial distribution, retail sale and home use) is considered, then in 1970, 1 kcal of food required an input of 9 kcal of fossil fuel energy, a magnitude that was only 2 kcal in 1925 (Steinhart and Steinhart 1974). In Australia, each joule of digestible food-energy eaten required at least 5 J expended in making it available. About 0·6 J are applied as far as the farm gate, 2 J from farm to retail store and 2–8 J from retail shop to consumer (Gifford and Millington 1975).

TABLE 10.3   Energy ratios of food-production systems

System	Location	Energy ratio (output per one unit input)
(A) NO FOSSIL-FUEL SUBSIDY		
Rice paddy	Yunan	53
Maize	Yucatan	13–29
Taro-Yam	New Guinea	20
Rice Paddy	Borneo	10
Pigs	New Guinea	2
(B) WITH FOSSIL-FUEL SUBSIDY		
Wheat	UK farm gate	2·2
White bread	Baker; from UK wheat	1·4
Barley	UK farm gate	1·8
Milk (+ calves, cow)	UK farm gate	0·33
Broiler hen	UK farm gate	0·11
Battery egg	UK farm gate	0·16
Fish	Freezer trawler, UK dockside	0·05

Source: Leach 1976.
Note: The statistics date from before the major oil-price rises of 1973–74.

280 *Productive resource processes underlain by biota*

Such calculations raise the question of whether DC agriculture is the model of efficiency which the LDCs are often exhorted to adopt. One way of comparing DC systems with other food-producing systems is given in Table 10.3, and Slesser (1975) compares them additionally with some more conventional systems in terms of energy input in megajoules per kilo of food harvested (Plate 36). Rice in Asia averages 2·0 MJ/kg; maize in the USA 2·0–6·0 MJ/kg; grass-fed beef in England 20 MJ/kg; feedlot beef in USA 48–70 MJ/kg; the industrial foods: single-cell protein from oil 100 MJ/kg; yeasts from molasses 27 MJ/kg; and synthesized lysine (an amino acid often lacking in protein-deficient diets) 90 MJ/kg. A further calculation suggests that if the US input of energy into agriculture were to be extended to the whole world, then 1200 litres of petroleum per head per year would be needed, which would equal 14 years' supply at current world production rates. Not too gloomy a view of these energy subsidies should be taken, however. The absolute gains of DC farmers, with their fossil-fuel inputs, are high and in

Plate 36 Intensive agriculture with a very high energy density is shown in this picture of a beef feedlot operation set in the midst of a highly mechanized farming area on the flat lands near the Gulf of Mexico in the USA.

the UK only 3 per cent of the national energy budget is expended in agriculture, a quantity which could easily be saved elsewhere. One of the more disturbing aspects of energy use in the DC food system is the quantities expended in unnecessary processing (probably to the detriment of people's health) and packaging (Mellanby 1978).

A more spatial view of energy subsidies can be taken by envisaging energy inputs per unit area for various production systems. This view provides us with a taxonomy of agriculture in energy terms (Table 10.4).

TABLE 10.4 An agricultural taxonomy in energy terms

System	Energy density (GJ/ha)*	Protein yields (kg/ha)
Andean Village	0·2	0·5
Scottish hill-sheep	0·6	1·0–1·5
Marginal farming, UK	4	9
Open-range beef, NZ	5	130
Mixed farm in a DC	12–15	500
Intensive crop production	15–20	2 000
Feed-lot animal production	40	300
Algae production	1600	22 000

Source: Slesser 1975.
*GJ (Gigajoules): $1 \text{ GJ} = 1 \times 10^9$ J.

TABLE 10.5 Actual and 'ghost' acreage for selected countries (10^6 ha)

	Tilled land	Fish	Trade	Total
China	13·7	4·4	0·5	18·6
India	31·7	1·5	1·2	33·4
Indonesia	15·9	3·0	0·7	19·6
Japan	5·8	22·3	15·9	44·0
UK	13·5	9·1	29·1	51·7
France	30·5	4·2	1·3	46·0
Netherlands	7·3	18·1	11·1	36·5

Source: Borgstrom 1969.

A number of policy-oriented conclusions can be drawn from these energy statistics. Firstly, the diet of the DCs is a consequence of high levels of energy input: about 10–12 per cent of the total national fuel flow goes into the food system in average DCs, more in special cases like Hong Kong and Israel. This dependence is particularly serious for small, densely populated countries since more energy must be invested per capita to maintain food supplies. If self-sufficiency through intensification is impossible, then DCs with buying power rely on 'ghost acreages' abroad to supply their needs, some areas being terrestrial, others marine. Table 10.5 gives the 'ghost acreages' for selected countries and Fig. 10.4 shows how they feed in to the energy flow of the UK food system. Inescapable, too, is the conclusion that the cultural preference for animal protein is a reason for a large proportion of 'ghost acreage' (to provide animal foodstuffs, e.g., from fish meal and imported cereals), and for the high industrial-energy input: the second law of thermodynamics is inexorable.

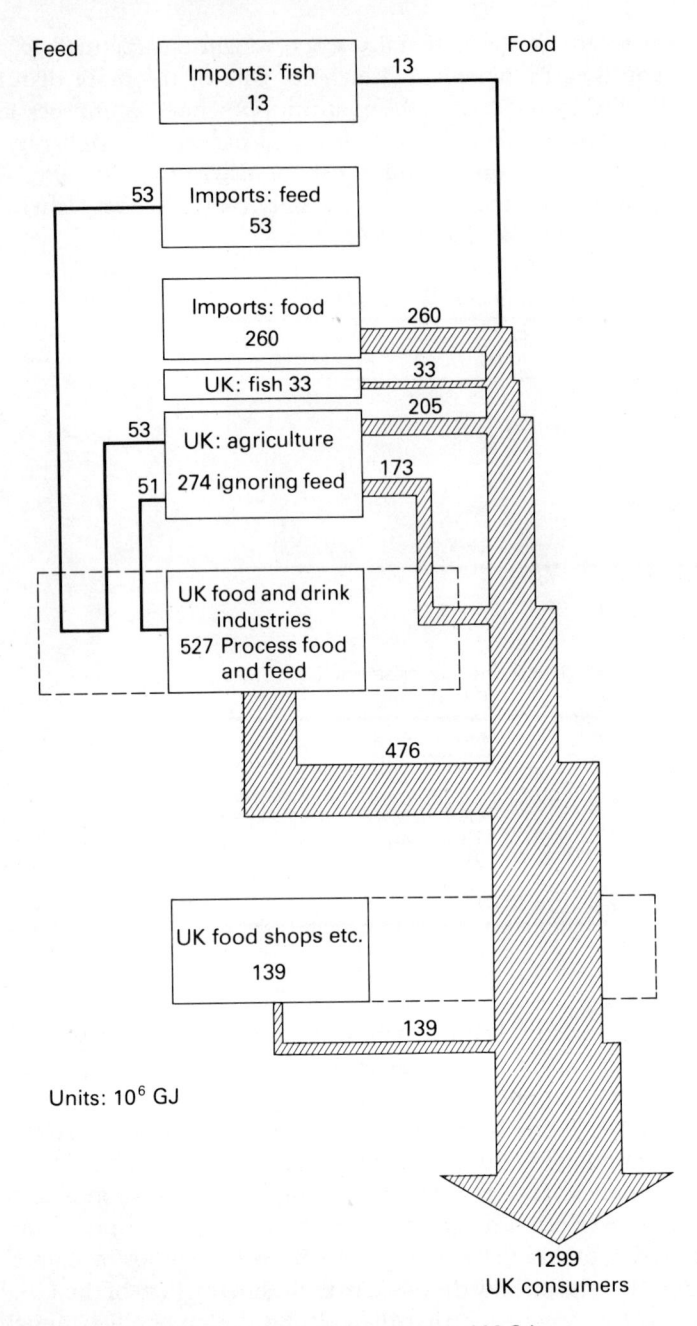

282 *Productive resource processes underlain by biota*

Fig. 10.4 Energy flow for the UK food system, 1968, values in 10⁶ GJ
Source: Leach, 1976.

Secondly, there seems little way in which the LDCs can aspire to current levels of DC diet: there is unlikely to be enough energy for them to get to that stage unless they have access to abundant sources of cheap energy. However, there is a great deal that can be done to improve LDC standards of diet in the near future by using widespread energy sources such as sun, wind, wood and excreta, on a renewable basis. There is the additional advantage that they are independent of the price structures of the industrial world's economy (Makhijani and Poole 1975).

Thirdly, it seems prudent for the DCs to reduce their dependence upon high fossil-fuel inputs by practices such as more use of organic manures, biological pest control, the use of rotations and direct drilling, energy from solar, wind and waste-fermentation sources, altogether less processing and packaging of food, and even fewer frostless refrigerators. We should note too that above certain intensities of energy input (*ca* 15 GJ/ha, see Table 10.4), agriculture begins to have an environmental impact that poses ecological problems (see below) which themselves might require extra energy to treat.

The intensification of agriculture inevitably reduces the quantity of wildlife in rural areas: land reclamation, improvement of accessibility to machines, intensive grassland management, drainage, modern harvesting methods, hedgerow and tree removal, pollution of surface waters and the effects of toxic pesticides, have all made inroads on wildlife populations (Davidson and Lloyd 1977). Sometimes some of these may act together: in the Netherlands the number of breeding pairs of barn owls (*Tyto alba*) fell from 3000 to 300–500 pairs in 1964–74. The factors involved were a combination of pesticides; the disappearance of the common shrew because of the loss of wooded banks and hedgerows, the storage of corn in places accessible only to mice but not to owls; and the shortage of nesting places in new agricultural buildings (Braaksma and De Bruyn 1974). The ploughing up of permanent pasture in the UK, leading to seasonal food shortages, has been held to be one of the factors in the decline of a common bird like the rook, *Corvus frugilegus* (Sage 1978).

Some data have been produced to show that 'organic' farming can be as productive and profitable as conventional high fuel-subsidy agriculture. Maize and orchard fruits seemed to be the only exceptions in these European studies (Hodges 1978). Such farming practices are also less prone to the eradication of the wildlife of the countryside by the eradication of, for example, hedges in England (Pollard *et al.* 1974), small pieces of 'waste' ground between fields and along streams, or small wetland areas, all of which can be important in providing food and cover for wild plants and animals.

Most accounts of agriculture emphasize the crop organisms, but these exist as part of a system in which there are many other biota and which is influenced from outside the limits of the crop area. In turn this exerts reciprocal influences beyond its borders (Fig. 10.5). The influence of non-crop biota is generally labelled the 'weed/pest problem', but as Van Emden (1965) has argued, too little is known about the possible effects upon crops of the biota of crop margins and their role in 'biological control' of pests (see p. 264). However, such management measures can do little about problems such as locusts or birds which fly onto a crop from a long distance away. In Africa, the red-billed quelea (*Quelea quelea*, a weaver-finch) will migrate in large flocks to areas where rain has been falling and feed on seed plants, among which are large areas of

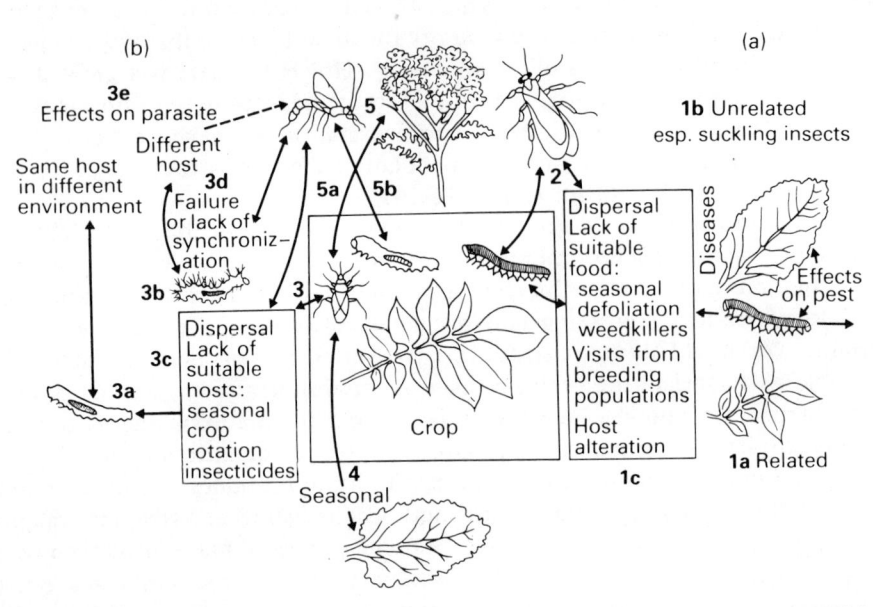

284 Productive resource processes underlain by biota

Fig. 10.5 Biological relationships between uncultivated land and both pest (a) and beneficial (b) insects. Injurious species feed on both crop and related (1a) or unrelated (1b) wild plants for a variety of reasons (1c); Pest insects (2), predators (5a) and parasites (5b) may also feed at flowers (5). (3) Beneficial insects may utilize alternative prey related (3a) or unrelated (3b) to the crop pests for several reasons (3c, d), and such alternative food may effect the physiology of the parasite (3e). Sometimes, carnivorous insects feed phytophagously on uncultivated land (4).
Source: Van Emden, 1974.

crops such as rice, millet and wheat. In Senegal, 9·72 ha of rice was eaten in 14 days by these birds. The eared dove (*Zenaida auriculata*) fills a similar role in South America, and many other examples of migratory birds which are agricultural pests are reported (Murton and Westwood 1976).

Other organisms in the system may consume or spoil part of the crop before it gets to the consumer: micro-organisms are particularly effective here, and the propensity of rats to eat stored cereals is well known from the nursery onwards. On a world-wide basis some 38 per cent of maize, 24 per cent of wheat and 47 per cent of rice are thought to be lost before harvest and smaller but not negligible proportions thereafter: in East Africa corn losses of 9–23 per cent of the harvest during 6–7 months' storage have been reported, and yearly storage losses of 50 per cent of the harvest of sorghum in the Congo and 25 per cent of rice in Sierra Leone have been quoted. An estimate of a world loss of 3·5 per cent of the wheat crop to rodents has been quoted for the 1960s: an absolute quantity of 33·5 million t/yr.

An external influence of the agricultural ecosystem is its production of wastes. Inedible carcasses, for example, may pose disposal problems because of the large numbers associated with intensive rearing operations: in the USA there are crop residues of *ca* 6·7 t/family/yr, and 58 million dead birds with unusable carcasses (Taganaides 1967). More serious perhaps are the effects of nitrogen and phosphate runoff from modern agricultural operations: the two main sources are chemical fertilizers and the sewage slurry of intensively reared animals, especially pigs, cattle and chickens. Apart from the eutrophication of fresh waters which generally results, nitrates may also cause nitrite poisoning of young children if they drink contaminated water. Disposal of excretory wastes (e.g. from feedlots of 10 000 cattle) clearly cannot be via normal water courses and various ways are being explored: pasture irrigation, decomposition in lagoons, composting, drying are among them. In the last-named, worms and flies fed on the manure may then be used as fodder for cattle, pigs and poultry.

Another possible use of wastes relates to the previous discussion of farm-energy use. Much farm waste can be processed relatively simply to yield energy: sewage can be converted to methane, for example, and plant wastes fermented to methanol. Leach (1975) calculates that the calorific value of one long ton of straw is $ca\ 3 \times 10^6$ kcal/acre ($1 \cdot 2 \times 10^6$ kcal/ha) and this is more than twice the energy used per unit area for all purposes except pig and poultry enterprises. If fermented the straw might yield $1 \cdot 5 \times 10^6$ kcal/acre ($0 \cdot 6 \times 10^6$ kcal/ha) of methanol and so with a judicious use of fermentation and combustion to generate power, many farms might become self-reliant for fuels and power, and would need to import energy only in the form of fertilizers (even these are less necessary if manures are put back on the land) and machinery. Asia, for example, generates $346 \cdot 6 \times 10^6$ t/yr of straw, out of an estimated world total of $394 \cdot 8 \times 10^6$ t/yr (Han and Anderson 1974).

The question of plants and energy has been given a new twist by the suggestion made in California that the milky latex of the shrub genus *Euphorbia*, being an emulsion of hydrocarbons in water, could be processed to yield a crude oil. This would be a renewable resource, and one preliminary calculation suggests that an area the size of Arizona would supply the USA's requirements for petrol (gasoline). Even if not a practicable energy source, such a plant might be an enduring source of materials for the petrochemical industry. The use of guayule shrub, sugar-cane wastes and maize wastes to provide rubber, turpenes and ethanol is another possibility under investigation (Lipinsky 1978).

The big question asked of agriculture is, inevitably, how well can it cope with increasing demands from a world population currently doubling in 35 years and in which rising expectations demand more varied diets often with more meat? The improvement of the crops themselves is obviously a primary target of concern and breeding programmes for all kinds of organisms are under way (pp. 258–61). These often involve a move towards genetic uniformity since a uniform crop with a predictable yield is required by the economies of industrially-linked agriculture (Table 10.6). A necessary corollary therefore is the conservation of genetic variety in plants and animals by means of seed banks, botanical and zoological gardens, and the protection of large areas of natural and semi-natural terrain. Other sources of genetic material include local varieties adapted to a set of conditions found only at a particular place; spontaneous mutations (Opaque-2 maize arose thus and was spotted by a farmer who took it to experimenters); and induced mutations (NAS 1972).

If energy requirements are met with a diet which averages 9 per cent protein, then there should be no overall protein deficiency, so that removing calorie deficiencies is

TABLE 10.6 Area of major US crops dominance by small numbers of varieties

Crop	Ha (×10^6)	Total varieties	Major varieties	% in major varieties
Cotton	4 536	50	3	53
Corn	26 851	197	6	71
Peanuts	567	15	9	95
Potato	567	82	4	72
Rice	729	14	4	65
Soybean	17 091	62	6	56
Wheat	17 941	269	9	50

Source: NAS 1972.

probably the most necessary agricultural strategy for the near and medium-term future and to achieve this, the upgrading of photosynthetic efficiency and biological nitrogen fixation are probably the two most important advances to be made. In this connection, understanding of the complexities of N_2 fixation in nature by nodulated legumes, nodulated non-leguminous woody species, free-living blue-green algae and blue-green algal associations with plants is an important item on the research agenda (Evans 1975; Nutman 1976; Evans and Barber 1977). Likewise, the intensive study of mineral nutrient recycling from wastes, sewage and refuse is essential, as are methods of preventing the considerable losses of stored foods, especially grain in the LDCs. Reduced tillage has favourable effects on the yield of some crops and it generally leads to enhanced soil and water conservation as well (Wittwer 1974). Where the economy and environmental conditions permit, the enrichment of the atmosphere in huge greenhouses with extra CO_2 (from power generation) may also be a viable way of obtaining extra productivity (Bassham 1977).

Remarkable advances can be made in productivity but every crop has its limits, beyond which no amount of breeding, fertilizer, pesticide and energy will raise the yield (Jensen 1978); and for most the economic limit comes well below the feasible maximum. As Slesser (1975) puts it, 'What we must not assume is that it is possible for any given crop to proceed ineluctably up the line of production versus input' (Figs. 10.6 and 10.7). Maize is one example: its yields at about 1×10^3 kcal/m^2 are about half the photosynthetic limit, which suggests that energy is not now a limiting factor in its productivity. The same is presumably true of protein production which has so far been (in DCs) a direct function of energy subsidy. In LDCs, medium-term changes are likely to come in the shape of an adaptation to lower intensities of fuel subsidy than the current style of 'development' which at present largely consists of extending high-energy-use agriculture at the expense of non-industrial farming. Yet increased output without linkage to an industrial nexus would enable many LDCs to achieve better standards of nutrition without mass rural unemployment (Makhijani and Poole 1975). Unless an LDC undergoes a significant degree of development, coupled with assured availability of energy, many of the unconventional sources of protein are unnecessarily expensive and the energy could be better spent at lower intensities where it might ensure better harvests. The use of 'appropriate technology' which uses little fossil-fuel derived energy is an important element of the new-style approach to 'development', along with a renewed recognition that many so-

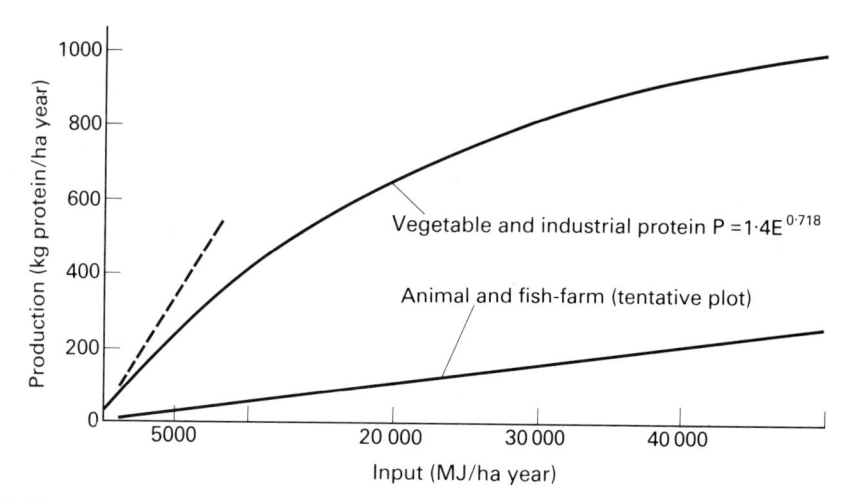

Fig. 10.6 The correlation between energy input and protein production for two different food-producing systems.
Source: Slesser, 1975.

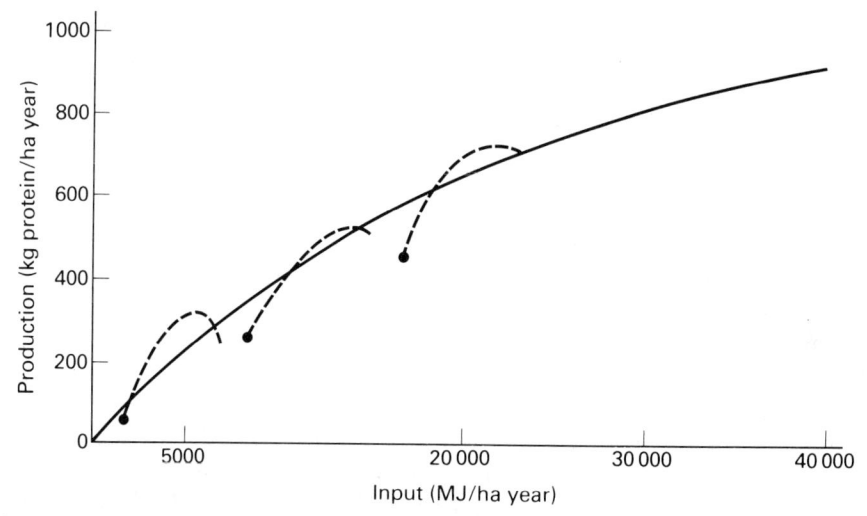

Fig. 10.7 The response of three different protein-producing crops to additional inputs of energy.
Source: Slesser, 1975.

called 'primitive' agricultural systems were adequate for thousands of years and have only become less so under the disruptive effects of western medicine and other forms of aid (Schumacher 1973, Slesser 1973).

And the ultimate carrying capacity of the agricultural resource process? There are many calculations, based on different assumptions, but perhaps a Dutch study (Buringh et al. 1975) is one of the most careful. On a regional temperature regime/soil type/water availability basis, the calculation is made that 40 times the present cereal crop production can be achieved, assuming that all areas can be made to yield at the level of the best so far accomplished, e.g. triple cropping of rice giving 22–26 t/ha/yr (cf. Table 10.2). The authors do not deal with social and political factors, nor the energy input needed for such a raising of yield, nor whether such a yield level would be constant. Neither do they examine the likely consequences to the ecology of the biosphere of converting more wild land into crop land and of the intensification of extant cropland. Beyond a certain intensity of energy subsidy (Table 10.4), we can see that the effect of agriculture has a strong local and regional environmental impact (De Soet 1974), and we must wonder if there is a point at which on a world scale, perturbations analogous to algal blooms might occur. Thus the case argued by Chancellor and Goss in 1976 for trying to bring about a largely solar-based food-producing system, preferably allied to a stable human population, is on all grounds a strong one.

Forestry

The biogeography and ecology of natural forests has been discussed earlier in the book; here we need to think of man's uses of them and the ecological consequences. Not all forests used by man are natural, and indeed the last four centuries' impact have meant that truly natural forests, or even areas in which man's effects are little detectable, are very rare. There exists a complete spectrum of modification, from the occasional removal of individual trees to the clear-felling of all or part of a forest and its replacement by a different forest species or even another form of land use altogether. Even though the forest ecosystem is dominated by trees, the other components of the ecosystem are clearly important in its ecology. The soil parent material affects the nutrient supply to the forest during its initial phases of establishment, and thereafter through weathering processes; according to their level of seed consumption, the small mammal populations may affect the regeneration of the trees, and large herbivores may consume young saplings but also provide a seed-bed by trampling out a competing ground flora. Nearly all these processes are changed when man uses the forest (Pesson 1974).

The human use of a forest can be of two types. The first is of a consumptive type, when the woody parts of the tree are extracted and transported elsewhere for processing and use; the second is non-consumptive and involves the use of the forest as a protector of watersheds in terms of water yield and quality; for outdoor recreation, nature conservation and aesthetic pleasure; and for a possible harvest of animals. The consumptive use usually implies the cutting down of the tree and removal of the trunk, although large branches may also be harvested from hardwood species. Leaves are rarely used (except for local purposes such as dyes and drugs) although experiments in processing them for human food have been made (pp. 267–69). The uses to which timber is put are well known: fuel is the most important product of the world's forests, although its per capita use declines as industrialization provides alternatives like coal or petroleum products (Fig. 10.8); thereafter industrial wood (lumber) is used in housing and many other industries and for a great variety of related purposes. The bark is not generally used directly with the exception of certain specialized species such as the cork oak. Wood may also be converted industrially and the fastest growing use is for pulp production

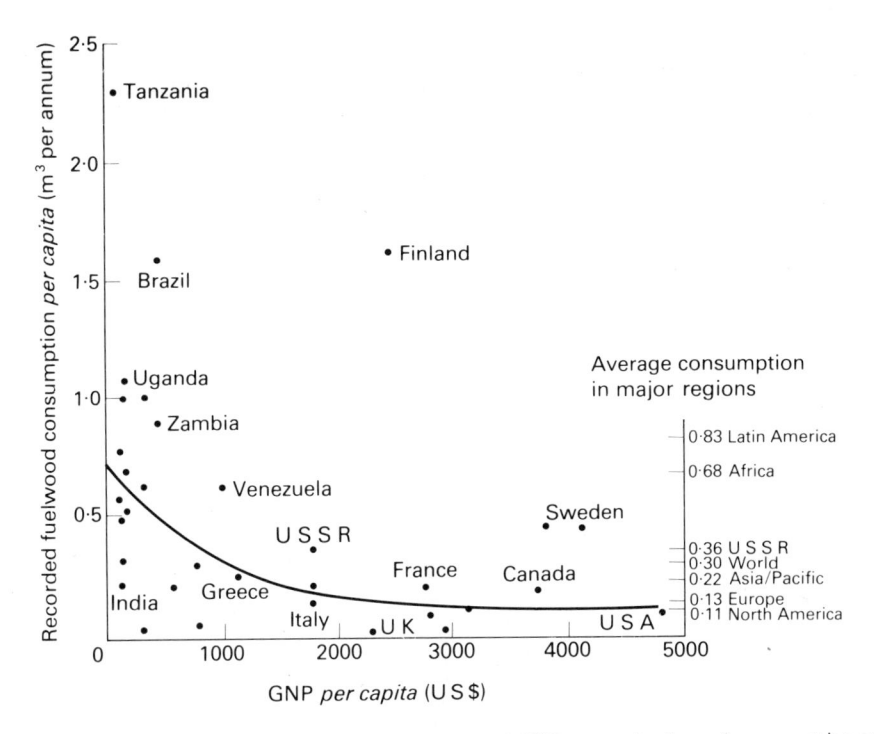

Fig. 10.8 Relationship between fuelwood consumption and GNP per capita for various countries and consumption for major regions in recent years.
Source: Earl, 1975.

TABLE 10.7   World roundwood production ($\times 10^3 m^3$)

	1961–65 average	1970	1976
WORLD	2 121 119	2 387 661	2 524 219
India	94 726	110 770	130 947
Sabah	3 756	7 072	16 624
UK	3 182	3 421	3 343
North and Central America	438 107	489 525	517 750
Developing countries	787 573	946 513	1 050 625

Source: *FAO Yearbook of Forest Products 1976*.

for paper; much of this goes into such articles as newsprint and packaging. Table 10.7 gives some sample figures for recent years and it is clear that the forest ecosystems are, like the agro-ecosystems, coming under pressure for more intensive use in order to raise an output demanded by rising human populations with increased expectations. An understanding of the ecology of forest use therefore becomes crucial.

In consumptive use the tree is felled and stripped of branches, leaves and sometimes bark, at or near its site of growth, and the trunk is then shipped off for further processing. Thus some of the nutrient content is removed from the ecosystem in the form of the trunk whereas another portion is available for return to the soil by the normal processes of organic decomposition. If the slash is burnt, the resulting mineral ash may leach nutrients into the soil quite quickly, although nitrogen is lost in the smoke. The proportion of nutrients lost by harvest is very variable but in the trunks and roots of a 100-year-old stand of deciduous hardwoods in Europe the proportion of calcium was 20 per cent of the total for the stand; for conifers other than pines the proportion was 19 per cent; for pines 30 per cent. For phosphorus (often a relatively scarce element) the same calculations were 28·5 per cent, 14 per cent and 27 per cent (Rennie 1955). In a 16-year-old loblolly-pine plantation it was estimated that by harvesting the aerial portion and larger roots of the trees 12 per cent of the total nitrogen, 8 per cent of the extractable phosphorus and 31 per cent of the extractable potassium from the site were removed. Overall it was discovered that one-third of the quantity of nutrients are removed when debarked pulpwood only was harvested (Jorgensen *et al.* 1975). Thus although a substantial measure of nutrients is lost via the trunks, a major proportion is left in the ecosystem. Clear-felling however causes nutrient loss from the whole ecosystem: transpiration is reduced so that the leaching effect of precipitation is increased; root surfaces able to remove minerals from leaching water are diminished in number and more rapid mineralization of organic matter may add to the loss. The order of increase of nutrient loss was $\times 3 - \times 20$ for various cations in New Hampshire experimental watersheds studied by Bormann and his co-workers (1968, 1969). Silt losses are also much increased, especially when heavy machinery churns up the soil at extraction sites and where logging roads are situated near to streams, leading to a recommendation that a strip of uncut forest should be left on both sides of watercourses.

In the case of the Hubbard Brook studies, where experimental deforestation and suppression of regrowth was carried out, the experimental watersheds began to regain normal values of hydrological and biogeochemical parameters 3–4 years after the vegetation was allowed to regrow, in conditions where wood products were not extracted and where the forest floor was essentially not disturbed. Parallel investigations of commercial clear-cutting in New Hampshire, where roads and skid trails are built, and where wood products are removed, showed that nutrient and particulate loss was much higher than in the experimental watershed and recovery of the ecological functions took in the order of 60–80 years. In this case, the normal 110–120 year rotation practised by the US Forest Service is sufficient to allow a recovery of the nutrient cycles and hence obviate any long-term downturns in biological productivity (Likens *et al.* 1978). The study has emphasized the important role of the pioneer phase of vegetative growth, especially of shrubs such as pin cherry (*Prunus pensylvanica*) and other Rosaceae, in stabilizing hydrology and nutrient flows after deforestation. Tropical forests generally lose a much higher proportion of their nutrients when large-scale felling takes place because most of the nutrients are in the trees. Subsequently, laterization may take place and savanna or grassland results. Temperate ecosystems have a higher proportion of the nutrients in the soil and so a reservoir is always left to fuel new forest growth. This balance might be changed if whole-tree utilization became more widespread. In Swedish forestry, for example, only about 65 per cent of the tree biomass is harvested, and the remaining parts include the green parts which hold a large part of the nitrogen, phosphorus and potassium of the tree. A much larger proportion of these nutrients is in the ground but rapid-cycle cropping might lead to the need for some form of soil fertilization (Nilsson and Wernius 1976; Kimmins 1977).

Palaeoecological studies show that natural forests are often characterized by a diversity of species. Man's activity usually reduces this diversity: some species may be preferentially used, as with oak being differentially extracted from the New Forest of England for shipbuilding during the pre-industrial period. Economic considerations may dictate that only homogeneous stands are worth cropping: this is particularly so with softwoods destined for pulp, as with spruces in Canada, Scandinavia and the USSR. Thus there is a tendency for cropped forests to become monocultures, just as the elimination of weeds makes monocultures of most agricultural crops. Predictably, one-species forests are prone to instabilities: insect pests such as the spruce budworm (*Choristoneura fumiferana*) can spread with great rapidity when one food source is contiguous with the next. Such forests, especially of conifers, are also particularly liable to damage by fire. Fire is a constant enemy of commercial forestry, much feared because it can mean the loss of a crop which will take another 30 years to grow to harvest. It is particularly likely in coniferous forests during a dry season, when a relatively harmless ground fire may become an all-consuming crown fire if fanned by a good wind. Thus foresters will accept the costs of maintaining an expensive fire prevention and extinction programme, and such precautions will in time affect the flora and fauna, especially in forests where fire has long been a part of the natural ecology of the forest. Some species of plants find their seedlings shaded out by competitors whereas under natural conditions repeated ground fires resulted in establishment of a mixture. The giant sequoia (*Sequoiadendron giganteum*) of the western slopes of the Sierra Nevada of California is one such species: unless the ground around the base of the adult trees is burned regularly, its seedlings are shaded out by other conifers (Vale 1975). In general, the accumulation of slowly decaying litter (normally mineralized by low-temperature ground fires but allowed to build up when fire-control programmes operate) may prevent the establishment of seedlings: young trees only a year or two old may root only in the humus and thus are deprived of water if desiccated by a dry spell. If an accidental fire burns off some of the humus then the next generation of seeds may be able to germinate and root in the mineral soil successfully. This process appears to be the case with the semi-natural forests of *Pinus sylvestris* in Scotland, where even-aged stands of trees have sprung up upon areas of known burns. The accumulation of humus in coniferous forests is said to lead to the greatest economic losses from fire, for the depths of humus that can accumulate in a fast-growing conifer stand mean that a fire once ignited has access to large stores of fuel and can reach high enough temperatures to ignite bark and resin drips on the trunks and so a crown fire is started which can devastate huge areas (Plate 37). The rational response by managers has been to burn forest floors in a controlled manner so as to keep the humus layers thin: this is called 'prescribed burning'. There are beneficial side-effects: the ground flora may become dominated by grasses and herbs which provide a forage resource for cattle or wild herbivores, and other wild life may also benefit from the food supplies and ground cover where there is only a thin layer of humus instead of perhaps 0·5–1·0 metre. Fire is also being tried in Australian jarrah

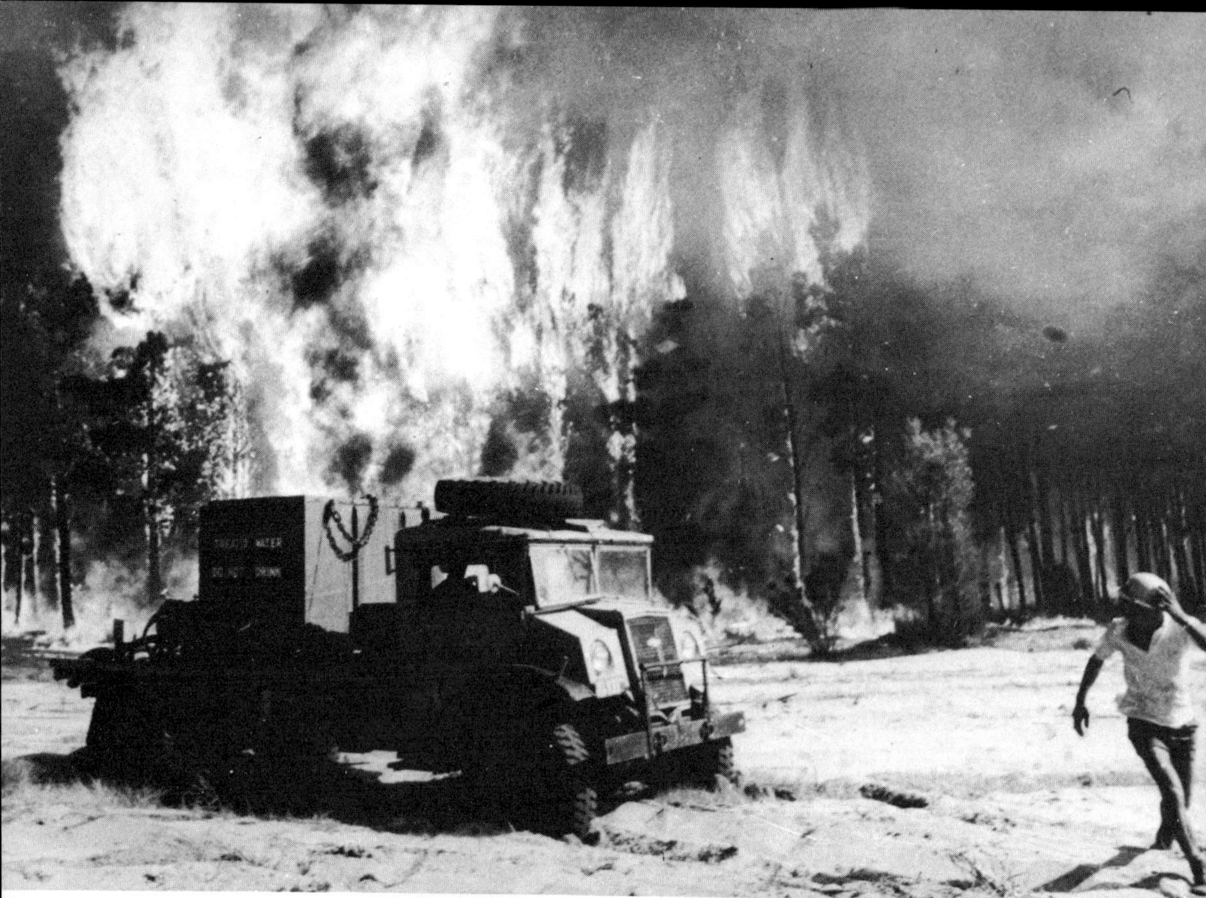

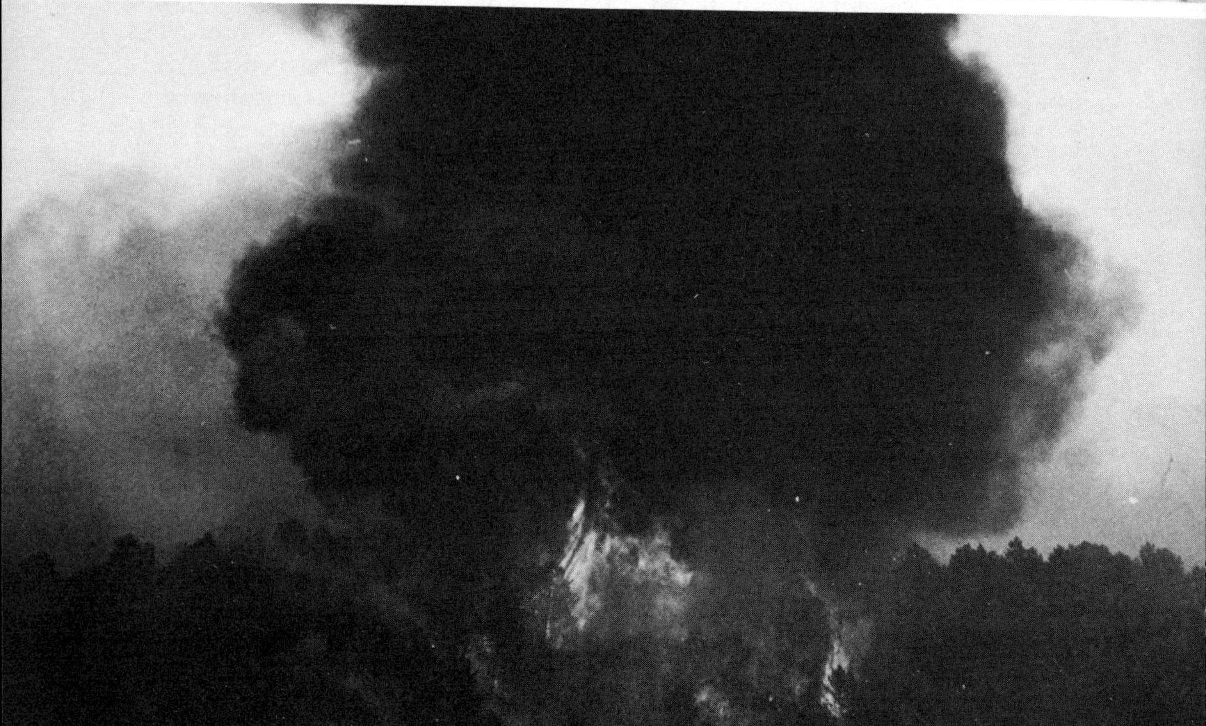

(*Eucalyptus marginata*) forests in order to control the spread of a fungus *Phytophthora cinnamoni* which is causing a widespread dieback in several parts of that continent (Anon. 1978).

The demand for wood and wood products, seen in the context of the results of the clearing of forests for many millennia, has led to some re-evaluations of their role in this century, especially in smaller industrial nations. So re-afforestation has taken place on terrains formerly forested or on marginal agricultural land. Such forests are usually of fast-growing conifers: in western Europe, exotic species from western North America such as Sitka Spruce (*Picea sitchensis*) and Lodgepole Pine (*P. contorta*) have been particularly suitable, along with more distant imports such as Japanese Larch (*Larix leptolepis*). These forests grow quickly and are harvested after 30–40 years, when re-seeding takes place, often accompanied by chemical fertilization of the soil: such forests are thus analogous to agriculture in ecological terms, though not yielding an annual harvest. Afforestation has not been confined to industrial countries. Protective forests to stop the spread of sand have been planted in various locations peripheral to major deserts, and commercial plantations have been made in some African savannas. These aim to increase the supply of all forms of timber for developing countries, including charcoal for domestic use. The tree species used for afforestation are, as in Europe, fast-growing exotics: in the case of the savannas species of *Eucalyptus* are often used. The most serious enemy of these plantations is the termite which is controlled with organochlorine insecticides. Fungi, fire, goats and elephants are also liable to be troublesome (Laurie 1974). Fuel and construction needs have been the main reasons for China's attempts to create large afforestation projects, using a variety of native and exotic species, including *Eucalyptus globulus* in the south (Woo 1974).

Commercial forests are also subject to a number of other influences which may affect their yield but which can be discussed only briefly. Close to urban-industrial areas, air pollution may affect growth rates of many species, especially conifers, and may eventually kill trees (see p. 233); this phenomenon has been most marked in western North America but is by no means confined to that region. Climatic change should also be mentioned: this is not normally considered by foresters except at forest limits, both altitudinal and latitudinal. But in a country like Finland, which relies on timber exports for the balance of its economy, a shrinking of the area of forest during a colder phase is a serious matter and responses such as the attempt to breed hardier trees must be undertaken.

Perhaps one of the most serious re-evaluations of forests which is just beginning is that of their role as energy producers, where their possibilities as converters of solar energy are now being re-assessed, particularly in LDCs (Earl 1975). In the USA, studies have suggested that densely grown hardwood plantations can compete with coal in terms of price per joule of recoverable energy, thus leading to the concept of 'energy plantations' (Evans 1974; Brown 1976; Burwell 1978). Some of the success of such a programme would probably depend upon a positive energy balance during harvesting and processing: at present the forest industries of America supply about 27 per cent of their own energy requirements from recycling of wood, bark and extractives (Grantham and Ellis 1974); genetic selection of appropriate trees would no doubt improve the output

Plate 37 (a) A forest fire in Australia. (b) A forest fire in southern France in 1969. Fanned by a wind reaching 70 kph, the fires burn not only the forest floor but run up the tree trunks and ignite the crowns.

of energy tremendously. A desk study showed that *Eucalyptus* woodland covering 200 million acres (809 375 km^2) in the western US could supply 18 900 Btu/yr (5·54 × 10^6 kwh) of electricity, 27 per cent of the US energy consumption in 1974 (Livingston and McNeil 1975). A calculation by Earl (1975) asserts that the world has a forest growing stock equivalent to 271 000 × 10^6 coal equivalent (CE), to which is added an increment of 770 × 10^6 t CE/yr. (In 1970 the world's total energy consumption was 7435 × 10^6 t CE.) In any terms, this is no minor resource, but it is particularly significant for LDCs: if they could 'farm' their energy from forests and scrubland it would free them from dependence upon external sources of fossil-fuel supply and expensive capital projects like HEP. A balance between forest sales for pulp, 'energy farming' and the use of forests for food such as leaf protein might well be the most rational use of many LDC forest areas, especially in the tropics. In the Developed Countries, forestry is not free from linkage to the energy flows of industrial economics for as in agriculture, fossil fuel goes into the forest before the crop comes out. In Sweden, one study (Nilsson 1976) showed that solar input of 178 × 10^6 million kWh into the forest area of 23·7 × 10^6 ha was aided by 4486 million kWh of industrial energy, both directly and indirectly, giving a yield of 178 000 million kWh of forest products. (At 1972 energy prices this meant that industrial energy costs were 10 per cent of the income from sales.)

The non-consumptive uses of forests are quite diverse but since they do not involve the removal of the trees they can often be made compatible with each other; indeed they may be reconcilable with extractive forestry if very large-scale clear-felling is not practised. The uses which yield a product are grazing, water catchment and game preservation. The grazing of domestic animals is facilitated where the forests have an understorey layer available to browsers and a ground flora suitable for grazers. Thus high altitude forests where the trees are becoming sparse may well bear a grassy ground layer or contain a proportion of shrub species which will support cattle or sheep during the summer season. The density of grazing animals is critical since beyond a certain threshold no natural regeneration of the forest trees will occur, but in some cases the role of the animals in churning up the soil may be a help in providing a seed-bed free of competitive species of plants. Wild animals of the forest may be preserved for game: no management may be necessary if the habitat is little manipulated, or alternatively various techniques may ensure that sufficient feed is available for the game species, particularly during severe weather. Large mammals are often browsers and so the management of the forest to provide browse species becomes important and here fire may be a help: in some North American deciduous forests it has been shown that 2–3 years after a burn, the browse species fed upon by deer are much higher in protein than they would otherwise have been (Dills 1970). As with domestic animals, the mammals need to be managed so that their numbers do not impinge on the regenerative capacity of the forests and so culling for sport may need to be supplemented by a cull of a different sector of the population: for example, if only deer stags with good heads of antlers are sought for sport, then a supplementary cull of hinds may be essential to prevent too rapid a population growth, with its attendant possibility of a 'crash' during the winter or some other period of environmental stress.

A forested watershed is generally a useful adjunct to the management of water supplies for irrigation and urban-industrial purposes. Yield is usually steadier than unforested catchments because of the 'sponge' effect of the root–soil system, and because in appropriate places the shade of the trees even in early spring retards snow-melt, releasing water steadily to the runoff. Unless clear-felled, a mature forest will yield little silt to the rivers. On the other hand, forests evaporate from their crowns and transpire from their leaves far more water than other vegetation types, so that the net water yield may be lower from forest lands. Given the other benefits of forests this disadvantage may not weigh particularly heavily except in areas of high water demand and low supply, such as the southwest of the USA.

Non-consumptive uses may include outdoor recreation, with which may be included aesthetic pleasure whether this be from being in the woods or looking at them as an element of scenery. Another such use is nature protection, which is not necessarily incompatible with recreation but there are places where strict preservation of plants and animals of the forest is essential and in such cases it may be necessary to exclude recreationists who may damage the biota. Otherwise, outdoor recreation is usually regarded as a suitable employment for forests and compatible with most other uses; a forest provides a setting for most recreations even if it is not a necessary element of all of them, and in general seems to have a higher carrying capacity for people and their vehicles than open land because of the screening effect of the trees.

The ecological effects of these non-consumptive uses are demonstrably less than those caused by timber extraction. The retention of forests for watershed protection may involve little or no management, although impoundment reservoirs may be constructed in forest areas and convert terrestrial systems to aquatic ones. If a rare plant or animal is characteristic of a **seral** stage of a developing forest then management may involve holding the forest at that stage in order to protect the organism's habitat. Recreational impact, here as in other ecosystems, shows the effects of trampling and vehicle use in bare ground and modified floral assemblages, depending upon the intensity of use. Animals if timid will absent themselves, temporarily or permanently, from areas frequented by visitors; if bold, they may become scavengers.

Attention must also be drawn to what may prove to be a significant non-consumptive role of forests. Recent experiments have involved the diversion of waste water and sewage into the forest ecosystem with interesting results. When the sewage was applied at a rate of 2·5–10 cm per week to forest plots the forested areas were highly efficient in removing phosphorus and nitrogen concentrations, and detergent residues were rapidly degraded into forest–soil horizons. The forest thus functioned as a living filter system but it did not seem to have suffered adverse effects from this 'sewage irrigation'. On the contrary, the foliage of vegetation was higher on irrigated plots than on control plots and the height of white spruce saplings was found to have increased significantly after application of the sewage (Sopper and Kardos 1972).

The carrying capacity of forests is scarcely calculable in the same terms as agro-ecosystems. To determine the future output of wood from the world's forests, given assumptions about biological productivity and the technology involved, would be possible but not very useful, given all the other values (usually not priced but subject to political evaluation, see, for example, Hasel 1971; Clawson 1975) which forests possess. Under pressure from competing land uses, usually those for food production, many forest areas are shrinking and so wood output may well decline. Many of the uses of wood products are ephemeral in terms of basic human needs (as are some of the products of agriculture) and so perhaps those demands will perforce fall away as the price of wood mounts. Substitution of other materials for wood may occur, if the price of synthetics is sufficiently low; more efficient use of the wood produced would be a better step forward, even if some of the products such as paper are of a poorer quality than the West is accustomed to using (Van der Meiden 1974).

Little is known for certain of the wider roles of forests in the functioning of the biosphere: forests are responsible for about half the photosynthesis of the world, which must mean a much higher proportion of the terrestrial photosynthesis; so to what extent is the balance of CO_2 and O_2 in the atmosphere regulated by such vegetation? What effect do forests have on regional climate in terms of evapotranspiration and albedo? There can be no shortage of practical reasons for protecting the remaining areas of forests and for further research into their role in the biosphere as a whole.

Fisheries

Until the advent of fuel-powered technology, man's impact upon fish and other aquatic populations was relatively low. The crop organisms are dispersed in large volumes of water and so only in confined places such as small lakes and accessible shorelines was pre-industrial man likely to deplete a population beyond its reproductive capacity. The

TABLE 10.8   Aquatic organisms used by man

1   Freshwater and Diadromous fishes
    (including shads, milkfish, salmon, trout, eels, pike, etc.)
2   Marine fish
    (flatfish, cods, basses, mullets, herring, anchovies, tunas, mackerels, sharks, etc.)
3   Crustaceans, molluscs, other invertebrates
    (crabs, lobsters, mussels, oysters, winkles, sea cucumbers, sea urchins, etc.)
4   Whales
    (blue, fin, sperm, Minke, pilot, etc.)
5   Seals and dolphins
    (porpoises, dolphins, seals, walruses, manatees, etc.)
6   Miscellaneous animals
    (e.g. turtles, frogs, corals, etc.)
7   Aquatic plants
    (large brown algae, red algae, etc.)

Source: Adapted from FAO, *Yearbook of Fishery Statistics*.
Note: Relatively few of the animal species harvested are herbivorous: most are either carnivores (the larger fish coming apparently from 4th or 5th trophic levels) or detritus feeders. The gregarious habits common to many marine creatures help make them a potential crop.

coming of the steam vessel, industrial processing and refrigeration meant a greater usage of the seas and a demand for a wide variety of species (Gulland 1972). Thus a very wide spectrum of aquatic animals, and a few plants, are now exploited (Table 10.8). The course of industrialization of fishing ('fishing' will be used to include exploitation of all aquatic animals, including those taxonomically not fish, such as molluscs and whales) has meant that the catch per unit of input effort increased greatly during the phase of industrialization. There is thus one thousand times difference between the 'efficiency' of simple harvesting of fish from man- or wind-powered boats with simple fishing equipment, and a modern purse-seine trawler catching an abundant fish such as the anchovy off Peru (Table 10.9). But as would be expected, the energy balance of such

TABLE 10.9   Production levels of fishing types

Catch (long tons/man/yr)	Fishing method
1	Lines, traps, spears or nets from normally operated boats
3	As above in small power boats
10	Small coastal vessels with lines, trawls or gill nets
30	Medium-large vessels with lines, trawls or purse-seiners
100	Modern trawlers and purse-seiners for moderate-high-value fish
300	Modern purse-seiners for low-value fish
1000	Purse-seiners for anchovies.

Note: 1 long ton=1·016 tonnes.

fishing has (as in agriculture) shifted towards a negative position, so that some forms of fishery are possible only because rich nations are willing to invest energy in producing high-protein food either for direct consumption or for feeding to terrestrial livestock. The energy-intensive types are characteristic of industrial nations like those of Europe and North America, the USSR and Japan. Most LDCs (with a few exceptions like Peru) are confined to lower-intensity types of fishing, with correspondingly less impact upon the stocks, although the rise in demand due to rapidly growing human populations is beginning to affect these fish populations as well. At a world scale, the markets for fish and fish products are characterized by a strong demand, supplies which in recent years have been either static or declining, and record prices. In recent years, the biggest single feature of production has been the decline of the anchovy fishery of Peru and Chile, but overall the reduction of stocks, fishing controls and production quotas, together with the implementation of national resource-management measures have all helped to bring about a near-stasis of world fisheries' production. As Fig. 10.9 shows, the years 1950–69 showed a rapid increase in landings at a rate of 6·6 per cent p.a. to the level of 60–70 million tonnes, between which levels catches have subsequently fluctuated. The plateau figure of *ca* 70 million tonnes may well show a decline as catch restrictions in the new 200-mile economic zone are applied, although a later rise, when the benefits of conservation policies are apparent, may be hoped for. The gross total is also affected in recent years by fluctuations in the Peruvian anchovy catch. Of the contributors to

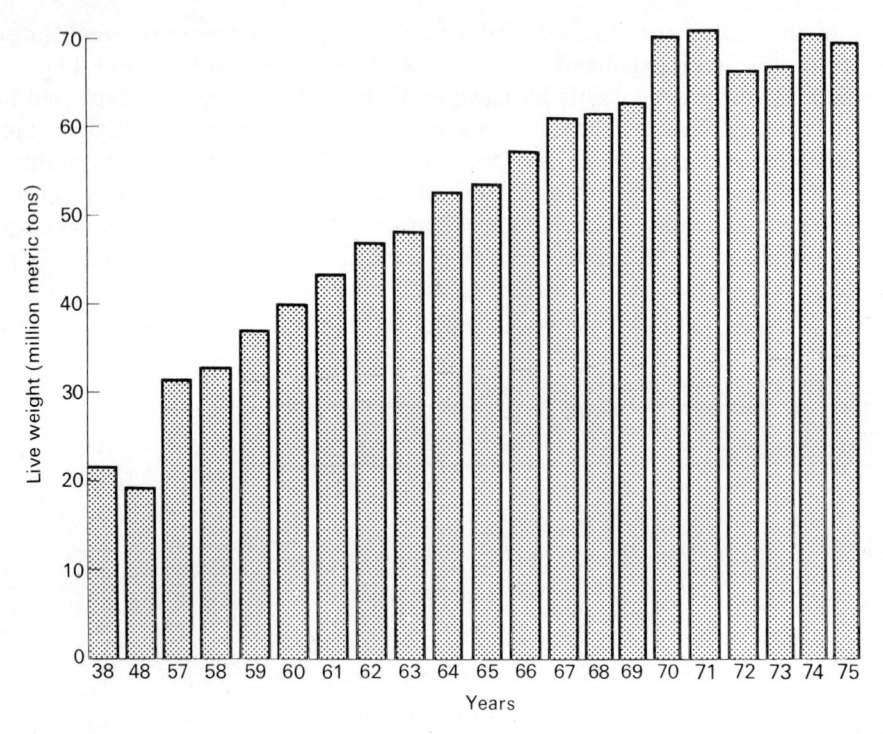

Fig. 10.9 World fish catch 1938–75. Note the fall-off in quantity after the rapid rise to 1971–72 levels. Source: *FAO Fisheries Yearbook*, 1977.

the rapid rise, the Pacific Ocean was outstanding with a rise in output of 9·3 per cent p.a., and inland waters rose by 8·7 per cent p.a. during 1950–60, although dropping to 2·7 per cent p.a. in 1960–69. In terms of continental landings, South America rose from 1 million t/yr in 1956 to *ca* 10 million t/yr in 1963 and a peak of *ca* 13 million tonnes in 1963, from which level a decline has taken place. Of the animals caught, the herring–sardine–anchovy group showed the most rapid rise of 8·4 per cent p.a. (e.g., the catches of anchovies rose from one to ten million tonnes between 1948 and 1966). The cod group of fish, which includes the haddock, pollock and hake of temperate waters, rose rapidly (3·7 million tonnes to 7·3 million tonnes) during the same period. So the great increase in world fish landings has been of cheap sources of protein rather than expensive delicacies. In part, though, this interpretation conceals the fact that a good proportion of the fish (an average of 12·75 per cent in recent years) are converted to meal and oil, much of which is fed to terrestrial animals, especially poultry and cattle. This aspect of food consumption is therefore underwritten by the protein yield of the oceans, and a significant part of the world catch does not help to nourish those who are said to be protein-deficient.

The ecology of fisheries must necessarily begin with the biological productivity of the seas. The whole question of the roles of various groups of organizations in the functioning of oceanic ecosystems is now being reconsidered (Pomeroy 1974) so estimates of biological productivity such as produced by the IBP are only for the visible elements of the food chains although there now seems to be a convergence of $50-55 \times 10^9$ t/yr dry matter as the total primary productivity of the oceans (Lieth 1975), distributed as shown in Table 3.2 (p. 134). Difficulties occur in translating such estimates into forecasts of sustainable yield from the oceans because of uncertainty about the length and complexity of marine food chains and the efficiency of energy transfer at each stage, which may in some cases be higher than expected (Gulland 1970; 1974) and because knowledge of the population biology of fish is still unreliable, especially with respect to normal cyclic variations in abundance: 'base-line' surveys taken from a single year, for example, are likely to be misleading (Longhurst et al. 1972). Further, at each stage of a food web there is competition for the production from animals which are not human resources.

Based on these types of data nevertheless, scientists such as Ryther (1969) and Ricker (1969) have made estimates of the ultimate sustainable yield of the sea, varying from 100–200 million t/yr. A recent statement by Gulland (1976) suggests that there is now general agreement that the total potential catch of fish of familiar kinds is at the lower end of this spectrum, i.e. about 100 million t/yr. The levelling-off of the yield curve in recent years suggests that any marginal increments will not be achieved easily: the rises in fossil-fuel prices of recent years were not of course foreseen by the predictors of large increases in catches. To achieve any sustainable increase will presumably mean not only more efficient fishing in technological terms, but also the rational management of known stocks and the exploitation of species currently outside the commercial nexus. There will be recourse to small crustaceans like the Antarctic krill (*Euphasia superba*) for which one estimate has even suggested that the potential yield is equal to the present total world fish catch (Kasahara 1975), and to various other unconventional species. These could include the pelagic phase of the California red crab (*Pleuroncodes planipes*), lantern fish (Mycophidae), deep-sea smelts (Bathylagidae) and open-sea squids. Some very peculiar looking fish indeed from over 1000 m depth are now about to find their way to the slab alongside the familiar flatfish and kippered herrings. The main group of organisms still restricted in their use are the cephalopods: octopus, squid and cuttlefish may yet be teamed with fried potatoes and wrapped with ecosystemic appropriateness in the *Sun*.

The attainment of the maximum sustained yield (which has usually been equated with the carrying capacity for the exploitation of this particular system) is dependent upon the rational management of the fisheries stocks and in this respect the record is not particularly hopeful. In 1883 T. H. Huxley averred that cod, herring and mackerel fisheries were inexhaustible, but since the appearance of the steam-powered trawler, the improvement of detection equipment and new types of nets, many individual populations of fish have been subjected to overfishing and have declined in abundance, in some cases catastrophically. Examples of fishery collapses include the Hokkaido herring in the 1890s, East Asian sardine (1946), Northwest Pacific salmon (1950), Atlanto-Scandinavian herring (1961) and Barents Sea cod (1962); a number are under strain including the flatfish of the Bering Sea, the yellow-fin tuna in the East Pacific and the marlin. In 1976 the North Sea herring seemed likely to last only a short while because of the unwillingness of the involved parties to accept catch limits although firm British action in 1977 eventually produced an EEC policy which banned herring fishing for one and possibly two years; a scramble to catch mackerel off Cornwall seemed also likely to result in overfishing since the small fish which should theoretically slip through the net are often fatally damaged by it (Allaby 1976). A determined fleet can rip its way through a stock in a short period, in the way that the Japanese have exhausted the tuna of the South Indian Ocean and West Pacific in the last two decades before transferring to the East Pacific and Atlantic oceans. The demand for some organisms has meant immense killing of associated or supposedly competitive species: 250 000–400 000 dolphins per year in the eastern tropical Pacific have been netted along with the yellow-fin tuna; grey seals are regularly culled on the Farne Islands of the North Sea because they are a predator of salmon. Even species which are caught by simple methods not involving deep-sea vessels and technology can be brought to low levels. Within the last few decades the mollusc *Haliotis* (abalone) was common in the intertidal zone of the Pacific Coast of the USA. Now it can be caught only by deep-sea divers well offshore. Marine turtles have suffered similar fates, either because of wide demands for their flesh or their shells, as with the green turtle (*Chelonia mydas*) and the Atlantic hawksbill (*Eretmochelys imbricata*). Protection programmes for these animals, as well as hatcheries, have been set up but enforcement of regulations is difficult (Parsons 1962; Bennett 1975).

The saddest depletion story of all is that of whaling. The industry is supposed to be regulated by the International Whaling Commission but the Commission is able to disregard the advice of its scientists because the IWC Convention provides that measures are to be taken for the benefit of the whaling industry (Myers 1975). Admittedly, knowledge of the social structures, population dynamics and ecology of whales is inadequate but by 1950 it was clear that most of the catch quotas were too

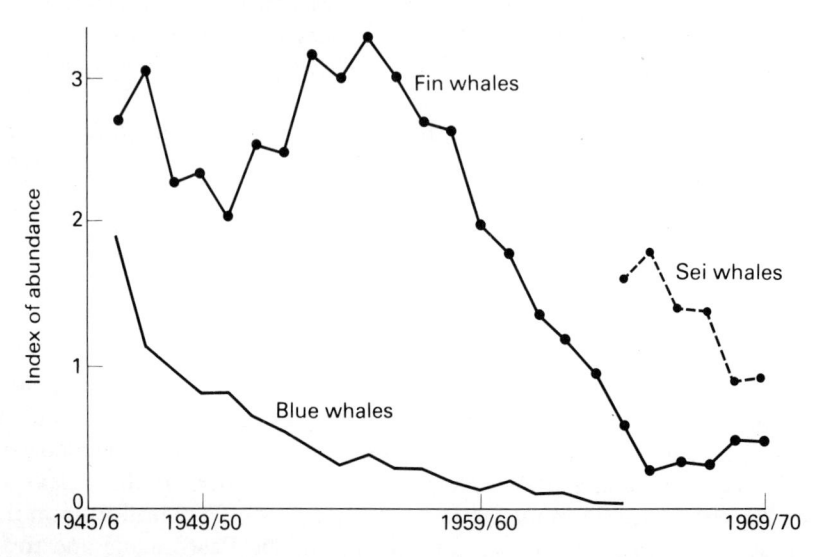

Fig. 10.10 The abundance of certain whales in the Antarctic. The index is estimated as the number caught per 10^3 catcher-tonne-days.
Source: Varley, 1974.

high (Fig. 10.10). The advice given in 1963 to reduce the catch of blue, fin and sei whales was ignored and all three fisheries collapsed. Because of slow rates of reproduction, whale populations are slow to recover: the Greenland right whale has been protected since 1935 but has still not recovered. As the 'great whales' have become depleted, so interest is shifting to the 'small whales': the pilot whale, bottle-nosed whale, killer whale and short-finned whale are now being exploited by Norway, the USSR and Japan (Burton 1975). It is argued that many whales are being caught at sustainable yield levels but even if this is true, the level must be far below what might have been possible had not a period of intensive culling occurred: the 1971 population of the blue whale in the Antarctic is believed to be 4 per cent of the population prior to whaling. The present situation is summarized by Gambell (1976):

> Many stocks have been so reduced by successive phases of the industry that there is only a small remnant of some of the major species left in the world's oceans. Modern whaling has the power to bring the remaining whale stocks to the point of economic if not biological extinction.... Yet this natural renewable resource could make an important contribution amounting to about 10 per cent of the total yield of marine products on a sustainable basis.

There have been suggestions too that the **maximum sustained yield** (MSY) concept is no longer a valid basis for fishery management (Talbot 1975). As far as the exploited population is concerned, MSY models do not make any allowance for impact of exploitation on behaviour and it assumes stable population levels affected only by human cropping. The ecosystem relations of the harvested species are also ignored: for example natural or induced changes in carrying capacity, responses within a trophic level or between trophic levels and impacts on symbiotic or commensal relationships. Harvest of one species to a theoretical MSY level may result in another species' occupation of its energy and nutrient pathways and continued harvesting at the theoretical MSY level will lead to rapid further depletion. Work upon the theory of population fluctuations (Beddington and May 1977) suggests as the catching effort increases, the predictability of the catch decreases, especially if there is over-exploitation. This is particularly so under a strategy which aims at a constant catch rather than a fixed level of effort. If fishing policy, too, is aimed at maximizing present level of net economic value, then populations will fluctuate more severely. In general, management policies like MSY which do not include sensitive feedback controls seem to be unsuited for biological populations in the oceans.

Another counterproductive influence upon fisheries is contamination of the oceans, especially enclosed seas and continental shelves near industrial countries (Ruivo 1972). Because of monitoring difficulties, the combined effects of industrial effluents, oil, sewage, fertilizer runoff and detergents are not known (H. A. Cole 1974). Radionuclides are more carefully checked and no effects upon the dynamics of fish populations have been reported. Some contaminants have been carefully followed through marine food chains: mercury is an outstanding example largely because of the poisoning of human consumers which occurred at Minamata in Japan; detrimental effects upon the fish and molluscs themselves are not reported. By analogy with fresh water, eutrophic 'blooms' might be expected in shallow seas receiving contaminant loads of nitrogen and phosphorus, and perhaps the explosions of dinoflagellates like *Gonyaulax* spp and *Gymnodinium brevis* known as 'red tides' may sometimes be attributed to such a cause, although they are also found where active unwelling of sea water takes place (Prakash 1974). The biological productivity of oligotrophic waters ought, if eutrophicated in a controlled fashion, to be increased: there is some evidence that the North Sea has benefited in this way, albeit by accident rather than design. Silt from rivers may also nourish productivity and so its loss (e.g. by damming the river) may cause the defection of the fishery. By contrast, a sudden influx of particulate matter may damage fisheries: the primary productivity of coral reefs which underlies their rich yield of fish can be wiped out by a heavy silt load resulting from e.g. mining operations on an island or the reclamation of mangroves (see p. 140).

One of the best documented histories of cultural stress upon a fishery is that of Lake Erie in North America which has endured 150 years of heavy fishing and contamination by wastes as well as the introduction of non-indigenous species, the alteration of shorelines by drainage and construction works, and the inflow of silt increased by the urbanization of the watershed, which now houses some 13 million people. A number of fish species were sequentially fished out: lake trout, lake sturgeon, lake whitefish, and lake herring were all exhausted between 1850 and 1950. In the mid 1950s emphasis shifted to walleye, blue pike and yellow perch, resulting in the depletion of the first two by 1960. Around 1931 the rainbow smelt invaded Lake Erie and predated upon the younger stage of many of the commercial fish and exacerbated the reduction in their stocks. In the meantime, cultural eutrophication from sewage caused algae blooms and deoxygenation during the summer, again reducing the survival chances of the fish as well as drastically curtailing the recreation value of the lake. Toxic pollutants from industry have also added to the disturbance of biotic patterns. The USA and Canada have jointly agreed a management programme to inject new life into Lake Erie. The diminution of pollution inputs and the restocking of the lake with Coho and Chinook salmon along with conservation of existing commercial stocks, are the first objectives, although implementation of the agreement of 1972 is slow because of the lack of cooperation by the fishing industry and a certain lagging in US interest and funding (Regier and Hartman 1973).

It can be rightly if reluctantly inferred that fishing is still at a Mesolithic level of man–nature relationship: wild populations are hunted. The Neolithic revolution in the shape of domestication or even in its probably precursor, the herding of semi-wild animals, is yet to come. There are signs that it is on the way in a limited sense for 'fish farming' (mariculture, aquaculture) is being practised on a small scale in many parts of the world and a large scale in a few, giving a yield from both fresh water and marine sources of about 5 million t/yr, although even in Japan, the entire freshwater catch only accounts for about 1 per cent of the national haul (Tamura 1970). Fish ponds, enclosures, beds and racks (for sessile organisms) and the partial control of open waters are all being used to manage the populations of amenable species such as carp, salmon, trout, the milk-fish *Chanos*, the African cichlid *Tilapia*, shrimp, oysters, mussels, the yellowtail fish (*Seriola quinqueradiata*), octopus, crabs and kelps. The abalones (*Haliotis* spp) and other molluscs such as scallops (*Agropecten* and *Patinopecten*) and squids are included, as are several genera of crabs. Flatfish possibilities run to five or six species, and there are a wide variety of possible fin fish, like the pompano (*Trachinotus carolinus*) and the grey mullet (*Mugil cephalus*). One problem with temperate-zone flatfish such as sole and turbot is that they are carnivorous and so consume more fish protein than they produce. They do, however, offer an opportunity to convert species such as sprats and sand-eels to more expensive foods for DCs (Windsor and Cooper 1977). The siganids or rabbit fishes are particularly promising for LDCs since they are herbivores with low costs of production (Webber and Riordan 1976). There is also a potential for the extension of farming edible seaweeds from East Asia to other parts of the world, especially the North Pacific (Hunter 1975). In general terms, the scope of fish culture seems to

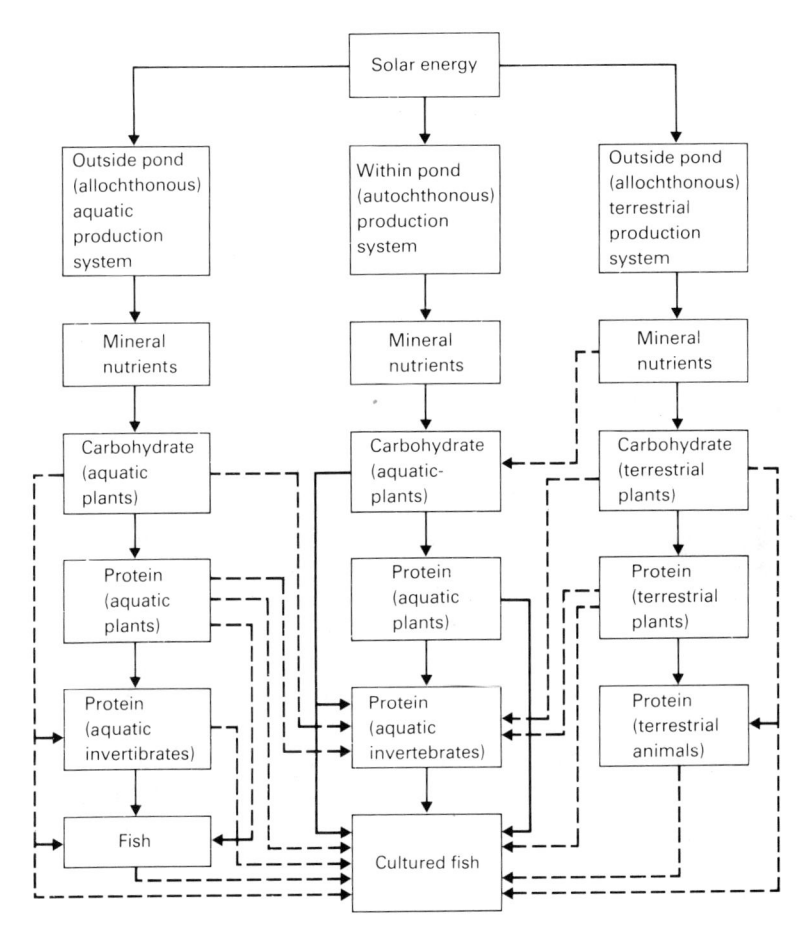

Fig. 10.11 The possible sources of food for cultured fish. The central column is the obvious line of production within a pond; the outer columns indicate potential food linkages from other aquatic systems (left-hand) and from terrestrial production systems (right-hand).
Source: Weatherley and Cogger, 1977.

be very wide since any unpolluted area of fresh, brackish or salt water is a potential space for fish rearing, and both 'waste' water and 'waste' organic matter could be used in a variety of trophic linkages both within aquatic systems and with terrestrial systems (Fig. 10.11).

As one step in the scientific development of fish culture, selective breeding is beginning to emerge since the physiology of spawning is well understood, the nutritive requirements of captive species are now known, and disease-control practice established. Breeding for selective characteristics has now been noted in carp, trout and bivalve molluscs, and the criteria by which successful candidates for aquaculture can be assessed are now more clear. Biologically these hinge around growth rate, feed-conversion efficiency and hardiness. Economic and cultural factors are also very important: eels grown in the waste water of a coal-fired power station in England fetched £2500–3000/t in 1976 (Royce 1972; Calaprice 1976; Purdom 1976; Rounsfell 1977; Weatherley and Cogger 1977).

TABLE 10.10 Estimated production of farmed fin fish, 1975, selected countries (tonnes)

China	2 200 000
Taiwan	81 236
India	490 000
USSR	210 000
Japan	147 291
Indonesia	139 840
Thailand	80 000
Bangladesh	76 458
Nigeria	75 000

Source: Windsor and Cooper 1977.

Mariculture for food is dominated regionally by eastern Asia (Table 10.10) and potential yields range from about 500 kg/ha/yr if herbivorous fish like the milkfish are used with no food added, to 5000 kg/ha/yr with fertilization by organic materials. Given completely optimal conditions with augmentation of water flow and food supply, 2 million kg/ha/yr are said to be possible (Royce 1972). Molluscs may produce 300 000 kg/ha/yr, and both these groups compare extremely well with protein yields from terrestrial systems. One of the most promising ways ahead seems to be a polyculture in which an enclosed area contains a relatively complex artificial ecosystem (Plate 38) containing herbivorous and carnivorous fish, bottom detritus feeders, and browsers, the whole being fed by wastes such as sewage and perhaps calefacted water (Noble 1975; Bardach 1976). For instance, polycultures of six species of carp in India gave yields of 8200 kg/ha/yr of fish provided supplemental feeding was undertaken (Chaudhuri et al. 1974).

Looking forward a few decades marine biologists have postulated additional methods by which open-sea mariculture could become viable commercially. The most promising species of oyster and mussel might well be harvested from structures closely akin to offshore drilling rigs or even from the rigs still engaged in oil production: the potential of the Gulf of Mexico and Californian coasts is already being assessed in this light. In addition, shellfish and even some species of turtle could be harvested from the lagoons

Plate 38 An experiment in fish-and-duck culture. These Peking ducks not only provide food and a marketable commodity but their excreta fertilize the water giving rise to a dense growth of plankton upon which carp feed.

of atoll-based Pacific Islands (Ribakoff *et al.* 1974). Money is probably better invested in mariculture than in building bigger trawlers, but even so the FAO view that even a five-fold increase in the output of mariculture by 1985 would be only marginal to the world situation has a gloomy air of reality. (If we assume a total world catch of *ca* 70×10^6 t/yr in the 1980s, then a probable output of 4×10^6 t/yr of farmed fish represents 7 per cent of world supplies but about 10 per cent of fish for direct human consumption.) The more important statistic is the sustainable yield of the conventional ocean fisheries and the contribution which they could make to the protein needs of the world's people, perhaps about 20 per cent on the basis of a catch of 100 million t/yr. This can only be realized by increased knowledge of the biology of the seas and more rational management of their biota: changes in the law of the sea need to be viewed in this light.

The grazing system

The husbandry of animals for their various products (Table 10.11) comes from agricultural systems as well as pastoral systems but in the latter there is a lack of conscious

TABLE 10.11   Grazing animals of the world and their products

	Species	Milk	Fibre	Meat	Fertilizer	Hide	Wool	Traction	Transport
Cattle	*Bos taurus*	+	+	+	+	+			
	Bos indicus	+	+	+	+	+		+	+
Sheep	*Ovis aries*	+	+	+	+	+	+		
Camel	*Camelus dromedarius*	+	+	+	+	+		+	+
Goat	*Capra hircus*	+	+	+	+	+			
Horse	*Equus caballus*	+	+	+		+		+	+
Ass	*Asinus spp.*	+		+		+		+	+
Mule	*E.asinus × E.caballus*				+			+	+
Buffalo	*Bubalus bubalis*	+		+		+		+	+
Yak	*Bos grunniens*	+	+	+		+			+
Llama	*Lama glama*		+	+					
Alpaca	*Lama pacos*		+	+					
Vicuña	*Lama vicugña*		+						
Eland	*Taurotragus oryx*			+					
Zebra	*Equus burchelli bohmi*			+					
Goose	*Anser cinereus*		+	+					
Kangaroo	*Megaleia rufa*			+		+			
Reindeer	*Rangifer tarandus*	+		+		+		+	+
Rabbit	*Oryctolagus cuniculus*			+		+			

Source: Spedding 1975b.

nutrient return to the soil except in very advanced versions in industrial countries. The animals of the grazing system are dominated, now as in the past (see p. 196), by a relatively small number of domesticates, of which the goat, sheep and cattle are the most important. These animals eat a variety of forage resources made up of either natural or near-natural vegetation beyond the margins of agriculture, in regions too high, hot, cold, dry or steep to grow crops. So grasslands of both long and short varieties, scrublands, steppes, semi-deserts, mountain vegetation, and forest understorey vegetation may all be cropped. Table 10.12 gives some examples of the NPP of vegetation types used for grazing systems.

TABLE 10.12   Examples of plant formations used in grazing ecosystems

	Latitudinal belt and plant formation	NPP (g/m²/yr)
BOREAL	Mountain meadows	975
SUB-BOREAL	Herbaceous prairie	1500
	Steppe	800
	Clay steppe	900
	Mountain meadow	1100
SUB-TROPICAL	Herbaceous prairie	1300
	Semi-desert	1000
TROPICAL	Forest and secondary savanna	1600
	Grass and shrub savanna	1200
	Desertlike savanna	400

Source: Simplified from Caldwell 1975.
Note: These data come from a source different from Table 2.4 but are quite comparable, often representing values for individual vegetation types whereas the mean of several is given in Table 22.4.

As in all ecosystems, the loss of energy at each transfer stage is large, and this is especially so with large mammal herbivores, where a considerable proportion of their food energy is 'lost' as faeces, urine, as inedible or unusable portions, in moving about, and in the gestation of young (Spedding 1975b). Table 10.13 shows the general loss of energy averaged for sheep and cattle on seasonally different parts of the range in Colorado, USA. So as in animal production in agro-ecosystems, the quantity of solar energy per unit area recovered as animal tissue for food is very small. At the same time, the influence of domesticated livestock on mineral nutrient flows may also be very small. One study in Colorado (Dean et al. 1975) showed that only 17·4 per cent of the nitrogen was removed as cattle-biomass from heavily-grazed areas. The cattle appear merely to divert above-ground plant biomass from the litter layer. On montane grasslands in North Wales, domestic sheep removed 14·5 per cent of the nitrogen and 37·5 per cent of the phosphorus in the above-ground vegetation (Perkins 1978).

TABLE 10.13   Energy budget of some range livestock

	Gross energy/acre (Mcal)	Gross energy consumed/acre (Mcal)	Gross energy voided/acre (Mcal) Faeces	Urine	Methane	Metabolizable energy/acre (Mcal)
SPRING (Seeded range) Sheep/Cattle average	1884	942	354	64	68	455
SUMMER mountain	2025	810	354	46	43	367
WINTER desert	655	328	154	29	16	129

Source: Cook 1970.

Occasionally, energy-budget studies yield surprises: again in the western-range ecosystems of the USA, Cook (1972) showed that one pound of animal biomass of jackrabbit had an energy input cost of 12 248 kcal or metabolizable energy (i.e. 1 kg had an input of 113×10^3 kJ), whereas the equivalent for sheep and cattle were 22 331 and 24 936 kcal respectively (i.e. 206 and 231×10^3 kJ). A wholesale shift in the American diet from prime-cut steak to rabbit pie seems unlikely, however. The converse seems to exist in the Scottish Highlands, where domestic sheep appear to need less food per unit of body weight than the wild red deer (Mitchell et al. 1977).

Some animals are normally efficient users of range resources (Plate 39) and even unimproved breeds are good meat producers, like some types of goat (Owen 1975), but even though the grazing system converts otherwise unused cellulose to meat and other animal products, the generally low energetic efficiency of the system has made it subject to calls for intensification, in a manner analogous to agriculture. The major methods of so doing in this case are to increase the forage supply, to remove inedible plants from the vegetation, and to augment the actual management of the animals themselves, including close control of genetic traits. A prime area for future treatment is the savanna zone of many LDCs where cattle productivity would be a useful part of the economy but

308 *Productive resource processes underlain by biota*

Plate 39 The wide range of food sources available to the goat is seen in this photograph from Upper Volta where even thorn trees are edible.

where calving and growth rates are low and pasture management is poor. One research organization in South America is experimenting with the plant genus *Stylosanthes*, which is leguminous (fixing nitrogen), tolerant of acid soils with a high aluminium content, can withstand long periods of drought, burning, trampling and yet be a good forage plant for cattle (Harrison 1977). Recent work on the North American grassland has suggested that the efficiency of energy capture could be tripled by the addition of nitrogen and the maintenance of an adequate water supply (Sims and Singh 1978).

The alteration of ecosystem components such as the plants, soils and consumer organisms offers most hope for the resource manager wishing to increase his output and at the same time consider the long-term biological productivity of the range. The decision of what to manipulate and the likely interaction of changes is made more rational by the use of computer simulations of the type described by Patten (1972). The management of the crop herbivores themselves is an obvious first stage: thus the numbers of animals, their species and strain, their proportions if pastured in mixed herds, and their distribution in space and time, all contribute to influencing the species of plants which are grazed, the time in their life cycle at which they are cropped, and the frequency with which they are eaten. The manager may control the distribution of the animals by providing water points, salt licks or specially fertilized patches of grassland, and above all he may be able to fence or herd them. Such measures are designed to spread the animals to the optimum density for utilizing the forage, since many plants are easily damaged by being grazed too soon after they have started their growth. Thus grazing systems must be elaborated, either by folk custom or by complex socio-legal arrangements. If a system other than continuous grazing is employed, then additional water developments are usually necessary, as is the fencing of the range into units with more or less equal stocking potential.

Predation by carnivorous animals and parasitism are also fought by the range manager with a wide variety of methods; he may also wish to eliminate other herbivores which he perceives as competitors with his flock: wild herbivores often come into this category although research may indicate that there are wide differences in diet selection. Nevertheless species of deer in the Great Plains of North America and the kangaroo in Australia have been hunted because of their alleged abduction of the forage of beef cattle.

The bringing about of changes in vegetation is another method of manipulating grazing systems. The aims are to produce a high proportion of palatable plants, preferably with a spread of maximum growth periods so as to provide a long grazing season, and to reduce the proportions of open ground, unpalatable shrubs, and poisonous or injurious plants (Vallentine 1971, Burton 1973). In nations like the USA, with a heavy and diverse battery of technological weapons, the conversion of sage-brush, juniper scrub and mesquite woodland can be attempted by chaining and cabling, with hormone herbicides, burning and seeding. Box (1974) thinks that the forage supply could be doubled in the southwest USA by such means although Vale (1974) notes that large-scale clearance reduces the populations of sage grouse (*Centrocerus urophasianus*), pronghorn antelope (*Antilocapra americana*) and mule deer (*Odocoileus hemionus*) which reduces the sporting value of the land. Some sage-bushes (e.g. *Artemisia tridentata*) may also provide forage for sheep during periods of heavy snow cover.

The soil itself is more difficult to alter than animals or plants but is not beyond treatment: improvement of the nitrogen status of the soil is one of the most usual desirabilities. Crops and tame pastures of the world contain about 200 leguminous species, whereas wild range and forest ecosystems have *ca* 12 000: thus the encouragement of leguminous species helps to counteract the loss of soil nitrogen; the secondary productivity of California range can be raised from 20 kcal/m^2/yr to 800 kcal/m^2/yr by the use of legumes together with fertilizer. Furthermore, some soils appear to be deficient in mineral elements such as copper, cobalt and molybdenum, which are essential for the nutrition of animals, so that fertilization of grassland may be necessary. In contrast, one edaphic element which cannot easily be directly manipulated is the soil flora and fauna. These biota are especially important in the detritus food chain which is critical to nutrient circulation and most other management techniques will affect them, some adversely.

The search for improved forage may lead to the import of exotic grassland species. The pastures to which the forests of South America are rapidly being converted are dominated by six African grasses, one of which (Guinea grass, *Panicum maximum*) was taken to the West Indies in the seventeenth century, and the latest (Pangola grass, *Digitaria decumbens*) some time after 1935. Their importance is highlighted by Parsons (1972) when he says that these six species are 'at the base of the new hope for the development of a viable commercial livestock industry in the low latitudes of the New World tropics'.

In the absence of deliberate management of pastures, changes in vegetation and subsequently in other ecosystem components are likely. The plants which prove most palatable to the stock are grazed frequently, and those weakened by cropping are replaced by resistant species. These are taxa capable of rapid immigration into areas of lessened competition and which escape being eaten because of low stature, a short season of growth, low palatability, poisonous properties, or spines (Moore and Buddiscombe 1964). Concomitant changes involve the reduction of the mulch cover of the soil. The microclimate then becomes drier and more severe, and so many invaders are plants of xeric habit; in semi-arid areas they are frequently plants of the desert biome such as succulents and thorny shrubs. The absence of humus cover may mean that the mineral soil surface is heavily trampled when wet, producing puddling of the surface layers, which in turn reduces the infiltration of water into the soil and accelerates its runoff, causing drought. These changes all contribute to the reduction of the rate of energy flow, and the disruption of the stratification and periodicity of the primary producers results in a breakdown of the biogeochemical cycles, especially those involving water, carbon and nitrogen. Where such 'overgrazing' is persistent, water and wind erosion may be the signs of total system breakdown. Recovery is invariably lengthy since a long fallow period has to be arranged during which succession from bare ground can take place. Technically complex erosion-control procedures are expensive and their application tends to be limited relative to the area of erosion.

The advent of domestic flocks to extensive pastures formerly grazed only by wild animals may alter the vegetation completely. The *Stipa* grasslands of North Africa may well be the result of centuries of selective grazing under selective pastoralism, and in New Zealand the aboriginal vegetation of the plains of South Island was tussock grassland dominated by species of *Poa* and *Festuca*. The introduction by Europeans of grazing animals such as sheep, rabbits, red deer and goats, pig and hare, and the use of fire as a management technique, has resulted in a more or less complete replacement of the native flora, and the forage is now mostly introduced grasses such as *Anthoxanthum odoratum*, *Festuca rubra* and *Holcus lanatus*. Introduced animals have also destroyed the natural stability of their new habitats, especially forests and alpine grasslands, apparently accelerating soil erosion (Howard 1964). As another instance we may quote the deforestation of Iceland in the course of sheep grazing, which has meant the denudation of 30–40 per cent of the soils; the annual loss of soil and vegetation cover is still greater than that regained through plant recolonization and management efforts (Thorsteinsson et al. 1971).

Intensive grazing which results in more areas of bare soil creates a new habitat in which burrowing animals may flourish. In North America, mice, jackrabbits, gophers and prairie dogs probably benefit from heavy grazing; predator control applied to wolves or coyotes may also have removed another element of natural control and so the levels of the rodents may reach epidemic proportions. In their increased burrowing they render sterile thousands of hectares of forage land, and control measures aimed only at killing them individually are clearly ecologically unsound. Although there is no direct evidence, we may also speculate about the connection between intensive grazing in semi-arid areas and locust plagues. The presence of loose sandy soil is necessary for oviposition in locusts and heavy grazing helps to produce such conditions.

The unstable nature of some contemporary pastoralism can be seen from recent events in the Sahel zone of Africa. Failure of the short rains in the autumns of 1973 and 1974 resulted in immense stock losses: estimates suggested that northern Niger suffered animal mortalities of 100 per cent of cattle, 70 per cent of goats and 60 per cent of camels in 1974. The problems were not helped by well-digging in previous years since uncontrolled use had allowed overgrazing and puddling near the wells. The loss of stock was also made greater when herdsmen took their cattle south to seek water, thus bringing them into contact with diseases to which the northern beasts were not immune (Omerod 1976). Concentration of nomadic pastoralists in areas with water or where emergency relief food is available during periods of crises accelerates their sedentarization. Such permanent settlement may bring increased rates of population growth and the pressure for the possession of more animals is then reflected in demands for new wells and better veterinary medicine so that larger herds live in a state of near-starvation, as has been forecast for the Twareg of Mali. If, alternatively, pastoralists are settled on the margins of their lands, whole areas may produce no meat and so large quantities are then imported, as in Saudi Arabia (Darling and Farvar 1972; Heady 1972; Talbot 1972; Swift 1975).

The whole complex of man-made changes in the vegetation of areas marginal to the great deserts of the world which results in their extension is called desertification. This is usually held to be the result of over-grazing by pastoralists or by settled nomads but can also be attributed to agricultural use of some arid and semi-arid areas like the Mediterranean fringe of the Sahara. Desertification (in which there may be a climatic causal factor as well) appears in the form of dune encroachment, sand drifting and gulley erosion. Most of these have their inception in a loss of vegetation cover together with physical processes such as crusting or sealing of the soil surface or its break-up for cultivation. Many features of desert soils (e.g. lack of a litter horizon, lime concentration near the surface, salinization, concentration of nutrients near the surface) render them easily prone to nutrient loss and physical degradation (Mabbutt 1977).

All these problems apart, extensive grazing can still play an important role in providing animal products from ecosystems which man is unlikely otherwise to be able to crop, and one which in general is capable of having its carrying capacity determined with some precision. Additionally, harvesting costs are low even in difficult terrain, the selective grazing habits of the animals allow the use of swards of mixed quality, and no more herbate is harvested than is required at any one time. The disposal of excreta costs little and may improve plant production. A traditional nomad culture exerts a light, flexible pressure on the plant cover which does it no permanent damage (Toupet 1975, Spedding 1975b). But the balance between the needs of man and animals, and the available resources of water and pasture is precious.

Biota and the quality of air and water resources

These two resources share the characteristic that, as far as consumption of them is concerned, the absence of biological material (as well as intrusive inorganic substances) is regarded as an indicator of high quality. One of the main forms of water contamination is eutrophication (p. 145); sewage also introduces bacteria and viruses into water. The general aim of management of water for consumption by people and domestic animals, and often for industrial purposes as well, is to reduce the biomass within the water, either by preventing contamination or by treatment. The commonest measure of contamination by unwanted biota is Biochemical Oxygen Demand (BOD) a parameter in which the amount of dissolved oxygen needed for oxidation of intrusive material by aerobic biochemical action is measured over a 5-day period, the results being expressed in mg/l. Thus water standards are often expressed in terms of a maximum BOD: fish require at least 5 mg/l of dissolved oxygen for survival and the normal saturation level of fresh water for O_2 is 9·2 mg/l at 20°C (and 6·6 mg/l at 40°C, so that calefaction will obviously increase the BOD level). Treatment of waste water with a high BOD is not usually economic once it gets into watercourses such as lakes, rivers and estuaries: only dilution will then reduce it. This anathema to living things does not extend to water used for other purposes; for boating, nature conservation, and aesthetic pleasure (though less so for fishing), the fringing vegetation of water bodies is a desirable adjunct and the normal biological processes ensure that the sudden release of large quantities of nutrients from the breakdown of organic matter are very rare (Hynes 1970).

In the case of the air, contamination is not usually with living materials except when people are sufficiently confined for the aerosol transmission of micro-organisms, normally from one respiratory system to another. As with water, desirable standards of air-contamination levels (e.g. for SO_2, CO, particulates, nitrogen oxides, hydrocarbons and lead), are being set in many industrial nations. Plants can be used positively in some places to ameliorate the effects of atmospheric contamination since certain species can exert a filtering effect while not themselves suffering greatly. This effect can be more or less purely physical: open areas in a city will, if tree-clad, reduce windspeed through them and cause particulate matter to settle out. Some trees are more effective than others in removing particulates from the air: effective trees include maple, lime, poplar and lilac. Other trees noted for their resistance to air pollution include birch, catalpa, elms, gingko, hawthorns, London plane, magnolia, tree of heaven and the tulip tree. Additionally there is evidence that tree vegetation can reduce the concentration of some gaseous contaminants.

In suburban areas, a dense screen of trees and shrubs between the road and the houses will reduce carbon monoxide levels as much as 54 per cent in the dwellings. Lead concentrations fall off rapidly away from highways in any case, but a screen of trees and shrubs will accelerate this distance-decay factor by absorbing some of the lead via leaves and roots. Thus the separation of busy highways from residential areas by earth banks with shrub cover will help to insulate people from both noise and contaminants which above certain concentrations would be a danger to health.

Further Reading

BARDACH, J. E. 1976: Aquaculture revisited. *J. Fisheries Research Board of Canada* **33**, 880–87.
BURINGH, P., VAN HEEMST, H. D. J., and STARING, G. J. 1975: *Computation of the absolute food production of the world*. Agricultural University Wageningen: Dept. of Tropical Soil Science.
DAVIDSON, J. and LLOYD, R. 1977: *Conservation and agriculture*. Chichester: Wiley.
EARL, D. E. 1975: *Forest energy and economic development*. Oxford: Clarendon Press.
GAMBELL, R. 1976: Population biology and the management of whales. *Applied Biology* **1**, 247–343.
GRIBBIN, J. (ed.) 1978: *Climatic change*. London: Cambridge University Press.
GULLAND, J. A. (ed.) 1972: *The fish resources of the ocean*. London: Fishing News (Books) Ltd.
— 1976: Production and catches of fish in the sea. In D. H. Cushing and J. J. Walsh (eds.), *The ecology of the seas*. Oxford: Blackwell, 283–316.
KINNE, O. and ROSENTHAL, H. 1977: Commercial cultivation (aquaculture). In O. Kinne (ed.), *Marine ecology* vol. III, part 3, 1321–98. Chichester, New York etc.: Wiley.
LEACH, G. 1976: *Energy and food production*. London: IPC Press.
LEWIS, G. M. 1969: Range management viewed in the ecosystem framework. In G. M. Van Dyne (ed.), *The ecosystem concept in resource management*. London and New York: Academic Press, 97–187.
LIETH, H. 1975: Primary productivity of the major vegetation units of the world. In H. Lieth and R. H. Whittaker (eds.), *Primary productivity of the biosphere*. Ecological Studies 14, 203–15. Berlin, Heidelberg and New York: Springer-Verlag.
MAKHIJANI, A. and POOLE, A. 1976: *Energy and agriculture in the third world*. Cambridge, Mass.: Ballinger Publishing Co.
MONOD, T. (ed.) 1975: *Pastoralism in tropical Africa*. London, Ibadan and Nairobi: OUP, for the International African Institute.
MURDOCH, W. W. (ed.) 1975a: *Environment, resources, pollution and society*, 2nd edn. Sunderland, Mass.: Sinauer, ch. 15.
REID, R. L. (ed.) 1975: *Proceedings of the III World Conference on animal production*. Sydney: Sydney University Press.
RUIVO, M. (ed.) 1972: *Marine pollution and sea life*. London: Fishing News (Books) Ltd.
SLESSER, M. 1975: Energy requirements of agriculture. In J. Lenihan and W. W. Fletcher (eds.), *Food, agriculture and the environment*. Glasgow: Blackie, 1–20.
SPEDDING, C. R. W. 1975a: *The biology of agricultural systems*. London: Academic Press.
— 1976: The biology of agriculture. *Biologist* **23**, 72–80.

11
Protective resources processes underlain by biota

The work of E. P. Odum referred to earlier in this book (p. 254) has given a conceptual basis to the idea that protected ecosystems in a natural or near-natural condition are as important a part of man's resources as the productive systems discussed in the previous section. The detailed reasons for protecting various systems from man-induced change are discussed below: here we will recall the argument that ecological stability on a global scale seems to be more assured if there is at any one time a mosaic of successional stages and mature communities, and it is usually these latter which are given protection. So we have landscapes and ecosystems which have been given a definite legal status and called by a special name: National Parks, Wilderness Areas, Protected Landscapes, Nature Reserves, Game Parks are all part of this complex. Many of them are relict in character since a great deal of their surroundings will have been converted to productive or inert systems: this is particularly so in densely populated areas. The formally protected lands are complemented by residual areas of 'unused' land where little or no manipulation of the ecosystems has taken place; such areas are often very harsh terrain where productive resource processes are discouraged by the natural conditions: Greenland, parts of the Sahara, and the Eurasian-North American tundra are examples. The present network of designated parks and reserves (excluding Greenland and Antarctica) covers about $1 \times 10^6 \text{ km}^2$ or 1·1 per cent of the Earth's land area. But they tend to concentrate on the spectacular and unique rather than on the representative and of the world's 193 terrestrial biotic provinces delineated by Udvardy (1975), more than one quarter contain no protected areas and a further 15 per cent only one (Myers 1976).

Nature protection

The role of biological organisms in the many types of protected landscapes and ecosystems is nearly always important. (There are protected areas of primarily physiographic interest such as geological exposures or coastal landforms but few of these will be biologically inert and it is common for interesting plants and animals to be present as well.) At one end of a continuous spectrum are reserves for the perpetuation of a single species of plant or animal. Such taxa are generally rare (and indeed may be threatened with extinction) or possess some intangible value, for example as a national emblem or symbol. Examples are the reserves to protect the last few Javan rhino; the sanctuary in Hokkaido for the *tsuru*, a sacred Japanese crane; the refuge of the California Condor (Plate 40), whose members are probably down to forty individuals (Miller *et al.* 1965); and the

Plate 40 A pair of the rare California Condor.

reserves protecting individuals stands of rare orchids on the chalklands of southern England. In the case of rarities, a sectoral approach may also be applied, prohibiting the taking of certain species whether they are found in a reserve or not: thus the Golden Eagle in Great Britain is in theory totally protected, as is the Bald Eagle in its North American habitats. Similarly in Britain many species of wild flowers are protected by legislation no matter where they grow.

A more complex category is that of the protection of habitats together with their characteristic assemblages of plants and animals. These are often larger in area than the previous category and may have some linked affinities: wildfowl refuges which supply food and shelter for migrating birds are one example of the general category of wetland whose value has been strongly emphasized in recent years (Bellamy and

Pritchard 1973). In northern Britain, an assemblage of arctic-alpine plants near the southern end of their ranges is protected in Upper Teesdale. These plants are common in Scandinavia and occur relatively frequently in the Scottish Highlands but are rare enough in England to merit preservation. Efforts are often made in densely populated nations to protect woodland, especially if this is thought to be virgin or nearly so. The relic pinewoods of Scotland, the mixed oak forest of Bialowieza in Poland and the beech spruce *urwald* of Boubinsky Prales in Czechoslovakia are instances of this practice. Even in England, where there is no virgin forest, the Nature Conservancy Council identifies 'primary woodland' (Plate 41)—i.e. stands which have been deciduous forest as long as historical evidence persists (Peterken 1974). Protected habitats may also include underwater areas where protected status has been applied to assemblages of marine plants and animals, in spite of opposition from fishermen and shell-collectors, for example. Coral reefs are often the nuclei of such marine parks as in the Caribbean, Hawaii, and Australia (Schultz 1967, Greden 1975).

Plate 41 Tourism and wild animals overlap in this specially constructed hide at a water-hole in the forest savanna of southern Africa at Mkuse in Zululand.

The size and configuration of such reserves is often determined as much by politics as biology, but certain groups may require minimum areas if they are to survive. Food sources are an obvious aspect of this, but a spatially isolated population is also genetically isolated and if rates of dispersion across unprotected terrain are low, then it may be essential to keep corridors of protected land connecting the main reserves in order not to interrupt the free exchange of populations. This has been suggested as especially necessary for tropical birds (Terborgh 1975). From such work reserves and reserve systems can be designed whose shape and spatial relations maximize the probability of survival of a threatened species. One large reserve is usually better than several smaller reserves, for example because it can hold more species at equilibrium diversity. A reserve should also be as nearly circular as possible to minimize dispersal distances within the reserve and to avoid 'peninsula effects' where local extinctions within the reserve are not replaced from a more central pool. These guidelines are subject to qualifications according to local circumstances, but represent a movement towards a theoretical base for nature-reserve needs (Diamond and May 1976).

The last category is the set of reserves that are large enough and sufficiently diverse to protect whole sets of systems and landscapes. If their primary purpose is to protect ecosystems then they are likely to be largely natural; if it is to protect landscapes then a considerable degree of human manipulation may be tolerated or even welcomed. But in densely populated regions any wild areas are valued because they provide open space and some wildlife. Nations with a lot of natural and near-natural terrain can set aside large zones (usually called National Parks) for the conservation of wild biota and their attendant ecosystems: the parks of eastern and central Africa which are intended to protect the highly diverse savanna fauna are perhaps the best known (Myers 1972a, 1972b). Also well loved is the spectacular scenery of some of the National Parks of western North America such as Yosemite, Crater Lake and Banff–Jasper. Even arid areas have their attractiveness, and Australia for example has an Ayers Rock–Mount Olga National Heritage Area which is sufficiently heavily used to require a careful management plan (Lacey and Sallaway 1975). Most industrial nations have National Parks consisting of cultural landscapes to protect their most valued scenery from over-development by industrial, agricultural or recreational users, even if the flora and fauna have been depleted by centuries of human occupation: examples are Italy, Japan and India.

At the far end of the protection spectrum from the small nature reserve is the wilderness. *De facto* wildernesses occur in the very sparsely inhabited zones of the earth but *de jure* examples are less frequent. Within a developed nation, the National Wilderness Preservation system of the USA is the outstanding example (Simmons 1966). Here 'natural areas' within National Parks and National Forests are set aside from all economic use and intrusion and allowed only a low level of recreational use. Similar 'wilderness zones' are designated in other National Parks, such as those of western and northern Canada, and the High Tatra Mountains in Czechoslovakia-Poland. We should note in passing that, as Vale (1977) shows for the Warner Mountains Wilderness of California, fire protection policies and a previous history of grazing make the idea of 'natural wilderness' conditions difficult to support in most parts of the Western world. Perhaps the most outstanding instance of a wilderness is Antarctica, where the ecosystems and the

318 *Protective resources processes underlain by biota*

inert terrain are both protected by the Agreed Measures appended to the Antarctic Treaty of 1959. Although the productive ecosystems of the offshore waters are excluded from the Measures, considerable protection is afforded to the native fauna and flora and steps are taken to prevent the introduction of alien species which might 'explode' in the biologically impoverished ecosystems of the fringes of the continent.

Outdoor recreation

The designation and use of land for rural outdoor recreation both cuts across and adds to the types of protected area discussed above. In the first place, many of the protected areas also house recreationists. Some may come to see the flora, fauna and physiographic features which are the main purpose of the reserve (Plate 42), but others will want to engage in one or more of the multitude of outdoor activities which are now popular; yet more will simply want to be in the open air, preferably in the company of an internal-combustion engine. Differences exist in the ability of the nature-protection areas to cope with recreationists. At one end of the scale, the small nature reserve is likely to be entirely devastated by more than a few visitors per year, especially if the ecosystems are fragile, though in a limited number of places devices such as walkways can allow

Plate 42 Erosion from recreational use can be clearly seen on the chalklands of southern England: Coombe Hill in the Chilterns.

visitors to pass over for example a mire without damaging it. At the other extreme, the wilderness is also likely to be popular with 'back-to-nature' recreationists but at the same time to consist of fragile environments: in mountains for example the growing season of the flora is very short and plants if grazed by pack animals or heavily trampled will soon be eradicated, leading to floral impoverishment and perhaps erosion (Harvey, Hartesveldt and Stanley 1972). So the management of recreation in wilderness areas tends to aim at a low density of people: the users themselves like to feel isolated and do not care to see a great number of other people. This also facilitates 'biocentric' wilderness management in which human influence is not allowed to alter significantly the natural energy flows of the ecosystems (Houson 1971; Hendee and Stankey 1973). In the case of Antarctica, visitors arrive by boat and make day-trips to the coast but do not remain there.

Some outdoor recreation does not conflict with nature protection. Even hunting and fowling may not be incompatible with the perpetuation of game species, if careful management of populations of both animals and men can be achieved. This is often the case in Europe where for some long time now, game bags have been taken without rendering extinct or even rare any of the principal hunted species (Table 11.1). However, popular areas such as National Parks experience a constant tension between those who would keep them inviolate against most if not all forms of human intrusion, and those who wish to see them available 'for the people' (Simmons 1974a; 1976). Most protected areas which have to cope with a heavy visitor pressure, receive much of it from people attracted by features other than the primary values and so interpretative services have the task of trying to instil some knowledge of the protective function (be it of wildlife or scenery) of the area visited (Burrell 1973).

TABLE 11.1 Annual game bags in some European countries

Country	Year	Pheasant	Partridge	Pigeons	Ducks and Geese	Grouse
Britain	Annual	5 500 000	750 000	4 000 000	500 000	2 000 000
West Germany	1971–72	1 387 500	381 000	473 697	393 350	—
France	pre-1939	500 000	4 000 000	—	—	—
Austria	1971	564 991	130 980	37 452	48 464	—
Switzerland	1971	6 561	1 270	16 253	19 421	—
Denmark	up to 1962	500 000	ca 300 000	ca 350 000	ca 500 000	—
Poland	1958–59	7 714	325 850	—	—	—
Jugoslavia	1967–68	294 000	294 000	—	—	—
Czechoslovakia	Av.	300 000	400 000	—	35 000	—

Country	Year	Roe	Red deer	Boar	Rabbits	Hares
West Germany	1971–72	542 000	25 960	36 385	926 000	1 330 000
East Germany	1971	106 000	9 627	34 000	—	118 600
France	pre-1939	10 000	3 000	20 000	10 000 000	1 000 000
Austria	1971	154 563	32 349	2 670	12 607	403 187
Switzerland	1971	28 832	1 300	100	369	21 367
Denmark	up to 1962	ca 28 000	—	—	—	max. 475 000
Poland	1958–59	11 693	6 593	21 228	—	536 084
Jugoslavia	1967–68	20 600	4 042	4 000	—	638 000
USSR	Annual	500 000–600 000 wild ungulates				
Czechoslovakia		60 000	7 000	—	—	600 000

Source: Bigalke 1975.

Outdoor recreation also adds to the total of protected areas in rural areas since many tracts of terrain may be attractive to recreation and protected against most other productive uses which are deemed incompatible with recreation. Thus another category of protected land is created and, like nature reserves, the areas concerned vary in size. In Britain, small areas called Country Parks are found close to the fringes of some cities; in the Netherlands special bodies manage individual recreation areas like large lakes; to the north of Toronto, a special body manages lands jointly for flood control and urban-fringe outdoor recreation; in the USA the large dams in the Western Cordillera are often the focus for National Recreation Areas which surround the water resource.

It may also be possible to schedule areas for nature conservation and outdoor recreation within the overall context of productive ecosystems of a low intensity such as grazing and water conservation. Careful management of scrub in the Mediterranean lands, for example, might result in multi-layered recreation and fodder forests where both types of use would benefit from the reafforestation of denuded slopes (Naveh 1974).

Management

A term which has been used in the above discussion without definition or explanation is 'management'. It might be thought that the essence of a protected ecosystem or landscape was a 'hands-off' attitude by the managers. Reflection brings the realization that the purposes of protection are only likely to be achieved in such a *laissez-faire* manner where the areas concerned are very large and ecologically self-contained: i.e. the wildernesses and large national parks. Elsewhere the perimeters of the designated area are unlikely to be the boundaries of the ecosystems containing the protected wildlife and so all kinds of human influences will be exerted.

A simple instance of the problem is that of wanton destruction: the poaching of protected animals for eggs, hides, meat or ivory, for example, the cutting of protected forests for fuel or construction timber and the uprooting of rare wild flowers to grace the collector's garden or herbarium, are all continuing problems and can be countered immediately only by wardening or other policing services. Visitors in large numbers bring other ecological problems even if they are not malevolently motivated. The most destructive of these is probably fire, which is often accidental in cause but which can destroy great areas of forest or other wild vegetation. In general it is a worse problem where vegetation is all of one type, as in coniferous forest or on the heather moors of the British uplands; more diverse communities tend to have their own natural firebreaks. Protection against fire may bring its own ecological shifts as discussed above (p. 291).

Another visitor-caused difficulty is trampling: concentrations of people lead to the eradication of tender plants and their replacement with tougher species, often generally normally thought to be 'weeds' and where pressure is intense nothing will grow. Thus paths and tracks and zones near car parks become bare and so are susceptible to erosion: 'tourist gulleying' is a familiar sight in many National Parks the world over (Plate 43). In forest campsites, the incidence of human feet compacts the soil and prevents litter breakdown and natural regeneration so that 'a period of rest' is necessary, with mulching and scarifying the soil essential to revive the soil flora and fauna. Large numbers of

Plate 43 A piece of 'primary woodland' in England: though not natural, woodland has been on this site as long as records permit investigation. Staverton Thicks in Suffolk.

people also cause shifts in animal communities. Some species are intolerant of human presence and so become scarcer, whereas others start to scavenge on recreationists' leavings and may become at least seasonally dependent upon them, gathering round picnic places like children at an ice-cream van. The brown bear, blue jay, yellow hammer, chipmunk, bighorn sheep, white-tail deer and domesticated hill sheep have all been observed filling this niche. Dangers to visitors are often present and may not be correctly perceived by them, as shown by Bryan and Jansson (1973) in their study of hazards from wildlife in the national parks of the Rocky Mountains in Alberta. The brown bear of North America, for example, is unpredictably bad-tempered and visitor-related sources of food brought injuries to people from grizzly bears (*Ursus arctos horribilis*) to the level of 4·4/year during 1963–69 in Yellowstone National Park (G. F. Cole 1974); even small and attractive mammals like chipmunks and ground squirrels can be dangerous because they may carry rabies. (A great deal of work has now been done on the ecological impact of outdoor recreation in Western countries; see, for example, Goldsmith, Munton and Warren 1970; Stankey and Lime 1973; La Page 1974; Mattyasovsky 1974; Rudolf 1974.) Natural variations in environmental conditions which also threaten species may also be ameliorated by management: the dry summer of 1976 in Britain desiccated the last remaining waterlogged peat-holes which are the habitat of the giant raft spider (*Dolomedes plantarius*) but deeper peat cuttings were made with the support of the World Wildlife Fund, so that enough water remained for the spiders to live in.

The population dynamics of protected species may be affected by processes outside the reserves. For instance, a reserve may be only part of a drainage basin so that events elsewhere in the catchment may affect it. At a simple level, a protected wetland may receive drainage from agricultural land: it may thus be subject to gradual eutrophication (with subsequent floral and faunal changes) from fertilizer runoff, or it may be suddenly devastated by an influx of pesticide from the surrounding lands. The last riverside stands of California Coast Redwood (*Sequoia sempervirens*) which contain the tallest trees in the world are dependent for their growth on silt inputs from flooding. Yet there are powerful pressures for upstream dams to control flooding outside the Redwood National Park and to supply water to urban-industrial California. Deprived of silt, the riverside trees of the next generation would no longer grow to such impressive proportions (Simmons and Vale 1975; Plate 44). Mobility of a protected species may also cause trouble: a migratory animal may spend part of its yearly cycle in a protected area and so perhaps flourish greatly and reproduce well, so that a large number of beasts come out of the park and compete with domestic animals or depredate upon crops. This syndrome is especially bad when the upper altitudinal levels of a region are protected but not the lower so that animals like deer are forced to spend the winter season in the valleys where there may be productive ecosystems. Not only are they unpopular with farmers and graziers but there may be insufficient feed for them and hence a high mortality rate. Such difficulties are encountered by the elk of Yellowstone National Park, and by the deer of some of the Nature Reserves of the Scottish Highlands.

Economic and recreational development of all kinds within or near to protected areas brings problems in the management of flora and fauna. Roads are sources of mortality to animals and disturb the ecology of rivers during their construction; there may be a temptation to re-seed bare roadsides, cuttings and embankments with fast-growing mixtures that contain species alien to the region (as was once done in the tundra area of Mount McKinley NP in Alaska) with the danger that one or more exotic species may become explosively invasive. As the numbers of people rise, manipulation of the environment takes on a quasi-urban character; paths are asphalted or gravelled, more areas are rendered biologically inert under roads, car parks and trampled soil; refuse dumps and sewage works become essential, and time and effort must be spent in catching scavenging bears and transporting them to the backwoods to get back to the berry-and-root habit. Small wonder that some managers of protected areas wish to acknowledge an upper limit of visitors with which they can cope and to try to 'de-develop' the areas under their control (Wagar 1974).

Plate 44 California Coast Redwoods: riverside groves in the Humboldt Redwoods.
'Among the scenes which are deeply impressed on my mind, none exceed in sublimity the primeval forests undefaced by the hand of man;... no one can stand in these solitudes unmoved, and not feel that there is more in man than the breath of his body.' (Charles Darwin, *The Voyage of the Beagle*.)

Not all the management problems come from direct human interference: natural succession will be taking place in the protected area and this may mean the replacement of one species by another by normal processes. If, however, a species of an early successional stage is valued for example because of its rarity or beauty then managers must try and manipulate the habitat so as to maintain the early-stage habitat: fire, flooding, deforestation, and sprays may all be considered legitimate techniques of manipulation (Stone 1965). Examples include the use of mowing machines or sheep to maintain the short-turfed grassland which is the habitat of several species of orchids on the chalklands of southern England; and the use of pumping to maintain a wet fenland in eastern England which supports a population of the large copper butterfly, *Lycaena dispar batava* (Duffey 1974; Fig. 11.1).

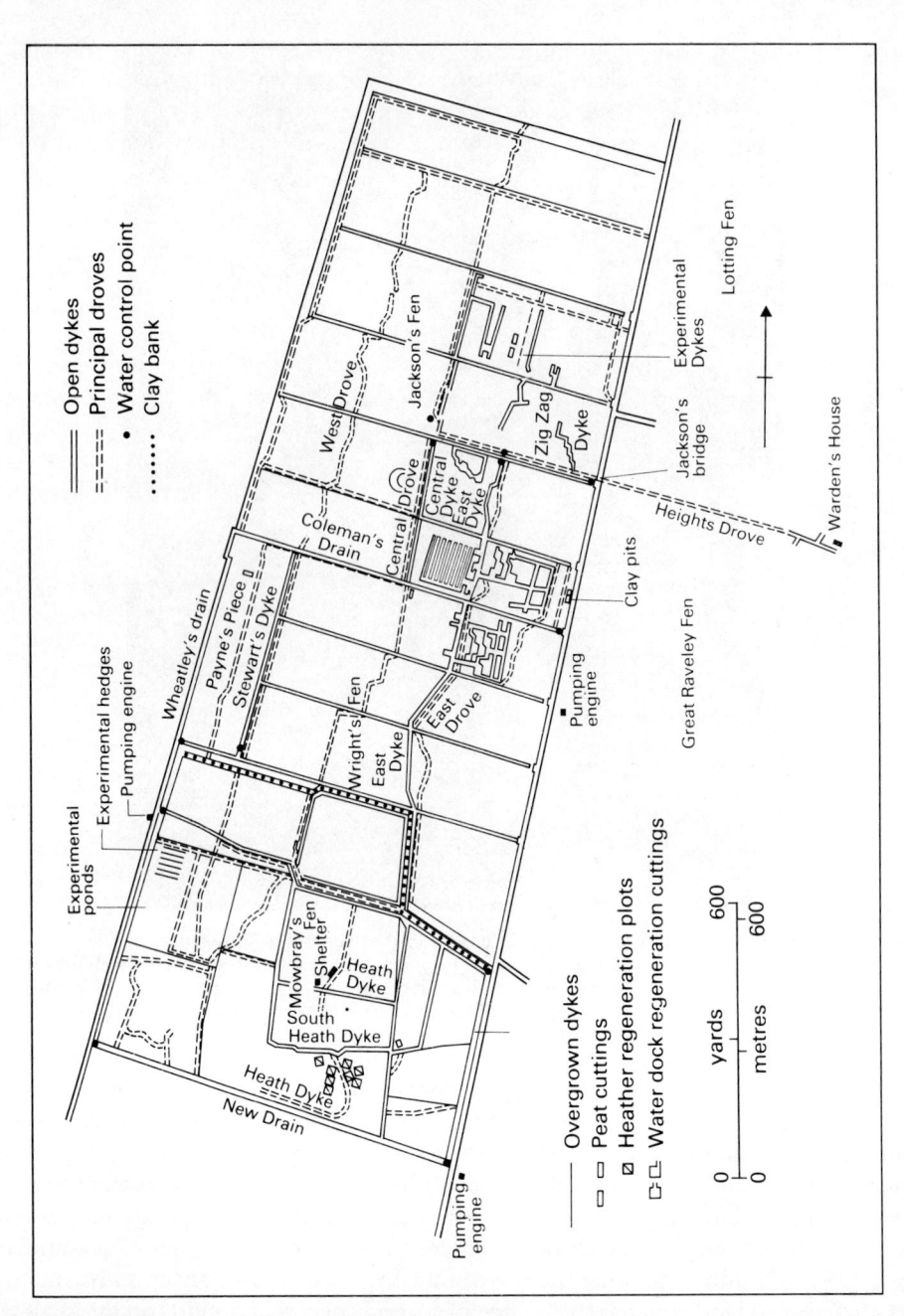

Fig. 11.1 The National Nature Reserve at Woodwalton Fen (Cambridgeshire, England), showing drains, compartments, peat cuttings and other management aids.
Source: Duffey, 1974.

From the initial thought of a totally 'natural' nature reserve we come to the opposite idea: that it is now customary for protected areas to have a Management Plan which states the objectives of the reserve and then proceeds to set out how these should be achieved, whether by complete seclusion and non-interference or by habitat or population management.

Purposes in protection

The reasons for protection fall into two groups (Ehrenfeld 1976). The first of these is scientific and starts with the need to preserve natural and near-natural ecosystems for research. Much of this is academic but increasingly it is apparent that the way in which natural systems function carries important lessons for human use of them: energy and matter pathways, and the ways in which these can be cropped without exceeding their carrying capacities, are one such piece of guidance. The IBP has once more emphasized the need for unmanipulated ecosystems which can act as base-line or datum-line studies against which to measure the effects of man's activities (Jenkins and Bedford 1973; di Castri and Loope 1977).

More obviously of importance to human affairs is the use of wild and protected areas as gene pools. They are not the only gene pools, since seed banks together with both botanical and zoological gardens can be pressed into service, but are the only ones in which the processes of natural selection can continue. Previous discussion (p. 250) has argued that the tendency of modern resource processes is to reduce biological diversity, both by eliminating species and by breeding from a very narrow range of genotypes within favoured taxa. The practice of cloning, which gives a uniform product and so is economically desirable, makes the organisms an easy prey to environmental changes such as climatic shifts or the introduction of new pests. So the role of wild places as a reservoir of new genetic material has an importance which will only be significantly diminished if the laboratory manipulation of genetic material becomes a widespread and well understood practice. (This argument applies, *mutatis mutandis*, to the preservation of the genetic resources of animals already domesticated, like cattle, where the preservation of stocks of variability in non-industrial societies or animal parks or as frozen semen or even perhaps in future as frozen embryos, is being strongly advocated (Rendel 1975b).)

A less easily documented reason is the role of natural areas in the general matrix of ecosystem types on the globe, discussed on pp. 254–5. It would be interesting to know at what scales the concept applies: whether it can be applied only at a global level, or whether smaller units (continents, perhaps, or major biomes) need to achieve the balance which is advocated.

The second group of reasons for protection are based on human values. In general terms, the Western world-view has seen nature as resources—i.e. as a subordinate world which can be altered to suit human ends. There is little evidence that this view is changing markedly (Tocher and Milne 1974; Shaw 1974), but more acceptance that what is left of wild nature might in some way be allowed to coexist with man on terms of equal importance. There is a growing view that natural objects such as trees, mountains, rivers and lakes should have rights as do people and business corporations. These might not then be infringed without recourse to the law being available to their protectors (Stone 1974). (The intellectual history of the Western world-view and the antecedents of present attitudes towards nature protection are not in detail relevant to this book but are very interesting: see Glacken 1967; Montefiore 1970; Barbour 1973; Passmore 1974.) So this view adds to the scientific reasons for both the creation of protected areas and for the protection of individual species outside any such sanctuaries. In some countries, nature is still sacred and areas are protected for religious purposes (e.g. Gadgil and Vartak 1976), a tradition with a long history (e.g. Schafer 1962).

Less intellectual but of equal if not greater importance is the appeal of some biota to what is generally called 'emotion' in human beings. Views differ on what notice should be taken of 'emotion' but it cannot be denied that it has played a key role in many countries in the steady build-up of groups of people devoted to wildlife protection. People thus motivated are without doubt selective of the objects of their concern, however. Beautiful scenery (both natural and cultural landscapes) is approved of, but interesting physiographic features which have little aesthetic appeal (like a fluvioglacial kame or an esker threatened with gravel working) attract little public attention. With animals the position is somewhat similar: birds qualify for almost universal approbation and people protecting them were in the forefront of most nineteenth-century citizen concern over biota (Fitter 1963; Sheail 1976). In his review *Dangerous to Man*, it is notable that Roger Caras (1978) devotes very little space to birds; unprovoked attack is confined to cassowaries (*Casuarius* spp), and defensive attack is common in only a restricted group like the swans and colonial seabirds. Compared with most other animal groups they are not dangerous to man. Many mammals, too, are looked on with favour, notably those with a furry appearance or with especially cuddly-looking young. It is doubtless no accident that from among all the rare and threatened species of the globe, the World Wildlife Fund chose the giant panda as its emblem. On the other hand, many species of considerable ecological importance excite disgust rather than acclamation: the societies for the protection of vultures, poisonous snakes and large black insects are few and far between (Kellert 1975). An attempt to show visually the curve of human concern and esteem for animals has been made by D. B. Luten (pers. comm.) and is given as Fig. 11.2. Legislation, of course, will cut across this curve: the United States' Endangered Species Act of 1973 halted the completion of a TVA dam because of the presence in a river of a rare species of fish. Plants evoke a narrower range of responses, although the more showy flowering plants such as orchids are a target of attention, as are very large plants *en masse* such as the riverside groves of the California coast redwood (*Sequoia sempervirens*) and the mountainside specimens of the giant sequoia (*Sequoiadendron giganteum*).

An extension of this view leads us to consider the importance of recreation in nature conservation. Although recreation may conflict with protection, as suggested above, the demand for recreation space in industrial nations may carry the protection movement along with it. People like to look at valued scenery and to be in what they perceive as beautiful places, and this often voluble desire can sometimes be linked with nature protection: a National Park may be created largely to cater to a recreational demand but it may serve to protect certain rare species as well: the Nikko NP in Japan, for example,

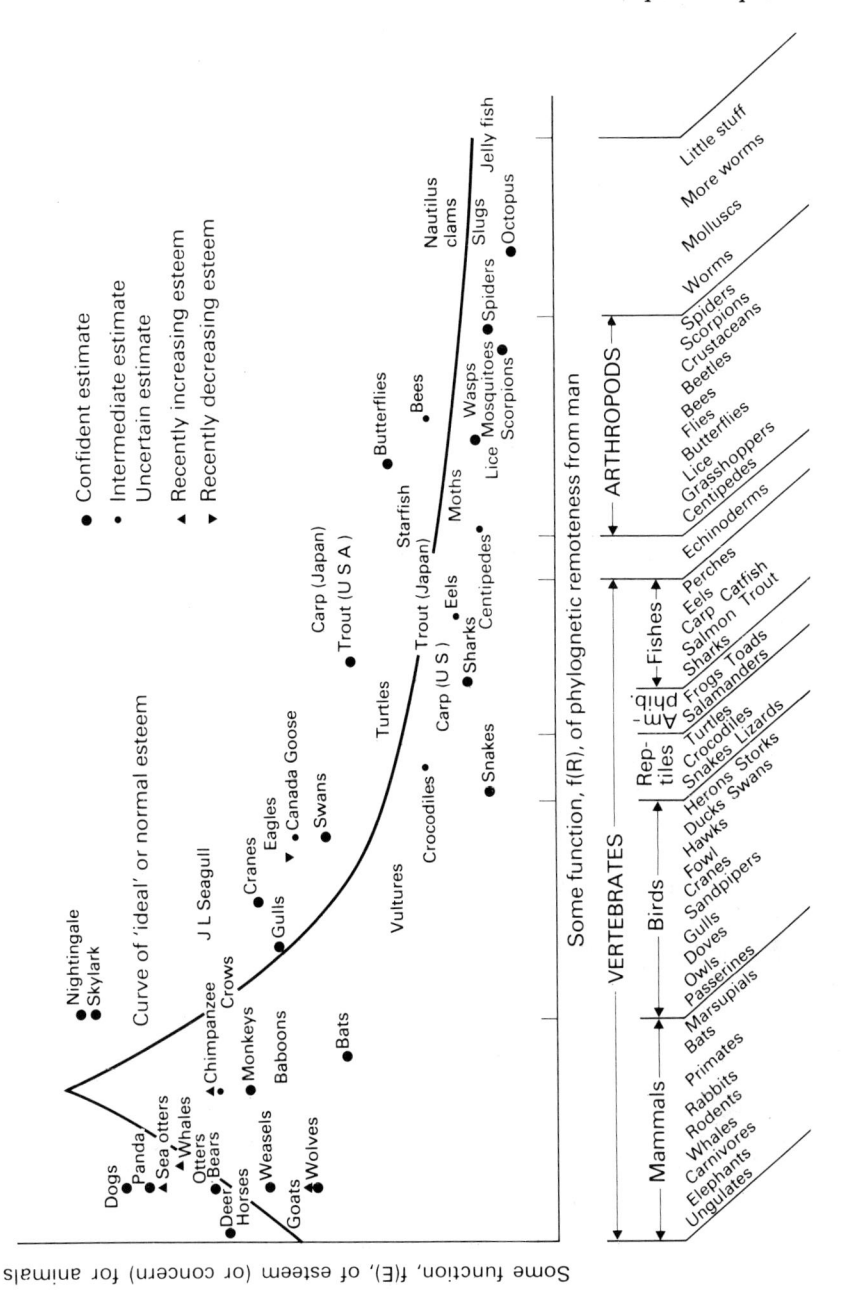

Fig. 11.2 An attempt to plot the pattern of human concern for animals together with deviations from the 'normal' or 'ideal' esteem.
Source: an original and unpublished idea by Dr D. B. Luten, reproduced by kind permission.

functions largely as a weekend recreation area for the Kanto but houses as well several areas of unusual wetland flora, the protection of which becomes all the more important in a nation where all the higher plants which have been extirpated are from marshy areas (Numata 1974). More directly, observation and photography of wildlife is an important recreation practised by growing numbers of people, and so demands to visit places of observation are high, whether these are man-made habitats like the London reservoirs at wildfowl-migration time, or wild like the African savannas. This recreation may lead to involvement in movements to protect nature and scenery, although sometimes a narrowly sectoral approach may cause problems: too many birdwatchers clustering round a storm-driven stray from another continent may frighten it away from a food source, and entomologists may devastate great swathes of vegetation in the great beetle hunt.

There is, too, the unprovable human value that wildlands are of importance spiritually to man, either in a directly religious or quasi-religious sense of places of refreshment of the soul, or in the more secular connotation of being 'a different kind of country' to quote Raymond Dasmann's (1968) phrase. In other words, the wild places form a contrast to the conditions and perceptions of everyday living and so to be in them constitutes a form of renewal and re-creation. The presence of nature in the city in the form of parks or simply tracts of undeveloped land probably has a similar value.

Whatever the multifarious reasons for designating protected landscapes and ecosystems, the practice is now common in most nations and a World List is issued and kept up to date by IUCN. As might be expected, coverage of major world ecosystem types as well as rare biota and outstanding scenery, is spotty and the International Union for the Conservation of Nature and Natural Resources would like to see all the natural regions of the world represented (Dasmann 1972). As a start, the detailed inventory compiled for Britain by the Nature Conservancy Council is worthy of imitation. It identifies 734 key sites, only about 12 per cent of which are now designated as National Nature Reserves. The document provides a basis for consultation and negotiation over the future of the other 88 per cent of the 950 000 ha (4 per cent of the surface of Britain) identified as the most important sites (Ratcliffe 1977).

Many other countries are reviewing their plans for systems of reserves in order to make sure that they are complete and representative of the various types of ecosystems. The USA, for example, hopes to extend its number of protected ecosystem reserves from 3153 in 1975 to a future total of 6000 (Sparrowe and Wight 1975; Darnell 1976), and to develop clusters of reserves for pure protection and for parallel experimental work (Johnson *et al.* 1977). Whatever the immediate aims of the designation and management of a protected area, however, its long-term aim must be to contribute to the maintenance of biological diversity on the face of the earth (Westhoff 1970).

Carrying capacity

Determination of the carrying capacity for people of the various types of protected ecosystem and landscape is not easy. In some cases of nature conservation, the answer *is* easy: the carrying capacity for people is virtually nil. A plant community may be very easily trampled out of existence: rare plants of wetlands or of narrow mountain ledges beyond the sheep and goats, for example; an animal may adopt an abnormal and non-survival-based behaviour pattern if it sees more than few people per year: the California condor is like this. Occasionally, an animal may be too dangerous to allow people near except under special conditions, as with the Asiatic tiger. But beyond these absolutes, the determination becomes difficult and capacity is usually based on some visible or easily measured parameter of environmental alteration.

Recreational carrying capacities are even more complex (Barkham 1973) since they are affected by at least two elements. One of these is the ecological tolerance of the area used in terms of how much alteration of the original ecosystems is deemed to be permissible in order not to compromise their chances of perpetuation; the other is the psychological attitudes of the visitors to such features as the presence of other people, the degree of crowding and other quasi-urban phenomena, and the amount of ecological manipulation away from a wild condition. People differ in their expectations and tastes: most wilderness users in the USA prefer to see no more than one other party during a day's progress, but vehicle-based campers and picnickers clearly like to be within sight and sound of other people, unless engaged in one of the more biological phases of the pair-bonding process. All of which complicates the determination of carrying capacity to the point where it is often determined by some artificial factor such as the number of developed camp-sites or car-park spaces. Without doubt the use of the private car in the West as the main means of access to recreation areas reduces the carrying capacity for people and so many experiments in recreation management, from Yosemite Valley in California through the Goyt Valley in Derbyshire to the Kamikochi area of the Chubu Sangaku NP in Japan, have started with the substitution of feet or public bus for the private automobile.

More sophisticated ways of determining carrying capacity will no doubt be determined. We might be struck by the suggestions made by Slesser (1975; see p. 281) that above a certain energy intensity, farming starts to have noticeable environmental impacts. By analogy it might be possible to establish a natural energy: man-introduced energy ratio for protected areas which would define the quantity of alteration by people that was permissible.

Further Reading

IUCN 1966: *Red Data Book*. Morges: IUCN, continuing.
— 1971: *United Nations list of national parks and equivalent resources*, 2nd edn. Brussels: Hayez.
JENKINS, R. E. and BEDFORD, W. B. 1973: The use of natural areas to establish environmental baselines. *Biological Conservation* **5**, 168–74.
MYERS, N. 1972b: *The long African day*. New York: Macmillan.
— 1976: An expanded approach to the problem of disappearing species. *Science* **193**, 198–202.
NELSON, J. G., NEEDHAM, R. D. and MANN, D. L. (eds.) 1978: *International experience with national parks and related reserves*. Waterloo, Ont.: University of Waterloo Dept. of Geography Publication Series 12.

SHEAIL, J. 1976: *Nature in trust*. Glasgow: Blackie.
WILSON, E. O. and WILLIS, E. O. 1975: Applied biogeography. In M. L. Cody and J. M. Diamond (eds.), *Ecology and evolution of communities*. Cambridge, Mass.: Balknap Press, 522–34.

12
The ecology of man in his environment

The focus of this book has not been directly upon man-resource-environment relationships but these have been implicit through most of it and undisguised in Chapters 10 and 11. What is apparent is the present-day dominance of the human species within the biosphere: from a place very much within the systems of nature at the beginning of the Pleistocene, man has risen to a position of such considerable power that the future of most species (including his own) is dependent upon human behaviour. This last chapter of the book will therefore consider some of the consequences for the biosphere of man's present activities and look at the alternative paths to the future.

Ecological demand and impact

It is not very helpful to imagine some Arcadian time when men lived in a totally harmonious relationship with nature and somehow existed without making any environmental impact beyond a little gathering of wild fruit. At some early stage in his evolution, man must have started to alter nature's patterns beyond what would have been achieved by any other predatory but diversivorous species. The process of man-induced change has gathered momentum as man's ability to transform natural systems has increased; and the means of alteration has been technology, which has been the channel through which man has been able to direct an incremental series of energy sources at his 'natural' environment in order to manipulate it (Table 12.1). The more the desired manipulation, the greater the control over energy sources that is needed. The management of fire is the first step of importance, since not only does it imply an ability to manipulate both plant and animal communities, but it allowed the movement of man out of the tropics into virtually all parts of the world except those of extremely harsh climate. From then onwards, the addition of each energy source has conferred an ability to effect change in ecosystems (very often in the process of extracting resources from them, since with more energy and more technology the cultural evaluation of what constitutes a resource

TABLE 12.1 Energy sources available to different types of culture

Early hunter-gatherer	Solar energy via plants and animals.
Advanced hunter-gatherer	Solar energy, stored solar energy as fire.
Early agriculturalist	Solar energy, fire, domestic animals (traction, dung).
Advanced agriculturalist	Solar energy, fire, domestic animals, wind power, water power.
Early industrialist	Solar energy, fire, domestic animals, wind power, water power, power from coal.
Advanced industrialist	As above, plus power from oil and natural gas, and geothermal power.
Contemporary	As above plus technological methods of trapping solar radiation, and nuclear power from fission and fusion. Tidal power.

encompasses more and more materials, both living and non-living) which has continued to grow in extent and intensity. Especially important phases are perhaps the perfection of the ocean-going sailing vessel, and without any doubt the Industrial Revolution. The former allowed the coalescence of the various parts of the inhabited world and the transmission on a relatively large scale of materials (including plants and animals) between them. The latter has facilitated the impact of the energy concentrated in fossil fuels upon most parts of the biosphere and permitted the alteration of nature on a large scale, making the Industrial Revolution a great discontinuity in man's relationship with nature (E. P. Odum 1971).

We must also remember that these alterations are in most cases cumulative. It has not been a question of a change being made to a new condition which has then lapsed into a former state, in a fashion analogous to shifting agriculture or the selective felling of trees within a forest. On the contrary, one phase of change has been imposed upon the last in a continuing process, so that the genes of a domesticated dog are far removed from that of its wild ancestors; indeed we might be surprised that there often is so much resemblance between them. Some natural ecosystems were changed in prehistoric times and have remained so: the effects of Mesolithic man on the uplands of Britain (p. 160) are a case of this. But in many places the earliest changes have been wiped away by the more radical developments of recent times, though often leaving relics of a biotic or cultural type which gives clues to the palaeoecologist or the historical geographer. Thus the ecosystem types of the present-day world are a mosaic of ages and of degrees of manipulation: some have functioned in their present fashion for millions of years, like the equatorial lowland forests; a great number have shuffled into place since the Pleistocene glaciations, such as the coniferous forests of Eurasia and North America; some represent old-occupied agricultural lands like those of Western Europe and Eastern Asia; other quite newly converted to agriculture by irrigation of semi-arid lands as in the Sudan and the west of the USA. Increasingly common are very new ecosystems which are the results of the imposition of man's wastes, personal and industrial, upon pre-existing plant and animal communities.

Other generalizations about the role of man in altering the functioning of nature might encompass, for example, the transformation of largely separate and closed ecosystems into a huge open web with substantial translocations of energy and matter. There is too the use of fossil fuels to enhance the productivity of the second and third trophic levels, where 8–10 per cent of the energy now comes from stored rather than current photosynthesis: an increasing proportion of the total energy flow in the biospheres is being channelled through the one species. Broadening the scale of concern, it is clear that the manipulative power of man makes the world's ecosystems increasingly dependent upon the integrity of human society (Jacobs 1975). The extreme example is the potential ecological effect of weapons of mass destruction if that societal integrity were to dissolve into large-scale thermonuclear war (Westing 1977).

A comprehensible integration of all the processes of human interaction with the biosphere at a world scale would be a measure which summarizes the utilization of resources (including non-living materials, the extraction of which frequently alters plant and animal communities) and the disposal of wastes—i.e. which reflects the sum of the intensity of interaction between the man-made and the natural worlds. The nearest approach to such a measure was proposed by the Study of Critical Environmental Problems sponsored by MIT (SCEP 1970). The statistic they proposed was the easily available (from UNO data) Gross Domestic Product minus services, on the attractive but not unassailable grounds that services are less likely to result in man-biosphere interactions. A plot (Fig. 12.1) of the movements of this index in recent years shows a rise in the

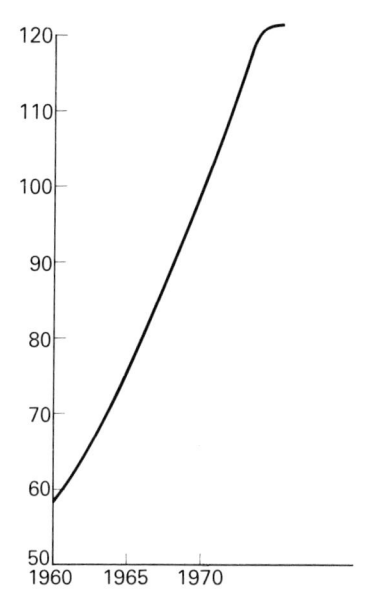

Fig. 12.1 Index of GDP minus services for 1960–75 (1970=100). A levelling-off marks the world 'recession' of recent years.
Source: *UNO Statistical Yearbook*, 1976

statistic of between 4 and 5 per cent per year, with the exception of the years following the oil price rises of 1973–74.

This value is over twice the rate of population growth and seems to accord with intuitive perceptions quite well: we know for example that in the developed nations the cultural demand for resources well exceeds that which would satisfy their purely metabolic needs, and in the LDCs rising expectations are added to often rapidly growing populations. So a doubling time of 14·3 years in this Index of Ecological Demand seems quite a reasonable statement of its trends: if human population doubles in the next 35 years then the IED will rise six-fold if the rates persist.

To examine the consequences of such rises is difficult for it involves forecasting where the actual interactions will take place and their differing intensities: will the mix of environmental change wrought in the causes of agriculture, settlement and industry, for example, be the same? Or will the threat of starvation mean an overwhelming concentration on increasing food supplies, with all that is implied for ecological systems?

The consequences of the extension of ecological demand: individual resource processes

With prediction comes uncertainty, except for death and taxes. We cannot be sure that the extrapolation of present trends into even the medium-term future will bring ecological impacts of the same kind as at present, amplified only in degree. But it seems worthwhile to look at some of the possible consequences of the rises in ecological impact which will be fuelled by greater demands upon both the productive and protective resource processes.

The food system is an obvious first candidate for consideration. The sources of food supply have been discussed earlier (pp. 273–88), and it seems clear that the emulation of Western agriculture practices will continue to be a cornerstone of development in the LDCs in spite of some suggestions that low-energy or 'intermediate' development would be more gainful in the long run. This basically means intensification via the application of considerable energy subsidies and the reduction of genetic diversity (Wilkes and Wilkes 1972). At one level, therefore, we must expect that the environmental consequences of intensive agriculture experienced by the DCs in recent decades will be transferred to the developing nations, a process which has already been seen in regions where 'Green Revolution' high-yield crop varieties have been widespread. As Alexander (1973) points out, there will be increasing applications of fertilizers and pesticides, the conversion of wild lands to agriculture, more irrigation projects, together with any difficulties which might arise from the disposal of solid wastes. Of these, the most likely to have heavy ecological impact are the fertilizers and pesticides. The eutrophication caused by runoff from nitrogenous fertilizers will disturb biological patterns and may diminish the supplies of fish which are so important a source of protein in many LDCs. High nitrogen levels in water are known to cause methomoglobinemia in some infants (alternative supplies such as bottled water are not likely to be available in LDCs) and apparently it may occur in domesticated ruminants. The effects of residual pesticides are well known but the loss of song-birds and spectacular predators is probably acceptable to people who are badly nourished or affected by malaria. Objectively, such acceptance does not alter the fact that elimination of members of ecosystems is always likely to bring shifts, among which may be explosions of other animals or plants, among which could be pest species; again, the poisoning of fish in LDCs is especially retrograde. Future scientific developments may ameliorate these two problems in the sense that more target-specific and non-persistent biocides (such as those that inhibit the synthesis of chitin in insects) are now becoming available, though at higher prices than for example the organochlorines, and eventually, techniques of inoculation of non-leguminous plants with the bacteria which will enable them to fix their own atmospheric nitrogen will probably become available. But even if DDT were withdrawn today, it would take at least twenty years for it to come into an equilibrium distribution in the biosphere and many more for it to degrade into non-toxic substances. To take a pessimistic view, it is possible that there are still surprises to be sprung by toxic residues in the form of unexpected environmental effects and it is hardly likely that they will be pleasant.

Wild land converted to agriculture is likely to be used mainly for high-protein calorie sources rather than the energetically wasteful meat-animal, so that the animal-waste problem is unlikely to be a feature of LDCs, although it will persist in the DCs. In both groups of countries, the sheer over-cropping of agricultural lands, combined in some places with the physical impact of machinery upon the soil, will continue to produce soil erosion, one of the commonest and most persistent signs of over-intensive land use. It is also one of the most devastating since it reduces the ecosystem virtually to a biologically inert condition and needs a long recovery time; there are places too where reclamation is scarcely feasible since an irreversible change has taken place: for example, deforestation in the tropics may mean laterization of the soils, which cannot then be cultivated but only grazed by animals like cattle.

At a higher level of generalization, we may speculate about the overall effects of the search for more food of a higher quality: the trend will be towards a greater extension of simpler systems with low biological diversity, certainly to the point of monocultural crops of low genetic variability. This may for some time yet be accompanied by a wide spin-off effect caused by runoff of fertilizers, organic wastes and pesticides, together with soil loss from present agricultural land and from areas deforested or otherwise cleared of native vegetation to make farmland.

The intensification of demand upon the forests (*qua* forests and not as sources of cleared land for other uses) tends to bring to them an agricultural air, so that more and more farming practices such as ploughing, seeding, fertilizing, weeding and pest control are applied. However, the environmental impacts associated with certain degrees of intensification in agriculture (see pp. 281–3) are not yet reported from forestry, whose environmental influence is generally thought to be benign. As in agriculture, greater demand leads to monoculture with all the problems posed thereby for the managers, especially diseases and fire. Nevertheless a strong demand for wood products is probably one way of keeping a large area under forest, which in times of low demand might well be cleared for conversion to grazing or arable crops.

It seems strange that a greater demand for water could in any way affect the workings of the hydrological cycles since the proportion of its flows open to man's intervention is so small. That the demand for water will be extended and intensified is in little doubt since irrigation forms an important part of many LDC schemes and in those countries also the provision of more and better-quality drinking water and sewage systems is a high priority. There is however a limit to the supply from the runoff and groundwater in the arid and semi-arid zones of the earth; much of the groundwater currently being used is 'fossil' water from aquifers charged during periods of wetter climate and so is for practical purposes a non-renewable resource. Desalination could be called upon to supply large quantities of fresh water, if sufficient cheap energy were available. More germane at this point is the unknown climatological effect of greatly increasing the evaporation over land in the dry areas, accelerating the hydrologocial cycle; at present the evaporation from irrigated lands is 1700 km^3/yr and agricultural extension might increase this 20-fold (Flöhn 1973). To these considerations must be added the continuing rises in demand for water in the DCs. Here the highest quality is also required to meet urban and industrial standards, and the implications for land-use patterns and for the energy consumption needed to purify and distribute the water are far reaching.

Any greater use of energy will also change ecological systems. At local and regional scales, its ability to do so through air contamination and calefaction of water is already apparent. Globally, there are other trends which if continued may bring considerable and unforeseen changes to all ecosystems. On the one hand, the greater amounts of CO_2 now present in the atmosphere because of the combustion of fossil fuels might be expected to produce a global warming, the so-called 'greenhouse effect' (Fig. 12.2). This

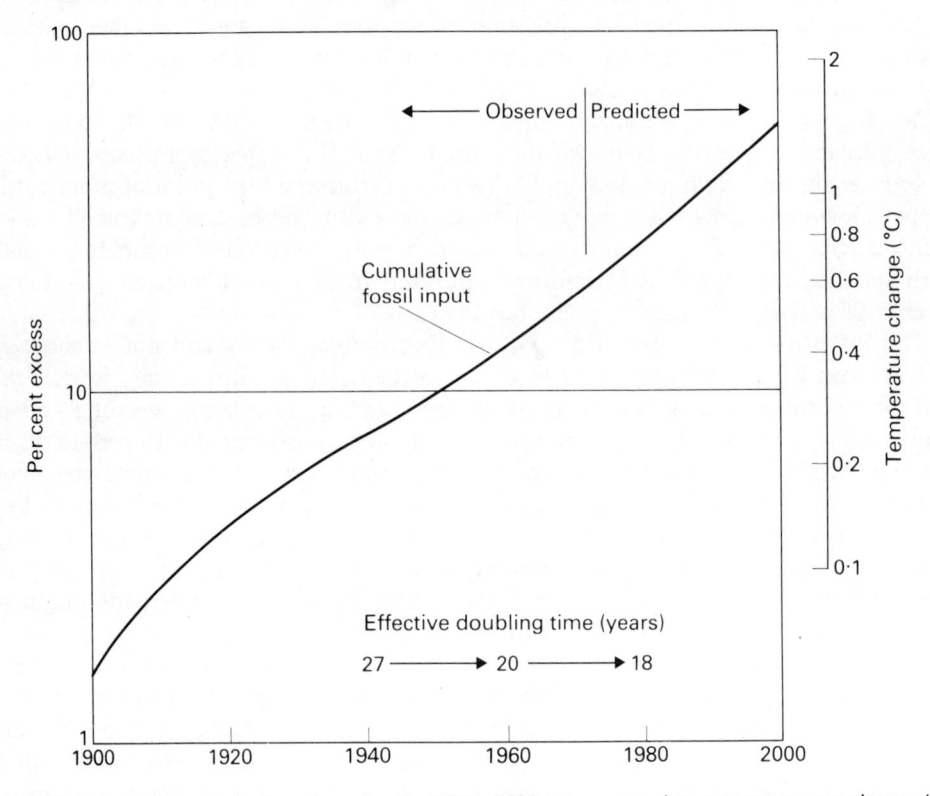

Fig. 12.2 Growth of atmospheric carbon dioxide 1900–2000 AD expressed as present excess above nineteenth-century base level. It assumes all fossil input remains in the atmosphere (but see Fig. 2.8). On the right is the associated temperature change with various levels of CO_2.
Source: Mitchell, 1975.

would be intensified by all lowerings of the biological productivity of ecosystems since their CO_2 intake would be lessened. The effect could be amplified if in turn the raised temperature of the oceans decreased their content of carbon dioxide since it is less soluble in warmer water. It would then be released in the atmosphere and cumulatively add to the raising of the global temperature. On the other hand, the particulate matter placed in the atmosphere from the same processes reflects light and would be expected to cause a cooling effect (Fig. 12.3). Both these trends must be superimposed upon natural climatic changes and the long-term outcome is not clear; but necessarily the higher the

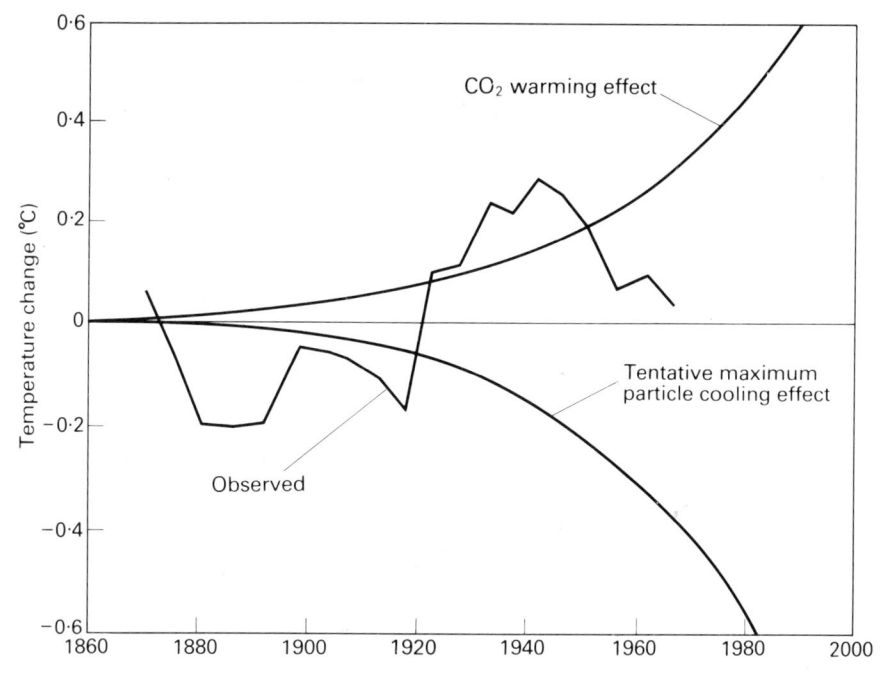

Fig. 12.3 Prediction of trends of mean temperature 1860–2000 AD. The upper curve is the warming effect of CO_2, the lower curve the cooling effect of particulate matter in the atmosphere. The broken curve is observed temperature in the northern hemisphere.
Source: Mitchell, 1975.

energy demand, the greater the potential for interference in normal climatic trends. Superimposed upon both of these is the idea of a heat limit: all energy used by biota and by man ends up as heat and the atmosphere has a finite capacity to radiate this heat back to space. The more fossil and nuclear energy that is used the greater the chances of heat 'injections' changing the circulation pattern of the atmosphere (Kellogg 1978).

Even local events during the extraction of energy have been credited with the possibility of bringing about large-scale climatic changes. A large oil spill in the Arctic Ocean, for example, would reduce albedo and might melt the Arctic Sea ice; once this has melted, the addition of non-saline ice-melt water to the ocean is replaced by more saline water making the re-freezing of the Arctic Ocean unlikely. This would cause all the climatic belts to shift southwards by several hundred kilometres: the consequences for the major food-producing areas of the world would appear to be disastrous.

Even a smaller cooling trend in the climate, to something like the climate of the nineteenth century, might make the winter wheat and rangelands of the northern half of the USA rather wetter, though this would not affect US food production very much. Such a change would also bring severe drought every 3–4 years in northern and northwest India, persistent drought in Sahelian Africa, and shorter growing seasons in Canada, northern Europe, northern Russia and northern China, together with more frequent monsoon failures in southeast Asia and the Philippines. Such a cooling trend (not universally accepted by all commentators) would have political as well as biotic implications (CIA 1974).

So far, man's use of non-biotic energy has now reached about 5 per cent of the magnitude of worldwide net production of plants (Woodwell 1974) and so secondary effects of its use are having serious effects on the world's biota, probably reducing the fixation of solar energy and its availability for life-support. Regionally, the ratio of industrial energy consumption to photosynthetic production is much higher: for the continental USA, it is about 13 per cent and for New York City it is 200 times the natural rate (Hall 1975). Growing demands for protected landscapes and ecosystems are not entirely free of a contribution to ecological demand, largely because their preservation may entail more intensive land- or water-use elsewhere. This is in addition to any erosion caused by over-use of recreation areas.

Last of these processes is that of the generation of wastes. Since his beginnings, man has created wastes but their volume and concentration since the Industrial Revolution has in some places exceeded the capacity of any ecosystem to absorb and process them. Additionally, scientific advance now allows the synthesis of compounds not present in nature and for which there are no natural degradation pathways in ecosystems. It is misleading to pretend that biota and man can be protected from toxins by setting standards based on threshold effects. There is no way of determining such thresholds for many of the substances that are developed and released, no way of controlling the releases, no proper way of monitoring most of the compounds and the meaning of threshold in terms of the variation inherent in systems with living components is indistinct.

One consideration, perhaps only marginally relevant at this point, is the demands of the human organism itself for resources such as space (given that crowding seems to intensify cultural characteristics, good or bad, already there), and the wider effects of population growth upon the functioning of the organism. In the first instance, it seems as if in many cultures, crowding beyond a certain point and in particular types of housing which diminish the sense of personal space as distinct from communal or no-man's space, induces crime, vandalism and neuroses like 'high-rise blues'. Put another way, nobody knows if man in a population irruption will remain particularly 'human' or whether behaviour patterns may be drastically modified (Catton 1976). In the second instance, population growth and ecosystem change bring different diseases to replace those which can be controlled in more advanced economies. So a suite of 'development diseases' can be recognized in LDCs, just as can a set of diseases most common in advanced industrial societies. In addition, we may recognize that diseases can now be transmitted overnight from continent to continent and be introduced to a cultural group which has had no chance to build up an immunity to them.

The consequences of the extension of ecological demand: the totality

If, as seems likely, all these systems are affected by increased demand from human societies, what likely general consequences may we expect in the medium-term future? The overall impact of this ecological demand and its extension seems to crystallize around two generalizations.

The first of these is the simplification of ecological systems. An easy example is that of converting a complex forest ecosystem to a relatively simple monocultural-crop ecosystem but the same process also happens downstream from a raw-sewage inlet or in a lake undergoing rapid eutrophication, or in the conversion of grassland to a biologically inert dump of industrial waste materials. Therefore the extension and intensification of productive resource processes will tend to simplify ecosystems and produce conditions more akin to the early stages of succession in natural systems than to self-sustaining mature systems (Whittaker and Woodwell 1972). This trend is reinforced by current practices which reduce the genetic diversity in domesticated plants and animals.

The second generalization is the acceleration by man of the mobilization of materials in the biosphere. The natural nutrient cycles, often spatially very tight, are changed to more open cycles since all kinds of materials are transported long distances. Furthermore, the natural mobility of material is subject to the imposition of a heavy loading due to human activities (Table 12.2). The increased role of mobility of materials and

TABLE 12.2 Mobilization rates of selected geological materials

Element	Geological rates of discharge in rivers	Mining rates
	$\times 10^3$ t/yr	
Iron	25 000	319 000
Nitrogen	8 500	9 800
Manganese	440	1 600
Lead	180	2 330
Phosphorus	180	6 500
Mercury	3	7

Source: SCEP 1970, p. 116.

the greater number of ecosystems into which they are cast as wastes, often in concentrated form or as chemical synthates of great molecular complexity, is another example of post-Industrial Revolution impact upon ecosystems.

Clearly the next question is, does it matter? Suppose that these processes continue for another doubling or two of the world's population (i.e. 70 years at current rates) will the ecological impacts necessarily be deleterious to the continued functioning of the biosphere as a life-support system for *Homo sapiens*?

The answer to such a question is not completely clear but some indication should be given of the present thinking of ecologists. It centres around the relationship between diversity and stability in ecosystems, and the response to perturbation by ecosystems of differing diversities (Woodwell and Smith 1969). Diversity in this instance is measured in terms of the number of species per unit area and hence high diversity implies the existence of complex food webs in diverse ecosystems with a large number of interlocking energy pathways. Stability has two meanings. The first refers to persistence: the constancy of species numbers in an ecosystem, or the numbers of individuals in a single-species population. The second relates to the ability of the system to return to its original condition after an external perturbation—i.e. its resilience (Fig. 12.4). Several writers (e.g. MacArthur 1955; Elton 1958) have suggested that both types of

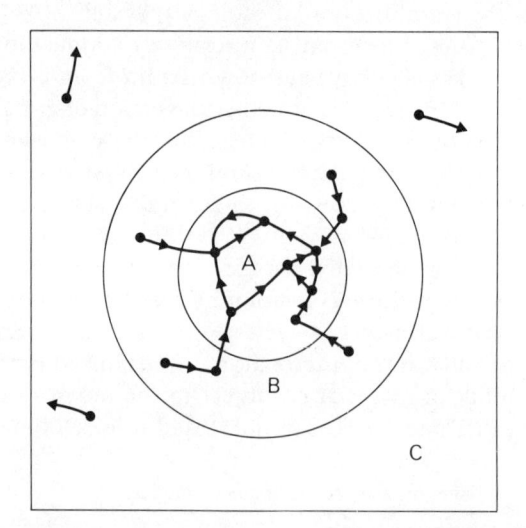

Fig. 12.4 Margalef's model of stability–resilience. An ecological system receives perturbations which force it to move to area B but its resilience enables it to return to A. However, severe stress may move it to C where it must seek a new equilibrium.
Source: Hill, 1975.

stability are enhanced in ecosystems with a high biotic diversity since the system is more resistant to outside invasions and has feedback mechanisms for damping down population oscillations within itself. The idea that the complexity of the food-webs provides a large number of interaction pathways which can absorb stress is supported by some empirical evidence that simple ecosystems such as agricultural areas and the tundra have much greater fluctuations in biotic populations than more complex ecosystems. Yet some of these, such as the tropical rainforest, which are well adapted to persist in the relatively stable environment in which they have evolved, seem less likely to be resistant to man-induced perturbations than relatively simple temperate ecosystems. Further, some natural monocultures such as bracken fern and *Spartina* marsh seem highly stable (May 1975).

The position is probably even more complex. Studies of individual systems reveal a wide variety of relationships between diversities of herbivores and carnivores and the numbers of competitors at any one trophic level, and cases have been recorded where a single species occupies a key role. Paine (1969) noted that the disappearance of a starfish (*Pisaster ochreus*) led to the collapse of the trophic structure of an intertidal community and the reduction of a fifteen-species system to a two-species system virtually monopolized by mussels. As Hill (1975) remarks, the equation of diversity and stability is now regarded more as an hypothesis than an axiom or even an article of faith. Even if the diversity-stability hypothesis is false, it is still quite likely that the disruption of patterns of evolved interaction in natural communities will have unforeseen (and sometimes untoward and catastrophic) consequences (Goodman 1975). Man-induced stresses may of course affect the abiotic parts of an ecosystem as well as the living parts and so produce perturbations which are initially unrelated to its trophic structure.

If, nevertheless, stability and diversity should be positively connected, then the effects of human activity in transforming ecosystems to simpler states can be seen as a set of processes which could produce instabilities. These might appear as greater susceptibilities to explosions of populations which would certainly include crop pests, or vulnerability to hitherto harmless organisms whose large numbers turned them into *de facto* pests. Other events might be the unpredictable disappearance of economic species such as fish; the more rapid transmission of plant and animal diseases through genetically uniform populations, and the lessened resistance of populations of domesticated organisms to minor fluctuations in climate or to stress from weather conditions. But it would be wrong to give the impression that nature always constitutes a very fragile set of webs: nature is very resilient and adaptable but the response to a perturbation is likely to be large-scale oscillations in some populations (pessimists would include man amongst these), and the steady state may take a long time to achieve because disturbance is likely to return ecosystems to an early stage of succession. Even when stability is again achieved, the species mixture may be less useful to human societies than before: small rodents and insects, for instance, are the animal groups most likely to flourish under such conditions.

Such possible fluctuations in plant and animal populations must necessarily be put into the context of other, abiotic, mutabilities which man might cause, especially those concerned with the workings of the atmosphere (Bryson 1973; Kellogg 1978). If human activity were to cause climatic shifts then all ecosystems would be affected, simple and diverse alike, although it could be argued that the more heterogeneous such as the oceans and the equatorial forests (if there are any left) will change less in response to the alterations in climate. Coping with ecosystem variability would be as nothing compared with trying to adjust to unpredictable vagaries of climate. In certain limited circumstances, instability may not be necessarily harmful. If closed forests of exotic conifers in Scotland were shown to be unstable to the point where only a single rotation could be run, then that single crop might be justified provided that it did not preclude the occupation of the site by a wide range of land uses thereafter (Mutch 1974).

One more aspect of diversity might be mentioned, though it is not strictly biogeographical. In nature, diversity of life-forms and of genotypes means that natural selection is likely to produce a system with survival capabilities. Most mature natural systems have undergone severe tests of survivability. Not all man-made systems have done so, especially the industrial life-style and its systems which are barely 150 years old. There would seem to be an excellent argument for encouraging the survival of pre-industrial economic modes, therefore, as a source of diversity for future human existence should the industrial way of life prove to be unviable: a variety of culture seems as desirable as a diversity of plants and animals. Another facet of this view is seen in the proposals to build on indigenous modes of production in LDCs rather than uncritically importing high-energy agricultural and other productive systems from DCs.

Yet another aspect of the resilience–stability aspect of ecosystems which are subject to human influence must be mentioned. It is increasingly evident that both cropping and contamination are stresses upon ecosystems which reduce the amount of energy available for self-maintenance (E. P. Odum 1972). But because of their history of manipulation of ecosystems (especially since the agricultural and industrial revolutions), human societies will inevitably go on trying to direct the energy and matter and pathways, no matter how high the amplitude of oscillations. The ability to manage means access to energy sources, and the more manipulation is needed, the more energy is required. So if rapid fluctuations of high amplitude are to be damped down or cropped for human use, then it is likely that vast quantities of energy will be needed.

The idea of limits

The implications of the discussion in the previous sections are simple: there is a limit to the extent to which people can manipulate ecosystems without losing control. Phrased differently, every ecosystem, and hence every resource process that depends upon ecosystems, has a carrying capacity for human activity. Stated thus, the argument seems self-evident, but to large numbers of people, it is not so. In the first place, the final limits are not clearly defined, in time, space or appearance. Secondly, many people are inadequately supported by the systems at present and see in further manipulation of them the chance of improving their or their descendants' material standards. Thirdly, there is greed: the continued use of resources ensures a high material standard for some, and they are rarely content to stay at a level of 'modest sufficiency'. The pressures therefore to 'develop' and 'grow' are sufficiently strong to provoke, for example, the somewhat hysterical debate that followed the publication of Meadows's *Limits to Growth* (1972). In this context, however, we must not forget that abstractions like 'growth' and 'development' really mean the manipulation of ecosystems, and that there must necessarily be thresholds beyond which the resilience of such systems breaks down. This has happened in many places: where soil erosion is manifest, where waste gases from a chimney have killed all the vegetation, where fertilizer runoff has caused algae blooms and killed all the fish, and as in the Sahel during the early 1970s where the economy collapsed because too many cattle were too concentrated spatially. 'Development' and 'growth' have the potential to multiply such instances and perhaps synergistic interactions of which we have as yet no knowledge: nobody, for example, predicted the effects of organochlorine pesticide residues when the compounds were introduced. Our overall attitude to ecological systems and their living components might then follow the words of G. M. Woodwell (1974):

> ... I suggest that the Earth's biota is our single, most important source. While protecting it will not assure wealth and grace for man, its decimation will assure increasing hardship for all.

Alternative man–biota–environment relationships

The discussion in the previous paragraph suggests both that an alternative path to the extrapolation of present trends in the use of ecosystems may be necessary, but that pressures for continued 'development' are strong. In that context it is therefore essential to consider the types of relationship of man and nature which might be viable in the medium and long-term future.

The extrapolation of present trends

The extension of present rates of growth of human population, and of the associated ecological demand is likely to bring either plenty to all, or famine and ecosystem collapse, depending upon the opinion of the type of observer. If it is to bring plenty, via the medium of technology, then the decisive factor must be the supply of energy, since this underpins the success of a technology-based existence (Odum and Odum 1976). The discussions has postulated however that the systems of the biosphere have an upper limit to the quantity of man-directed energy that they can absorb without breakdown. Since these systems function as the life-support system for man (via food, oxygen and the processing of wastes) as well as suppliers of culturally-generated demands, it follows that if man's demands continue beyond the breakdown point of the biosphere's systems, then nature's system must be replaced by totally man-made life-support organizations, i.e. technology must replace nature for all practical purposes. It is difficult to envisage such a world although its endpoint might be something like Fremlin's (1964) conception of a population of $60\,000 \times 10^{12}$ (890 years at 2 per cent p.a.) living in a 2000-storey building covering the entire surface of the planet; each person having $7 \cdot 5\,m^2$ of floor space, and the life-support machinery occupying 1000 of the storeys. The limiting condition would be the technological ability to radiate into space all the heat generated by the people and their machines: the outer skin temperature of the building would be $1000°C$. The key technological element in even a reduced version of such a lifestyle would be the availability of cheap and safe energy, and presumably atomic fusion is the main hope. Opinions differ about the likelihood and timing of this latter development but without it, the total technology option which would replace biosphere life-support systems will not be feasible. An intermediate stage based on fission reactors of the fast-breeder type is technically within reach but seems something of a Faustian compact since the disposal of wastes and the creation of a plutonium economy with all its dangers are a high price to pay for the power thus generated. Whatever the source of the energy, it is difficult to overstress man's access to energy sources as a mediating factor in and perhaps even a determinant of man–biota relationships.

The isolation of the econosphere

If there is a clash between the demands of human society and the carrying capacity of the biosphere's systems, then perhaps it would be useful to investigate the possibility of uncoupling them so that the 'econosphere' is functionally sealed off from the 'ecosphere'. The most complete visualization of this view is to be found in Nigel Calder's book *The Environment Game* (1967) where everybody lives under transparent domes (some on ice-islands in the seas) and eats industrially-processed foods which start as intensively produced algae. Outside the domes the natural environment predominates and a person's status in society is related to their effectiveness in maintaining these ecosystems. Living close to nature (and perhaps because of the transparent domes), everybody will see the necessity for keeping family size small. The basic idea of the book, that of keeping man and all his works separate from nature, is attractive and some premonitions of similar thinking can now be seen from time to time. The burial of solid wastes has always represented this way of thought but more modern developments include the proposals for siting nuclear power stations offshore from urban areas or setting them underground. Such proposals are usually too expensive to be viable alternatives at present but changes in the economics of energy are by no means impossible as we have seen in recent years. The substitution of mining for quarrying in areas of high aesthetic value is another example.

It is relatively simple to denounce this course as unviable: the system might still produce wastes that had to be led off into the environment and unless a population-resource stasis were achieved, these could still produce stress in receiving ecosystems. The problem of heat is not overcome and so the necessity for a limit to ecological demand is still present: in fact the whole idea can perhaps be interpreted as one form of acknowledgement of limits to growth, symbolized by the domes that contain the people. Nevertheless, many of the subsystems are attractive and in any future, the idea of isolating natural systems even more than is done now with protected landscapes and ecosystems deserves careful consideration.

The dynamic equilibrium

This alternative postulates that the systems of the biosphere must continue to be the life-support systems for man and so ecological demand has to be adjusted to as not to exceed the carrying capacity of these ecosystems. Thus the biosphere could continue to play its public-service function which is currently free of charge to society: water is stored, soil structure and productivity are maintained, the chemical equilibrium of air and water is maintained, epidemics are damped down via natural predators and diverse communities, and biological metabolism 'cleanses' soils, air and water. (The economic, social and political consequences are not appropriate matters for this book but are considerable; see, for example, Caldwell 1971.) In terms of the role of biota, the role of productive systems would continue to be vital for they will still provide food and fibre, and would have to be interlinked with the energy and inorganic matter-producing systems to the benefit of both rather than the detriment of the biological systems. Increased understanding of the nature of processes like photosynthesis may make possible the revaluation of many living plants as a renewable source of energy and materials via such processes as fermentation: sugar-cane, kelp and the rubber tree appear to be extremely good converters of sunlight and may provide pioneering examples (Calvin 1976). But it will still have to be realized that there is a limit to production capacity and hence to the number of people that can be supported (Simmons 1974b). A major change would have to come in the valuation of the protected systems. No longer would they be regarded as purely cosmetic (i.e. as pleasant things to have especially in societies rich enough to afford them), but as integral parts of a mosaic of different ecosystem types, both productive and protective, which keep the biosphere in a stable condition (Odum 1969; Odum and Odum 1972). They thus become seen as a necessity, just as much as food-producing systems, in their function as 'anchormen' in controlling the rates of important processes like the exchange of O_2 and CO_2, in acting as a source of biological diversity, and in the provision of refuges not only for the rare, beautiful and the strange, but in seeming to play an important if intangible role in providing habitat diversity for many people. At government level, therefore, the destruction of the world's wild places and wildlife needs to be halted, particularly in fragile areas such as islands, mountains and estuaries and both national and international programmes of management for the oceans worked out.

In the man-modified lands, protection would be given to people who wish to pursue a non-industrial way of life: they have a right to do so, and are in any case a source of survival value. Productive systems need all to be managed on a sustained-yield basis, bearing in mind that the energy subsidies (pp. 276–83) which underpin much of their productivity may not continue for more than the medium-term future. (What is the future for flying out flowers from California to New York at an energy cost of $27 \cdot 33 \times 10^6$ J/tonne-km?) The economic development of poor countries must continue, but a more serious analysis of total costs (environmental, social and economic) should accompany the calculation of benefits; a longer view will undoubtedly encompass the value of the natural environment to such countries. We need also to know the conditions for long-term stability of man–environment relations within inherently fragile regions such as oceanic islands (UNESCO 1973b; Winslow 1977).

Research is clearly needed on the linkages and couplings between the three major energy systems of the world: the solar-powered natural systems; the solar-powered but energy-subsidized agricultural food- and fibre-producing systems; and the fuel-powered urban-industrial systems. We need to know the optimum mixes and spatial arrangements of these elements, a field in which ecosystem modelling has much to contribute (Hall and Day 1977). The study of the 'interface' between physical and cultural systems clearly needs a lot of elucidation if more rational management structures are to be found (Bennet and Chorley 1978). As the level of individual people, a commitment to population control is clearly necessary, with all that is implied in terms of changes in cultural attitudes in many nations. A set of moves towards decentralization of settlement, particularly in industrial nations, would also be a step towards greater viability of the biosphere since the man-made concentrations of materials, particularly wastes, which have so strong an impact upon ecosystems would be lessened (Dasmann 1975).

The future of recombinant DNA

This is a joker in the pack for all the alternatives outlined above. The control over genetic material which may be conferred by the development of research into the fundamental structure of the gene (pp. 256–8 probably exceeds our present imagination, although at present it is at a relatively early stage of development, and is carefully controlled by enforceable guidelines (Harris 1977). Many of the properties of productive ecosystems will be altered and even the genetics of man might be subject to choice rather than chance. Like nuclear fission, the bargain seems Faustian, since the potential for evil seems every bit as great as that for good.

The choice between alternatives

Which of the above alternative paths to take (or none, by simply working on an *ad hoc* basis) is determined by human values, a tangled and thorny stand of undergrowth for most biogeographers. What should perhaps concern us most is whether a particular set of values is positive in an evolutionary sense: will it increase the probability of survival of the species in the long term? And is survival worthwhile, indeed, if it is some version of the Fremlin-type world (p. 343)? No blanket answer can be given to these questions since by definition the answers lie in a largely unpredictable future. But to this author it looks very much as if the dynamic-equilibrium alternative has more chance of supporting a human population on a long-term basis than its main competitor. Put negatively, the technology-dependent case is an act of faith, a placing of trust in a way of life dependent upon an energy source not yet harnessed. By contrast, the equilibrium state is attainable with present knowledge, does not foreclose the options of a different kind of development later in time, and it is perhaps also the path of the prudent caution which academics are temperamentally inclined to adopt.

Whether all this would matter in the context of a swift ending of the present interglacial is another consideration. Just how the world's biota would emerge from a glaciation as widespread as the Devensian but with a starting population of perhaps 6×10^9 *Homo sapiens* is the subject for an essay in science fiction rather than a book on biogeography.

Further Reading

BENNETT, R. J. and CHORLEY, R. J. 1978: *Environmental systems. Philosophy, analysis and control.* London: Methuen.

BISWAS, A. K. and BISWAS, M. R. 1976: State of the environment and its implications to resource policy development. *BioScience* **26,** 19–25.

DASMANN, R. 1975: *The conservation alternative.* New York and London: Wiley.

HILL, A. R. 1975: Ecosystem stability in relation to stresses caused by human activities. *Canadian Geographer* **19,** 206–19.

KELLOGG, W. W. 1978: Global influences of mankind on the climate, In J. Gribbin (ed.), *Climatic change.* London: Cambridge University Press, 205–227.

MEADOWS, D. H., MEADOWS, D. L., RANDERS, L., and BEHRENS, W. W. 1972: *The limits to growth.* London: Earth Island Press.

MESAROVIC, M. and PESTEL, E. 1975: *Mankind at the turning point.* London: Hutchinson.

ODUM, E. P. and ODUM, H. T. 1972: Natural areas as components of man's natural environment. *Transactions 37th N. American Wildlife and Natural Resources Conference,* 178–89.

ODUM, H. T. and ODUM, E. P. 1976: *Energy basis for man and nature.* New York and London: McGraw-Hill.

SCEP 1970: *Man's impact on the global environment.* Cambridge, Mass.: MIT Press.

SCHUMACHER, E. M. 1973: *Small is beautiful.* London: Blond and Briggs.

SIMMONS, I. G. 1974b: *The ecology of natural resources.* London: Edward Arnold.

WOODWELL, G. M. 1974: Success, succession and Adam Smith. *BioScience* **24,** 81–7.

Glossary

Abiotic The non-living factors influencing an ecosystem.
Acidophile ('Acid-lover') A plant which is tolerant of, and grows best on, acid soils.
Adaptive radiation Evolution from a common ancestor of divergent forms adapted to distinct modes of life, in order to exploit new habitats.
Allogenic An allogenic succession is one where the stimulus for progressive change is external to the ecosystem.
Amino acid Amino acids are organic compounds containing nitrogen which link together to form proteins.
Aphotic A zone of permanent darkness, usually used of the oceans or very deep lakes beyond about 300 m depth.
Assimilation Assimilation is the incorporation of food materials into the cells and fluids of an organism.
Autecology The study of an individual organism or species, including its life-history and behaviour.
Autogenic An autogenic succession is where the stimulus for progressive change comes from within the ecosystem.
Autotrophic ('self-nourishing') An organism is autotrophic if it is independent of external sources of organic substances and can manufacture its own food from inorganic materials.
Biocide A toxic substance capable of killing plants or animals.
Biogeochemical cycle A circulation of chemical elements in the biosphere from the biotic to the abiotic element.
Biological productivity The rate at which growth processes occur in an organism or ecosystem. Usually expressed as the quantity of dry matter/unit area/unit time, e.g. kg/ha/yr; or less frequently as grams of carbon/unit area/unit time, e.g. gC/m²/day. A rough conversion from weight C to dry weight is $\times 2 \cdot 2$.
Biomass The weight of living material in all or part of an organism, population or ecosystem. Usually expressed as dry matter/unit area, e.g. kg/ha or g/m².
Biome A major regional ecological community of plants and animals extending over a large area, with a uniform climate and vegetation.
Biota Groups of plants and animals occupying a place together.
Biotic The living part of an ecosystem and any influences that arise from the activity of living organisms.
Calcicole A plant tolerant of, and which grows best on, lime-rich (calcareous) soils.
Calcifuge A plant which is tolerant of, and grows best on, acid soils.
Calorific value A measure of the amount of heat or energy yielded when a substance undergoes combustion.
Carpels Female reproductive organ of flowering plants, containing ovules (which

become seeds) and a receptive surface for pollen, the stigma, which is often at the end of a stalk, the style.

Carrying capacity The number of individuals that the resources of a habitat can support.

Carnivore An animal which eats mainly flesh.

Chelation The mobilization of certain metals (e.g. iron, manganese, aluminium) by polyphenols derived from organic sources such as plants and plant litter.

Chlorinated hydrocarbon A set of organic compounds used as insecticides, such as DDT, aldrin and dieldrin. They have very slow rates of breakdown.

Chlorophyll A green pigment found in most plants which traps the energy of sunlight and uses it to manufacture foodstuffs in the process of photosynthesis.

Chromosome A microscopic thread-shaped structure, consisting largely of DNA and proteins, numbers of which occur in the nucleus of every plant or animal cell. All nuclei of a given species have the same number and type of chromosomes.

Climax The final stage of the successional sequence which consists of a relatively stable ecosystem which is in equilibrium with its abiotic environment.

Cloning A process which produces a group of genetically identical organisms asexually from a single ancestor.

Club mosses A group of small evergreen plants which reproduce by means of spores. They were more prominent during the Carboniferous when they took the form of trees.

Commensalism Members of different species living in association where one of them benefits from the association while the other is not affected.

Community A group of organisms inhabiting a common environment and often interrelated by food chains.

Competition The process by which species with similar requirements vie for a resource such as water, light or food.

Cuticle A waxy outer coating of the aerial parts of plants which is impervious to water and gases.

DNA (Deoxyribonucleic acid) A complex compound found in the chromosomes of plants and animals which is the basic hereditary material due to its ability to replicate itself.

Decomposition The breakdown of organic material, releasing nutrients into the environment.

Density dependence A factor whose influence on a population depends on the population density (number of individuals/unit area) is density dependent.

Detritus Freshly dead or partially decomposed organic debris.

Diadromous A fish which lives part of its life-cycle in fresh water and part in the oceans, e.g. eels, salmon.

Diploid *Of a nucleus:* Having the chromosomes in pairs so that twice the haploid number is present. Characteristic of almost all animal cells except the gametes.

Diversivore An organism which eats both plants and animals.

Dominant *In genetics:* When a pair of chromosomes contain two different genes for the same character in the genotype (e.g., eye colour or flower colour) then the dominant gene is the one that is expressed.
In ecology: The organism which by its size or other key role exerts a controlling influence over other living components of the ecosystem, e.g. trees in forest ecosystems.

Dysphotic   A zone in deep waters of oceans or a few lakes, beyond 200 m depth, where there is some light but of the wrong wavelength for photosynthesis.
Ecology   The study of the relationships of plants and animals to each other and to their environment.
Ecosystem   A unit of space-time containing living organisms interacting with each other and with their abiotic environment by the interchange of energy and materials.
Ecotone   The boundary between two types of ecosystems, often a transitional zone.
Ecotype   A group of plants within a species adapted genetically to a particular habitat but able to breed freely within other ecotypes of the same species.
Endemism   Restriction of a particular species or other taxonomic group to a particular locality, due to factors such as soil, climate or isolation.
Enzyme   A protein contained in cells which acts as a catalyst in facilitating specific biochemical reactions.
Eury-   The converse of **steno-**. A prefix used to denote that an organism can tolerate a wide range of a specific environmental factor.
Euphotic   The upper zone of aquatic biomes where there is sufficient light penetration of the correct wavelength for photosynthesis to take place.
Eutrophication   The addition to an ecosystem of mineral nutrients, generally raising the NPP. It is usually used of man-made additions of, e.g., nitrogen and phosphorus, to waters (fresh and salt) low in those elements but it can occur in terrestrial systems and also occurs naturally.
Evolution   The process of gradual and continual change in the life-form or behaviour of the successive generations of an organism.
Exotic   A plant or animal introduced to a place where it is not native.
Fecundity   The reproductive potential of an organism, i.e. the number of offspring it could produce in the whole of its reproductive life if given ideal conditions.
Feedback loop   *In ecology:* An interaction mechanism within an ecosystem which often acts to restore the equilibrium of the system.
Fertility   The rate at which an organism produces offspring.
Food chain/web   The series of organisms through which food energy is transferred. All the food chains in an ecosystem form a food web, an interlocking system of food relationships.
Gamete   A reproductive cell having a haploid number of chromosomes which when combined with another gamete leads to the formation of a new individual.
Gene   A unit of genetic material localized in the chromosome which cannot be subdivided and which influences a particular set of characters in the phenotype.
Genetic engineering   The production of new genes by the manipulation of genetic material, e.g. by gene surgery.
Genetics   The study of heredity and the variation of organisms.
Genotype   The sum total of genetic information contained in the chromosomes.
Habitat   The place where a plant or animal normally lives, characterized by its biotic or physical characteristics, e.g. a lake or a forest.
Half-life   The time taken for half the atoms of a given quantity of a radioactive element to disintegrate through radiation.
Haploid   Having a single set of unpaired chromosomes in each nucleus. Characteristic of gametes.
Herbicide   A toxic chemical compound used to kill plants.
Herbivore ('Plant-eater')   An animal which eats mainly plants.

Heterotrophic ('outside nourishing') An organism unable to manufacture its own food (cf autotrophic) and which must obtain it from other organisms, living or dead.

Homeostasis The process whereby a change in the environment stimulates an equilibrium-seeking response so that constant conditions are maintained in the face of a varying environment.

Homoiothermic Maintaining a constant body-temperature, usually raised above the environmental level, i.e. 'warm-blooded'.

Hybrid A plant or animal resulting from a cross between two unalike individuals, often the offspring of two different species or varieties.

Hydrophyte A plant whose habitat is in or near water and which is adapted to life under such conditions.

Limiting factor An environmental factor limiting the growth or reproduction of an individual or community.

Linkage The occurrence of genes on the same chromosome so that they tend to stay together during inheritance.

Maximum Sustained Yield (MSY) The highest level of crop which can be taken from a population for an indefinite length of time without affecting its viability.

Mesophyte A plant which grows under average conditions of water supply.

Mineral nutrients Inorganic substances (e.g. nitrogen, phosphorous, calcium) required by organisms for their growth, acquired in solution from the soil or water by autotrophs and from their food source by heterotrophs.

Mortality The rate at which the individuals of a population die.

Mutation A sudden and rare change in the chromosomal DNA. It introduces a random element into the processes of evolution.

Mutualism An association between two or more species in which all derive benefit and none can normally survive without the other(s).

Natality The birth rate of a population.

Natural selection The way in which the environment of a population winnows out those individuals least suited to it and allows the 'fittest' to survive and reproduce, thus passing on their characteristics to the next generation.

Net Primary Production (NPP) The rate at which plants store energy as organic matter in excess of that used in respiration. Expressed usually as dry weight of organic matter/unit area/unit time, e.g. $g/m^2/day$ or $kg/ha/yr$.

Niche The status or position occupied by an organism in a community due to its structural, physiological or behavioural adaptations to its particular environment and way of life.

Omnivore See **diversivore**.

Optimal range The range of environment factors which are most favourable for the growth and reproduction of an organism.

Osmosis The movement of solvent molecules (usually water) from a dilute to a more concentrated solution through a selectively permeable membrane.

Parasite A plant or animal living in or upon another (the host) and drawing nutrients directly from it and thereby causing it harm.

Pesticide A poisonous chemical designed for use against animals considered to be pests.

Phenotype The sum of characters manifested in an actual organism, produced by the interaction of the individual's genetic potential (genotype) and its environment.

Pheromone A chemical substance produced by one organism that influences the development or behaviour of another.

Photoperiod The length of a period of light (daylength) needed to stimulate a light-regulated response, such as flowering in plants and reproduction in some animals.

Photosynthesis The manufacture of organic compounds by green plants from water and carbon dioxide using the energy from sunlight trapped by chlorophyll.

Phytoplankton Microscopic plants of aquatic habitats; the main primary producers of open water, especially in the ocean.

Placenta In mammals, the organ in the uterus (womb) which enables food, oxygen and waste products to pass between the embryo and its mother.

Ploidy See **haploid, diploid, polyploidy**.

Poikilothermic Animals which maintain a varying body temperature which approximately follows that of the surroundings, i.e. 'cold-blooded'.

Polychlorinated biphenyls (PCBs) An organic compound with a structure similar to that of chlorinated hydrocarbons. Similarly, they are toxic to some organisms (though not used as biocides) and are persistent in residual form in the environment.

Polyploidy Having three or more times the haploid (*q.v.*) number of chromosomes. Not commonly found in animals, in plants the condition may confer a selective advantage since polyploid individuals of a species tend to be larger than those with fewer chromosomes.

Population The total number of organisms of a particular taxonomic group inhabiting a particular area.

Predation One organism's use of another for food by direct ingestion.

Production ecology The study of energy flow through ecosystems involving studies of photosynthesis, uptake and dissipation of energy, accumulation of biomass and food relationships.

Pyrophytic Plants having an adaptation to withstand fire, e.g. a very thick bark.

Radionuclide An atom which emits radiation.

Recessive A recessive gene is only expressed in the phenotype when it is present in a doubled pair on the chromosomes of the genotype.

Recombination Any process which gives rise to cells or individuals with a combination of two or more genes which is not present in the parent.

Respiration The breakdown of organic compounds, e.g. glucose in living cells to release energy, usually involving the uptake of oxygen by the organism and the release of carbon dioxide and water.

Roentgen A measure of the intensity of radiation emitted by a radioactive source, in terms of the ionization of a given volume of gas.

Saprophyte An organism which obtains organic substances in solution from dead and decaying tissue of plants and/or animals.

Saprovore See saprophyte; includes animals as well as plants.

Seed-ferns A group of plants present in the Lower Carboniferous to Jurassic periods, with fern-like fronds bearing structures very similar to seeds.

Sere A series of stages during succession involving changes in the composition of the community in a particular place.

Steno- The converse of **eury-**. A prefix used to designate an organism with only a narrow tolerance range of a particular environmental factor.

Stomata The pores in the epidermis of the aerial parts of plants which allow gaseous exchange to occur.

Succession The regular and progessive change in the components of an ecosystem from the initial colonization of an area to a stable state or mature ecosystem.

Symbiosis (adj. **symbiotically**) A condition where two dissimilar organisms are associated to the mutual benefit of both.

Synecology The study of groups of organisms associated as a unit and their relationships to one another.

Taxonomy The classification of organisms in logical and natural groups along evolutionary lines. A particular group (e.g. species, genus) is a **taxon**.

Thermocline A zone of air or water where the temperature changes rapidly in contrast to the relatively uniform temperatures above and below it.

Transpiration The evaporation of water from the shoot, particularly the leaves, of a flowering plant.

Trophic level The designation of groups of organisms in an ecosystem according to their food sources: e.g., the autotrophs, herbivores (first heterotrophic level), carnivores (second heterotrophic level), saprovores (decomposer level). Each level is a distinct number of energy transfer steps away from the primary producers, i.e. the green plants.

Xeric Organisms having adaptations to withstand arid conditions, e.g. water retention in fleshy tissues in plants (i.e. succulence).

Xerophyte A plant of dry habitats ('xeric') able to withstand drought due to special modifications.

Zooplankton Very small animals which drift in the waters of seas or lakes, feeding on phytoplankton.

Zygote A cell resulting from the fusion of two gametes.

Zoogeographical realm A large region with a distinctive fauna. The world is often divided into six such realms.

Bibliography

ABELSON, P. (ed.) 1974: *Energy: use, conservation and supply.* Washington: AAAS.
AGRICULTURAL ADVISORY COUNCIL (OF ENGLAND AND WALES) 1970: *Modern farming and the soil.* London: HMSO.
ALEXANDER, M. 1973: Environmental consequences of increasing food production. *Biological Conservation* 5, 15–19.
— 1974: Environmental consequences of rapidly rising food output. *Agro-Ecosystems* 1, 249–64.
ALLABY, M. 1976: Cornwall's mackerel war. *New Scientist* 69, 610–12.
ALLEGRO, J. M. 1970: *The sacred mushroom and the cross: a study of the nature and origins of Christianity within the fertility cults of the ancient Near East.* London: Hodder and Stoughton.
ALTSHUL, A. and ROSENFIELD, D. 1970: Protein supplementation: satisfying man's needs. *Progress* 3, 72–80.
AMERICAN CHEMICAL SOCIETY [ACS] 1969: *Cleaning our environment. The chemical basis for action.* Washington DC: ACS.
— 1976: The rising clamour about PCBs. *Environmental Science and Technology* 10, 122–3.
ANTONOVICS, J., BRADSHAW, A. D. and TURNER, R. G. 1971: Heavy metal tolerance in plants. *Adv. Ecol. Res.* 7, 1–85.
ANON. 1978: A destructive forest fungus. *Ecos* 15, 3–14.
ARCHER, B., GRENVILLE, H. W., JAGO, M., JOHNSON, C. B. 1976: *Understanding biology.* London: Mills and Boon.
ARKCOLL, D. B. 1971: Agronomic aspects of leaf protein production in Great Britain. In Pirie, N. W. (ed.), *q.v.*, 9–18.
ARMSTRONG, P. H. 1973: Changes in the land-use of the Suffolk Sandlings: a study of the disintegration of an ecosystem. *Geography* 58, 1–8.
ASCHMANN, H. 1973: Man's impact on the several regions with Mediterranean climates. In F. di Castri and H. A. Mooney (eds.), *q.v.*, 363–72.
ASKEW, R. R., COOK, L. M., and BISHOP, J. A. 1971: Atmospheric pollution and melanic moths in Manchester and its environs. *J. appl. ecol.* 8, 247–56.
ATHWAL, D. S. 1975: Present and future research on rice. *Cereal Foods World*, August 1975, 354–61.
BABB, T. A. and BLISS, L. C. 1974: Effects of physical disturbance on Arctic vegetation in the Queen Elizabeth Islands. *J. appl. Ecol.* 11, 549–62.
BÄCKSTRAND, G. and STENRAM, H. 1971: *Air pollution across national boundaries. The impact on the environment of sulfur in air and precipitation.* Stockholm: Royal Ministry for Foreign Affairs/Royal Ministry of Agriculture.
BAKER, H. G. 1970: *Plants and civilization*, 2nd edn. Belmont, Calif.: Wadsworth.

BALL, M. E. 1974: Floristic changes on grasslands and heaths on the Isle of Rhum after a reduction or exclusion of grazing. *J. Environmental Management* **2,** 299–318.
BARBOUR, I. G. 1973: *Western man and environmental ethics.* Reading, Mass.: Addison-Wesley Publishing Co.
BARDACH, J. E. 1976: Aquaculture revisited. *J. Fisheries Research Board of Canada* **33,** 880–87.
BARKHAM, J. P. 1973: Recreational carrying capacity: a problem of perception. *Area* **5,** 218–22.
BARNARD, C. and FRANKEL, O. H. 1964: Grass, grazing animals and man in historical perspective. In C. Barnard (ed.), *Grasses and Grasslands,* London and Melbourne: Macmillan, 1–12.
BASSHAM, J. A. 1977: Increasing crop production through more controlled photosynthesis. *Science* **197,** 630–38.
BATCHELDER, R. B. and HIRT, H. F. 1966: *Fire in tropical forests and grasslands.* US Army Material Command Earth Sciences Division Technical Report 67–41–ES, Natick, Mass.
BATES, R. P. and HENTGES, J. F. 1976: Aquatic weeds: eradicate or cultivate? *Economic Botany* **30,** 39–50.
BEDDINGTON, J. R. 1975: Economic and ecological analysis of Red Deer harvesting in Scotland. *J. Environmental Management* **3,** 91–103.
BEDDINGTON, J. R. and MAY, R. M. 1977: Harvesting natural populations in a randomly fluctuating environment. *Science* **197,** 463–5.
BELLAMY, D. J. and PRITCHARD, T. 1973: Project Telma: a scientific framework for conserving the world's wildlife peatlands. *Biological Conservation* **5,** 33–40.
BENDER, B. 1975: *Farming in prehistory.* London: John Baker.
BENDER, A. E., KINLBERG, R. LOFQVIST, B. and MUNCK, L. (eds.) 1970: *Evaluation of Novel Protein Products.* Oxford: Pergamon Press.
BENNETT, C. F. 1968: Human influences on the zoogeography of Panama. *Ibero-Americana* **51.** Berkeley and Los Angeles: University of California Press.
— 1975: *Man and earth's ecosystems.* New York and London: Wiley.
BENNETT, R. J. and CHORLEY, R. J. 1978: *Environmental systems. Philosophy, analysis and control.* London: Methuen.
BENSON, A. A. and LEE, R. F. 1975: The role of wax in oceanic food chains. *Scientific American* **232**(5), 76–86.
BEST, R. H. 1965: Recent changes and future prospects of land-use in England and Wales. *Geographical Journal* **131,** 1–12.
BEST, R. and WARD, D. 1956: *The garden controversy.* Wye College Papers in Agricultural Economics.
BICCHIERI, M. G. (ed.) 1972: *Hunters and gatherers today.* New York: Holt, Rinehart and Winston.
BIGALKE, R. C. 1975: Technological problems associated with the utilization of terrestrial wild animals. In R. L. Reid, *q.v.,* 36–46.
BILLINGS, W. D. 1972: *Plants, man and the ecosystem,* 2nd edn. Belmont, California: Wadsworth.
— 1974: Environment: concept and reality in B. R. Strain and W. D. Billings (eds.), *Vegetation and environment.* Handbook of Vegetation Science, part VI. The Hague: Junk.

BISWAS, A. K. and BISWAS, M. R. 1976: State of the environment and its implications to resource policy development. *BioScience* **26,** 19–25.
BLAXTER, K. L. 1973: The limits of agricultural improvement. *J. Agricultural Society Newcastle-upon-Tyne* **25,** 3–12.
— 1975a: The energetics of British agriculture. *Biologist* **22,** 14–18.
— 1975b: Conventional and unconventional farmed animals. *Proc. Nutrition Society* **34,** 51–6.
BOESCH, D. F., HERSHNER, C. H. and MILGRAM, J. M. 1974: *Oil spills and the marine environment.* Cambridge, Mass: Ballinger.
BOLIN, B. 1970: The carbon cycle. In *Scientific American, q.v.,* 47–56.
BORGSTROM, G. 1969: *Too many.* New York: Collier–Macmillan.
BORMANN, F. H., LIKENS, G. E., FISHER, D. W. and PIERCE, R. S. 1968: Nutrient loss accelerated by clear-cutting of a forest ecosystem. *Science* **159,** 882–4.
BORMANN, F. H., LIKENS, G. E. and EATON, J. S. 1969: Biotic regulations of particulate and solution losses from a forest ecosystem. *BioScience* **19,** 600–10.
BOUGHEY, A. S. 1973: *Ecology of populations,* 2nd edn. New York: Macmillan.
BOWMAN, J. C. 1977: *Animals for man.* Studies in Biology 78. London: Edward Arnold.
BOX, T. W. 1974: Increasing red meat from rangeland through improved range management practices. *J. Range Management* **27,** 333–5.
BRAAKSMA, S. and DE BRUYN, O. 1974: Kerkuilen, wel en wee. *Het vogeljaar* **22,** 694–8.
BRANHAM, J. M. 1973: The crown of thorns on coral reefs. *BioScience* **23,** 219–26.
BRETH, S. A. 1975: Durum wheat: new age for an old crop. *CIMMYT Today* **2.**
BRONSON, B. 1975: The earliest farming: demography as cause and consequence. In S. Polgar (ed.), *Population, ecology and social evolution.* Paris and The Hague: Mouton, 53–78.
BROWN, A. W. A. 1971: Pest resistance to pesticides. In White-Stevens, R. (ed.), *Pesticides in the Environment,* Vol. 1, part II, 458–552. New York: Dekker.
BROWN, C. L. 1976: Forests as energy sources in the year 2000. *J. Forestry* **74,** 7–12.
BRUBAKER, S. 1972: *To live on earth.* New York: Mentor.
BRYAN, R. B. and JANSSON, M. C. 1973: Perception of wildlife hazard in national park use. *Trans 38th N. American Wildlife and Natural Resources Conference,* 281–95.
BRYSON, R. 1973: *Climatic modification by air pollution.* Madison, Wisconsin: Institute for Environmental Studies Report 9.
BRYSON, R. and ROSS, J. E The climate of the city. In Detwyler and Marcus, *q.v.* 51–68.
BULLETIN OF THE ATOMIC SCIENTISTS, 1972: Can man control his biological evolution? A symposium on genetic engineering. *Bulletin of the Atomic Scientists* **28,** 12–28.
BUNNELL, F. L., MACLEAN, S. F. and BROWN, J. 1975: Barrow, Alaska, USA. In T. Rosswall and O. W. Heal, *q.v.,* 73–124.
BUNT, J. S. 1975: Primary productivity of marine ecosystems. In H. Lieth and R. H. Whittaker (eds.), *q.v.,* 169–83.
BURINGH, P., VAN HEEMST, H. D. J. and STARING, G. J. 1975: *Computation of the absolute food production of the world.* Agricultural University of Wageningen: Dept. of Tropical Soil Science.
BURNETT, J. H. (ed.) 1964: *The vegetation of Scotland.* Edinburgh: Oliver and Boyd.
BURRELL, T. S. 1973: National Parks. The big three—conservation, recreation and education. *J. Environmental Management* **1,** 201–5.

BURTON, G. W. 1973: Breeding better forages to help feed man and preserve and enhance the environment. *BioScience* **23,** 705–10.
BURTON, J. A. 1975: The future of small whales. *New Scientist* **66,** 650–51.
— 1976: Illicit trade in rare animals. *New Scientist* **72,** 168.
BURWELL, C. C. 1978: Solar biomass energy: an overview of US potential. *Science* **199,** 1041–8.
BUTZER, K. W. 1974: *Environment and archaeology*. 2nd edn. London: Methuen.
CALAPRICE, J. R. 1976: Mariculture—ecological and genetical aspects of production. *J. Fisheries Research Board of Canada* **33,** 1068–87.
CALDER, N. 1967: *The environment game*. London: Secker and Warburg.
CALDWELL, L. K. 1971: *Environment: a challenge to modern society*. New York: Doubleday, Anchor Books.
CALDWELL, M. M. 1975: Primary production of grazing lands. In J. P. Cooper (ed.), *Photosynthesis and productivity in different environments*, IBP Studies 3, 41–73. London: Cambridge University Press.
CALVIN, M. 1976: Photosynthesis as a resource for energy and materials. *American Scientist* **64,** 270–8.
CARAS, R. A. 1978: *Dangerous to man*. Harmondsworth: Penguin.
CARLSON, P. S. and POLACEO, J. C. 1975: Plant cell cultures: genetics. Aspects of crop improvement. *Science* **188,** 622–5.
CAST (Council for Agricultural Science and Technology) 1973: *Energy in agriculture*. Report 14.
CATTON, W. R. 1976: Can irrupting man remain human? *BioScience* **26,** 262–7.
CENTRAL INTELLIGENCE AGENCY [USA] 1974: *Potential implications of trends in world population, food production and climate*. Washington DC: CIA Report OPR-401.
CHADWICK, M. J. 1975: The cycling of materials in disturbed environments. In M. J. Chadwick and G. T. Goodman (eds.), *The Ecology of resource degradation and renewal*. Symposia of the British Ecological Society 15, 3–16. Oxford: Blackwell Scientific Publications.
CHANCELLOR, W. J. and GOSS, J. R. 1976: Balancing energy and food production 1975–2000. *Science* **192,** 218–28.
CHAUDHURI, H., CHAKRABARTY, R. D., SEN, P. R., RAO, N. G. S. and JENA, S. 1975: A new high in fish production in India with record yields by composite fish culture in freshwater ponds. *Aquaculture* **6,** 343–55.
CHANG, T-T., 1976: The rice cultures. *Phil. Trans. Soc. Lond.* **B 275,** 143–57.
CHEMURGIC COUNCIL 1972: *Waste utilization for environmental quality and profit*. New York: Chemurgic Council.
CLAWSON, M. 1975: *Forests for whom and for what*. Baltimore: Johns Hopkins Press, for RFF.
CLAPHAM, W. B. Jr 1973: *Natural Ecosystems*. New York: The Macmillan Co., London: Collier-Macmillan Ltd.
CLARK, D. 1976: Mesolithic Europe: the economic basis. In G. de G. Sieveking, I. H. Longworth, and K. E. Wilson (eds.), *Problems in Economic and Social Archaeology*. London: Duckworth, 449–81.
CLARKE, J. G. D. 1972: *Star Carr: an essay in bioarchaeology*. Reading, Mass.: Addison-Wesley Modules in Anthropology.
CLARKE, J. G. D. and HUTCHINSON, J. (eds.) 1976: The early history of agriculture. *Phil. Trans. Roy. Soc. Lond.* **B 275,** 1–213.

CLOUDSLEY-THOMPSON, J. L. 1975: *Terrestrial environments*, London: Croom Helm.
— 1977: *Man and the biology of arid zones*. London: Edward Arnold.
COHEN, M. N. 1977: *The food crisis in prehistory*. New Haven and London: Yale University Press.
COHEN, S. N. 1975: The manipulation of genes. *Scientific American* **233,** 24–33.
COLE, G. F. 1974: Management involving grizzly bears and humans in Yellowstone National Park 1970–73. *BioScience* **24,** 335–8.
COLE, H. A. 1974: Marine pollution and the UK fisheries. In F. R. Harden-Jones (ed.), *Sea Fisheries Research*, New York: Wiley; London; Elek Ltd, 277–303.
COLLIER, B. D., COX, G. W., JOHNSON, A. W. and MILLER, P. C., 1973: *Dynamic Ecology*. Englewood Cliffs, NJ: Prentice-Hall.
COLLINS, N. J., BAKER, J. H. and TILBROOK, P. J. 1975: Signy Island, maritime Antarctic. In T. Rosswall and D. W. Heal, *q.v.*, 345–74.
COLLINSON, A. S. 1977: *Introduction to World Vegetation*. London: George Allen and Unwin.
CONNELL, J. H. 1978: Diversity in tropical rain forests and coral reefs. *Science* **199,** 1302–10.
CONWAY, G. 1976: Man versus pests. In R. M. May (ed.), *q.v.*, 257–81.
COOK, C. W. 1970: *Energy budget of the range and range livestock*. Colorado Agricultural Experimental Station Technical Bulletin 109.
— 1972: *Energy budget for rabbits compared to cattle and sheep*. Colorado State University Range Science Dept. Science Series 13.
COOK, L. M. and WOOD, R. J. 1976. Genetic effects of pollutants. *Biologist* **23,** 129–39.
COOPER, J. P. (ed.) 1975: *Photosynthesis and productivity in different environments*. IBP Studies 3. London: Cambridge University Press.
COULSON, J. and MONAGHAN, P. 1978: Herring gulls move into town. *New Scientist* **79,** 456–8.
COURTENAY, W. R. and ROBINS, C. R. 1975: Exotic organisms: an unsolved, complex problem. *BioScience* **25,** 306–13.
COX, C. B., HEALEY, I. N. and MOORE, P. D. 1976: *Biogeography. An ecological and evolutionary approach*, 2nd edn. Oxford: Blackwell Scientific Publications.
CRAWFORD, M. A. 1975: A re-evaluation of the nutrient role of animal products. In R. L. Reid (ed.). *q.v.*, 21–35.
CUSHING, D. H. and WALSH, J. J. (eds.). 1976: *The ecology of the seas*. Oxford: Blackwell.
DALE, D. and WEAVER, T. 1974: Trampling effects on vegetation of the trail corridors of North Rocky Mountain forests. *J. appl Ecol* **11,** 767–72.
DAMAS, D. 1972: Environment, history and central Eskimo society. In D. Damas (ed.), *Contributions to Anthropology: Ecological Essays*. Ottawa: National Museums of Canada Bulletin 230, Anthropological Series, 40–64.
DANSEREAU, P. 1957: *Biogeography. An ecological perspective*. New York: Ronald Press.
— (ed.) 1970: *Challenge for survival. Land, air and water for man in megalopolis*. New York and London: Columbia University Press.
DARLING, F. F. and FARVAR, M. 1972: Ecological consequences of sedentarization of nomads. In M. Taghi Farvar and J. P. Milton (eds.), *q.v.*, 671–82.
DARLINGTON, P. J. 1957: *Zoogeography: the geographical distribution of animals*. New York: Wiley.
DARNELL, R. M. 1976: Natural area preservation: the US/IBP conservation of ecosystems program. *BioScience* **26,** 105–8.

DASMANN, R. 1968: *A different kind of country.* London: Macmillan.
— 1972: *Environmental conservation,* 3rd edn. New York and London: Wiley.
— 1975: *The conservation alternative.* New York and London: Wiley.
DAUBENMIRE, R. 1968: Ecology of fire in grasslands. *Adv. Ecol. Res.* **5,** 209–66.
DAVIDSON, J. and LLOYD, R. 1977: *Conservation and agriculture.* Chichester: Wiley.
DEAN, R., ELLIS, J. E., RICE, R. W. and BEMENT, R. E. 1975: Nutrient removal by cattle from a shortgrass prairie. *J. appl. Ecol.* **12,** 25–9.
DE BACH, P. 1964: *Biological control of insect pests and weeds.* New York: Reinholt.
DECOURT, N. 1974: Sur quelques rôles des arbres et forêts dans l'environment urbain. In P. Pesson, q.v., 43–53.
DEEVEY, E. S. 1960: The human population. *Scientific American* **203,** 195–204.
— 1970: Mineral cycles. *The biosphere.* San Francisco: W. H. Freeman, 81–92.
DELWICHE, C. C. 1970: The nitrogen cycle. In *Scientific American,* q.v. 71–80.
DETWYLER, T. R. 1971: Summary and prospect. In T. R. Detwyler (ed.), *Man's impact on environment.* New York: McGraw-Hill, 695–700.
— 1972: Vegetation of the city. In T. R. Detwyler and M. G. Marcus (eds.), *Urbanization and environment,* Belmont, Calif: Duxbury Press, 229–59.
DETWYLER, T. R. and MARCUS, M. G. 1972: Urbanization and environment in perspective. In T. R. Detwyler and M. G. Marcus (eds.), *Urbanization and environment,* Belmont, Calif.: Duxbury Press, 3–25.
DE SOET, F. 1974: Agriculture and the environment. *Agriculture and Environment* **1,** 1–15.
DE VOS, A. 1969: Ecological conditions affecting the production of wild herbivorous mammals on grasslands. *Adv. Ecol. Res.* **6,** 237–83.
DIAMOND, J. M. and MAY, R. M. 1976: Island biogeography and the design of natural reserves. In R. M. May (ed.), *Theoretical ecology. Principle and applications.* Oxford: Blackwell, 163–86.
DI CASTRI, F. and LOOPE, L. 1977: Biosphere reserves: theory and practice. *Nature and Resources* **13,** 2–7.
DI CASTRI, F. and MOONEY, H. A. 1973: *Mediterranean-type ecosystems. Origin and structure.* Ecological Studies 7. Berlin, Heidelberg and New York: Springer-Verlag.
DILLS, G. G. 1970: Effects of prescribed burning on deer browse. *J. Wildlife Management* **34,** 540–45.
DIMBLEBY, G. W. 1976: Climate, soil and man. *Proc. Roy. Soc. Lond.* **B 275,** 197–208.
DIX, R. L. 1964: A history of biotic and climatic changes within the North American grassland. In D. J. Crisp (ed.), *Grazing in terrestrial and marine environments,* Oxford: Blackwell, British Ecological Society Symposia 4, 71–89.
DOBBEN, W. H. VAN, and LOWE-McCONNELL, R. H. (eds.) 1975: *Unifying concepts in ecology.* The Hague: Junk.
DODDS, K. S. 1965: The history and relationships of cultivated potatoes. In Hutchinson, q.v., 123–41.
DOE [Department of the Environment] 1976: *Report of the Working Party on dogs.* London: HMSO, DoE list 1/157, chairman, W. J. S. Batho.
DORST, J. 1970: *Before nature dies.* London: Collins.
DOUGHTY, R. W. 1975: *Feather fashions and bird preservation.* London, Berkeley and Los Angeles: University of California Press.
— 1978: *The English sparrow in the American landscape: a paradox in nineteenth-*

century wildlife conservation. University of Oxford School of Geography Research Paper 19.
DUFFEY, E. 1974: *Nature reserves and wildlife*. London: Heinemann.
EARL, D. E. 1975: *Forest energy and economic development*. Oxford: Clarendon Press.
ECKHOLM, E. 1978: *Disappearing species: the social challenge*. Washington DC: Worldwatch Paper 22.
EDWARDS, C. A. (ed.) 1973: *Environmental pollution by pesticides*. New York and London: Plenum Press.
EDWARDS, P. J. 1977: Studies of mineral cycling in a Montane Rain Forest in New Guinea. II. The production and disappearance of litter. *J. Ecol*. **65**, 971–92.
EGLER, F. E. 1942: Indigene versus alien in the development of arid Hawaiian vegetation. *Ecology* **23**, 14–23.
— 1970: Ecology and management of the rural and suburban landscape. In P. Dansereau (ed.), *q.v.*, 81–98.
EHRLICH, P. and EHRLICH, A. 1972: *Population, resources and environment*. San Francisco: Freeman.
EHRENFELD, D. W. 1976: The conservation of non-resources. *American Scientist* 648–56.
ELIAS, T. S. 1977: An overview. In G. T. Prance and T. S. Elias (eds.), *Extinction is forever*. New York: New York Botanical Garden, 13–16.
EL TITI, A. and STEINER, H. 1975: Möglichkeiten und Grenzen des integrierten Pflanzenschutzes in Gemüsebau. *Qual. Plant*. **25**, 57–75.
ELTON, C. 1958: *Ecology of invasions by animals and plants*. London: Methuen.
ENGSTRÖM, A. 1971: *Air pollution across national boundaries. The impact on the environment of sulfur in air and precipitation*. Report of the Swedish preparatory committee for the UN conference on human environment. Stockholm: Kungl. Bogtryckeriet.
EVANS, G. M. 1976: Rye. In N. W. Simmonds (ed.), *q.v.*, 108–11.
EVANS, H. J. (ed.) 1975: *Enhancing biological nitrogen fixation*. Washington DC: National Science Foundation.
EVANS, H. J. and BARBER, L. E. 1977: Biological nitrogen fixation for food and fiber production. *Science* **197**, 333–9.
EVANS, R. S. 1974: *Energy plantations: should we grow trees for powerplant fuel?* Report VP-X-129, Dept. Environment, Canadian Forest Service, Vancouver, BC.
EYRE, S. R. 1968: *Vegetation and soils. A world picture*. 2nd edn. London: Edward Arnold.
— (ed.) 1971: *World vegetation types*. London: Macmillan.
— 1978: *The real wealth of nations*. London: Edward Arnold.
FALK, J. H. 1976: Energetics of a suburban lawn ecosystem. *Ecology* **57**, 141–50.
FAO (Food and Agriculture Organization [of the UN]) *Production yearbook* (annually). Rome: FAO.
FARVAR, M. T. and MILTON, J. P. (eds.) 1972: *The careless technology. Ecology and international development*. Garden City, New York: Natural History Press.
FELDMAN, M. 1976: Wheat. In N. W. Simmonds (ed.), *q.v.*, 120–8.
FISCHER, D. W., LEWIS, J. E. and PRIDDLE, G. B. (eds.) 1974: *Land and Leisure. Concepts and methods in outdoor recreation*. Chicago: Maaroufa Press.
FISHER, J., SIMON, N. and VINCENT, J. 1969: *Wildlife in danger*. London: Collins.
FITTER, R. S. R. 1963: *Wildlife in Britain*. Harmondsworth: Pelican.
FITTKAU, E. J. and KLINGE, H. 1973: On biomass and trophic structures of the central Amazon rain forest ecosystem. *Biotropica* **5**, 2–14.

FLANNERY, K. V. 1965: The ecology of early food production in Mesopotamia. *Science* **147,** 1247–56.
— 1969: Origins and ecological effects of early domestication in Iran and the Near East. In Ucko and Dimbleby (eds.), *q.v.,* 73–100.
FLÖHN, H. 1973: Der Wasserhaushalt der Erde. *Naturwissenschaften* **60,** 310–48.
FORESTRY COMMISSION 1957: *Exotic forest trees in Great Britain.* Forestry Commission Bulletin 30. London: HMSO.
FORMAN, R. T. T. 1975: Canopy lichens with blue-green algae: a nitrogen source in a Colombian rain forest. *Ecology* **56,** 1176–84.
FOSBERG, F. R. 1963: The island ecosystem. In F. R. Fosberg (ed.), *Man's role in the island ecosystem,* Honolulu: Bishop Museum Press, 1–6.
— (ed.) 1963: *Man's role in the island ecosystem.* Honolulu: Bishop Museum Press.
FRANKEL, O. H. and HAWKES, J. G. (eds.), 1975: *Crop genetic resources of today and tomorrow.* IBP Studies 2. London: Cambridge University Press.
FREEDMAN, R. 1978: Gene manipulation: a new climate. *New Scientist* **79,** 268–69.
FREMLIN, J. H. 1964: How many people can the world support? *New Scientist* **24,** 285–7.
FRISON, G. C. (ed.) 1974): *The Casper site. A Hell Gap bison kill on the High Plains.* New York: Academic Press.
FRYER, J. D. and CHANCELLOR, R. J. 1970: Our changing weeds. In F. Perring (ed.), *The flora of a changing Britain.* London: BSBI, 105–16.
GADGIL, M. and VARTAK, V. D. 1976: The sacred groves of Western Ghats in India. *Economic Botany* **30,** 152–60.
GAMBELL, R. 1976: Population biology and the management of whales. *Applied Biology* **1,** 247–343.
GEERTZ, C. 1964: *Agricultural involution: the processes of ecological change in Indonesia.* Berkeley and Los Angeles: University of California Press.
GELUSO, K. N., ALTENBACH, J. S. and WILSON, D. E. 1976: Bat mortality: pesticide poisoning and migratory stress. *Science* **194,** 184–6.
GIFFORD, R. M. and MILLINGTON, R. J. 1975: *Energetics of agriculture and food production with special emphasis on the Australian situation.* CSIRO Melbourne: Bulletin 288.
GILBERT, O. L. 1970: Further studies on the effect of sulphur dioxide on lichens and bryophytes. *New Phytologist* **69,** 605–27.
— 1971: Some indirect effects of air pollution on bark-living invertebrates. *J. Appl. Ecol.* **8,** 77–84.
GILL, D. and BONNETT, P. 1973: *Nature in the urban landscape: a study of city ecosytems.* Baltimore: York Press.
GILLESPIE, R., HORTON, D. R., LADD, P., MACUMBER, P. G., RICH, T. H. and WRIGHT, R. V. S., 1978: Lancefield Swamp and the extinction of the Australian megafauna. *Science* **200,** 1044–8.
GIMINGHAM, C. H. 1975: *An Introduction to heathland ecology.* Edinburgh: Oliver and Boyd.
GLACKEN, C. J. 1967: *Traces on the Rhodian shore.* Berkeley and Los Angeles: University of California Press.
GOLDSMITH, F. B., MONTON, R. J. C. and WARREN, A. 1970: The impact of recreation on the ecology and amenity of semi-natural areas: methods of investigation used in the Isles of Scilly. *Biol. J. Linnean Society* **2,** 287–306.

GOLLEY, F. B. (ed.), 1977: *Ecological succession.* Benchmark Papers in Ecology vol. 5. London: Wiley.
GOLLEY, F. B. and MEDINA, F. (eds.) 1975: *Tropical ecological systems. Trends in terrestrial and aquatic research.* Ecological Studies 11. Berlin, Heidelberg and New York: Springer-Verlag.
GOOD, R. 1974: *The geography of the flowering plants.* 4th edn. London: Longmans.
GOODMAN, D. 1975: The theory of diversity–stability relationships in ecology. *Quart. Rev. Biol.* **50**, 237–66.
GOODMAN, M. M. 1976: Maize, in N. W. Simmonds (ed.), *q.v.* 128–36.
GOODWIN, H. A. 1973: Ecology and endangered species. *Trans. 38th N. American Wildlife and Natural Resources Conference*, 46–54.
GRANTHAM, J. B. and ELLIS, T. H. 1974: Potentials of wood for producing energy. *J. Forestry* **72**, 522–6.
GREDEN, G. A. 1975: Managing marine national parks: conflicts in resource exploitation. *Proc. Ecological Society of Australia* **8**, 147–55.
GREEN, M. B. and MCCULLOCH, A. 1976: Energy considerations in the use of herbicides. *J. Science of Food and Agriculture* **27**, 95–100.
GRIBBIN, J. 1976: Climatic change and food production. *Food Policy* **1**, 301–12.
— (ed.) 1978: *Climatic change.* London: Cambridge University Press.
GRIGG, D. B. 1974: *The agricultural systems of the world: an evolutionary approach.* London: Cambridge University Press.
GUENZI, W. D. (ed.) 1974: *Pesticides in soil and water.* Madison, Wisconsin: Soil Science Society of America.
GUGGISBERG, C. A. W. 1970: *Man and wildlife.* London: Evans Brothers.
GULLAND, J. A. 1970: Food chain studies and some problems in world fisheries. In J. H. Steele (ed.), *q.v.*, 296–315.
— (ed.) 1972: *The fish resources of the ocean.* London: Fishing News (Books) Ltd.
— 1974: Fishery science, and the problems of management. In F. R. Harden-Jones (ed.), *Sea Fisheries Research.* New York: Wiley, 413–29.
— 1976: Production and catches of fish in the sea. In D. H. Cushing and J. J. Walsh (eds.), *q.v.*, 283–316.
GUNN, D. L. and STEVENS, J. G. R. 1977: *Pesticides and human welfare.* Oxford: Oxford University Press.
HAAG, R. W. and BLISS, L. C. 1974: Energy budget changes following surface disturbance to upland tundra. *J. appl. Ecol.* **11**, 355–74.
HALL, C. A. S. 1975: Look what's happening to our earth. *Bulletin of the Atomic Scientists* **31**, 11–21.
HALL, C. A. S. and DAY, J. W. 1977: *Ecosystem modeling in theory and practice.* New York and London: Wiley.
HALL, D. O. and RAO, K. K. 1974: *Photosynthesis.* Studies in Biology 37. London: Edward Arnold.
HAN, Y. W. and ANDERSON, A. W. 1974: The problem of rice-straw waste. A possible feed through fermentation. *Economic Botany* **28**, 338–44.
HANSON, E. D. 1964: *Animal diversity*, 2nd edn. Englewood Cliffs, NJ: Prentice-Hall.
HANSON, W. C. 1967: Radioecological concentration processes characterizing Arctic ecosystems. In B. Aberg and F. P. Hungate (eds.), *Radioecological concentration processes.* New York and London: Academic Press, 183–91.

HARLAN, J. R. 1975: *Crops and Man*. Madison, Wisconsin: Crop Science Society of America.
— 1976a: The plants and animals that nourish man. *Scientific American* **235,** 89–97.
— 1976b: Barley, in N. W. Simmonds (ed.), *q.v.*, 93–8.
HARPER, J. L. 1961: Approaches to the study of plant competition. *Soc. Exper. Biol. Symp.* **15,** 1–39.
HARPSTEAD, D. D. 1971: High lysine corn. *Scientific American* **225,** 39–42.
HARRIS, D. R. 1962a: The distribution and ancestry of the domestic goat. *Proc. Linn. Soc. Lond.* **173,** 79–91.
— 1962b: The invasion of oceanic islands by alien plants: an example from the Leeward Islands, W.I. *Trans. Institute of British Geographers* **31,** 67–82.
— 1966: Recent plant invasions in the arid and semi-arid southwest of the United States. *Ann. Assoc. Amer. Geogr.* **56,** 408–22.
— 1969: Agricultural systems, ecosystems and the origins of agriculture. In P. J. Ucko and G. W. Dimbleby (eds.), *q.v.*, 3–16.
— 1972: Swidden systems and settlement. In P. J. Ucko, R. Tringham and G. W. Dimbleby (eds.), *Man, settlement and urbanism*. London: Duckworth, 245–62.
— 1978: Alternative pathways towards agriculture. In C. A. Reed (ed.), *Origins of agriculture*. The Hague: Mouton, 179–243.
HARRISON, P. 1977: Beyond the Green Revolution. *New Sci.* **74,** 575–8.
HARRISS, R. J. C. 1977: Common sense in genetic engineering. *Biologist* **24,** 185–93.
HARVEY, H. T., HARTESVELDT, R. J. and STANLEY, J. T. 1972: *Wilderness impact study report*. San Francisco: Sierra Club.
HASEL, K. 1971: *Waldwirtschaft und Umwelt*. Hamburg and Berlin: Paul Parey Verlag.
HAVLICK, S. W. 1974: *The urban organism. The city's natural resources from an environmental perspective*. London: Collier-Macmillan.
HAWKES, J. G. 1969: The ecological background of plant domestication. In P. Ucko and G. W. Dimbleby (eds.), *q.v.*, 17–29.
HAWKSWORTH, D. L. and ROSE, F. 1970: A qualitative scale for estimating sulphur dioxide air pollution in England and Wales using epiphytic lichens. *Nature, London* **227,** 145–8.
HEADY, H. F. 1972: Ecological consequences of Bedouin settlement in Saudi Arabia. In Farvar and Milton (eds.), *q.v.* 683–93.
HEAL, O. W. and MACLEAN, S. F. 1974: Comparative productivity in ecosystems: secondary productivity. In W. H. van Dobben and R. H. Lowe-McConnell (eds.), *Unifying concepts in ecology*. The Hague: Junk, 89–108.
HEIZER, R. F. 1955: *Primitive Man as an Ecologic Factor*. Kroeber Anthropological Society Papers 13. Berkeley: University of California Press.
HENDEE, J. C. and STANKEY, G. H. 1973: Biocentricity in wilderness management. *BioScience* **23,** 535–38.
HIGGS, E. S. and JARMAN, M. R. 1972: The origins of animal and plant husbandry. In E. S. Higgs (ed.), *Papers in Economic Prehistory*. London: Cambridge University Press.
HILL, A. R. 1975: Ecosystem stability in relation to stresses caused by human activities. *Canadian Geographer* **19,** 206–19.
HILLS, T. L. 1976: *The savanna biome: a case study of human impact on biotic communities*. McGill University Dept. of Geography, Savanna Research Series 19.
HODGES, L. 1973: *Environmental pollution*. New York: Holt, Rinehart and Winston.

HODGES, R. D. 1978: The case for biological agriculture. *The Ecologist Quarterly* **2**, 122–43.
HOPKINS, B. 1965: Observations on savanna burning in the Olokemeji Forest Reserve, Nigeria. *J. appl. Ecol.* **2**, 367–82.
HORN, H. S. 1976: Succession. In R. M. May (ed.), *q.v.*, 187–204.
HOULT, D. P. (ed.) 1969: *Oil on the sea.* New York and London: Plenum Press.
HOUSE, P. R. and WILLIAMS, E. R. 1971: *The carrying capacity of a nation.* Lexington, Mass.: Lexington Books.
HOUSON, D. 1971: Ecosystems of national parks. *Science* **172**, 648–51.
HOWARD, N. J. 1964: Introduced browsing animals and habitat stability in New Zealand. *J. Wildlife Management* **28**, 431–9.
HOWELL, R. K. and KREMER, D. F. 1972: Ozone injury to soybean cotyledonary leaves. *J. Environmental Quality* **1**, 94–7.
HUFFAKER, C. B. (ed.) 1971: *Biological control.* New York: Plenum Press.
HUFFAKER, C. B. and MESSENGER, P. S. (eds.) 1976: *Theory and practice of biological control.* London: Academic Press.
HUGHES, M. K. 1974: The urban ecosystem. *Biologist* **21**, 117–26.
HULSE, J. H. and SPURGEON, D. 1974: Triticale. *Scientific American* **231**, 72–80.
HUNTER, C. J. 1975: Edible seaweeds—a survey for the industry and prospects for farming the Pacific Northwest. *Marine Fisheries Review* **37**, 19–26.
HUNTER, J. M. 1966: Ascertaining population carrying capacity under traditional systems of agriculture in developing countries; notes on a method employed in Ghana. *Professional Geographer* **18**, 151–4.
HUSTICH, I. 1966: On the forest-tundra and the northern tree-lines. *Ann. Univ. Turku* **A.II: 36**, 7–47.
HUXLEY, A. 1978: *Plant and planet.* Harmondsworth: Penguin.
HYAMS, W. 1970: *English cottage gardens.* London: Nelson.
HYNES, H. B. N. 1970: The ecology of flowing waters in relation to management. *J. Water Pollution Control Federation* **42**, 418–24.
IRRI 1974: *Research highlights for 1974.* Los Banos, Philippines: IRRI.
ISAAC, E. 1970: *Geography of domestication.* Englewood Cliffs, NJ: Prentice-Hall.
IUCN [International Union for the Conservation of Nature and Natural Resources] 1966: *Red Data Book.* Morges: IUCN, continuing.
— 1971: *United Nations list of national parks and equivalent resources*, 2nd edn. Brussels: Hayez.
IVES, J. D. 1974a: The impact of motor vehicles on the tundra environments. In J. D. Ives and R. G. Barry (eds.), *Arctic and Alpine Environments*, London: Methuen, 907–10.
— 1974b: The impact of man as a biped. In J. D. Ives and R. G. Barry, *q.v.*, 921–24.
IVES, J. D. and BARRY, R. G. (eds.) 1974: *Arctic and Alpine environments.* London: Methuen.
JACOBI, R. M., TALLIS, J. H. and MELLARS, P. A. 1976: The southern Pennine Mesolithic and the ecological record. *J. Archaeological Science* **3**, 307–20.
JACOBS, J. 1975: Diversity, stability and maturity in ecosystems influenced by human activities. In W. H. van Dobben and R. H. Lowe-McConnell (eds), *q.v.* 187–207.
JACOBSEN, J. S. and HILL, A. C. (eds.) 1970: *Recognition of air pollution injury to vegetation.* Pittsburgh, Pa.: Air Pollution Control Association.

JANSSON, B.-O. and WULFF, F. 1977: Baltic ecosystem modelling. In C. A. S. Hall and J. W. Day (eds.), *q.v.*, 324–43.
JARMAN, M. R. 1972: European deer economies and the advent of the Neolithic. In E. S. Higgs (ed.), *Papers in Economic History*, London: Cambridge University Press, 125–45.
JENKIN, J. F. 1975: Macquarie Island, Sub-antarctic. In T. Rosswall and O. W. Heal, *q.v.*, 375–97.
JENKINS, R. E. and BEDFORD, W. B. 1973: The use of natural areas to establish environmental baselines. *Biological Conservation* 5, 168–74.
JENSEN, N. F. 1978: Limits to growth in world food production. *Science* 201, 317–20.
JEWELL, P. A. 1969: Wild animals and their potential for domestication in P. J. Ucko and G. W. Dimbleby (eds.), *The domestication and exploitation of plants and animals*. London: Duckworth, 101–9.
JOHANNESSEN, C. L. 1963: Savannas of interior Honduras. *Ibero-Americana* 46. Berkeley and Los Angeles: University of California Press.
JOHNSON, W. C., OLSON, J. S. and REICHLE, D. E. 1977: Management of experimental reserves and their relation to conservation reserves: the reserve cluster. *Nature and Resources* 13, 8–14.
JORDAN, C. F. and KLEIN, J. R. 1972: Mineral cycling: some basic concepts and their application in a tropical rain forest. *Ann. Rev. Ecol. System* 3, 33–50.
JORGENSEN, J. R., WELLS, C. G. and METZ, L. J. 1975: The nutrient cycle in continuous forest production. *J. Forestry* 73, 400–403.
JOSHI, R. N. 1971: The yields of leaf protein that can be extracted from crops of Aurangabad. In N. W. Pirie, *q.v.*, 19–28.
KASAHARA, H. 1975: State of marine fishery resources—a general review. *Agriculture and Environment* 2, 205–18.
KELLERT, S. R. 1975: Perception of animals in American society. *Trans. 40th N. American Wildlife and Natural Resources Conferences*, 533–46.
KELLOGG, W. W. 1978: Global influences of mankind on the climate. In J. Gribbin (ed.), *q.v.*, 205–227.
KELSALL, J. P. 1968: *The migratory barren-ground caribou of Canada*. Ottawa: Queen's Printer, Dept of Indian Affairs and Northern Development Canadian Wildlife Service Monograph 3.
KIMMINS, J. P. 1977: Evaluation of the consequences for future tree productivity of the loss of nutrients in whole-tree harvesting. *Forest Ecology and Management* 1, 169–83.
KINNE, O. and ROSENTHAL, H. 1977: Commercial cultivation (aquaculture). In O. Kinne (ed.), *Marine ecology*. Vol. III, part 3, 1321–98. Chichester, New York etc.: Wiley.
KING, J. 1977: A science for the people. *New Sci.* 74, 634–6.
KIRITANI, K. 1972: Strategy in integrated control of rice pests. *Review Plant Protection Research* 5, 76–104.
KLAGES, K. W. H. 1942: *Ecological crop geography*. New York: Macmillan (7th printing, 1961).
KLEIN, W. 1976: Environmental pollution by pesticides. In R. L. Metcalfe and J. J. McKelvey (eds.), *The future for insecticides*. New York and London: Wiley, 65–92.
KLINGE, H., RODRIGUES, W. A., BRUNIG, E. and FITTKAU, E. J. 1975: Biomass and structure in a central Amazonian rain forest. In F. B. Golley and E. Medina (eds.), *q.v.*, 115–22.

KNIGHT, C. B. 1965: *Basic concepts of ecology*. New York: Macmillan.
KOZLOWSKI, T. T. and AHLGREN, C. E. (eds.) 1974: *Fire and ecosystems*. London and New York: Academic Press.
KREBS, C. J. 1972: *Ecology*. New York: Harper and Row.
KREBS, C. J. and MYERS, J. M. 1974: Population cycles in small mammals. *Adv. Ecol. Res.* **8**, 268–399.
KRUMBEIN, W. E. (ed.) 1978: *Environmental biogeochemistry and geomicrobiology*. London and New York: Wiley, 3 vols.
LACEY, J. A. and SALLAWAY, M. M. 1975: Some aspects of the formulation of a planning policy for the management of the Ayers Rock–Mount Olga National Heritage Area. *Proc. Ecological Society of Australia* **9**, 256–66.
LAMOTTE, M. 1975: The structure and function of a tropical savanna ecosystem. In F. B. Golley and E. Medina (eds.), *q.v.* 179–222.
LA PAGE, W. F. 1974: Campground trampling and ground cover response. In Fisher, Lewis and Priddle (eds.), *q.v.*, 237–44.
LASLEY, J. F. 1972: *Genetics of livestock improvement*, 2nd edn. Englewood Cliffs, NJ: Prentice-Hall.
LAURIE, M. V. 1974: Tree planting practices in African Savannas. *FAO Forestry Dept. Paper* 19. Rome: FAO.
LEACH, G. 1975: The energy costs of food production. In F. Steele and A. Bourne, *The man/food equation*, London: Academic Press.
LEACH, G. 1976: *Energy and food production*. London: IPC Press.
LEARMONTH, A. T. A. and SIMMONS, I. G. 1977: *Man–environment relationships as complex ecosystems*. OU Course D204 Unit 8.
LEDOUX, L. (ed.) 1975: *Genetic manipulation with plant material*. London: Plenum Press.
LEE, R. B. 1968: What hunters do for a living, or, how to make out on scarce resources. In R. B. Lee and I. De-Vore *q.v.*, 30–48.
LEE, R. B. and DE VORE, I. (eds.) 1968: *Man the hunter*. Chicago: Aldine Press.
LEE, R. B. and DE VORE, I. 1968: Problems in the study of hunters and gatherers. In R. B. Lee and I. De Vore (eds.), *q.v.*
LEEDS, A. and VAYDA, A. P. (eds.) 1965: *Man, culture and animals*. Washington DC: AAAS Publication 78.
LEVER, C. 1977: *The naturalized animals of the British Isles*. London: Hutchinson.
LEWIN, R. 1978: Galapagos: the endangered islands. *New Scientist* **79**, 168–71.
LEWIS, G. M. 1969: Range management viewed in the ecosystem framework, in G. M. Van Dyne (ed.), *The ecosystem concept in resource management*. London and New York: Academic Press, 97–187.
LEWIS, H. T. 1972: The role of fire in the domestication of plants and animals in Southwest Asia: a hypothesis. *Man* **7**, 195–222.
LEWIS SMITH, R. I. and WALTON, D. W. H. 1975: South Georgia, Subantarctic. In T. Rosswall and O. W. Heal, *q.v.*, 399–423.
LIETH, H. 1972: Modelling the primary productivity of the world. *Nature and Resources* **8**, 5–10.
— 1975: Primary productivity of the major vegetation units of the world. In H. Lieth and R. H. Whittaker (eds.), *q.v.*, 203–15.
LIETH, H. and WHITTAKER, R. H. (eds.) 1975: *Primary productivity of the biosphere*. Ecological Studies 14. Berlin, Heidelberg and New York: Springer-Verlag.

LIKENS, G. E. 1975: Primary production of inland aquatic ecosystems. In H. Lieth and R. H. Whittaker (eds.), *q.v.*, 185–202.
LIKENS, G. E., BORMANN, F. H., JOHNSON, N. M. 1972: Acid rain. *Environment* 14, 33–40.
LIKENS, G. E. and BORMANN, F. H. 1975: An experimental approach in New England landscapes. In A. D. Hasler (ed.) *Coupling of land and water systems*. Berlin, New York and Heidelberg: Springer-Verlag, 7–29.
LIKENS, G. E., BORMANN, F. H., PIERCE, R. S., EATON, J. S. and JOHNSON, N. M. 1977: *Biogeochemistry of a forested ecosystem*. New York, Berlin, and Heidelberg: Springer-Verlag.
LIKENS, G. E., BORMANN, F. H., PIERCE, R. S., and REINERS, W. A. 1978: Recovery of a deforested ecosystem. *Science* 199, 492–6.
LINDEMAN, R. I. 1942: The trophic-dynamic aspect of ecology. *Ecology* 23, 339–418.
LIVINGSTON, R. S. and MCNEILL, B. (eds.) 1975: *Beyond Petroleum. Vol 1: Biomass Energy chains*. Palo Alto: Stanford University Institute for Energy Studies.
LIPINSKY, E. S. 1978: Fuels from biomass: integration with food and materials systems. *Science* 199, 644–51.
LONGHURST, A., COLEBROOK, M., GULLAND, J., LE BRASSEUR, R., LORENZEN, C. and SMITH, P. 1972: The instability of ocean populations. *New Scientist* 54, 500–2.
LOOMIS, R. S. and GERAKIS, P. A. 1975: Productivity of agricultural environments. In J. P. Cooper (ed.), *q.v.*, 145–72.
LOOSLI, J. K. 1974: New sources of protein for human and animal feeding. *BioScience* 24, 26–31.
LUGO, A. 1974: Tropical ecosystem structure and function. In E. G. Farnworth and F. B. Golley (eds.), *Fragile ecosystems*, Berlin, Heidelberg and New York: Springer-Verlag, 67–111.
LUNDHOLM, B. 1975: Interactions between oceans and terrestrial ecosystems. In S. F. Singer (Ed.) *The changing global environment*. Dordrecht: Reidel, 329–36.
MACARTHUR, R. H. 1955: Fluctuations of animal populations and a measure of community stability. *Ecology* 36, 533–6.
— 1972: *Geographical ecology*. New York: Harper and Row.
MCTAGGART COWAN, I. 1976: Biota Pacifica 2000. In R. F. Scagel (ed.), *Mankind's future in the Pacific*, Vancouver: University of British Columbia Press, 86–98.
MABBUTT, J. A. 1977: Climatic and ecological aspects of desertification. *Nature and Resources* 13, 3–9.
MAKHIJANI, A. and POOLE, A., 1976: *Energy and agriculture in the Third World*. Cambridge, Mass: Balinger Publishing Co.
MANGELSDORF, P. C. 1961: Biology, food and people. *Economic Botany* 15, 279–88.
MARTIN, P. S. 1967: Prehistoric overkill. In P. S. Martin and H. E. Wright (eds.), *Pleistocene extinctions: the search for a cause*, New Haven: Yale University Press, 75–120.
— 1973: The discovery of America. *Science* 179, 969–74.
MATTYASOVSKY, E. 1974: Recreation area planning: some physical and ecological requirements. In D. W. Fischer, J. E. Lewis and G. B. Priddle (eds.), *Land and leisure. Concepts and methods in outdoor recreation*. Chicago: Maaroufa Press, 221–36.
MAUGH, T. H. 1977: Guayale and jojoba: agriculture in semi-arid regions. *Science* 196, 1189–90.
MAY, R. M. 1975: Stability in ecosystems: some comments. In W. H. van Dobben and R. H. Lowe-McConnell (eds.), *q.v.*, 161–8.

— (ed.) 1976: *Theoretical ecology. Principles and applications.* Oxford: Blackwell.
MEADOWS, D. H., MEADOWS, D. L., RANDERS, L and BEHRENS, W. W. 1972: *The limits to growth.* London: Earth Island Press.
MELLANBY, K. 1972: *The biology of pollution.* Studies in Biology 38. London: Edward Arnold.
— 1978: Crops–oil = not enough food. *The Countryman* **83**, 45–50.
MELLARS, P. 1976: Fire ecology, animal populations and man: a study of some ecological relationships in prehistory. *Proc. of the Prehistoric Society* **42**, 15–45.
MERRITT, M. L. and FULLER, R. G. 1977: *The Environment of Amchitka Island, Alaska.* Springfield, Va.: National Technical Information Service catalogue no. TD-26712.
MESAROVIC, M. and PESTEL, E. 1975: *Mankind at the turning point.* London: Hutchinson.
METCALF, R. L. 1972: Agricultural chemicals in relation to environmental quality: insecticides today and tomorrow. *J. Environmental Quality* **1**, 10–14.
METCALF, R. and LUCKMAN, W. H. 1975: *Introduction to insect pest management.* New York: Wiley.
MILLER, A. H., MCMILLAN, I. I. and MCMILLAN, E. 1965: *The current status and welfare of the California condor.* New York: National Audubon Society Research report 6.
MILLER, G. R. and WATSON, A. 1973: Some effects of fire on vertebrate herbivores in the Scottish Highlands. *Proc. 13th Tall Timbers Ecology Conf.* 39–64.
MILLER, R. S. and BOTKIN, D. B. 1974: Endangered species: models and predictions. *American Scientist* **62**, 172–81.
MITCHELL, B., STAINES, B. W. and D. WELCH, 1977: *Ecology of red deer.* Cambridge: Institute of Terrestrial Ecology.
MITCHELL, J. M. 1975: A reassessment of atmospheric pollution as a cause of long-term changes of global temperature. In S. F. Singer (ed.), *The changing global environment.* Dordrecht: Reidel, 149–73.
MONTEFIORE, H. 1970: *Can man survive?* London: Fontana.
MONOD, T. (ed.) 1975: *Pastoralism in tropical Africa.* London, Ibadan and Nairobi: Oxford University Press, for the International African Institute.
MONTGOMERY, G. G. and SUNQUIST, M. E. 1975: Impact of sloths on neotropical forest energy flow and nutrient cycling. In F. B. Golley and E. Medina (eds.), *q.v.*, 69–98.
MOONEY, H. A. 1974: Plant forms in relation to environment. In B. R. Strain and W. D. Billings (eds.), *Vegetation and environment.* Handbook of Vegetation science part VI, 113–22. The Hague: Junk.
MOONEY, H. A. and PARSONS, D. J. 1973: Structure and function of the California chaparral—an example from San Dimas. In F. di Castri and H. A. Mooney (eds.), *q.v.*, 83–112.
MOORE, P. D. 1975: Origin of blanket mires. *Nature, London* **256**, 267–9.
MOORE, R. M. and BUDDISCOMBE, E. F. 1964: The effects of grazing on grasslands. In C. Barnard (ed.), *Grasses and grasslands.* London and Melbourne: Macmillan, 221–35.
MOORE, R. W. 1975: Penoxalin, a new selective herbicide for weed control in peas, beans and other crops. *Proc. 28th New Zealand Weed and Pest Control Conference*, 185–8.
MOSIMANN, J. E. and MARTIN, P. S. 1975: Simulating overkill by Palaeoindians. *American Scientist* **63**, 304–13.

MUDD, J. B. and KOZLOWSKI, T. T. (eds.) 1975: *Responses of plants to air pollution*. London and New York: Academic Press.
MUELLER-DOMBOIS, D. 1975: Some aspects of island ecosystem analysis. In F. B. Golley and E. Medina (eds.), *q.v.*, 353–66.
MÜLLER, P. 1974: *Aspect of Zoogeography*. The Hague: Junk.
MÜLLER-WILLIE, L. 1974: The snowmobile, Lapps and reindeer herding in Finnish Lapland. In J. D. Ives and R. G. Barry (eds.), *q.v.*, 915–20.
MURDOCH, W. W. (ed.) 1975a: *Environment. Resources, pollution and society*. 2nd edn. Sunderland, Mass.: Sinauer.
— 1975b: Diversity, complexity, stability and pest control. *J. appl. Ecol.* **12,** 795–807.
MURDOCK, G. P. 1965: Human influences on the ecosystems of high islands of the tropical Pacific. In F. R. Fosberg (ed.), *q.v.*, 145–52.
MURTON, R. K. and WESTWOOD, N. J. 1976: Birds as pests. *Applied Biology* **1,** 89–181.
MUTCH, W. E. S. 1974: Land management—an ecological view. *J. Environmental Management* **2,** 259–67.
MYERS, N. 1972a: National parks in savannah Africa. *Science* **178,** 1255–63.
— 1972b: *The long African day*. New York: Macmillan.
— 1975: The whaling controversy. *American Scientist* **63,** 448–55.
— 1976: An expanded approach to the problem of disappearing species. *Science* **193,** 198–202.
NAS, 1972: *Genetic vulnerability of major crops*. Washington DC: NAS.
— 1975: *Underexploited tropical plants with promising economic value*. Washington DC: NAS.
NAVEH, Z. 1974: The ecological management of non-arable Mediterranean uplands. *J. Environmental Management* **2,** 351–71.
NAVEH, Z. and DAN, J. 1973: The human degradation of Mediterranean landscapes in Israel. In F. di Castri and H. A. Mooney (eds.), *q.v.*, 373–90.
NELSON, D. J. and EVANS, F. C. (eds.) 1969: *Symposium on Radioecology*. Springfield, Va.: US Dept. of Commerce.
NELSON, J. G., NEEDHAM, R. D. and MANN, D. L. (eds.) 1978: *International experience with National Parks and related reserves*. Waterloo, Ont.: University of Waterloo Dept. of Geography Publication Series 12.
NELSON-SMITH, A., 1977: Estuaries. In R. S. K. Barnes (ed.), *The Coastline*. London: Wiley, 123–46.
NILSSON, P. O. 1976: The energy balance in Swedish forestry. In C. O. Tamm (ed.), *q.v.*, 95–101.
NILSSON, P. O. and WERNIUS, S., 1976: Whole-tree utilization. In C. O. Tamm (ed.), *q.v.*, 131–6.
NOBILE, P. and DEEDY, J. (eds.) 1972: *The complete ecology fact book*. Garden City New York: Anchor books.
NOBLE, R. 1975: Growing fish in sewage. *New Scientist* **67,** 259–61.
NUMATA, M. 1974: Conservation of flora and vegetation in Japan. In M. Numata (ed.), *The Flora and Vegetation of Japan*. Tokyo: Kodansha Ltd.; Amsterdam: Elsevier, 269–77.
NUTMAN, P. S. (ed.) 1976: *Symbiotic nitrogen fixation in plants*, IBP vol. no. 7. London: Cambridge University Press.

O'CONNORS, H. B., WURSTER, C. F., POWERS, C. D., BIGGS, D. C. and ROWLAND, R. G. 1978: Polychlorinated biphenyls may alter marine trophic pathways by reducing phytoplankton size and production. *Science* **201**, 737–9.
ODUM, E. P. 1969: The strategy of ecosystem development. *Science* **164**, 262–70.
— 1971: *Fundamentals of Ecology*, 3rd edn. Philadelphia, London and Toronto: W. B. Saunders Company.
— 1972: Ecosystem theory in relation to man. In J. A. Wiens (ed.), *q.v.*, 11–24.
— 1975: *Ecology*, 2nd edn. New York and London: Holt, Rinehart and Winston.
ODUM, E. P. and ODUM, H. T. 1972: Natural areas as components of man's natural environment. *Transactions 37th N. American Wildlife and Natural Resources Conference*, 178–89.
ODUM, H. T. (ed.) 1970: *A tropical rain forest*. Washington DC: Div. Tech Info. US Atomic Energy Commission.
— 1971: *Environment, power and society*. London and New York: Wiley.
ODUM, H. T. and ODUM, E. P. 1976: *Energy basis for man and nature*. New York and London: McGraw-Hill.
OLKOWSKI, W., OLKOWSKI, H., VAN DEN BOSCH, R., and HORN, R. 1976: Ecosystem management: a framework for urban pest control. *BioScience* **26**, 384–9.
OLMO, H. P. 1976: Grapes. In N. W. Simmonds (ed.), *q.v.*, 294–8.
OLSEN, S. J. and OLSEN, J. W. 1977: The Chinese wolf, ancestor of New World dogs. *Science* **197**, 533–5.
OMEROD, W. E. 1976: Ecological effect of control of African trypanosomiasis, *Science* **191**, 815–21.
ORME, B. 1977: The advantages of agriculture. In J. V. S. Megaw (ed.), *Hunters, gatherers and first farmers beyond Europe*. Leicester: Leicester University Press, 41–49.
OSBURN, W. S. 1974a: The snowmobile in Eskimo culture. In J. D. Ives and R. G. Barry (eds.), *q.v.*, 911–13.
—1974b: Large-scale examples [of the impact of technology on the tundra]. In J. D. Ives and R. G. Barry (eds.), *q.v.*, 925–51.
— 1974c: Radioecology. In J. D. Ives and R. G. Barry (eds.), *q.v.*, 875–903.
ØSTBYE, E. (and 18 others) 1975: Hardangervidda, Norway. In T. Rosswall and O. W. Heal (eds.), *q.v.*, 225–64.
OUDEJANS, J. H. M. 1976: Date palm. In N. W. Simmonds (ed.), *q.v.*, 229–31.
OVINGTON, J. D., HEITKAMP, D., and LAWRENCE, D. B. 1963: Plant biomass and productivity of prairie, savanna, oakwood and maize field systems. *Ecology* **48**, 515–24.
OWEN, J. E. 1975: The meat-producing characteristics of the indigenous Malawi goat. *Tropical Science* **17**, 123–38.
PAINE, R. T. 1969: A note on trophic complexity and stability. *American Naturalist* **103**, 91–3.
PARKER, I. S. C. and GRAHAM, A. D. 1971: The ecological and economic basis for game ranching in Africa. In E. Duffey, and A. S. Watts (eds.), *Scientific Management of animal and plant communities for conservation*. Oxford: Blackwell, 393–404.
PARSONS, D. J. 1976: Vegetation structure in the Mediterranean scrub communities of California and Chile. *J. Ecol.* **64**, 435–47.
PARSONS, D. J. and MOLDENKE, A. R. 1975: Convergence in vegetation structure along analogous climatic gradients in California and Chile. *Ecology* **56**, 950–7.
PARSONS, J. J. 1962: *The green turtle and man*. Gainesville, Fla.: University of Florida Press.

— 1972: Spread of African pasture grasses to the American tropics. *J. Range Mgmt.* **25**, 12–17.

— 1976: Forest to pasture: development or destruction? *Rev. Biol. Trop.* **24** (Suppl. 1), 121–38.

PASSMORE, J. 1974: *Man's responsibility for nature.* London: Duckworth.

PATTEN, B. C. 1972: A simulation of the shortgrass prairie ecosystem. *Simulation* **19**, 177–86.

PEAKALL, D. B. and LINCER, J. L. 1970: Polychlorinated biphenyls: another long-life widespread chemical in the environment. *BioScience* **20**, 958–64.

PEARS, N. 1977: *Basic biogeography.* London: Longmans.

PEARSALL, W. H. 1968: *Mountains and moorlands,* 2nd edn, revised by W. Pennington. London: Collins.

PEDEN, D. G., VAN DYNE, G. M., RICE, R. W. and HANSEN, R. M. 1974: The trophic ecology of *Bison bison* L. on shortgrass plains. *J. appl. Ecol.* **11**, 489–98.

PERKINS, D. F. 1978: The distribution and transfer of energy and nutrients in the Agrostis-Festuca grassland ecosystem. In D. W. Heal and D. F. Perkins (eds.), *Production ecology of British moors and mortane grasslands,* Ecological Studies 27, 375–95. Berlin, Heidelberg and New York: Springer-Verlag.

PERRING, F. H. and MELLANBY, K. (eds.) 1978: *Ecological effects of pesticides.* London: Academic Press. Linnean Society Symposium 5.

PESSON, P. (ed.), 1974: *Écologie Forestière.* Paris: Gauthier-Villais Éditeur.

PETERKEN, G. 1974: A method for assessing woodland flora for conservation using indicator species. *Biological Conservation* **6**, 239–45.

PHILLIPS, L. L. 1976: Cotton. In N. W. Simmonds (ed.), *q.v.*, 196–200.

PHILLIPSON, J. 1966: *Ecological energetics.* Studies in Biology 1. London: Edward Arnold.

PIMENTEL, D., HURD, L. H., BELLOTTI, A. C., FORSTER, M. J., OKA, I. N., SHOLES, O. D. and WHITMAN, R. J. 1974: Food production and the energy crisis. In P. Abelson (ed.), *q.v.*, 41–47.

PIRIE, N. W. 1969: *Food resources conventional and novel.* Harmondsworth: Penguin.

— (ed.) 1971: *Leaf protein: its agronomy, preparation, quality and use.* IBP Handbook 20. Oxford: Blackwell.

— 1975: Using plants optimally. In J. Lenihan and W. W. Fletcher (eds.), *Food, agriculture and the environment,* Glasgow and London: Blackie, 48–70.

PIRES-FERREIRA, J. W., PIRES-FERREIRA, E. and KAULICKE, P. 1976: Pre-ceramic animal utilization in the Central Peruvian Andes. *Science* **194**, 483–90.

POLLARD, E., HOOPER, M. D. and MOORE, N. W. 1974: *Hedges.* London: Collins.

POLUNIN, N. 1960: *Introduction to Plant Geography.* London: Longmans.

POMEROY, L. R. 1974: The ocean's food web: a changing paradigm. *BioScience* **24**, 499–504.

POMEROY, D. E. 1978: The abundance of large termite mounds in Uganda in relation to their environment. *J. appl. Ecol.* **15**, 51–64.

POTTS, G. R. and VICKERMANN, G. P. 1974: Studies on the cereal ecosystem. *Adv. Ecol. Res.* **8**, 107–197.

PRAKASH, A. 1974: Dinoflagellate blooms—an overview. *Proc. First International Conference on Toxic Dinoflagellate Blooms.* Wakefield, Mass.: Mass. Science and Technology Federation, 1–6.

PRUITT, W. O. 1978: *Boreal ecology.* Studies in Biology 91. London: Edward Arnold.

PULLAN, R. A. 1974: Farmed parkland in West Africa. *Savanna* **3**, 119–51.

PURDOM, C. E. 1976: Genetic techniques in flatfish culture. *J. Fisheries Resource Board of Canada* **33**, 1088–93.
RACKHAM, D. 1975: *Hayley Wood. Its history and ecology* Cambridge: Cambridgeshire and Isle of Ely Naturalists' Trust Ltd.
RAPPAPORT, R. A. 1965: Aspects of man's influence upon island ecosystem: alteration and control. In F. R. Fosberg (ed.) *q.v.*, 155–70.
— 1971: The flow of energy in an agricultural society. *Scientific American* **224**, 116–32.
RATCLIFFE, D. A. 1977: *A nature conservation review.* London: Cambridge University Press, 2 vols.
RAVEN, P. H. 1976: Ethics and attitudes. In J. B. Simmonds, R. I. Beyer, P. E. Brandham, G. Ll. Lucas and V. T. H. Parry (eds.), *Conservation of threatened plants.* London and New York: Plenum Press, 155–79.
RAY, A. J. 1975: Some conservation schemes of the Hudson's Bay Company 1821–50: an examination of the problems of management in the fur trade. *J. Historical Geography* **1**, 49–68.
REED, C. A. 1969: The pattern of animal domestication in the prehistoric Near East. In P. J. Ucko and G. W. Dimbleby (eds.), *q.v.*, 361–81.
— 1970: Extinction of mammalian megafauna in the Old World late Quaternary. *BioScience* **20**, 284–8.
REGIER, H. A. and HARTMAN, W. L. 1973: Lake Erie's fish community: 150 years of cultural stress. *Science* **180**, 1248–55.
REICHLE, D. E. (ed.) 1973: *Analysis of temperate forest ecosystems*, Ecological Studies 1. Berlin, Heidelberg and New York: Springer-Verlag.
REICHLE, D. E., O'NEILL, R. V. and HARRIS, W. F. 1975: Principles of energy and material exchange in ecosystems. In W. H. van Dobben and R. H. Lowe-McConnell (eds.), *q.v.*, 27–43.
REID, R. L. (ed.) 1975: *Proceedings of the III World Conference on animal production.* Sydney: Sydney University Press.
RENDEL, J. 1975a: Animal breeding in the future. In R. L. Reid (ed.), *q.v.*, 145–57.
— 1975b: The utilization and conservation of the world's animal genetic resources. *Agriculture and Environment* **2**, 101–19.
RENNIE, P. J. 1955: The uptake of nutrients by mature forest growth. *Plant and Soil* **7**, 49–95.
REVELLE, R. 1976: The resources available for agriculture. *Scientific American* **235**, 164–78.
REYNOLDS, E. R. C. and WOOD, P. J. 1977: Natural versus man-made forests as buffers against environmental deterioration. *Forest Ecology and Management* **1**, 83–96.
RIBAKOFF, S. B., ROTHWELL, G. N. and HANSON, J. A. 1974: Platforms for open sea mariculture. In J. A. Hanson (ed.), *Open sea mariculture.* Stroudsburg, Pennsylvania: Dowden, Hutchinson and Ross.
RICE, T. R. and WOLFE, D. A. 1971: Radioactivity: chemical and biological aspects. In D. W. Hood (ed.), *Impingement of man on the oceans*, London: Wiley, 325–79.
RICHARDS, P. W. 1952: *The tropical rain forest.* London: Cambridge University Press.
— 1973: The tropical rain forest. *Scientific American* **229**(6), 58–67.
RICKER, W. E. 1969: Food from the sea. In P. E. Cloud (ed.), *Resources and man.* San Francisco: Freeman.
RICKLEFS, R. E. 1973: *Ecology.* London: Nelson.

RODIN, L. E. and BAZILEVICH, N. I. 1967: *Production and mineral cycling in the terrestrial vegetation*. Edinburgh and London: Oliver and Boyd (English translation ed. G. E. Fogg).
RODIN, L. E., BAZILEVICH, N. I. and ROZOV, N. N. 1975: Productivity of the world's main ecosystems. In NAS, *Productivity of world ecosystems*. Washington DC: NAS, 13–26.
RODRIGUEZ, G. 1975: Some aspects of the ecology of tropical estuaries. In F. B. Golley and E. Medina (eds.), *q.v.*, 313–33.
ROGERS, E. S. 1972: The Mistassini Cree. In M. G. Bicchieri, *q.v.*, 90–137.
ROSSWALL, T. and HEAL, O. W. 1974: *Structure and function of tundra ecosystems*. Stockholm: Swedish Natural Science Research Council Ecological Bulletins No. 20.
ROSTLUND, E. 1952: *Freshwater fish and fishing in native North America*. University of California Publications in Geography, Vol. 9.
ROUNSFELL, G. A. (ed.) 1977: *Handbook of Marine Science: 3. Mariculture*. Oxford: Blackwell, 2 vols.
ROYCE, W. F. 1972: *Introduction to the fishery sciences*. New York: Academic Press.
RUDD, R. L. 1964: *Pesticides and the living landscape*. Madison: Univ. Wisconsin Press.
RUDOLF, P. O. 1974: Silviculture for recreation area management. In Fischer, Lewis and Priddle, *q.v.*, 252–7.
RUIVO, M. (ed.) 1972: *Marine pollution and sea life*. London: Fishing News (Books) Ltd.
RYTHER, J. H. 1969: Photosynthesis and fish production in the sea. *Science* **166,** 72–6.
SAFRANY, D. R. 1971: Nitrogen fixation. *Scientific American* **225,** 64–80.
SAGE, B. 1978: The rook in Britain. *New Scientist* **78,** 898–9.
SALAMAN, R. N. 1949: *The history and social influence of the potato*. London: Cambridge University Press.
SALISBURY, E. 1961: *Weeds and aliens*. London: Collins.
SALMON, J. T. 1975: The influence of man on the biota. In G. Kuschel (ed.), *Biogeography and Ecology in New Zealand*. The Hague: Junk, 643–61.
SAN JOSE, J. J. and MEDINA, E. 1975: Effect of fire on organic matter production and water balance in a tropical savanna. In F. B. Golley and E. Medina (eds.), *q.v.*, 251–64.
SARMIENTO, G. and MONASTERIO, M. 1975: A critical consideration of the environmental conditions associated with the occurrence of savanna ecosystems in tropical America. In F. B. Golley and E. Medina (eds.), *q.v.*, 223–50.
SARRE, P. 1978: The diffusion of Dutch elm disease. *Area* **10,** 81–5.
SAUER, C. O. 1960: Maize into Europe. *Akten des 34. internationalen Amerikanisten-kongresses*, Wien, 777–88.
— 1969: *Agricultural origins and dispersal*, 2nd edn. Cambridge, Mass.: MIT Press.
SAUER, J. D. 1950: The grain Amaranths; a survey of their history and classification. *Annals Missouri Bot. Gdn.* 561–632.
SCEP 1970: *Man's impact on the global environment*. Cambridge, Mass.: MIT Press.
SCHAFER, E. H. 1962: The conservation of nature under the T'ang dynasty. *J. Econ. Soc. Hist. Orient* **5,** 282–308.
SCHALL, J. J. and PIANKA, E. R. 1978: Geographical trends in numbers of species. *Science* **201,** 679–86.
SCHMID, J. A. 1975: *Urban vegetation: a review and Chicago case study*. University of Chicago Dept. of Geography research paper 161.

SCHNEIDER, S. H. and TEMKIN, R. L. 1978: Climatic changes and human affairs. In J. Gribbin (ed.), *Climatic change*. London: Cambridge University Press, 228–46.
SCHULTZ, P. E. 1967: Public use of underwater resources. *Towards a new relationship of man and nature in temperate lands*, IUCN Publ., NS, 7, 153–9.
SCHULTZ, V. and KLEMENT, A. W. 1963: *Radioecology*. Washington DC: American Institute of Biological Sciences; London: Chapman and Hall.
SCHUMACHER, E. M. 1973: *Small is beautiful*. London: Blond and Briggs.
SCIENTIFIC AMERICAN 1970: *The biosphere*. San Francisco: Freeman.
SCOTTER, G. W. 1970: Reindeer husbandry as a land use in Northern Canada. In W. A. Fuller and P. G. Kevan (eds.), *Productivity and conservation in northern circumpolar lands*, IUCN Pubs, NS, 16. Morges: IUCN.
SEDDON, B. 1971: *Introduction to Biogeography*. London: Duckworth.
SHEAIL, J. 1976: *Nature in trust*. Glasgow: Blackie.
SHAW, W. W. 1974: Meanings of wildlife for Americans: contemporary attitudes and social trends. *Trans. 39th N. American Wildlife and Natural Resources Conference*, 151–62.
SIEGLER, D. S. (ed.) 1977: *Crop Resources*. London: Academic Press.
SIGURBJORNSSON, M. 1971: Induced mutations in plants. *Scientific American* **224**, 86–95.
SILBERBAUER, G. B. 1972: the G/wi bushmen. In M. G. Bicchieri (ed.) *q.v.*, 271–326.
SIMMONDS, N. W. (ed.) 1976: *Evolution of Crop Plants*. London: Longmans.
SIMMONS, I. G. 1966: Wilderness in the mid-twentieth century USA, *Town Planning Review* **36**, 249–59.
— 1974a: National parks in developed countries. In A. Warren and B. Goldsmith (eds.), *Conservation in practice*. London: Wiley, 393–421.
— 1974b: *The ecology of natural resources*. London: Edward Arnold.
— 1975: Towards an ecology of Mesolithic man in the uplands of Great Britain. *J. Archaeological Science* **2**, 1–15.
— 1976: Protection and development in the national parks of England and Wales: the role of the physical environment. *Geografia Polonica* **34**, 379–90.
SIMMONS, I. G. and DIMBLEBY, G. W. 1974: The possible role of ivy (*Hedera helix* L.) in the Mesolithic economy of Western Europe. *J. Archaeological Science* **1**, 291–6.
SIMMONS, I. G. and VALE, T. 1975: Problems of the conservation of the California coast redwood and its environment. *Environment Conservation* **2**, 29–38.
SIMOONS, F. G. 1961: *Eat not this flesh*. Madison, Wisconsin: University of Wisconsin Press.
SIMS, P. L., SINGH, J. S. and LAUENROTH, W. K. 1978: The structure and function of ten western North American grasslands. I. Abiotic and vegetational characteristics. *J. Ecol.* **66**, 251–85.
SIMS, P. L. and SINGH, J. S. 1978: The structure and function of ten western North American grasslands. III. Net primary production, turnover and efficiencies of energy capture and water use. *J. Ecol.* **66**, 573–97.
SINGER, M. 1977: Scientists and the control of science. *New Sci.* **74**, 631–4.
SLESSER, M. 1973: Energy subsidy as a criterion in food policy planning. *J. Science of food and agriculture* **24**, 1193–207.
— 1975: Energy requirements of agriculture. In J. Lenihan and W. W. Fletcher (eds.), *Food, agriculture and the environment*. Glasgow: Blackie, 1–20.

SMITH, N. J. H. 1974: Destructive exploitation of the S. American river turtle. *Yrbk. Association of Pacific Coast Geographers* 36, 85–102.

SMITH, V. L. 1975: The primitive hunter culture, Pleistocene extinction and the rise of agriculture. *J. Political Economy* 83, 727–55.

SMITH, R. L., BOUTON, J. H., SCHANK, S. C., QUESENBERRY, K. H., TYLER, M. E., MILAM, J. R., GASKINS, M. H., LITTELL, R. C. 1976: Nitrogen fixation in grasses inoculated with *Spirillum lipoferum*. *Science* 193, 1003–05.

SÖDERLUND, R. and SVENSSON, B. H. 1976: The global nitrogen cycle. In B. H. Svensson and R. Söderlund (eds.). *Nitrogen, phosphorus and sulphur-global cycles*. Stockholm: Swedish National Science Research Council Ecological Bulletin No. 22, 23–73.

SOLOMON, M. E. 1969: *Population Dynamics*. Studies in Biology 18. London: Edward Arnold.

SOPPER, W. E., and KARDOS, L. T. 1972: Effects of municipal waste water disposal on the forest ecosystem. *J. of Forestry* 70, 540–45.

SOUTHWOOD, T. R. E. 1976: Bionomic strategies and population parameters. In R. H. May (ed.), *q.v.*, 26–48.

SPARROWE, R. D. and WIGHT, H. M. 1975: Setting priorities for the endangered species program. *Trans. 40th N. American Wildlife and Natural Resources Conf.*, 142–56.

SPECHT, R. L. 1973: Structure and functional response of ecosystems in the Mediterranean climate of Australia. In F. di Castri and H. A. Mooney (eds.), *q.v.*, 113–20.

SPEDDING, C. R. W. 1975a: *The biology of agricultural systems*. London: Academic Press.
— 1975b: Grazing systems. In R. L. Reid (ed.) *q.v.*, 145–57.
— 1976: The biology of agriculture. *Biologist* 23, 72–80.

STANHILL, G. 1974: Energy and agriculture: a national case study. *Agro-Ecosystems* 1, 205–17.

STANKEY, G. H. and LIME, D. W. 1973: *An annotated bibliography of selected references relative to recreational carrying capacity decision-making*. US Dept. of Agriculture Forest Service General Technical Report INT-3.

STEARNS, F. 1972: The city as a habitat for wildlife and man. In T. R. Detwyler and N. G. Marcus (eds.), *q.v.*, 261–77.

STEELE, J. H. (ed.) 1970: *Marine food chains*. Edinburgh: Oliver and Boyd.

STEENSBERG, A. 1976: The husbandry of food production. *Proc. Roy. Soc. Lond.* **B 275**, 43–54.

STEINHART, J. S. and STEINHART, C. E. 1974: Energy use in the US food system. In P. Abelson (ed.), *q.v.*, 48–57.

STEWART, O. C. 1956: Fire as the first great force employed by man. In W. L. Thomas (ed.), *Man's Role on Changing the Face of the Earth*. Chicago: Chicago University Press, 115–33.

STEWART, W. D. P., 1972: Estuarine and brackish waters—an introduction. In R. S. K. Barnes and J. Green (eds.), *The estuarine environment*. London: Applied Science Publishers, 1–9.

STONE, C. D. 1974: *Should trees have standing? Towards legal rights for natural objects*. Los Altos: Wm. Kaufman Inc.

STONE, E. C. 1965: Preserving vegetation in parks and wilderness. *Science* 150, 1261–7.

STRUGNELL, R. G. and PIGOTT, C. D. 1978: Biomass, shoot-production and grazing of two grasslands in the Ruwenzori National Park, Uganda. *J. Ecol.* 66, 73–96.

STURDY, D. A. 1975: Some reindeer economies in prehistoric Europe. In E. S. Higgs (ed.), *Palaeoeconomy*. London: Cambridge University Press, 55–95.
SUKACHEV, V. N. and DYLIS, N. V. 1964: *Fundamentals of forest biogeocoenology*. Edinburgh and London: Oliver and Boyd (English translation by J. M. Maclennan).
SUSSMAN, R. W. and RAVEN, P. H. 1978: Pollination by lemurs and marsupials: an archaic coevolutionary system. *Science* **200**, 731–6.
SWIFT, I. 1975: Pastoral nomadism as a form of land use: the Twareg of the Adrar n Iforas. In T. Monod (ed.), *q.v.*, 443–54.
TAGANAIDES, E. P. 1967: The animal waste disposal problem. In N. C. Brady (ed.), *Agriculture and the quality of our environment*. Washington DC: AAAS Publication 85, 385–94.
TAIT, R. V. 1968: *Elements of marine ecology*. London: Butterworth.
TALBOT, L. M. 1972: Ecological consequences of rangeland development in Masailand, East Africa. In Farvar and Milton (eds.), *q.v.*, 694–711.
— 1975: Maximum sustainable yield: an obsolete management concept. *Trans. 40th N. American Wildlife and National Resources Conference*, 91–6.
TALLIS, J. H. 1975: Tree remains in southern Pennine peats. *Nature, London* **256**, 482–4.
TAMM, C. O. (ed.) 1976: *Man and the boreal forest*. Stockholm: Swedish Natural Science Research Council Ecological Bulletin 21.
TAMURA, T. 1970: *Marine Aquaculture*. Washington: National Technical Information Service PB 194 051T, 2 vols. (Translation of 2nd edn. of *Suisan Zoshokugaku*. Tokyo: Tuttle & Co.)
TANSLEY, E. G. 1968: *Britain's green mantle*, 2nd edn. London: Allen and Unwin. M. C. F. Proctor (ed.).
TARZWELL, C. M. 1972: An argument for the open ocean siting of coastal thermal electric plants. *J. Environmental Quality* **1**, 89–91.
TATSUKAWA, R. 1976: PCB pollution of the Japanese environment. In K. Higuchi (ed.). *PCB poisoning and pollution*. Tokyo: Kodansha Ltd.; London: Academic Press, 147–79.
TAYLOR, D. 1976: Rabies and Britain's pets. *New Scientist* **72**, 166–7.
TAYLOR, I. J. and SENIOR, P. J. 1978: Single cell proteins: a new source of animal feeds. *Endeavour* NS 2(1), 31–4.
TEMPLE, S. A. 1977: Plant-animal mutualism: co-evolution with Dodo leads to near extinction of plant. *Science* **197**, 885–6.
TEN HOUTEN, J. G. (ed.) 1969: *Air pollution. Proc. first European Congress on the influence of air pollution on plants and animals*. Wageningen: Centre for Agricultural Publications and Documentation.
TERBORGH, J. 1974: Preservation of natural diversity: the problem of extinction prone species. *BioScience* **24**, 715–22.
— 1965: Faunal equilibria and the design of wildlife preserves. In F. B. Golley and E. Medina (eds.) *q.v.*, 369–80.
THOMPSON, C. R., TAYLOR, O. C., RICHARDS, B. L. 1970: Effects of air pollution on Los Angeles basin cities. *Citrograph* **55**, 165–6, 190–92.
THORSTEINSSON, I., OLAFSSON, G. and VAN DYNE, G. M. 1971: Range resources of Iceland. *J. Range Management* **24**, 86–93.
TINKER, J. 1971: One flower in 10 faces extinction. *New Scientist* **50**, 408–13.
TIVY, J. 1971: *Biogeography. A study of plants in the biosphere*. Edinburgh: Oliver and Boyd.

TOCHER, R. and MILNE, R. 1974: A cross-cultured comparison of attitudes towards wildlife. *Trans. 39th N. American Wildlife and Natural Resources Conf.* 145–50.

TOUPET, C. 1975: Le nomade, conservateur de la nature? L'exemple de la Mauritanie centrale. In T. Monod (ed), *q.v.*, 455–67.

TRUDGILL, S. T. 1977: *Soil and vegetation systems*. Oxford: Clarendon Press.

TUDGE, C. 1976: Last animals at the zoo. *New Scientist* **71**, 134–6.

TURNER, J. and SINGER, M. J. 1976: Nutrient distribution and cycling in a sub-alpine coniferous forest ecosystem. *J. appl. Ecol.* **13**, 295–301.

UCFTF (University of California Food Task Force) 1974: *A hungry world: the challenge to agriculture*. Berkeley: UC division of Agricultural Sciences.

UDVARDY, M. D. F. 1975: *A classification of the biogeographical provinces of the world*. Morges, Switzerland: IUCN Occasional Paper 18.

UCKO, P. J. and DIMBLEBY, G. W. (eds.) 1969: *The domestication and exploitation of plants and animals*. London: Duckworth.

UNESCO 1973a: Impact of human activities on mountain ecosystems. *Programme on Man and the Biosphere* (MAB) report series 8. Paris: UNESCO.

— 1973b: Ecology and the rational use of island ecosystems. *Programme on Man and the Biosphere* (MAB) report series 11. Paris: UNESCO.

VALE, T. R. 1974: Sagebrush conversion projects: an element of contemporary environmental change in the Western United States. *Biological Conservation* **6**, 274–84.

— 1975: Ecology and environmental issues of the Sierra Redwood (*Sequoiadendron giganteum*) now restricted to California. *Environmental Conservation* **2**, 179–88.

— 1977: Forest changes in the Warner Mountains, California. *Ann. Assoc. Amer. Geogr.* **67**, 28–45.

VALE, T. R. and VALE, G. R. 1976: Suburban bird populations in west-central California. *J. Biogeography* **3**, 157–65.

VALLENTINE, J. F. 1971: *Range development and improvements*. Provo, Utah: Brigham Young University Press.

VAN BATH, S. B. H. 1963: De oogstopbrengsten van verschillende gewassen, voorhamelijk granen, in verhouding tot het zaairaad ca 810–1820. *A.A.G. Bijdragen*, 9–125.

VAN DEN BOSCH, R. and MESSENGER, P. S. 1973: *Biological control*. New York: Intertext Publications.

VAN DER MEIDEN, H. A. 1974: Forests and raw material shortage. *Agriculture and Environment* **1**, 139–52.

VAN EMDEN, H. F. 1965: The role of uncultivated land in the biology of crop pests and beneficial insects. *Scientific Horticulture* **27**, 121–36.

— 1974: *Pest control and its ecology*. Studies in Biology 50, London: Edward Arnold.

VARLEY, M. E. (ed.) 1974: Open University course S323 (Ecology) Block A, unit 5— *Whole ecosystems*. Milton Keynes: Open University Press.

VAVILOV, N. I. 1951: The origin, variation, immunity and breeding of cultivated plants. *Chronica Botanica* **13**, nos. 1–6.

VOGL, R. L. 1969: The role of fire in the evolution of the Hawaiian flora and vegetation. *Proc. Ann. Tall Timbers Fire Ecology Conf.* **9**, 5–60.

VON RUMKER, R., LAWLESS, E. W. and MEINERS, A. F. 1974: *Production, distribution, use and environmental impact potential of selected pesticides*. Washington DC: US Environmental Protection Agency.

WADE, N. 1975: Go ahead for recombinant DNA. *New Scientist* **68**, 682–4.

WAGAR, J. A. 1974: Recreational carrying capacity reconsidered. *J. Forestry* **72**, 274–8.

WAGNER, K. A., BAILEY, P. C. and CAMPBELL, G. H. 1973: *Under siege*. New York: Intertext books.
WALLWORK, K. 1974: *Derelict land. Origins and prospects of a land-use problem*. Newton Abbot: David and Charles.
WASSINK, E. C. 1975: Photosynthesis and productivity in different environments—conclusions. In J. P. Cooper (ed.) *q.v.*, 675–87.
WATANABE, H. 1966: Die sozialen Funktionen des Bärenfestes der Ainu und die okologischen Faktoran in seiner Entwicklung. *Anthropos* **61**, 708–26.
— 1972a: The Ainu. In M. G. Bicchieri, *q.v.*, 448–84.
— 1972b: *The Ainu ecosystem. Environment and group structure*. Seattle and London: University of Washington Press.
WATERHOUSE, D. F. 1974: The biological control of dung. *Sci. Amer.* **230**(4), 100–9.
WEATHERLEY, A. H. and COGGER, B. M. G. 1977: Fish culture: problems and prospects. *Science* **197**, 427–30.
WEAVER, J. E. 1954: *North American prairie*. Lincoln, Nebr.: Johnsen.
WEBBER, H. H. and RIORDAN, P. F. 1976: Criteria for candidate species for aquaculture. *Aquaculture* **7**, 107–23.
WEBBER, P. J. 1978: *High altitude geoecology*. Boulder, Colo.: Westview Press.
WEIR, J. S. 1969: Importation of nutrients into woodlands by rooks. *Nature, Lond* **221**, 487–8.
WENT, F. W. and STARK, N. 1968: The biological and mechanical role of soil fungi. *Proc. Natl. Acad. Sci.* **60**, 497–504.
WESTHOFF, V. 1970: New criteria for nature reserves. *New Scientist* **46**, 108–13.
WESTING, A. H. 1976: *Ecological consequences of the second Indochina war*. Stockholm: SIPRI/Almqvist and Wiksell.
— 1977: *Weapons of mass destruction and the environment*. London: Taylor and Frances, for SIPRI.
WHEAT, J. B. 1972: *The Olsen-Chubbuck site: A Paleo-Indian bison kill*. Society for American Anthropology Memoir 26.
WHITMORE, T. C. 1975: *Tropical rain forests of the Far East*. Oxford: Oxford University Press.
WHITTAKER, R. H. and LIKENS, G. E. 1975: The biosphere and man. In H. Lieth and R. H. Whittaker (eds.), *q.v.*, 305–28.
WHITTAKER, R. H. and WOODWELL, G. M. 1972: Evolution of natural communities. In J. A. Wiens (ed.), *q.v.*, 137–56.
WHITTLE, T. 1975: *The plant hunters*. London: Picador Books.
WIENS, J. A. (ed.) 1972: *Ecosystem structure and function*. Corvallis, Ore.: Oregon State University Press.
WILCOX, B. A. 1978: Supersaturated island faunas: a species–age relationship for lizards on post-Pleistocene land-bridge islands. *Science* **199**, 996–98.
WILKES, H. G. and WILKES, S. 1972: The green revolution. *Environment* **14**, 32–9.
WILLARD, B. and MARR, J. 1970: Effects of human activities on alpine tundra ecosystems in Rocky Mountain National Park, Colorado. *Biological Conservation* **2**, 257–65.
WILLARD, B. and MARR, J. 1971: Recovery of alpine tundra under protection after damage by human activity in the Rocky Mountains of Colorado. *Biological Conservation* **3**, 131–90.
WILSON, E. O. and WILLIS, E. O. 1975: Applied biogeography. In M. L. Cody and

J. M. Diamond (eds.), *Ecology and Evolution of Communities.* Cambridge, Mass.: Balknap Press, 522–34.

WILSON, R. D., MONAGHAN, P. H., OSANIK, A., PRICE, L. C. and ROGERS, M. A. 1974: Natural marine oil seepage. *Science* **184,** 857–65.

WINDSOR, M. and COOPER, M. 1977: Farmed fish, cows and pigs. *New Sci.* **75,** 740–2.

WINSLOW, J. H. (ed.) 1977: *The Melanesian environment.* Canberra: ANU Press.

WITHERS, J. and ASHTON, D. H. 1977: Studies on the status of unburnt Eucalyptus woodland at Ocean Grove, Victoria. I. The structure and regeneration. *Aust. J. Bot.* **25,** 623–37.

WITTWER, S. H. 1974: Maximum production capacity of food crops. *BioScience* **24,** 216–24.

WOLMAN, A. 1965: The metabolism of cities. *Scientific American* **218,** 179–190.

WOO, M. 1974: Afforestation: a facet of the changing landscape of mainland China. *Yrbk. Association of Pacific Coast Geographers* **36,** 113–23.

WOODBURN, J. 1968: An introduction to Hadza ecology. In R. B. Lee and I. De Vore, *q.v.*, 49–55.

WOODWELL, G. M. 1962: Effects of ionizing radiation on terrestrial ecosystems. *Science* **138,** 52–7.

— 1970: Effects of pollution on the structure and physiology of ecosystems. *Science* **168,** 429–33.

— 1974: Success, succession and Adam Smith, *BioScience* **24,** 81–7.

WOODWELL, G. M. and SMITH, H. H. 1969: *Diversity and stability in ecological systems.* Upton, New York: Brookhaven Symposia in Biology 22.

WOODWELL, G. M., CRAIG, P. P. and JOHNSON, H. A. 1971: DDT in the biosphere: where does it go? *Science* **174,** 1101–7.

WOODWELL, G. M., WHITTAKER, R. H., REINERS, W. A., LIKENS, G. E., DELWICHE, C. C. and BOTKIN, D. B. 1978: The biota and the world carbon budget. *Science* **199,** 141–6.

WOOLHOUSE, H. W. 1978: Light-gathering and carbon assimilation processes in photosynthesis: their adaptive modifications and significance for agriculture. *Endeavour,* NS, 2(1), 35–46.

WYATT, T. 1976: Food chains in the sea. In D. H. Cushing and J. J. Walsh (eds.), *The ecology of the seas.* Oxford: Blackwell, 341–58.

WYNNE-EDWARDS, V. C. 1962: *Animal dispersion in relation to social behaviour.* Edinburgh and London: Oliver and Boyd.

YANCHINSKI, S. 1978: Air pollution evidence stacked against Sweden. *New Sci.* **78,** 731.

YARNELL, R. A. 1971: Early woodland plant remains and the question of cultivation. In S. Streuver (ed.), *Prehistoric agriculture,* American Museum Sourcebooks in Anthropology. Garden City, New York: Natural History Press, 550–4.

YEN, D. E. 1976: Sweet potato. In N. W. Simmonds (ed.), *q.v.*, 42–4.

YERMANOS, D. M. 1974: Agronomic survey of jojoba in California. *Economic Botany* **28,** 160–74.

YOUNG, E. 1975: Technological and economic aspects of game management and utilization in Africa. In R. L. Reid, *q.v.*, 132–41.

ZELITCH, I. 1975: Improving the efficiency of photosynthesis. *Science* **188,** 626–32.

ZEUNER, F. E. 1963: *A history of domesticated animals.* London: Hutchinson.

ZISWILER, V. 1967: *Extinct and vanishing animals.* New York: Springer-Verlag.

ZOHARY, D. 1969: The progenitors of wheat and barley in relation to domestication and agricultural dispersal in the Old World. In P. J. Ucko and G. W. Dimbleby, *q.v.*, 47–66.

Index

aardvark, 36
abalone, 198, 300, 302–3
Abies spp, 80, 117; *A. grandis*, 120
abrasion, by sand, 108
Abutilon avicennae, 199
Acacia spp, 100, 104
Acanthaster planci, 140, 267
Acarina spp, 49
Acer spp, 113–14, 312–13
acidophiles, 49
Adansonia digitata, 99
adaptive radation of species, 52
addax, 181–2
Adenostoma spp, 103; *A. fasciculata*, 20, 21; *A. sparsifolium*, 20, 21
Aedes aegypti, 247
Aegilops spp, 183; *A. comosa*, 260; *A. speltoides*, 260; *A. umbellata*, 260
Aepyornis sp., extinction of, 158
aerosol sprays, effects in atmosphere, 232, 237, 238, 247, 312
Africa: animals of desert areas, 111; character of basic agricultural systems, 185–6; co-evolution of man, fire, grasslands, animals, 2; consumption of fertilizers, 207; depletion of fauna by hunting, 100; diversity, distribution of biota, 36; domestication in, 168, 169, 178, 179; ecological significance of fire, 102; extinction of species in, 220; nature protection in, 317; origins in primitive continent, 30; spread of crops, animals, to, from, 193–4; unstable ecosystem of pastoralism, 311; wild game as source of food, 264; *see also individual areas and types of terrain*
Africa, central, 100

Africa, east, 32, 99–100, 101
Africa, north, 158, 196
Africa, south, 42, 46, 106
Agaricus bitorquis, 270
agave cactus, 269; *A. sisalina*, 199
agricultural land, *see* cultivated land
agricultural waste, *see* waste
agriculture: beginnings of, 162; consumption of fossil fuels, 206; early, energy sources available to, 331; ecological effects, 92, 97, 104, 116, 130, 205–9, 247; ecosystems, food chains based on, 192–3; establishment of different types, 185–6; future trends, 285–6, 334–8; intensive, 246, 283, 331; pre-industrial, 193–7, 201–2; problems, use of wastes, 284–5; productivity of major crops compared, 275–6; proportion of NPP contributed by, 88; shifting cultivation systems, 97, 185–6, 195, 196, 202; transition from biological to non-biological processes, 274; transition to, complexity of processes, 162, 163–4; weed/pest problems, controls, 265–6, 283–4; *see also* selective breeding *and individual crops*
Agropecten spp, 302–3
Agropyron spp, 171
Agrostis spp, 171, 247; *A. tenuis*, 237
Ainu people of Japan, bear cult, 161
air pollution, *see* atmosphere
Alaska, 120–1
albatross, 125, 135–6
Alces alces, 21, 119, 262
alcohol, 199
Aldabra Island, 52
aldehydes, 237
alder trees, 114, 120

aldrin, 208–9
alewives, 245
alfalfa, 60, 169, 171, 193
algae, 8, 11, 18, 28, 29, 30, 62, 86–7, 95–6, 138–9, 245, 269: blue-green, 11, 28, 30, 95–6, 141, 143, 145, 269; brown, 18, 87; calcareous, 29; freshwater, 142, 143, 145, 146, 239; green, 11, 141, 143; marine, 28, 86, 113, *see also* seaweeds; of Antarctic tundra, 124; of deciduous forests, 113; of mangrove swamps, 140; of salt marshes, 141; red, 140, 215; unicellular, 11
alkali, 267
alleopathy, 20
Allium spp: *A. cepa*, 169, 171, 184, 199; *A. porrum*, 171; *A. sativum*, 171, 199
allspice, 199
almond, 171, 199, 200
almond oil, 199, 200
Alnus glutinosa, 114, 120
alpaca, 128, 168, 178, 181, 185, 186, 197, 306
Alpine regions, *see* mountainous regions
Alps, 37
altitude, in ecostems, 127–8
Amanita muscaria, 200
Amaranthus spp, 188
Amazon river, region, 92–7
Ambrosia spp, 238
America
spread of crops, animals, to, from, 193–4; *see also individual areas and types of terrain*
America, central: character of basic agriculture, 185–6; consumption of fertilizers, 207; diversity, distribution of biota, 37; domestication in, 168, 169, 174–5, 178; extinction of species in, 220; spread of domesticates from, 187–8

380 Index

America, north: animals invading cities in winter, 227; burrowing animals in overgrazed areas, 311; character of basic agriculture, 185–6; consumption of fertilizers, 207; diversity, distribution of biota, 36, 37, 41, 44, 45, 46–7, 48, 106; domestication in, 168, 169, 178; ecology of grazing systems, 197; evidence of hunting peoples' use of fire, 156; extinction of species in, 158–60, 220; introduction of conifers from, 210; nature protection in, 315, 317–18, 320, 338; spread of domesticates from, 187–8; spread of starlings, 214

America, south: character of basic agriculture, 185–6; consumption of fertilizers in, 207; depletion of fauna by hunting, 100; diversity, distribution of biota, 36, 37, 42, 106; domestication in, 168, 169, 174–5, 178, 181; dominant tree species, 95; extinction of species in, 220; origins in primitive continents, 30; spread of domesticates from, 187–8

Amherstia nobilis, 210
amino acids, 196, 259, 270
ammonites, 28, 29
ammonium sulphate, 233
Amoeba spp, 11
amphibians, 12, 16, 28, 29, 37, 51, 95, 143, 327
Anastatica spp, 110
anchovy, 135, 297, 298
Andes Mountains, birds of, 129
Andropogon spp, 99, 106; *A. gerardi*, 106; *A. scoparius*, 106
Anemone nemorosa, 114
Anethum graveolens, 171, 199
Angelica archangelica, 199
angiosperms (flowering plants): distribution patterns, 42, 97; in classification of species, 12; in evolutionary history, 28, 29, 30, 34, 112; interrelation of species richness and latitude, 16; protected species, 315, 316; *see also* plants
angler fish, 137
anise, 171, 199
annelid worms, 12, 143

Anopheles spp: *A. culicifacies*, 247; *A. gambiae*, 247
Anser spp: *A. anser*, 178, 181, 188; *A. cinereus*, 306
Antarctica, 8, 124–7, 220, 317–18
ant-eaters, 95, 158
antelopes: habitats, 100, 106, 128; pronghorn, 106, 309; Saiga, 106, 159
Anthoxanthum odoratum, 310
Antidorcas marsupialis, 263
anti-knock additives, 232, 237
Antilocapra americana, 106, 309
ants, 12, 49, 112; Hymenoptera spp, 62, 101
apes, 13, 31
Aphelinus mali, 265
aphids, 227, 265, 267; elm, 227; linden, 227, woolly, 265
apple, 169, 171
Apium spp: *A. graveolens*, 199
apricot, 171
aquaculture, potential of, 270, 296–305
aquatic environments, waste in, 238–45; *see also* freshwaters, seas
Arabaena spp, 11
Arachis hypogaea, 169, 188
Arachnida, 12
Araucaria heterophylla, 222
arbutus, 103
Archaeopteryx, 31
Arctic regions, 37, 42, 43, 44–5, 47, 125–7, 145
Arctomys marmota, 128
Arctostaphylos spp, 103
Ardea cinerea, 81
Argyroxiphium spp, 130
Aristida spp, 110; *A. pungens*, 109
armadillos, 158
arrow-worms, 133
arsenic, 208
Artemisia spp: *A. dracunculus*, 199; *A. tridentata*, 309
arthropods, 12, 28, 29, 111, 245, 327
Artiodactyla, 262–3
ash trees, 114, 234
Asia, 111, 188, 207, 220
Asia, central, 168, 169, 178
Asia, east, 37, 41
Asia, southeast, 92, 97, 168, 169, 174, 178, 179, 193–4
Asia, southwest, 92, 163–4, 167, 168, 169, 170–4, 179–82, 185 193–4, 196
Asinus spp, 306

aspen, 117
asses, 111, 168, 177, 178, 185, 188
Astragalus spp, 183
Atacama Desert, 108
atmosphere: 'greenhouse effect', 73, 74, 336–7; pollution of, 232–8, 243, 247, 293, 312–13, 336–7
Atropa belladonna, 171, 200
auk, little, 62
aurochs, 179, 180
Australia, 36, 37, 42, 45, 100, 159, 194, 211–13, 316
Australopithecus spp, 32, 33
Austria, 208
autecology, defined, 54
autotrophs, 4, 28, 60–2, 72, 87–8
Avena spp, 169, 170, 183, 185, 194–5, 196: *A. sativa*, 169, 170, 185; *A. strigosa*, 170
Aves, 13
Avicennia nitida, 140
avocado, 169, 175, 185, 227

baboons, 327
bacteria: *B. aureus*, 80; biogeographical functions, 8, 62, 114, 115; density in urban store, forest, compared, 228; distribution, ecological role in, 95–6, 124, 133, 135, 136, 138, 141, 142, 143–5; in classification of species, 7, 11; in evolutionary history, 28, 29; rates of evolutionary change, 23, 27, 80; sources, control of manipulation, 256–8, 312; strains resistant to biocides, 209, 247, 262, 264; use of *Spirillium lipoferum* in plant breeding, 261
bacterial blight, breeding to combat, 267
badgers, 107, 157
bait, fisherman's, 145, 146
Baja California, 108
balm, 199
balsam, 80
Baltic Sea, effects of salinity, 18
bananas, 95, 169, 174, 185, 194, 258
bandicoot-rat, 184
Bangladesh, fish farming in, 304
Banksia spp, 104
banteng, 177, 178
baobab, 99
bark of trees, commercial use of, 288

Index

bark beetle, western, 233
barley: domestication of *H. vulgare*, 165–6, 168, 169, 170, 172, 185; economic importance, 185, 194–5, 196; energy input–output ratios, 280; selective breeding, 182–3, 260; yields, 278, 279
barnacles, 12, 140
Barro Colorado Island, 96–7
Basileuterus spp, 45
basswood, *see* lime trees
Bathylagidae, 299
bats, 13, 37, 327: Mexican free-tailed, 242
beans: broad, 169, 170; castor, 200; common, 168, 169; domestication of *Phaseolus* spp, 185; economic importance, 185–6, 187–8, 194–5, 258; jack, 185; lima, 168, 169; scarlet runner, 168, 169; sieva, 169; soya, 169, 185, 200, 261, 267, 270, 275–6, 286; tepary, 169
bears, 80, 157, 161, 327: Black, 114, 227; Brown, 119, 321; Grizzly, 104, 119, 125, 321; Malayan, 221; Polar, 62, 124, 125, 127, 161, 221
Bear Island, food web, 62
beaver, 80, 119, 157, 161
bedstraw, 183
beech trees, 20, 30, 37, 42, 113–14, 126
beech–spruce *urwald*, 316
beef, 191, 280; *see also* cattle
bees, honey, 22, 180, 181, 327
beet, *see Beta vulgaris*
beetles, 16, 111, 114, 115, 211–12, 327: bark-, 233; dung-, 112, 211–12; sugarcane root, 212
belladonna, 171, 200
Bellis perennis, 16
bentgrasses, 171, 237, 247
benthos, 133, 136
Bering land bridge, 37
Beta vulgaris, 169, 170, 275–6, 278, 279
BHC-Lindane, 208–9
Betula spp, 80, 114, 117, 120, 312–13; *B. nana*, 122
beverage plants, 198–200; *see also individual plants*
bezoar, 179
bicarbonate, 132, 136
Biochemical Oxygen Demand (BOD), 312

biocides: ecological effects, 104, 107, 207–9, 232, 241–2, 247; early, 208; release to environment, 231; resistant strains of organism, 209, 247, 262, 264; target-specific, 334; use regulated, 208–9; *see also* pests
biomass, 58–60
biomes, 42, 68, 90–3
birch trees, 80, 114, 117, 120, 312–13; dwarf, 122
Bird of Paradise, 37, 220
birds, 13, 28, 29, 31, 52, 161, 327: distribution patterns, 36, 37, 45, 46; evolution, adaptation, 95, 101, 104, 107, 111, 114, 119, 124, 125, 128, 135–6, 140, 141, 225, 227–8, 247; extinction of species, 158, 218–20; 221; fish-eating, in freshwater habitats, 143; flightless, 52, 130, 138, 158; insectivorous, agents of selection, 25; intraspecific competition, 20; marine, 124, 125, 135–6, 138, 243; migrant, 124, 222, 283–4; of prey, absorption of toxins, 242; protected species, 314–15, 315–16; protection of colonial habitats, 221; speciation on islands, 52
bison, 154, 157, 185, 220–1: Steppe-, 153; *see also* buffaloes
Biston betularia carbonaria, 2, 25–6, 27, 238
bittervetch, 170
bivalve molluscs, 29, 135, 137, 141, 143
blackfly, 124
bladderwrack, 11
blast disease of plants, 267
bleak fish, 146
blesbok, 263
blood worms, 143
bluebird, fairy, 36
bluestem grasses, 106: big bluestem, 106; little bluestem, 106
boar, 319
Boehmeria nivea, 199
bog mosses, 11, 117
bogs, 142
Bombacaceae, 199
Bordeaux mixture, 208–9
boreal (coniferous) forests, 86, 88, 91, 93, 96, 116–20
Borneo, 95

boron, 132, 136
Bos spp: *B. frontalis*, 128, 167, 178; *B. grunniens*, 128, 163, 177, 178, 186, 187, 306; *B. indicus*, 306; *B. primigenius*, 179, 180; *B. taurus*, 306
Bouteloua gracilis, 106
bowerbirds, 37
Brachiaria deflexa, 186
brachiopods, 28
bracken fern, 11, 20, 48, 114, 340
Bradypus infuscatus, 96–7
Brassica spp: *B. alba*, 199; *B. campestris*, 170; *B. nigra*, 170, 199; *B. oleracea*, 171; *B. rapa*, 170
Brazil, 94, 96
bread, 280
breadfruit, 169
British Isles: agriculture in, 278, 281, 282; diversity, distribution of biota, 41, 43–4, 49; ecological effects of late Mesolithic destruction of vegetation, 84; effect on lichens of atmospheric pollution, 233–4; evidence of prehistoric use of fire, 156, 160; nature protection in, 314–15, 316, 320, 323, 324, 328; species naturalized in, 215–18
broccoli, 171
bromegrass, smooth, 171
bromide, 132, 136
bromophos, 209
Bromus inermis, 171
brown earth soil, 114–15
brown plant hooper, 267
browsing animals, ecological effects of, 119, 146
Bryophytes, 11
Bubalus bubalus, 306
budgerigars, 216, 226–7
Buchloe dactyloides, 106
buckwheat, 169
budworm, spruce, 119, 247, 291
buffalo grass, 106
buffaloes, 107, 168, 177, 178: Asian, 188; *Bubalus bubalus*, 306; North African, 158; North American, 106; *Syncerus caffer*, 262; water, 185
Bufo marinus, 212
bulldozers, 63
bullrushes, 143
bunting, snow, 62
bush babies, 95
Bushmen, 153, 161
'butter of the poor', 175

butterflies, 327: copper, 323
button cactus, 200

cabbage, 12, 124–5, 168, 169, 171
cabbage looper, 209
cabbage worm, 209
cacao, 169
cactus, 108–11: button, 200; saguaro, 108, 111
caiman, 221
calcifuges, 49
calcioles, 49
calcium, 75–8, 96, 132, 136, 155, 290–1
calcium arsenates, 208
calefaction, 142
Calligonum comosum, 110
Callorhinus ursinus, 198
Calluna vulgaris, 49
calories, conversion to Joules, 3
Calvaria major, 222
Cambrian period, 28, 29
camels, 111, 112, 168, 177, 178, 181, 185, 197: bactrian, 168; *Camelus dromedarius*, 181, 188, 306
Canada, nature protection in, 317, 320; *see also* America, north
Canada goose, 215, 217, 327
canary grass, 171
Canavalia kauensis, 130–1
Cancer, 238
Canis spp: *C. familiaris*, 157, 168, 177, 180, 185; *see also* Dogs; *C. lupus*, 21, 104, 107, 114, 119, 123, 124, 125, 180, 311; *C. lupus chanco*, 180
Cannabis sativa, 169, 171, 199, 200
capercaillie, 216
Capra spp: *C. hircus*, 168, 177, 178, 179, 185, 306; *C. hircus aegagrus*, 179
caprines, 157
caraway, 171, 199
carbamate group of biocides, 209
carbon, 2, 69–70, 73, 74, 96; *see also* carbon cycle, carbon dioxide
carbonate of potassium, 108
carbonate of sodium, 108
carbon cycle, 73, 74
carbon dioxide, 18, 44, 55–60, 73, 74, 142, 232, 233, 238, 312–13, 336–7
Carboniferous period, 28, 29, 30, 34
carbon monoxide, 233

Caretochelys spp, 37
Carex spp, 122–3: *C. aquatilis*, 122
caribou, 119, 123, 124, 125, 126, 127, 161, 197, 248, 249
Carnegia gigantea, 108, 111
carnivores, 8, 327: evolution, adaptation, 95, 111, 114, 115, 119, 125; prehistoric extinction of, 158; roles in ecosystems, 60–2, 63, 88, 114, 115, 125
carob, 103, 171
carp, 198, 217, 302–3, 327
Carpinus betulus, 114
carrot, 170
carrying capacity of ecosystems, 84, 328–9, 342
Carthamus tinctorius, 170, 200
Carum carvi, 171, 199
Carya spp, 113–14
Caryophyllaceae, 124, 246
cassava, 169, 258
Cassia senna, 109
Cassiope tetragona, 122
cassowary, 32, 326
Castanea spp, 113–14
castor bean, 200
Castor fiber, 80, 119, 157, 161
Casuarina spp, 104
Casuarius spp, 37, 326
Catalpa spp, 312–13
caterpillars, 112, 115
catfish, 217, 327
Catharus spp, 44
cats, 8, 10, 140, 157, 177, 183, 185, 211, 216, 221, 226; wild, *see under individual families, species*
cattle: adaptation to forage sources, 197; breeding to improve, 262–3; dairy, potential sources of food for, 270; diseases, control of, 262; domestic (*Bos indicus, B. taurus*), products of, 306; domestication, 168, 177, 178, 179, 183, 184, 185, 211; evidence of role in pre-agricultural economy, 157; extinction of *Bos primogenius*, 179, 180; feral, 221; importance in basic southwest Asian agriculture, 185; longhorn, spread in Americas, 194; medical, scientific uses, 197; semi-domestication in southwest Asia, 167; subsistence ecosystems based on, 196–7; wild, 184

cauliflowers, 171
cave-dwellers of Pyrenees, 154
Caenothus spp, 103
cedar trees, 103, 161
Celebes, extinction of species in, 158
celery, 199
cellulose, 58
Centaurea diluta, 246
centipedes, 12, 327
Centrocerus urophasianus, 309
cephalopods, 28
Ceratonia siliqua, 103, 171
cereal crops, 92, 165–6, 170–3, 182–3, 258–61
see also individual crops
Ceriatola spp, 46
Cervus elephas, 154, 157–8, 160, 165–6, 197, 264, 310, 319, 322
cesium-137, 248, 249
Chaetognathae, 133
chalk soils, 114
Chamaea fasciata, 50
Chamaenerion angustifolium, 224
chameleon, 80
chamise, 20, 21, 103
Chanos spp, 185, 198, 302–3
Chaoborus, 144
chaparral, 91, 93, 103–4
chard, 170
cheetah, 181–2
Chelonia mydas, 300
chemical fertilizers, 70–2, 73, 206–7, 236, 247; *see also* agriculture *and individual substances*
chemicals: alteration of energy relations by, 63–6; contamination by, 142, 240–2; effects of persistence, 247; in man-made waste, 231–2; *see also* pollution, waste *and individual substances*
cherry trees: domestication of *P. avium*, 171; role of *P. pensylvanica* following deforestation, 75–6, 290
chestnut trees, 113–14
chestnut tree blight, 214, 267
chickens, 178, 185, 190, 191, 193, 280
chick-pea, 169, 170
Chihuahua Desert, 108
chimpanzees, 327
China: agriculture, 185, 281; domestication in, 168, 169, 178, 181, 188; fish farming, 304
chinchilla, 128

Index

chipmunk, 47, 48, 227, 321
Chironomids, 144, 146
chloride, 132, 136
chlorinated hydrocarbons, 138, 208–9, 241–2, 247
chlorine levels, 131–2
chlorofluoromethanes, 238
chlorophyll, 55–60; see also photosynthesis
chocolate, source of, 95, 199
Choctaw Indians, conservation practices, 161
Choloepus hoffmanni, 96–7
Chordata, 8, 12
Choristoneura fumiferana, 119, 247, 291
Chortoicetes terminifera, 16
Chromaphis juglandicola, 265
chromium, 76
chromosomes, 22–4
Chrysanthemum cinerariaefolium, 208–9
Chrysolina spp, 81
Cicer arietinum, 169, 170
cichlid fish, 302–3
cinnamon, 199
Citellus leucurus, 111
cities, 222–231
citrus fruits, 169, 193, 237
Cladocera, 144
clams, 80, 141, 327
classification of species, 7–13
climate: changes in, 274, 336–8; interaction with topography, 68; relation to biota, 36, 42–3, 46–7, 48, 53, 91–2; role in creation of soil, 90–1, 92
cloning, 262, 325
Clorella pyrenoidosa, protein yield, 269
clovers, 171, 183: see also alfalfa
cloves, 199
clubmosses, 11, 16, 28, 30, 35
Clupea spp, 135, 137
coal forests, 28
coal tips, 79
cobalt, 309
Coccosphaeres, 133
cochineal, 200
cockles, 141
cockroaches, 246
cocksfoot, 171
cocoa tree, 95, 199
coconut, 169, 258
coconut moth, 265
Cocos nucifera, 199
cod, 12, 298, 299
codeine, 171
Coelenterates, 12, 29

Coelodonta antiquitatis, 158
coffee, 169, 194, 198–200
coir, 199
Collembola spp, 49, 62, 114, 115
Colobanthus crassifolius, 124
Colorado potato beetle, 208
commensalism, 21
common cold, 11
communications, alteration of biotic patterns by, 204
competition, interspecific, intraspecific: defined, 19–22; effect on diversity, distribution of biota, 42–3, 48, 81; factors affecting in different regions, 47; in island ecosystems, 52, 130–1; interrelation with domestication, 182–3, 183–4
Compositae, 12, 106, 246
condor, 129; California, 220–1, 314, 315
Congo river, region, 92–7
conifers, 11, 28, 29, 30, 34, 102–3, 210, 213–14, 248–9, 290, 291, 293
conservation, see nature conservation, protection
contamination, see pollution
continental shelf, 86, 87, 131, 133, 134
continents: concepts of primitive, 30, 40–2, 53; distribution patterns of biota, 43–7
cooking, energy used in, 202
coot, 52
copepods, 133
Copernicia cerifera, 200
copper, 309
Copper acetoarsenite, 208
coppicing, effects of, 104, 116
coral reefs, 29, 138–40, 316
corals, 12
coriander, 171, 199
cormorant, 125, 181
corn, 23; see also maize, wheat
Cornus florida, 76
Corvus frugilegus, 115, 283
Corylus spp, 170; *C. avellana* (hazel), 114, 116, 156, 160, 170
cotton, 169, 185, 188, 189, 193, 194, 199, 286
cottongrass, 122
cottonseed, 200, 207
counting, in history, 32–3
cowpea, 169
coyote, 107, 112, 227, 311
coypu, 215–18
crabs, 12, 140, 299, 302–3: California red, 299; Chinese

mitten, 215; horseshoe, size and generation time, 80; Red Sea, 214–15
cranes, 181, 314, 327
Cree Indians, diet, 161
creosote bush, 108, 109, 110
cress, garden, 171
Cretaceous period, 28, 29, 30, 31, 34
Cricetus cricetus, 215
crickets, 112
crinoids, 29
Cro Magnon man, 32, 33, 35
crocodiles, 13, 31, 327: esturine, 221; see also caiman
crocodilians, 43
crop yields, 58–60, 193, 223, 264–7
crown of thorns starfish, 140, 267
crows, 52, 327
Cruciferae, 12, 246
crustaceans, 12, 133, 135, 137, 147, 327
cucumber, 169, 171
Cucumis spp: *C. melo*, 170; *C. sativus*, 169, 171
cultivated land, estimated NPP, 86
cumin, 171
curare, 171, 200
cuttlefish, 299
cycads, 28
Cydonia oblonga, 170
Cynodon spp, 183
Cyperus spp: *C. diffusus*, 247; *C. papyrus*, 86–7
cypress trees, 103
cyprinid fish, 36, 215
Cyzenis spp, 115
Czechoslovakia, 316

Dactylis glomerata, 171
Dactylopius sp, 200
daft-lamb disease, 262
daisy, English, 16
daisy family, see Compositae
Damaliscus spp, 263
Dansereau, P., 1–3
Daphnia, 12, 80, 142
darkness, respiration during, 56–7
dates, 169, 171, 184, 196
Daucus carota, 170
daylength, significance of, 17, 44–5
DDD, 208–9
DDE, 208–9
DDT, 62, 63, 138, 208–9, 242, 243, 247

decaying matter, role in food web, 62
deciduous forests, 91, 93, 112–16
deciduousness, tropical, 99
decomposer organisms, 60–2, 88, 101–2, 114–15; *see also* food chains, webs
deer, 13, 80, 95, 104, 114, 115, 119, 321; chamois, 157; Chinese water, 216, 217; fallow, 216, 217; mule, 104, 309; musk-, 200; red, 154, 157–8, 160, 165–6, 197, 264, 310, 319, 322; roe, 157, 319; sika, 217; white-tailed, 227, 321
deforestation: as management technique, 323; by introduced species, 212–13; ecological effects, 75–8, 140; following grazing, 310–11; for fuel for cooking, 202; recolonization following, 117, 290–1
Dendroica spp, 49, 50
deoxyribonucleic acid (DNA), 23–4, 256–8, 345
derris root, 208–9
desalination, 335
Deschampsia spp: *D. antarctica*, 124; *D. flexuosa*, 15
desertification, 112, 311
deserts, 86, 93, 107–12, 130, 306; *see also* desertification, semi-desert regions
dessication, toleration of, 18
detergents, 231, 243
detritus, 60–2, 68–9, 72, 75, 146
devil fish, 137
Devonian period, 28, 29, 30
diatoms, 11, 14, 143
diazinon, 208–9
dicotyledons, 12
Dicrostonyx spp, 124, 125
Didinium, 80
dieldrin, 208–9, 247
digitalis, 171
Digitaria spp; *D. decumbens*, 310
D. iburna, 186
dill, 171, 199
Dimorphandra spp, 95
dinoflagellates, 133, 143, 239, 301–2
dinosaurs, 28, 34
Dipodomys spp, 111
diptera, role in food web, 62
Dipterocapaceae, 95
disease, 230–1, 246, 258–64, 338 *see also individual diseases*

diver, red-throated, role in food web, 62
diversity, *see* species diversity
diversivores (omnivores), 8
dodo, 52, 222
dogs, 157, 211, 226, 327: domestication from *Canis familiaris*, 157, 168, 177, 180, 183, 185; prairie, 37, 107, 311
dog's mercury, 114
dogwood, 76, 80
Dolomedes plantanus, 322
dolphins, 136, 300
domestication, domesticates: complexity of processes, 162, 163–4; definitions, 164; diffusion of, 186–91; effect of cultural, religious factors, 187, 188–91; effects on local biogeography, 183–4; genetic and somatic changes, 182–3; of animals, 52, 176–82; of plants, 167–76, 198–200; sources, nature of evidence, 164–5
dom palm, 99
donkeys, 168, 177, 178, 185, 188, 211; wild, 111
dormouse, 181, 216
doves, 327: eared, 284
dredging, and coral reefs, 140
Drepaniidae, 52
drill, American oyster, 215
drinn grass, 109
dromedary, 111, 112, 168, 177, 178, 181, 185, 188, 197, 306
Drosophila, 80
drugs, 171, 200, 262
dry-weight, conversion to grams of carbon, 2
ducks, 13, 52, 62, 178, 181, 319, 327: Carolina (wood), 216, 217; long-tailed, 62; mandarin, 216; Muscovy, 178; northern eider, 62; ruddy, 216, 217
duckweed, 143
dune systems, 108, 109
dung, role in food webs, 62
dung-beetles, 112, 211–12
Dupontia spp, 122–3; *D. fischeri*, 122
durum wheat, 260–1
dust fall in cities, 237–8
Dutch elm disease, 214, 218, 267
dyes, 200

eagles: bald, 315; golden, 128, 220–1, 247, 315, 327
earthworms, 12, 30, 49, 114, 115, 128

Echeneis naucrates, 21
Echidna spp, 37
Echinochloa turnerana, 269
echinoderms, 12, 29, 135, 327
ecotypes within a species, 44–5
edentates, 158
eels, 12, 239–40, 327
eelgrass, marine, 269
eelworms, 12
eggplant, 169
eggs, 190, 191, 280
Egypt, ancient, 164–5
Eichornia crassipes, 268–9
einkorn, 166, 168, 170, 172, 182–3, 185
eland, 262–3, 306
elder, marsh, 186
elephant grass, 99
elephants, 36, 80, 158, 177, 185, 221, 264, 293, 327
elk, 80, 157, 185, 322: giant Irish, 158
elm aphid, 227
elm trees, 113–14, 214, 218, 267, 312–13
emigration of species, 79–85, 202
emmer, 166, 168, 170, 182–3, 185
Empetrum spp, 117
emu, 37
encinar, 103–4
endemism, 47, 52
endrin, 208–9
energy: and ecology, basis in photosynthesis, 55–60; demand for, 331–2, 336–8, 341–2, 345; flow, 68–9, 79; input–output ratios, 201–2, 223–4, 225, 273–88, 293–6, 298–305, 307–8, 310; relations in ecosystems, 62–6, 88; sources available, 331; transfer systems, *see* food chains, webs
Enteromorpha spp, 141
Entomostraca, 62
Eocene period, 28, 30
Eperna spp, 95
epiphytes, 21, 94–5, 113, 140
Equisetum spp, 11, 28
Equus spp, 111, 157, 168, 177, 178, 185: *E. asinus*, 111; *E. asinus x E. caballus*, 306; *E. burchelli bohmi*, 306; *E. caballus*, 168, 177, 178, 185, 306; *E. hemionus*, 111; *E. przewalskii*, 106, 159
Eragrostis tenella, 130
Eretmochelys imbricata, 221
Erica spp, 103
Eriocheir sinensis, 215

Eriophorum spp, 122–3; *E. scheuzcheri*, 122
Eriosoma lanigerum, 265
erosion, see soil erosion
Escherichia coli, 80, 256–8
Eschscholzia, 46–7
estuaries, 86, 138, 141–2
ethanol, 285
Ethiopia, 201
Ethiopian zoogeographical realm, 36–7
Eucallipterus tiliae, 227
Eucalyptus spp, 99, 102, 103, 104, 199, 291–3, 294; *E. calophylla*, 99; *E. globulus*, 293; *E. marginata*, 99, 291–3
Euchlaena mexicana, 174
Eugenia caryophylla, 199
Euglena, 80
Eugonia subsiguaria, 265
Eulaema meriana, 22
Euphasia superba, 299
Euphorbia spp, 108–9, 285
Eurasia, 92, 106, 158–60
  see also *individual types of terrain*
Europe: character of basic agriculture, 185; consumption of fertilizers, 207; diversity, distribution of biota, 37, 49; domestication in, 168, 169, 178, 179, 181, 188; evidence of prehistoric use of fire, 156; extinction of species in, 220; invaders of derelict land, 246–7; pH levels in precipitation, 235; pre-industrial subsistence ecosystems, 194–6; spread of crops, animals, to, from, 193–4
eutrophication: causes, effects of, 71–2, 73, 79, 239, 312, 322; in seas, 138, 301–2; natural, cultural in freshwater habitats, 145
evergreen forests, 112, 117–18; see also boreal forests
evolution: and speciation, 27, 52; characteristics of island species, 50–3; effect on interspecific, intraspecific competition, 22; long-term history of, 27–31; rates in tropics, temperate, arctic regions, compared, 47; through genetic mutation, 23–4; through genetic recombination, 24; see also under *individual biota*
extinction of species: factors

affecting rates on islands, 50–3; in evolutionary history, 33–5, 158–60; natural, man-made distinguished, 158, 218–19; of plants, 222; prevention by establishment of reserves, 314–15; see also *individual species*

Fagus spp, 20, 113–14, 116; *F. sylvatica*, 114
falcons, 181–2: peregrine, 247
families, in classification of species, 8
fatigue plants, 171
feathers, trade in, 220
fecundity, 79–80
Felidae, 8, 10
Felis spp: *F. geoffroyi*, 221; *F. lynx*, 119, 124; *F. pardalis*, 10, 221; *F. wiedii*, 221; *F. tigrina*, 221
fenitrothion, 209
Fennecus zerda, 111–12
fennel, 171, 199
fenoprop, 208
fenthion, 208–9
fenugreek, 171
ferns, 11, 16, 29, 30; seed, 28, 30, 34; spore-bearing, 28; tree, 29
ferrets, 181, 185, 216, 217
fertility, 79–80, 164
fescues, 171, 247, 310; *F. arundinacea*, 171; *F. ovina*, 15, 49; *F. rubra*, 310
fibres, 171, 198, 199
Ficus spp, 169, 170, 185
finches, 13, 52, 129
fir trees, 80, 117; Douglas, 120; grand, 120
fire: accidental, destruction by, 293, 320; as management technique, 291–3, 310, 323; co-evolution in Africa with man, grasslands, animals, 2; ecological effects, 72, 84, 98, 102, 104, 105–6, 116, 119, 120, 126, 127, 130, 155–6, 160, 291, 292; historical importance of man's control of, 331; impact of grazing combined with, 155–6; use by primitive, pre-agricultural communities, 154–6, 160, 184; vulnerability of forests to, 291
fireweed, 224
fish, fishes, 12, 28, 29, 327: adaptation to deep sea, 135, 137; as source of protein, 197–8, 270, 303; commercial,

dependence on salt marshes, estuaries, 141; distribution, 18, 36, 37, 133, 135, 136, 139; effects of contaminated waters on, 239–40, 242, 243, 245; effects of increase in freshwater algae on, 239; energy input–output ratios, 280; farming, 303, 304; in freshwater habitats, 43, 142, 143, 144, 145, 146; in mangrove swamps, 140; interrelation of species richness and latitude, 16; in tundra food webs, 125; potential source of food, 299, 303; problems of over-exploitation, 198; selective breeding of, 304; species used by man, 296, 297; tropical, as pets, 226–7; see also *individual species*
fishing, fisheries: as source of food in pre-agricultural communities, 153, 154, 161; depletion of stocks, 299–300; effect of 'red tides', 239; future direction of policies, 301; methods, production levels, 297; world catch levels, 297–8
flat fish, 135, 299, 303
flavours, plant sources of, 199, 200
flax, 169, 170, 171, 185, 199
fleas, 12; water, 12, 80, 142
flies, 112, 246, 327; horse, 80; house, 80
flooding, as management technique, 323
flounders, 245
flowering plants, see angiosperms, plants
fluoride, 132, 136, 236
fluorocarbons, 232
fluorine, 236
fluorosis, 236
flycatcher, Old World, 52
Foeniculum vulgare, 171, 199
forage crops, domestication of, 171
foramidines, 209
Foraminifera, 139
forestry, 116, 120, 288–96, 291, 335, 291, 335; see also forests
Forestry Commission, British, 213–14
forests: as productive resource, 288–96; acid rainfall and growth rates, 246; and air pollution, 293; and fire, 72,

forests—contd.
155–6, 160, 291, 292; and litter accumulation, 291; man's use of, 2, 116, 120, 288–96; estimated NPP, 86, 88; in evolutionary history, 28, 30; protected habitats, 316, 321; moisture levels and distribution, diversity of biota, 45; soil-types and nutrient circulation, 78; vulnerability to pests, 291
fossil fuels: and agricultural processes, 202, 274, 276, 277, 278; as basis of industrialization, 204–5; combustion, in carbon cycle, 74; effect of consumption on human population dynamics, 84–5; importance of harnessing of, 332; manipulation of energy relations by use of, 63–6; rates of consumption, 88–9; reduction of dependence on, 283; role in food production, 206
fossils, 27–31
fowl, see chicken
foxes, 27, 80, 107, 111–12, 157, 227, 247: arctic, 62, 125; fennec, 111–12; red, 125; silver, 27
France, agricultural acreage, 281
Fraxinus spp, 114, 234; *F. excelsior*, 234
freshwaters: distribution, character, life-zones, 142–7; pollution of, 236, 237, 238–45; species from, 296, 297; *see also individual species*
frigate bird, 135–6
frogs, 12, 37, 43, 80, 143, 327; edible, 217; tree, 217
fruits: domestication of, 169, 170–1; private garden produce, 224–5; spread of crops to Americas, 194; *see also individual crops*
Fucus spp: *F. serratus*, 18; *F. spiralis*, 18; *F. vesiculosus*, 11
fulmar petrel, 62
fungi: control by fire, 293; control by fungicides, 208, 242; ecological functions, 8, 95–6, 114, 115, 133; in classification of species, 7, 11; in evolutionary history, 28, 29; of Antarctic tundra, 124;

source of drugs, 200; *see also* biocides
fungicides, 208, 242; *see also* biocides
fur, 123, 221

Galapagos Islands, 52, 129, 211
Galinsoga parviflora, 210
Gallinule, speciation on Hawaii, 52
Gallium spp, 183
game, 294, 319, 320–5, 326–8
gametes, 24
gardens: energy input–output ratios, 224–6; private, 224–5; roof, 224, 225; specialist, 225–6
garlic, 171, 199
garrigue, 103–4
gas, natural, 126–7; *see also* fossil fuels
gases: concentration of, 18; effects of release to environment, 231, 232–8
gastropods, 29, 135
gathering, 33, 151–60, 160–1, 165–6; *see also* hunting
gazelles, 181–2, 184
General Systems Theory, 55
generating stations, 245
generation time, 23, 27, 80–1
genes, 22–4
genetic engineering, 23, 256–8, 264, 285, 286, 325, 345; *see also* selective breeding
genetic inheritance, 22–4, 345
Genista spp, 103
genotypes, defined, 23
genus, genera, in classification of organisms, 8
gerbil, Mongolian, 215, 216, 217
ginger, 199
gingko trees, 312–13
giraffes, 100, 101
glanders, 262
Glycine maxima, 169, 185, 261, 267, 270, 275–6
Glycyrrhiza glabra, 171
goats: adaptage to forage sources, 197, 308; domestication from *Capra hircus*, 168, 177, 178, 179, 183, 184, 185; ecological effects of grazing, 112, 130–1, 211, 212–13, 293, 310; feral, 221; in southwest Asian agriculture, 185; in pattern of human concern for animals, 327; mountain, 167; products, 306; semi-domestication of,

167; spread in Americas, 194; wild, 128, 216, 217
Gobi Desert, 108
'golden fruit of the Andes', 186
goldfish, 217
Gonyaulax spp, 302
goose, geese: *Anser anser*, 178, 181, 188; *A. cinereus*, 306; Canada, 215, 217, 327; Egyptian, 216, 217; role in food web, 62; speciation on Hawaii, 52; wild, extent of hunting of, 319
gophers, 107, 311
Gossypium spp, 169, 188, 199; *G. arboreum*, 188; *see also* cotton
gourds, 168, 185
grama, blue, 106
Gramineae, 12, 106
grapes, 169, 171, 174, 185, 194, 199
grasses: adaptation to contaminated environment, 246–7; bluestem, 106; breeding to improve, 261; buffalo, 106; Canary, 171; distribution, habitats, 49, 124, 130; domestication, 165–6, 170–2; dominance in tropical savannas, 98–9; drinn, 109; ecological impact of fire, 155; elephant, 99; environmental factors affecting behaviour, 15; guinea, 310; Harding-, 171; in classification of species, 12; introduction for grazing, 310; in tundra areas, 122–3; June, 106; marsh, 340; mat-, 49; needle-, 106; orchard-, 171; pangola, 310; *Phalaris* spp, 171, 183; productivity of, 276; *Spartina* spp, 141; tropic, 130; weed, emergence of, 183, 247; wild, in selective breeding of wheat, 260; *see also* grasslands
grasshoppers, 101, 111, 327
grasslands, temperate: character, distribution, 49, 93, 105–7, 130; co-evolution with man, fire, animals, 2; diversity, distribution of biota, 45; ecological effects of fire, 72, 155, 310; ecological effects of intensive, prolonged grazing, 2, 310–11; efficiency of decomposers, 88; estimated NPP, 86, 306; in evolutionary history, 28, 30; interrelation of

temperature, precipitation and, 91–2; NPP of grazing systems, 306; of north Africa, 310; recycling of mineral nutrients, 75; *see also* grazing, tropical savannas
grass pea, 170
grassy stunt, 267
gravel soils, 114
grazing, grazing systems: animals, behavioural adaptations to avoid overgrazing, 100–1; diversity in tropical savannas, 99–102; in freshwater ecosystem, 146; in phosphorus cycle, 72; ecological effects, of fire combined with, 155–6, of intensive, 310–11, of introduced domesticates, 104, 106, 112, 155, 211; ecology of systems based on, 196–7; economic virtues, 311–12; energy input–output ratios, 307–8; introduction of new grassland species, 310; methods of management, 107, 294, 307–10; resistant species, 310; soil improvement, 309; threat to plant species from, 222; *see also* grasslands
Great Basin Desert, 108
Great Plains, 105–7
greenfly, 12
greenhouse effect, 73, 74, 336–7
Greenland, *see* Arctic regions
greenleaf hopper, 267
Gross Primary Productivity, 61, 68–9
groundnuts, *see* peanuts
groundsel: sticky, 210; *S. vulgaris*, 128
grouse: extent of hunting, 319; sage, 309
guanaco, 106–7, 181, 186
guano, 130
guayule, 269–70, 285
guillemots, 62
guinea fowl, 178, 181
guinea pig, 168, 178, 185
Gulo gulo, 119, 125
gulls, 124, 125, 228, 327: glaucus, 62; herring, 228
gums, 200
guppy, 217
Guyana, 95
Gymnodinium brevis spp, 302

Gymnosperms, 11; *see also* Conifers
Gypsophila porrigens, 246

habitat, 42, 47–9, 218–19
haddock, 298
Hadza people of Tanzania, diet, 153
hair moss, 11
hake, 298
Haliotis spp, 198, 300, 302–3
Haplochromis spp: *H. angustifrons*, 144; *H. nigripinnis*, 144
Hardinggrass, 171
hardwoods, 290
hares, 161, 310, 319: arctic, 123; snow-shoe, 123, 124
hatchet fish, 137
Hawaii: adaptive radiation by hone creepers, 52; diversity of ecosystems, 130; endemic organisms, 52; evolution of *Lobelia* spp, 129; extinction of bird species, 220; flightless birds, 52
hawks, 52, 119, 125, 327
hawthorn, 312–13
hazel, 114, 116, 156, 160; *C. avellana*, 116, 156, 160, 170
heat, effects of release to environment, 232, 238, 245
heather, 49
Hedera helix, 113, 166–7
hemlock, western, 120
hemp: domestication of *Cannabis sativa*, 169, 171, 199, 200; manila, 199
henbane, 171
heptachlor, 208–9
herbicides, 208–9, 247–8; *see also* biocides
herbivores: behavioural adaptations to avoid overgrazing, 100–1; ecological impact of firing of habitat, 156; ecological role, 60–2, 63, 88, 125, 136; efficiency of NPP, 88; in Arctic tundra, 125; in deciduous forests, 113, 114, 115; in mountainous areas, 128; in seas, 136; interrelation with grassland ecosystems, 107; migratory habits, 107; protection afforded by herd, 107; source of food for hunting communities, 154
herbs (forbs), 28, 155

heroin, 200
herons, 81, 220, 327: black-crowned night, 52; night, 216
Herpestes auropunctatus, 140, 181, 211, 221
herring, 12, 135, 298, 299–300, 302; North Sea, 137
heterotrophs, 7, 8, 60–62, 87–8
Hevea braziliensis, 95, 194, 198, 210, 344
hibernation, 114
Hibiscus insularis, 222
hickory, 113–14
High Tatra Mountains, 317
Himalaya Mountains, 37, 45
hippopotamus: distribution, 36; pygmy, extinction of, 158
Holcus lanatus, 130, 310
holly, 114
holocene period, 28
holoplankton, 135
Homo spp: in evolutionary history, 28, 31–3; *H. erectus*, 32–3; *H. erectus pekinensis*, 32; *H. sapiens neanderthalensis*, 32, 33; *H. sapiens sapiens*, 32, 33
Homoiceros antiquus, 158
Homoiothermic animals, 13
Honduras, tropical savanna of, 99
honeycreepers of Hawaii, 52
honeyeaters of Hawaii, 52
hooded ladies' tresses, 37, 41
hookworms, 12
hops, 199
Hordeum spp (barley): domestication of *H. vulgare*, 165–6, 168, 169, 170, 172, 185; economic importance, 185, 194–5, 196; energy input–output ratios, 280; selective breeding, 182–3, 260; yields, 278, 279
horn, 69
hornbeam, 114
horse fly, 80
horseradish, 199
horses, 197, 202, 205, 262, 306, 327; domestication of *Equus caballus*, 168, 177, 178, 183, 185; wild, 106, 159
horsetails, 11, 28
house fly, 12, 23, 80
Hubbard Brook ecological studies, 75–8, 115, 290
Hudsonia spp, 46
humidity, 56–7; *see also* moisture, water resources
Humulus lupulus, 199

hunting: cause of extinction of species, 158, 218–19; depletion of fauna by, 100; ecological effects on caribou, 127; of game, continuing practice in Europe, 319; *see also* hunting-gathering communities
hunting-gathering communities: decline, transition to agriculture, 162; ecological impact, 151–6; energy sources available to, 331; evidence of selective killing, 154, 157, 160; in evolutionary history of man, 32, 33; herd-following economies, 156–8; population density, land ratios, 151, 152, 163–4, 202; problems of overkill, 154, 158
hurricanes, 129
hybridization, 260
Hydra spp, 12
hydrocarbons: derived from oil, 271; effects of release to atmosphere, 232, 236–7, 312–13; use of chlorinated, 138, 208–9, 241–2, 247
Hydrodamalis stelleri, 198
hydrogen, 69–70
hydrogen sulphide, 232, 233
hydrophytes, 8
hyena, spotted, 88
Hymenoptera, 62, 101
Hyoscyamus muticus, 171
Hyparrhenia, 99
Hypericum perforatum, 81
Hyphaene thebaica, 99

ice deserts, estimated NPP, 86
Iceland, deforestation following grazing, 310–11
Iceland moss, 11
ichthyosaurs, 28
iguana, 221
Ilex spp: *I. aquifolium*, 114; *I. paraguariensis*, 199
immigration of species, 50–3, 79–85
impala, 101
Imperata cylindrica, 99
India: domestication in, 168, 169, 178, 179, 181, 188; fish farming, 304; nature protection in, 317; use of pesticides, 208; *see also individual types of terrain*
Indians of North America, 116, 127, 161
indigo, 200

Indonesia, fish farming in, 304
industrialization: ecological effects, 204–5, 236, 249–50; energy sources, 331, 332; role of communications in, 204; *see also* pollution
industrial melanism, 2, 25–6, 27, 238
industrial waste, *see* waste
insecticides, *see* biocides
insects: evolution, adaptation to different environments, 49, 95, 101, 111, 112, 114, 115, 124, 128, 130, 140, 143, 145, 146; habitats, 45–6, 49, 124, 130; in classification of species, 12; in evolutionary history, 28, 29; pests, 227, 284; plant breeding to control, 267
International Biological Programme, 55
International Board for Plant Genetic Resources, 250–1
International Whaling Commission, 300
interspecific competition, *see* competition
intraspecific competition, *see* competition
Inuit peoples of North America, 154, 161, 197
invertebrates: in classification of species, 12; in evolutionary history, 28, 29, 30; in freshwater ecosystems, 146; in tundra, 125; mortality rates caused by predation on, 88; Production/Assimilation rates, 88
iodofenphos, 209
ionising radiation, 231
Ipomoea batatas, 169, 188, 193 194, 195, 196, 201–2
Irenidae, 36
iron, 96, 98, 104, 339
iron-55, 249
Iroquois Indians, 161
irradiation, 265
irrigation, 104, 183–4
Isatis tinctoria, 200
islands: distribution, character of flora, fauna, 129–31; ecological effects of introductions, 211, 212; extinction of species, 222; factors affecting species density, distribution, 50–3; threatened species, 221
Isle Royale, Lake Michigan, 21
Isoberlinia spp, 99

isoprenes, 236–7
Isoptera, 101
Israel, energy input–output ratios in agriculture, 276–9, 281
Italy: nature protection in, 317; use of pesticides, 208
Ivaannua, 186
Ivory Coast: ecological role of termites, 101–2; tropical savanna, 99
ivy, 113, 166–7

jackal, 180
jackdaws, 227
jackrabbits, 107, 111, 311
jaeger, 125
jaguars, 220, 221
Japan: domestication in, 181; fish farming in, 304; nature protection in, 314, 317, 326–8
jarrah, 291–3
Java, extinction of species in, 158
Java Man, 32–3
jays, demand for feathers, 220; blue, 321
jellyfish, 12, 28, 133, 327
Johnsongrass, 171
jojoba, 200, 269
Joule, conversion to calories, 3
Juglans regia, 171
Juncus squarrosus, 43–4
June grass, 106
Jurassic period, 28, 29, 30, 34
jute, 199

Kalahari Desert, diet of Bushmen, 163
kale, 171
kangaroo: habitat of *Macropus* spp, 100; products of *Megaleia rufa*, 306; red, 37, 38
kangaroo rat, 37, 111
kapok, 199
kaska, conservation of animal resources, 161
kelp, 80, 344
kerguelen cabbage, 124–5
ketones, 237
Kew, Royal Botanical Garden at, plant collections, 210
Kew weed, 210
Khazakstan, causes of crop failure in steppes, 107
kite, red, 230, 231
kittiwake, 62
Koeleria cristata, 106
kohlrabi, 171
krill, 299

kudu, 263
!Kung Bushmen, diet of, 153

Labiatae, 103
Lacerta spp, 43
Lactuca sativa, 171
lagomorphs, 157
Lake Erie, over-fishing on, 302
lakes: causes, effects of pollution, 239, 240, 243; character, distribution, 142, 143–5; estimated NPP, 86
Laminaria spp, 18, 87
land: cultivated, estimated NPP, 86; derelict, ruderal species, 246; effects of man-made waste, 245–9; reclamation, ecological effects of, 138, 140–2; use, modern intensive, 205–6
landscape, protection of, 317, 219–20, 326
lantern fish, 299
larch trees, 11, 117: Japanese, 293; *L. dahurica*, 117
Larrea divaricata, 108, 109, 110
Larus argentatus, 228
latex, 108–9, 285
Lathyrus sativus, 170, 171
Latin, in classification of species, 8
latitude and altitude, 127–8; and diversity, 45–6
latosols, of tropical forests, 95, 98, 99
lawns, energy ratios, 226
lead, 237, 312–13, 339
lead arsenates, 208
leaf-beetle, 81
leafbirds, 36
leaf rust, 260
leatherjackets, 49
Lecanora conizaeoides, 233
leeches, 12
leeks, 171
legumes, 174, 185, 200, 276
Leguminosae, 70–2, 106
lemmings: *Dicrostonyx* spp, 124, 125; *Lemmus* spp, 80, 124, 125
Lemna spp, 143
lemons, and pollution, 237; *see also* citrus fruits
lemurs, tree pollination by, 95
lentic habitats, 142, 143–5
lentils, 168, 169, 170
leopards: *P. pardus*, 8, 221; snow, 128
Lepidurus, 62

Leptinotarsa decemlineata, 208
Lepus spp: *L. americanus*, 123, 124; *L. arcticus*, 123
lettuce, 171
Leucobryum glaucum, 20
Levuana irridescens, 265
lianas, 94–5
lice, 327
lichens, 11, 86, 233–4, 248, 249; habitats, 96, 113, 114, 117, 120, 122–3, 124
light, 17, 56–60; *see also* sunlight
lilac trees, 312–13
Liliaceae, 12
lily, water, 143
lime trees, 113–14, 116, 161, 312–13
limestone soils, species associated with, 49, 114
limpets, 12
Lindemann models of ecosystems, 60–2
linden trees, *see* lime trees
linden aphid, 227
Linné, Karl von (C. Linnaeus), 8
linseed, 169, 170, 171, 185, 199, 200
lion, 8, 10, 227; mountain, 104, 220–1; *Panthera leo*, 8, 10
Liriodendron spp, 113–14
litter breakdown: and accumulation in forest ecology, 291; ecological effects of fire on, 104; function of termites, 101; in various habitats, 95–6, 98, 112, 114–15, 118
liverflukes, 12
liverworts, 11, 28, 29, 124
lizards, 13, 31, 37, 43, 45, 111, 327; *Lacerta* spp, 43; monitor, 221
llama spp: *L. glama*, 128, 168, 177, 178, 181, 185–6, 306; *L. guaniacöe*, 106–7, 181, 186; *L. pacos*, 128, 168, 178, 181, 185, 186, 197, 306; *L. vicugna*, 181, 186, 306
loam soils, and deciduous trees, 114
Lobaria amplissima, 233
Lobelia spp, 128, 129
lobsters, 12, 135, 245
locusts, 12, 16, 311; desert, 112; *Chortoicetes terminifera*, 16
Lolium spp, 171, 183
Lophophora williamsii, 200
lotic habitats, defined, 142
Loxodonta africana, 221

lucerne, *see* alfalfa
lupin, 170
Lycaena dispar batava, 323
Lycopersicum esculentum, 169, 174, 175
Lycopodium spp, 11, 16, 28, 30, 33
lye, 208
lynx: *Felis lynx*, 119, 124; *Lynx lynx*, 10, 124, 157
lyre birds, 27
lysine, 259, 270, 280

machinery, 63–6
mackerel, 300
Macquarie Island, 127
Macrocystis spp, 87
Macropus spp, 100; *M. rufus*, 37, 38
Madagascar: extinction of species in, 158; flightless birds, 52; *see also individual types of terrain*
madder, wild, 16, 17
magnesium, 132, 136, 155
magnolia trees, 312–13
magpie larks, 37
maize: agricultural ecosystems based on, 194–5, 196; annual primary production, respiration rates, 60; crop dominance by few varieties, 286; domestication, 168, 169, 174–5, 185; economic importance, 185–6, 258; energy input–output ratios, 279, 280; potential use of waste from to produce ethanol, 285; productivity, 275–6; selective breeding of, 182–3, 259, 261; source of vegetable oil, 200; spread from Americas, 187–8, 193
malaria, 11, 200, 208, 209
malathion, 208–9
Malaya, dominant tree species, 95
mallee scrub, of Australia, 103–4
Mammalia, 8, 13, 28, 29, 31, 327: arboreal, 95; ecological role in various habitats, 37, 107, 114, 115, 119, 125, 128, 135, 136, 140; extinction of species, 158–60, 218–20; in derelict land, 247; protection of colonial nesting habitats, 221; ungulate, intraspecific competition for territory, 20
mammoth, woolly, 28, 158
manganese, 339

Index

mangoes, 169
mangrove swamps, 138, 140–1, 247–8
Manila hemp, 199
manioc, 193–4, 195, 196, 275–6
manzanita, 103
maple trees, 113–14, 312–13
maquis, 103–4; see also scrublands
margay, 221
marine parks, 316
marlin, 299
marmot, 128, 157
marsh elder, 186
marsh grass, 340
marsh lands, estimated NPP, 86; see also salt marshes
marsupials, 31, 37, 38–9, 219–20, 327
martens, 157, 161
mate, 199
mat-grass, 49
matorral, of Chile, 103–4; see also grasslands
Mauritius, flightless birds, 52
mayfly, 16, 143
meadowlarks, 107
measles, 11
Medicago spp, 171, 183; *M. sativa*, 60, 169, 171, 193
mediterranean area: character of basic agriculture, 185, 196; domestication in, 170–1, 173–4, 178, 181; effects of human acitivities, 104; see also individual types of terrain
Megaleia rufa, 306
Megaloceros giganteus, 158
meiosis, 24
Melanerpes formicivorus, 83–4
melanism, 2, 25–6, 27, 128, 238
Melilotus spp, 171
Melissa officinalis, 199
melons: domestication of *Cucumis* spp, 170, 171; water-, 169
Mendel, G., 23
menhaden, 135, 245
Mentha piperata, 199
Mercurialis perennis, 114
mercury, 231, 242, 301, 339
merino sheep, see sheep
Meriones unguiculatus, 215, 216, 217
meroplankton, 133–5
Mesolithic period, destruction of vegetation in, 84
mesophytes, 8
mesquite, 108

metals, contamination by, 237, 242, 246
methane, 236–7
methanol, 271
methoxychlor, 208–9
methyl mercaptan, 240
Methylophilus methylotrophus, 271
Mexico, domestication of maize in, 174
microbenthos, 136
microflagellates, 133
midges, freshwater, 143
migration, seasonal: danger to species resulting from, 222; factors affecting, 17; of birds, 114, 124; of fish, 135
military activities, in tundra areas, 126–7
milkfish, 185, 198, 302–3
millet: as source of food, 269; Channel, 269; common, 169; depradation by pests, 283–4; domestication of, 168, 169, 182–3, 185; finger, 169; foxtail, 168, 169, 185; importance in basic agriculture, 185–6; lesser, 186; pearl, 169, 182–3, 261; selective breeding, 182–3, 261
millipedes, 12, 114, 115
Milvus milvus, 230, 231
Mineral nutrients: addition to grazing land, 309; availability, 13; biogeochemical cycles, 70; causes, effects of eutrophication by, 239; conservation within ecosystems, 68–9; cycling, function of micro-organisms, 8, in ecosystem development, 62, 68–9, interdependence with energy flow, 79, times, 70, 73, 74, 75, 78; ecological impact of fire on, 155; effect of forestry on local levels, 289–91; essentials for life, 69–70; function of termites in recycling of, 101–2; in various habitats 75–9, 95–7, 98, 99, 108, 114–15, 118–19, 120, 123, 124, 130, 131–2, 135–6, 139–40, 141, 145, 307, 311; recycling disturbed by man, 78–9; relation of soil type to circulation of, 78; scientific approach to use in agriculture, 206–7; trends in mobilization rates, 339

mineral oil, early use as insecticide, 208
minerals, ecological effects of extraction in tundra areas, 126–7
mink, 161, 216
Miocene period, 28
mirex, 208–9
Mistassini Cree Indians, diet, 161
mites, 62, 114, 115; *Acarina* spp, 49; spider-, problems of control, 247
mithan, 128, 167, 178
mitosis, 24
moa, 52, 158
moisture, and distribution of biota, 13, 16–17, 45, 48, 49
Mojave Desert, 108, 109
moles, 49, 115
Molinia caerulea, 49
molluscs, 12, 29, 245, 327; bivalve, 29, 135, 137, 141, 143; in mangrove swamps, 140; intertidal, 16; in sea, 135, 137, 139
mongongo tree, 153
mongoose, 140, 181, 211, 221
monkeys, 31, 36, 37, 327; in mangrove swamps, 140; rhesus, urban scavenger in India, 228; tree-dwelling, 95
monocotyledons, 12
Monodon monoceros, 127, 161
moorgrass, blue, 49
moor rush, 43–4
moose, 21, 119, 262
Mora spp, 95
morphine, 171, 200
mortality, 79–85, 236
Moschus moschiferus, 200
mosquitoes, 12, 327: *Anopheles* spp, 247; biocides in control of, 208, 209, 247; increase near waste dumps, 246; in tundra, 124; strains resistance to biocides, 247
mosses, 11, 28, 29: in boreal forests, 117; Iceland, 11; in beech woods, 113; in deciduous forests, 113, 114; in island habitats, 130; in tundra, 122–3, 124; role in food web, 62
moths, 49, 327; coconut, 265; diamond-back, 209; peppered (*Biston betularia*), 2, 25–6, 27, 238; winter, 115
mouflon, 179
mountain lion, 104, 220–1

mountains, mountainous areas: desert areas in lee of, 108; distribution, character, 127–9; effect on distribution of organisms, 45, 53; estimated NPP, 86; NPP of grazing systems, 306; vegetation in North African, 128
mouse, mice, 23, 80, 115, 197, 227–8, 311; desert, 37, 39; field, 216, 217; house, 216; yellow-necked, 216
mud flats, intertidal, 141–2
mudskipper, 140
mule-deer,
mules, 306
mullet, grey, 303
muntjac: Chinese (Reeve's), 216; Indian, 216
Musa spp, 95, 169, 174, 185, 194, 258; *M. textilis*, 199
mushrooms, 11, 270; Shiitake, 270
musk (perfume), 200
musk-deer, 200
musk-ox, 123, 158, 161, 185, 197
musk-rat, 161, 214
mussels, 141, 302–3; freshwater, 143; *Mytilispedulis* spp, contamination of, 239
mustard: black, 170, 199; yellow, 199
mustela, 157
mutation, 23–4, 285
Mycaster coypus, 215–18
mycorrhizae, 95–6
Mycophidae, 299
Myriapoda, 12
Myristica fragrans, 199
Myrmecobius fasciatus, 37, 39
myxomatosis, 81, 211
Mytilis edulis, 239

Namib Desert, 108
nanoplankton, 133
narcotics, plants yielding, 171, 200
Nardus stricta, 49
narwhal, single-tusked, 127, 161
natality, of species, 79–85
National Parks, 314, 317, 319, 320–5, 326–8
natural extinction, 34–5
natural selection, defined, 25–6
naturalization of species: ecological effects, 209–18 in Britain, 215–18
Nature Conservancy Council, British, 328

nature conservation, protection: by pre-agricultural peoples, 153, 154, 160, 161; carrying capacity of systems, 84, 328–9, 342; co-existence with recreation, 316, 318–20, 326–8; in reserves, 315–19; management problems, policies, 320–5, 326–8; purposes, 325–8
nature reserves: aims, distribution, 314, 317, 319; management problems, policies, 320–5, 326–8
nautiloids, 29
nautilus clams, 327
Neanderthal man, 32, 33
nearctic zoogeographical realm, 36–7
Near East, *see* Asia, southwest
needlegrass, 106
nekton, 133, 135
nematodes, 12, 62, 140
Neotoma fuscipes, 104
neotropical zoogeographical realm, 36–7
Neptunus pelagicus, 214–15
Nesokia indica, 184
Netherlands, recreation areas, 320
Net Primary Productivity (NPP): defined, 58; effects of mechanization in agriculture, 206; efficiency in photosynthesis, 85; estimates for various biome types, 85–8; expression, measurement of, 85; impact of fire, 155, 156; in boreal forests, 117–18; in continental shelves, 141; in deciduous forests, 115; in desert areas, 111, 112; in ecosystem development, 68–9; in energy transfer systems, 61; in freshwaters, 145–7; in intertidal areas, 138–42; in plant-man food chains, 192–3; in salt marshes, 141; in seas, 133–5, 141; in temperate grasslands, 106; in temperate woodlands, scrublands, 103–4; in tropical rain forests, 97; in tropical savannas, 99, 100; in tropical seasonal forests, 98, 141; in tundra, 121–2; in vegetation types used for grazing, 306; rates of autotrophs, heterotrophs compared, 87–8; world levels, 85, 88–9

Netsilik Inuit people of Canada, food resources, 154
neuston, 133, 135
New Guinea: energy input–output ratios of agriculture, 201–2; tropical seasonal forests, 98
New Zealand: changes in vegetation of grasslands, 310; ecological effects of introductions, 211, 212–13; extinction of moa, 158; flightless birds, 52, 158; spread of domesticates to, 187–8, 194
newt, 12, 80
nickel, accumulation in plants, 76
nicotine, 200, 208
Nigeria, fish farming in, 304
nitrate, levels following deforestation, 77
nitrogen: atmospheric, fixation by blue-green algae, 30; consumption in grazing systems, 307; cycle, 62, 70–2; ecological impact of fire, 104, 155; effect of forestry on levels of, 290–1; essential nutrient for life, 69–70; fixation in soil, 104, 155, 256, 261, 334; in various habitats, 95–6, 99, 123, 131–2, 135, 136, 145; mobilization rates, 339; relation to productivity of grasslands, 308, 309; scientific approach to needs in agriculture, 206–7; world consumption of fertilizers based on, 207; *see also* eutrophication
nitrogen oxides: control of levels in air, 312–13; effects of release to atmosphere, 232, 236–7, 238
nomadism, ecology of, 196–7; *see also* shifting agriculture
Nostoc spp, 11
Nothofagus spp, 30, 37, 42
nuclear devices, 127, 248–9
nuclear power, energy derived from, 63–6
numbat, 37, 39
nutmeg, 199
nutrients, *see* mineral nutrients
nuts, 116, 153, 160, 169, 170–1; pistachio, 171, 184
Nymphaea spp, 143

oak trees, 8, 9, 103, 104, 113–14, 115, 116, 316; evergreen, 103

392 Index

oak–pistachio ecosystem, 184
oats, 169, 170, 183, 185, 194–5, 196
Oceania (Pacific islands): consumption of fertilizers, 207; domestication in, 168, 169; extinction of species, 220; spread of crops, animals to, 194; tropical seasonal forests, 97
oceans, *see* seas
ocelot, 10, 21
octopus, 12, 299, 327
Odocoileus hemionus, 309
oil, oils: crude, 8; industry, 126–7, 231, 270–1; pollution of aquatic environments by, 142, 242–3; use, effects of, as biocide, 208–9; vegetable, crops yielding, 170, 194, 200; *see also* fossil fuels
oil palm, 194, 200
oil-seeds, 267
Olduvai Gorge, Tanzania, 32, 33
Olea europaea, 103, 169, 170, 173–4; *see also* olive trees
Oligocene period, 28
oligochaetes, 62, 144
olive trees: domestication of *Olea europaea*, 169, 170, 173–4; importance in basic Mediterranean agriculture, 185; source of vegetable oil, 200; subsistence ecosystems based on, 196
omnivores (diversivores), 8
onager, 185
Ondatra zibethica, 161, 214
onions, 169, 171, 184, 199
Onobrychis viciifolia, 171
Onothophagus gazella, 211–12
opencast mining, 240
opium, 171, 200
opossum, 212–13
Opuntia spp, 109, 211
orange trees, 227, 237
orchardgrass, 171
orchids: mutualistic interaction with bees, 22; rate, 210, 314–15, 323
orders, in classification of species, 8
Ordovician period, 28, 29
organisms, 7–13, 21: distribution, 13–18, 36–50; effects of low density on species survival, 82–3; eurytopic, stenotopic distinguished, 14; in ecosystem development, 68–9; interspecific competition, 20, 22; interspecific differential co-existence, 20; intraspecific competition, 19–20, 22; mutualistic interactions, 22; parasitism, 21–2; plants, animals distinguished, 7; population dynamics, 79–85; predation, 20–1; significance of generation time, 80
organochlorine insecticides, 293; *see also* biocides
organo-mercury compounds, 208–9; *see also* biocides
organophosphorus insecticides, 208–9; *see also* biocides
organ pipe, *see* saguaro cactus
oriental zoogeographical realm, 36–7, 45
Ornithorhynchus anatinus, 37
orthoptera, 101
Oryctolagus cuniculus: damage caused by, 127, 211, 228, 310; domestication, 181, 185; effects of myxomatosis, 81, 211; habitats, 49, 227; in pattern of human concern for animals, 327; naturalization in Britain, 216, 217; products, 306
oryx, 181–2
Oryza spp: *O. glaberrima*, 169, 175–6, 193, 194; *O. sativa*, 169, 174, 175–6, 193, 275–6
osmosis, in adaptation to freshwater habitats, 142–3
ostracods, 144
Ostrea spp: *O. edulis*, 315; *O. gigas*, 215
ostriches, 220; *see also* birds, flightless
otters: habitats, 140; importance in pre-agricultural economies, 157; in diet of hunting peoples, 161; in pattern of human concern for animals, 327
Ovibus moschatus, 123, 158, 161, 185, 197
ovis spp: *O. aries*, domestication of, 168, 177, 178, 179, 184, 185, 306; *O. orientalis*, 179
owls, 13, 220, 327: barn, 283; habitats, 107, 115, 119, 125; little, 216, 217; short-eared, speciation on Hawaii, 52; tawny, 115
ox, replacement by horse, 205
oxygen, causes, effects of depletion in aquatic systems, 239–40; concentration of, 18; effects of release to atmosphere, 232; essential nutrient for life, 69–70; levels, in freshwater habitats, 142, 145
Oxyria digyna, 44–5
oysters, 12, 80, 215
ozone, 233, 238

Pacific islands, *see* Oceania
Palaearctic zoogeographical realm, 36–7, 45
Palaeolithic Man, 33
pale, French, red-legged, 216
palm trees, Brazilian, 200; date, 169, 171, 184, 196; dom, 99; in evolutionary history, 30
pampas, of south America, 105–7; *see also* grasslands
Panama Canal, spread of species through, 215
panda, giant, 326, 327
pangola grass, 310
pangolins, 158
Panicum spp: habitat in savannas, 99; invasion of sugarcane, 247; *P. fasciculatum*, 247; *P. maximum*, 261, 310; *P. virgatum*, 106
Panthera spp, 8, 10, 221: *P. leo*, 8, 10, 227; *P. onca*, 220, 221; *P. pardus*, 8, 221; *P. tigris*, 8; *P. uncia*, 128
Papaver somniferum, 170, 200
papaya, 169
parakeet, ring-necked, 216
Paramecium, 80
paraquat, 208
parasitism, parasites: diseases caused by, 11; factors affecting density of, 82; in biological control of pests, 264–7; in classification of species, 8; in food web of deciduous forests, 115; in interspecific relationships, 21–2; transference with host species, 211, 215
parathion, 208–9
parks: urban, 224, 226; wildlife, 220–1, 314, 317, 319, 320–8
Parmelia saxatilis, 234
parrot, ring-necked, 216
parsley, 171, 199
Parthenium argentatum, 269–70, 285

partridges: chukar, 216; European rock, 216; extent of hunting of, 319; red-legged, 217
Passer domesticus, 227, 265
passerines, 327
pastoralism: agricultural ecosystems based on, 196–7; dominant species, 305–6; ecological effects, 92, 107, 112; population/land ratios, 202; unstable nature in Africa, 311; *see also* grasslands, grazing
pathogens, 264–7
Patinopecten spp, 302–3
peas: chick-, 169, 170; domestication of, 168, 169; garden, 170; pigeon, 169; wild, 222
peaches, 169
peacocks, 181, 210
peanuts, 169, 188, 193–4, 200, 258, 267, 286
pears, 169, 171; *see also* avocado
peaty soils, flora, fauna of, 49
peccaries, 95
Peking Man, 32
pelagic organisms, defined, 133
Pellia spp, 11
penguins, 13, 125
Penicillium spp, 11
Pennisetum spp: habitat in savannas, 99; *P. americanum*, 261; *P. purpureum*, 276
penoxyalin, 209
pentachlorophenyl, toxicity to shellfish, 240
peppermint, 199
perches (fish), 327; yellow, overfishing of, 302
perfumes, sources of, 199, 200
periodicity, of animals of boreal forests, 119
Perissodactyla, 262
Permian period, 28, 29, 34
peroxyacetyl nitrate (PAN), 237
Persea americana, 169, 175, 185, 227
pests: control by biological methods, 264–7; of forests, 291–3; r-pests, k-pests distinguished, 266–7; selective breeding for resistance to, 260; *see also* biocides *and individual pests*
petrels, 125, 135–6; fulmar, 62
petroleum combustion, effects in atmosphere, 237

Petroselinum sativum, 171, 199
pets, 215, 226
peyote, 200
Phacochoerus aethiopicus, 83
Phalaris spp: interaction between domesticates and, 183; *P. arundinacea*, domestication of, 171
Phasianus spp: domestication, 185; introduction to Britain, 215–18; *P. colchicus*, 217; *P. colchicus satscheuensis*, 216
Phaseolus spp, 185
pheasants: Chinese ring-necked, 216; demand for feathers, 220; domestication, 185; extent of hunting of, 319; golden, 216; introduction, naturalization in Britain, 215–18; Japanese, 216, 217; Lady Amherst's, 216; mongolian, 216; Pallas's, 216; Prince of Wales's, 216; Reeves's, 216; silver, 216
phenols, 208–9
phenotypes, defined, 23
Phleum pratense, 171, 193
Phoenix dactylifera, 169, 171, 184, 196
phosphorus: consumption in grazing systems, 307; cycle: 72–3, 231, 239; levels in ecosystems, 96, 104, 123, 131–2, 135, 136, 139–40, 145, 155, 290–1; mobilization rates, 339; world consumption of fertilizers based on, 207
photosynthesis: basis of ecological energy, 55–60; calculations of efficiency, 85; causes, effects of inhibition of, 237–8, 239, 240, 245; characteristic of plant life, 7; fundamental importance of light, 17; importance of balance between respiration and, 58–60; in evolutionary history, 30; in freshwaters, 142; in salt marshes, 141; in seas, 132, 136; plant breeding to improve, 259–60; role of forests in, 296
phyla, in classification of species, 8
Phytophthora cinnamoni, 291–3
phytoplankton: biogeographical function, importance, 8–13, 14, 59, 132–46; environmental factors affecting distribution, 18; freshwater, 143–5, 146; in

carbon cycle, 74; levels of primary production, 87; marine, 132–7, 141
Picea spp: dominance in boreal forests, 117; in classification of species, 11; pest of, 119, 247, 291; *P. sitchensis*, 120, 210, 213–14, 293
picnic areas, scavengers of, 321
pigeons, 181, 227–8, 319
pigs: breeding to improve, 183, 262; diseases of, 263; domestication from *Sus scrofa*, 163, 168, 177, 178, 179–80, 185, 188; ecological effects, 130, 211, 212–13, 310; efficiency in food chains, 193; energy input–output ratios, 202, 279; feral, 221; importance, 157, 185; wild, habitats, 95; *see also* pork
pike, blue, 302
pilchards, 135
Pimelea suteri, 76
Pimenta officinalis, 199
Pimpinella anisum, 171, 199
Pinaceae, 200; *see also* pine trees
pineapples, 169, 185
pine trees: annual primary production, respiration rates, 60; closed-cone, 156; ecological effect of forestry processes, 290; habitats, 99, 103, 117, 291; in classification of species, 11; lodgepole, 210, 293; Norfolk Island, 222; *P. pinea*, 103; *P. pinaster*, 103; ponderus, 233; *P. radiata*, 213; relic woods in Scotland, protection of, 316; *P. sylvestris*, 291
pine sawfly, 82, 265
pin mould, 11
Pipilo fuscus, 50
Pisaster ochreus, 340
Pisces, 12
pistachio nuts, 171, 184
pitcher-plant family, 37, 41
Planaria spp, 12
Planera spp, 46
plane trees, 312–13
plankton: biogeographical functions, 8–13; effects of thermal shock on, 245; in freshwaters, 62, 143, 145; in seas, 132–7, 139; saprophytic, 136–7
Plantago spp, 183; *P. psyllium*, 171

plant breeding: basis, achievements of, 258–61; limitations, 286; to improve nitrogen fixation, 256, 261, 334
plastics, useful waste from, 271
Platyhelminthes, 12
platypus, duck-billed, 37
Pleistocene period, 28, 30, 32–3, 35, 37, 53, 92, 156–7, 158–60
Pleiosaurs, 28
Pleurococcus spp, 22; *P. viridis*, 233
Pleuroncodes planipes, 299
Pleuro-pneumonia of cattle, 262
Pliocene period, 28, 30, 32
plum trees, damage to, 227; domestication of *Prunus domestica*, 171
Poa spp, 310
poaching, 320
Podocnemis expansa, 198
podsols, distribution of, 95–6, 97, 119
poikilothermic animals, 13
Poland, protected oak forests, 316
polecat-ferret, 216, 217
pollination, 22, 95
pollock, 298
pollution: effect of depletion of oxygen by, 18; of atmosphere, 232–8, 293; of freshwaters, 238–45, 250, 312; of land, 245–9; of rivers, 250; of salt marshes, estuaries, 142; of seas, 138, 238–45, 301–2, 312; overall effects of, 251
polychaete worms, 133
polychlorinated biphenyls (PCBs), 138, 241–2
Polytrichum spp, 11
pomegranates, 169, 171
pompano, 303
ponds, 67, 142, 143; *see also* freshwaters
pondweed, 143
poplar trees: aspen, 117; resistance to air pollution, 312–13
poppies: California, 46–7; narcotics derived from *Papaver somniferum*, 171, 200
population: density of pre-agricultural communities, 151, 152, 163–4, sustainable by subsistence agriculture, 202; dynamics, controlling factors, 79–85; defined, 54–5; effects of increase following industrialization, 204–5; increase and demand for space, 338, and spread of disease, 338, following disease control, 209; trends, problems, 151
Populus tremuloides, 117; *see also* poplar porcupines: crested, 216, 217; Himalayan, 216; in diet of hunting peoples, 161
pork, approaches to eating of, 189, 191
Portulaca oleracea, 171
possum, greater glider, 37, 38
Potamogeton spp, 143
potash, *see* potassium
potassium: constituent of sea water, 132, 136; levels in ecosystems, effects of fire, 155, effects of forestry, 290–1, following deforestation, 77; world consumption of fertilizers based on, 207
potato beetle, Colorado, 208
potato blight, 11
potatoes: crop dominance by few varieties, 286; domestication from *Solanum tuberosum*, 169; economic importance as basic food, 185–6, 258; in agricultural systems based on vegeculture, 195; pests, diseases of, 11, 12, 208; spread from, back to Americas, 188, 193–4; yields, 275–6, 278, 279
potatoes, sweet: domestication from *Ipomoea batatas*, 169, 188, 195, 196; economic importance as basic food, 258; energy input–output ratios of farming based on, 201–2; in agricultural ecosystems based on vegeculture, 195, 196; spread from Americas, 193, 194
poultry, breeding to improve, 262; *see also* chickens
power stations, effects on aquatic environments, 245
prairie chicken, 107
prairie dogs, 37, 107, 311
prairies, 105–7, 306
prawn, Red Sea, 215
Precambrian period, 28, 29
precipitation: acidification, effects on land biota, 245–6; interrelation with, other factors in determining distribution of organisms, 42, temperature and biome type, 91–2; levels, and intraspecific competition, 19; pH levels, 235, 236; seasonal, effect in deserts, 111–12; significance in, biomass of boreal forests, 120, development of temperate grasslands, 105–6, tropical savannas, 99, tropical rain forests, 96
predation, predators: effect of absence on island organisms, 52; effect on diversity, distribution, density of species, 48, 81, 82; in biological control of pests, 264–7; in food web of seas, 136; in interspecific relationships, 20–1; introduction of, a cause of extinction, 218–19; mortality rates caused among invertebrates, 88; protection against, afforded by herd, 107
prickles, ecological functions, 114
primates, 20, 31, 197, 327
Pringlea antiscorbutica, 124–5
Processa aequimana, 215
production ecology, defined, 54–5
Prosopis spp: interaction with domesticates, 183; *P. juliflora* (mesquite), 108
Proteaceae, 30
protected landscapes, 314, 317, 319, 320–5, 326–8
proteins: aquatic animals as source of, 197–8; breeding of plants with improved content, 259; content, in plant/animal/man food chains, 192–3; crops producing, response to energy input, 287; levels in shifting agriculture, 195, 196; potential sources, 267–71; single-cell (SCP), derived from industrial waste, 270–1, 280; sources in meat, 196; trends in search for supplies of, 334–5
protista, in classification of species, 11
protozoa, 11, 29: distribution, ecological functions, 62, 114, 115, 133, 142, 143
Prunus spp: domestication, 171, 199; habitats in deciduous forests, 114; *P. amygdalus*, 171, 199, 200; *P. armeniaca*, 171; *P. avium*, 171; *P. pensylvanica*, 75–6, 290
Psammomys obesus, 111
Pscocids, 234

Pseudomonas, 80
Pseudotsuga menziesii, 120
psyllium, 171
ptarmigan, 62, 123–4, 128
Pteridium spp: *P. aquilinum*, 11, 20, 48, 114, 340
Pteridophyta, 11
pteridosperms, 28, 30, 34
Ptychomyia remota, 265
Puccinia striiformis, 260
puffins, 62
pulses, domestication of, 170
Punica granatum, 169, 171
purslane, 171
Pyrenees, Pleistocene cave-dwellers of, 154
pyrethrum, 208–9
pyinkado tree, 98
pyrites, 240
pyrophycic trees, 99
Pyrus spp: *P. communis*, 169, 171; *P. malus*, 169, 171

quail, bobwhite, 216, 217
Quaternary periods, 29, 30
quelea, red-billed, 283–4
Quercus spp, 8, 9, 103, 113–14, 115, 116, 316; evergreen, 103; *Q. dumosa*, 8; *Q. gambelli*, 9; *Q. ilex*, 8, 103, 104; *Q. petraea*, 9, 114; *Q. robur*, 8, 113, 114; *Q. virginiana*, 9
quince, 170
quinine, 171, 200

rabbits: adaptation to urban environment, 227; damage caused by, 12, 127, 211, 228, 310; distribution, habitats, 49; domestication, 181, 185; effects of myxomatosis, 81, 211; extent of hunting of, 319; naturalization in Britain, 216, 217; products, 306
rabies, 226, 228, 262, 321–2
racoons, 12, 227, 228; crab-eating, 140
radiation: causes, effects, control of, 127, 248–9; in tundra, 127; release into aquatic systems, 243–5; ultra-violet, 238
radio-nuclides, release into sea, 244–5
radish, 170
ragweed, North American, 238
ragworts: effects of myxomatosis on density of *S. jacobea*, 81; spread of Oxford, 210

rails (birds), 52
rainfall, *see* precipitation
Ramalina maciformis, high NPP, 86
ramie, 199
Rana spp, 43, 143
Rangifer spp: *R. tarandus*, 119, 123, 124, 125, 126, 127, 156–7, 159, 197, 218, 306; *R. tarandus fennicus*, 197; *R. tarandus groenlandicus*, 197
rapeseed, 170
Raphus cucullatus, 52, 222
rats: bandicoot-, 184; black, 216, 217; brown, 216, 217; brown Norway, 211; depredations on cereal crops, 284; ecological effects of predation by, 130; effects of introduction to new habitats, 52, 130, 211; increase near waste dumps, 246; in derelict land, 247; kangaroo-, 37, 111; medical, scientific uses, 197; sand, 111; size and generation time, 80; urban, treated as pest, 227–8; wood, 104
rattlesnakes, 107
recreation and nature protection, 316, 318–20, 321; and modern forestry, 295; ecological effects, 112, 120, 126–7, 128–9; management problems, policies, 320–5, 326–8, 329
Red Sea, effect of salinity on species, 18
red shank, 20, 21
'red tides', cause of, 239, 301–2
redwood trees: California coast, 40, 48, 80, 120, 156, 322, 323, 326; sierra, 120, 291, 326
reindeer: as food resource, 197, 218; domestication, 177, 178; habitats, 119, 123, 124, 125, 126, 127; pre-agricultural man's dependence on, 156–7; products, 306; survival through environmental changes, 189
religion, and domestication of animals, 187, 188–91
remora fish, 21
Reproduction of species: fertility, fecundity distinguished, 79–80; function of genes, chromosomes, 22–4; rates, mechanisms for regulation of, 83–4; seasonal factors affecting, 17

reptiles (*reptilia*), 13, 16, 28, 29, 30–1, 34, 37, 43, 51, 327
resins, 200
resistant species: bacteria, to biocides, 262, 264; in grazing systems, 310; plants, to air pollution, 312–13; strains of malarial mosquitoes, 247
Resource processes: productive, agriculture, 273–88, concept defined, 273; fisheries, 296–305, forestry, 288–96, grazing, 305–12; protective, concept defined, 273
respiration: in carbon cycle, 74; in photosynthesis, 56–8
Rhacomitrium languinosum, 130
rhea, demand for feathers, 220
rhinoceroses: Javan, threatened species, 220, 221, 314; size and generation time, 80; woolly, extinction of, 158
Rhizobium spp, manipulation of, 256
Rhizocarpon spp, 11
Rhizopora spp: *R. candelaria*, 140; *R. mangle*, 140
rice: African, 169, 175–6, 193, 194; agricultural ecosystems based on, 194–5; Asian, 169, 174, 175–6, 193, 275–6; crop dominance by few varieties, 286; depredations by pests, 283–4; domestication of, 163; economic importance as basic food, 258; energy input–output ratios, 279, 280; factors affecting yields, 58, 275–6; importance in basic agriculture of southeast Asia, 185; population/land ratios, 202; response to chemical fertilization, 207; selective breeding of, 182–3, 259, 260, 261, 267; spread of, 193, 194
Ricinodendron rautanenii, 153
rinderpest, control of, 262
rivers: causes, effects of contamination, 239–40, 241, 243, 250; character, distribution, 142
roach, 146
roads, construction of, effect in tundra areas, 126–7
robins, 227
rock deserts, estimated NPP, 86
Rocky Mountains, diversity of biota, 37

rodents: evolution, adaptation to environment, 95, 111, 119, 128; giant, prehistoric, 158; in pattern of human concern for animals, 327
ronnel, 209
rooks, 115, 283
root crops, domestication of, 170; *see also individual crops*
Rosa spp, habitats in deciduous forests, 114
Roseaceae, 12, 114, 290
rose of Jericho, 110
rotenone, 208–9
Rotifera, 62
Royal Botanical Gardens, Kew, plant collection, 210
Rubber: commercial exploitation of *Hevea braziliensis*, 194, 198, 210; efficiency as converter of sunlight, 344; habitat, 95; production from waste, 285
Rubia peregrina, 16, 17
Rubus spp, habitat in deciduous forests, 114
ruderal plants: in cities, 224; on derelict land, 246; recolonization by, 238; response to radiation, 248–9
Rumex acetosa, 15
rushes, 141
rust diseases of wheat, 260
rye: agricultural ecosystems based on, 194–5; breeding to improve, 182–3; domestication of, 169, 170, 172–3; importance in basic agriculture of Europe, 185; *S. cereale*, 170, 173; *S. montanum*, 173
rye-grasses, 171, 183

Saccarinum spp, productivity, 275–6
Safari parks, 227
safflower, 170, 200
sage, 199
sage-bush, 309
sage grouse, 309
saguaro cactus, 108, 111
Sahara Desert, 108, 109
saiga antelope, 106, 159
sainfoin, 171
St Helena, adaptive radation of sunflowers, 52
St John's wort, 81
salamanders, 12, 37, 43, 80, 327
Salicornia spp, 141
salinification, 183–4

salinity: breeding of plants to withstand, 259, 260; diversity, in seas, 131–2; effect on distribution of organisms, 18; interaction with temperature in control of oxygen, 142
Salix spp, 114, 120
  S. arctica, 122
salmon, 299, 327
salt marshes: distribution, character, 141–2; reclamation, 138
Salvelinus tontinalis, 18
Salvia officinalis, 199
sandpipers, 327; purple, 62
saprovores, 8
sardines, 135, 298, 299
Sarraceniaceae, 37, 41
sarsparilla, 199
savanna *see* grasslands, tropical savannas
savanna grasses, 99, 276
Saxifraga oppositifolia, 40
sawflies, 82, 119
Scabiosa columbaria, 15
scallops, 80, 302–3
scavengers, natural, 230–1
Schistocerca gregaria, 112
Schoinobates volans, 37, 38
scorpions, 8, 327
Scotland, protected pinewoods, 316
scrubland: distribution, character, 102–5, 130; estimated NPP, 86; NPP of grazing systems, 306; on derelict land, 246
sea anemones, 12, 133, 135
sea-bed trenches, 131
sea birds, *see* birds *and individual species*
sea-cow, Steller's, 198
sea cucumbers, 12
sea depth, significance of, 132, 133–5
seagulls, *see* birds, gulls
sea-lettuce, 141
sea lilies, 28
seals, 62, 124, 161, 198, 242; Antarctic, 125, 242; bearded, 161; crab-eater, 125, 242; elephant, 125; grey, culling of, 300; leopard, 125; northern fur, 198; ringed, 161; Weddell, 125
sea otters, 327
seas, biota, 18, 132–42; concentrations of mineral nutrients, 131–2; currents,

131; dysphotic, aphotic, euphotic zones, 132; environmental character contrasted with land, 131; estimated NPP, 86, 87; function, importance of plankton, 8–13; geological frame, 131; inshore, open sea compared, 134, 135; interaction with island ecosystems, 130; introductions, ecological effects of, 214–15; land animals linked to food webs of, 124; movements of floor, 132; pollution of, 138, 238–45, 301–2, 312; proportion of NPP contributed by, 88; role in ecosystems, nutrient cycles, 62, 73, 74; significance of depth, 133–5; species from, used by man, 296, 297; upwelling zones, 131, 133, 135; seasonal change, effects of, in various habitats, 97–8, 98–9, 112, 114, 119, 128, 132, 133; on distribution of organisms, 17; on migration of fish, 135
sea urchins, 12, 29, 139
sea water, mineral constituents, 131–2
seaweeds, 11, 133, 303; *see also* algae
Secale spp: domestication, 169, 170, 172–3; *S. cereale*, 170, 173; *S. montanum*, 173
sedges: high NPP of *Cyperus papyrus*, 86–7; invasion of sugarcane by *Cyperus diffusus*, 247; of tundra areas, 122–3
seed cultivation, agricultural ecosystem based on, 194–5
seed dispersal, in deciduous forests, 113
selective breeding: effect on biotic diversity, 250–1; genetic and somatic changes resulting from, 182–3; limitations of, 286; of animals, 262–4; of fish, 304; of plants, 258–61; *see also under individual species*
semi-desert areas: distribution, characteristics, 130; estimated NPP, 86; NPP of grazing systems, 306
Senecio spp, 130, 210; *S. jacobea*, effect of myxomatosis on density, 81; *S. squalidus*,

spread of, 210; *S. viscosus*, 210; *S. vulgaris*, 128
senna, 109
sequoia:California Coast, 40, 48, 80, 156, 322, 323, 326; sierra, 120, 291, 326
sesame, domestication of, 169
sevin, 209
sewage: cause of eutrophication, 239; diversion to forests, 295; effect on water supplies, 142, 145, 312; function of microorganisms in, 8; in food web of seas, 136, 138; *see also* waste
sharks, 12, 21, 28, 327
sheep, 130, 167, 184, 185, 194, 196–7, 262, 306, 310, 321; bighorn, 321; domestication from *Ovis aries*, 168, 177, 178, 179, 184, 185; Guinea, 194; merino, 194, 197; Soay, 216, 217; wild, 127, 167, 178
sheep's fescue, 15, 49
shellfish, 239, 240
Shorea robusta, 98
shrews, 36, 115, 125; Whitetoothed, 216
shrimps, 140, 245
shrubland, estimated NPP, 86
shrubs, diversity, distribution, 113, 114, 115, 117, 120
Siberia, boreal forests, 117
silica, cycling of, in tropical savannas, 99
silk-moth, 181
Silurian period, 28, 29
silversword, 130
Simmondsia chinensis, 200, 269
single-cell protein (SCP), 270–1, 280
skate, 12
skuas, 62, 124, 125
skunk, 227
skunk-cabbage, 37, 41
sloths, 95, 96–7, 158
slugs, 11, 49, 111, 327
smallpox, 11
smelt, 299; rainbow, invasion of Lake Erie, 302
Smilax spp, 199
Sminthopsis spp, 37, 39
smog, 236
smoke, effects of release to atmosphere, 232
snails, 12, 16, 49
snakes, 13, 31, 37, 43, 80, 327; freshwater, 143; of desert areas, 111; tesselated,

naturalization in Britain, 217; *Vipera* spp, 43
snow bunting, 62
soap, 208
sodium, 132, 136
sodium pentachlorophenate, 240
soil erosion: caused by recreational use of land, 318, 319, 320; continuing problems, 335; effects of fire on rates of, 104; effects on flora and fauna, 238–9; following over-intensive use of land, 184, 202, 310
soils: changes following domestication of biota, 184; effects of machinery, fertilizers, on structure of, 205–6; ferrous, in Mediterranean area, 104; improvement for grazing, 306–7; pH levels, 245–6; roles of vegetation, climate, in creation of, 90–1; saline, breeding of plants to grow in, 259, 260; types, 13, 49, 78, 95–6, 98, 108, 114–15, 119, 311
Solanum spp: *S. quitoense*, 186; *S. tuberosum see* potatoes
solar energy, 64–6, 196, 201, 204
sole, farming of, 303
solvents, 231
sorghum: breeding to improve, 182–3; domestication, 169, 171; economic importance as basic food, 258; importance in basic agriculture of Africa, 185–6; Johnsongrass, 171; productivity, 276
sorrel: Alpine, 44–5; common, 15
soybeans, *see* beans
Spanish moss, 21
sparrow, English, 227
Spartina spp, 141
speciation, 27, 52
species: adaptive radation, 52; autecology, synecology distinguished, 54; causes of extinction, 218–20; diversity, 47, 68–9, 156, 160, 250–1, 325; dominance, 67, 95, 97–8, 103–4, 106–7, 109, 113–14, 117–18; ecotypes within, 44–5; endemism, endemics, defined, 47; in classification of organisms, defined, 7–8; interrelationships, 13; optimal range of environmental conditions, 14, 15; population

size, density, 79–85; resilience, stability of, 339–40; threatened, 220–2, 314–17
Speech, in evolutionary history of man, 32
Spermatophyta, 11
Sphagnum spp, 11, 117, 121–2
spiders, 12, 62, 111, 115, 327; giant raft, 322
spines, function of, 108–9, 114
Spiranthes romanzoffiana, 37, 41
Spirillum lipoferum, 261
Spirochaeta, 80
Spirogyra spp, 11
Spirulina platensis, 269
sponges, 139
spraying with chemicals, 323
springbok, 263
springtails, 49, 62, 14, 115
spruce budworm, 119, 247, 291
spruce, Sitka, 120, 210, 213–14, 293
squashes, 168, 185–6, 187–8
squids, 12, 29, 299, 302–3
squirrels: antelope ground, 111; grey, 215, 217; ground, 37, 104, 111, 125, 321–2; in diet of hunting peoples, 161; reservoir of rabies, 228
Stanhopea grandiflora, 22
Staphylococcus aureus, 11
starch, 58, 171
starfish, 12, 80, 327; crown of thorns, 140, 267; *Pisaster ochreus*, 340
starlings, 83, 214, 227–8
Steller's sea-cow, 198
stem rot, 267
Stentor, 80
steppes: character, distribution, 105–7; desertification, 112; ecological impact of fire, 155; NPP of grazing systems, 306; *see also* grasslands
steppe-bison, 153
sticklebacks, 237
stilt, speciation on Hawaii, 52
Stipa spp, 184, 310; *S. spartea*, 106
storks, 327
stratification, in ecosystem development, 68–9
strawberry tree, 103
Streblorrhiza speciosa, 222
Strepsiceros strepsiceros, 263
strontium, 132, 136; radioactive (strontium-90), 243, 244, 248, 249
sturgeons, 302

398 Index

Sturnus vulgaris, 214
Stylosanthes spp, 308
sub-boreal forests, NPP of grazing system, 306
succulent plants, 108–9, 112
Suez Canal, 214–15
sugar, sugars, 58, 171, 267
sugar-beet, 169, 170, 275–6, 278, 279
sugarcane: domestication, 169, 174; efficiency as converter of sunlight, 344; problems of pest control, 212, 247; productivity, 275–6; spread in Europe, Americas, 193, 194; use of waste from, to produce turpenes, 285
sulphates: constituent of sea water, 132, 136; levels in desert areas, 108
sulphite liquor, 240
sulphur: as biocide, 208–9; compounds, 232, 233–6, 240; cycling, recycling intensified by man, 231; impact of fire on levels in ecosystems, 155
sulphur dioxides, 232, 233–6, 312–13
sulphuric acid, 233, 240
sunflowers, 52, 169, 200
sunlight: adaptation to, 128; and photosynthesis, 56–60, 61
swamps: estimated NPP, 86; freshwater, character, distribution, 142; mangrove, 138, 140–1, 247–8
swans, 326, 327
swifts, 227
switchgrass, 106
Sylviidae, 114
symbiotic relationships, 11, 75
Symplocarpus foetidus, 37, 41
Syncerus caffer, 262
synecology, defined, 54
synergisms, 232

Tadaria brasiliensis, 242
tadpoles, 12
tannin, 108–9
tanning, 200
Tanzania, 32, 33, 153
tapeworms, 12, 21–2
Tardigrada, 62
taro: domestication of, 169; energy input–output ratios, 201–2, 280; importance in basic agriculture, 185, 195; spread to New Zealand, 194
tarpon, 215

tarragon, 199
Taurotragus oryx, 262–3, 306
tea, 169, 199
teak tree, 98
tegu, 221
temperate forests: distribution, character, 67, 93, 112–16; ecological effects of human activities, 92, 116; estimated NPP, 86, 88; interrelation of temperature, precipitation and, 91; recycling of mineral nutrients, 75
temperate grasslands, 105–7; *see also* grasslands
temperate sclerophyll woodland, 102–5
temperature: and distribution of organisms, 13, 14–16, 17, 42–3, 49; ecological significance, 142, 143–7, 120–1; fluctuations in deserts, 107; in energy environment of plants, 56–60; interrelation with precipitation and biome type, 91–2
Terminalia spp, 98
termites, 12, 101–2
tern, 135–6
terpenes, 236–7
terracing for cultivation, 104
territorialism, 20, 83
tertiary period, 29, 30, 31
Testudo spp, 221; *T. graeca*, 221; *T. horsefieldi*, 221
tetradifon, 208–9
tetraethyl lead, 237
Tetrahymena, 80
Thailand: fish farming in, 304; tropical seasonal forests, 98
Thalarctos maritimus, 62, 124, 125, 127, 161, 221
Thallophyta, 11
Thea sinensis, 199
Theobroma cacao, 95, 199
thermal radiation from atmosphere, 56
Thermocyclops hyalinus, 144
Thermodynamics, Second Law of, 62–3, 88
thorns, function of, 108–9
thrasher, California, 50
threadworms, 12, 62, 140
threatened species, 220–2, 314–17; *see also* individual species
thrushes, 44, 227
Thymus spp, 103
ticks, 12
tigers, 8, 329; *Panthera tigris*, 8, 220–1

Tilapia spp, 217; *T. nilotica*, 144
Tilia spp, 113–14, 116, 161
Tillandsia usneoides, 21
timber, 288–96
timberlines, 42
time, and evolution rates, 47
timothy, 171, 193
Tinocallis platani, 227
titmouse, 115
toads, 212, 327; yellow-bellied, 217
toadstools, 11
tobacco, 194, 200
tomatoes, 174, 175
topography, 46–7, 68
tortoises, 52, 111, 211, 221
Tortrix spp, 115
tourism, *see* recreation
towhee, brown, 50
toxaphene, 208–9
Toxostoma redivivum, 50
Trachinotus carolinus, 303
tractors, 205–6
trampling of plants, 320, 328–9
Trans-Alaska pipeline, 126
transpiration in photosynthesis, 56–60
tree of heaven, 312–13
trees: in cities, 224, 226; in control of air pollution, 312–13; in evolutionary history, 28; introduction of exotic, to Britain, 213–14; of deciduous forests, 113–14, 115; of tropical rain forests, 94–5, 96–7; of tropical seasonal forests, 97–8
Triassic period, 28, 29, 31
Trichosurus vulpecula, 212–13
Trifolium spp, 171; *T. repens*, 19
Trigonella foenum-graecum, 171
trilobites, 28, 29, 34
Tripsacum spp, 174
Triticum spp, 166, 168, 169, 170, 172, 173, 182–3, 185, 260–1, 275–6; *T. aestivum*, 170, 172, 173; *T. dicoccum*, 166, 168, 170, 185; *T. monococcum*, 166, 168, 170, 172, 182–3, 185; *T. timopheevii*, 170, 173; *T. turgidum*, 170, 173; *T. turgidum durum*, 260–1; *T. vulgare*, 275–6
tropical coal forests, 28
tropical rain forests, 28, 45–6, 60, 66–7, 86, 91–7, 130
tropical savannas, 86, 98–102, 112, 130, 306, 307–8

Index

tropical seasonal forests, 86, 97–8, 130
tropic grass, 130
trout, 18, 237, 239–40, 302, 327; brook, 18; rainbow, 237
trypanosomiasis, 266
tryptophan, 250
tsetse fly, 266
Tsuga heterophylla, 120
tsunami, 129
tuberous crops, 170, 174
tubifex worms, 239
tulip tree, 312–13
Tuna spp, 300; yellow-fin, 299, 300
tundra, 86, 88, 91, 92, 93, 120–7, 130, 156, 159, 197, 248–9
turbot, 303
turkeys, 164, 178, 185
turnips, 170, 195, 278, 279
turpentine, 208
turpenes, 285
turtles, 28, 31, 37, 43, 80, 135, 198, 221, 300, 327; Atlantic hawksbill, 221, 300; *Caretochelys* spp, 37; freshwater species, 143; green, 300; marine, 135, 221, 300; river, 198
Typha spp, 143
typhus, 208, 209
Tyrannosaurus rex, 34
Tyto alba, 283

Ulex spp, 103
Ulmus spp, 113–14, 214, 218, 267, 312–13
ungulates, 156, 158, 327
Union of Soviet Socialist Republics (USSR), 207, 304
United States of America, 208; *see also* America, North
urbanization, 104, 251–4
Ursus spp: *U. arctos arctos*, 119, 321
  U. arctos horribilis, 104, 119, 125, 321
  U. malayanus, 221
Usnea articulata, 233

Vaccinium spp, 117
vanilla, 199
Varanus spp, 221
vegeculture, 185–6, 195
vegetables, 171, 224–5; *see also* individual crops
vegetation: and air pollution, 312–13; impact of fire on, 84, 154–6, 160; and mineral retention, 75–8; NPP in grazing, 306; and soil creation, 90–1
vertebrates, 12–13, 28, 29, 88, 99–102, 104, 327
vetches, 171; bitter, 170
Vicia spp, 171: *V. ervilia*, 170; *V. faba*, 169, 170
vicuna, 181, 186, 306
Vietnam War, 208, 247–8
vines: *Vitis vinifera*, 169, 171, 174, 185, 194, 199; *Canavalia kauensis*, 130–1
Vipera spp, 43
viruses, 7, 141, 265, 312
vitamin B$_{12}$, 141
Vitis spp: *V. vinifera*, 169, 171, 174, 185, 194, 199; wild, 113
volcanoes, 66, 67, 129
voles, 115, 125: Guernsey, 215, 216, 217; Orkney, 215, 216, 217
vultures, 327

wallaby, red-necked, 216, 217
walleye, 302
walnut, English, 171
walruses, 124, 161
warblers, 45, 49, 50, 114: bay-breasted, 50; blackburnian, 50; black-throated green, 50; Cape May, 50; myrtle, 50; yellow, 48
warthogs, 83
wasps, 12, 112, 265, 327
waste, man-made: concentrations in modern times, 231; creation, disposal, in urban ecosystems, 223–4; disease caused by, 230–1; ecological effects, 2, 231–49; from agriculture, 284–5; future problems, 338; potential uses of, 270–1, 284–5; role of natural scavengers, 230–1; synergism caused by, 232; water, sewage, effects of diversion to forests, 295
water, 13, 16–17, 19; *see also* water resources
water deer, Chinese, 216, 217
water fleas, 12, 80, 142
water fowl, 227
water hyacinth, 268–9
water lily, 143
water resources: compatiblity with modern forestry, 294–5; contamination of, 138, 142, 238–45, 250, 301–2, 312, 336; ecological effects on, 126–7, 145–7; management of, 312; trends in demand for, 335
wavy hair grass, 15
waxes, 200
weasels, 107, 115, 125, 327
weeding, 162, 163
weeds, 52, 183–4
West Indies, 51, 220
whales, 13, 80, 82–3, 138, 161, 300–1, 327: baleen, 136; beluga, 161; blue, 300–1; bottle-nosed, 301; fin, 300–1; Greenland right, 301; killer, 301; pilot, 301; sei, 300–1; short-finned, 301; toothed, 136
wheat: agricultural ecosystems based on, 194–5; bread, 170, 172, 173; crop dominance by few varieties, 286; depradations by pests, 283–4; domestication, 166, 168, 169, 170, 172, 173, 182–3, 185; durum, 260–1; energy input-output ratios, 279; productivity, 275–6; response to chemical fertilizers, 207; selective breeding of, 182–3, 258–9, 260–1; tetraploid, 170, 173; yields in Britain, 278, 279
whitefish, 302
wildebeest, 88, 101
wilderness, protection of, 314, 317–18, 319, 329
wildlife: effects of grassland management on, 309; and forestry, 294; impact of intensive agriculture, 283; management problems, policies, 320–5; reasons for protection of, 325–8; and tourism, 316, 321; urban, encouragement of, 225, 229
willow trees, 114, 117, 120; Arctic, dwarf, 122
wind, 48, 56, 113
wine, 198–200
winter sports, 128–9
woad, 200
wolverine, 119, 125
wolves: Chinese, 180; and predation, 21, 107, 311
wolves: Chinese, 180; habits, habitats, 104, 114, 119, 123,

400 Index

wolves—*contd*
124, 125, 180; importance in pre-agricultural economies, 157; in pattern of human concern for animals, 327; and predation, 21, 107, 311
women, as gatherers, 154
wood: demand for, uses of, 288–9, 293–6; and GNP, 289; world production figures, 289; *see also* forestry
woodlands: ecosystems changed by man, 2; estimated NPP, 86; protected habitats, 316, 321; secondary, around cities, 224; *see also* Forestry *and particular types of woodland*
woodlice, 12, 114, 115
woodpecker, Californian, 83–4
wood rat, 104

Woodwalton Fen, Cambridgeshire, nature reserve, 324
woolly mammoth, 158
World Wildlife Fund, 326
worms, 12, 29, 62, 111, 245, 327: Annelida, 12, 143; blood, 143; freshwater, 143; marine, 133; *Tubifex*, 239
worm tracks, burrows, 28, 29
wrentit, 50

xerophytes, 8
Xylia xylocarpa, 98

yak, 128, 163, 177, 178, 186, 187, 306
yams, 169, 185, 195, 201–2, 280
yeasts, 11, 271, 279
yellow fever, 247

yellow hammer, 321
yellow perch, 302
yew, 11
yorkshire fog, 130, 310

zebras, 101, 262, 306
zebu cattle, 168, 185
Zenaida auriculata, 284
zinc deficiency, 267
Zingiber officinale, 199
Zinjanthropus boisei, 35
zoogeographical realms, 36–7
zooplankton: distribution, ecological functions, 132–7; freshwater, 143, 144, 146; in carbon cycle, 74; marine, 139–40
zoos, 227
Zostera marina, 269
zygotes, 24